ABOUT ISLAND PRESS

Island Press, a nonprofit organization, publishes, markets, and distributes the most advanced thinking on the conservation of our natural resources—books about soil, land, water, forests, wildlife, and hazardous and toxic wastes. These books are practical tools used by public officials, business and industry leaders, natural resource managers, and concerned citizens working to solve both local and global resource problems.

Founded in 1978, Island Press reorganized in 1984 to meet the increasing demand for substantive books on all resource-related issues. Island Press publishes and distributes under its own imprint and offers these services to other nonprofit organizations.

Support for Island Press is provided by The Geraldine R. Dodge Foundation, The Energy Foundation, The Ford Foundation, The George Gund Foundation, William and Flora Hewlett Foundation, The James Irvine Foundation, The John D. and Catherine T. MacArthur Foundation, The Andrew W. Mellon Foundation, The Joyce Mertz-Gilmore Foundation, The New-Land Foundation, The Pew Charitable Trusts, The Rockefeller Brothers Fund, The Tides Foundation, Turner Foundation, Inc., and individual donors.

ABOUT ISEE

Ecological economics is concerned with integrating the study and management of nature's household (*ecology*) with humanity's household (*economy*). Ecological economics acknowledges that, in the end, a healthy economy can only exist in symbiosis with a healthy ecology. Ecological economics is the name that has been given to the effort to transcend traditional disciplinary boundaries in order to address the interrelationship between ecological and economic systems in a broad and comprehensive way. Ecological economics takes a holistic worldview with human beings representing one component (albeit a very important one) in the overall system. Moreover, human beings play a unique role in the overall system because they can consciously understand their role in the larger system and manage it for sustainability. Ecological economics seeks to constitute a true marriage of ecology and economics so as to give meaning and substance to the idea of **sustainable development**. Mechanisms for achieving sustainable development will enable modern societies to develop and prosper within a natural world that is safeguarded from ecological destruction.

The International Society for Ecological Economics (ISEE) is a not-for-profit organization with more than 1600 members in over 60 countries. Regional chapters have been established in Russia and Canada, with plans being made for chapters in China, Africa, and Southeast Asia. ISEE promotes the integration of ecology and economics by providing information through its membership journal, *Ecological Economics,* and the Quarterly Newsletter, and encourages the exchange of ideas by supporting major international conferences and smaller regional meetings on topics of interest to members, as well as research and training programs in ecological economics.

Investing in Natural Capital, a video providing an overview of ecological economics, and membership information can be obtained by contacting the ISEE Secretariat, P.O. Box 1589, Solomons, MD 20688 USA.

Investing in
Natural Capital

Investing in Natural Capital

*The Ecological Economics
Approach to Sustainability*

Edited by
AnnMari Jansson, Monica Hammer,
Carl Folke, and Robert Costanza

Technical Editor, Sandra Koskoff

*Foreword by Olof Johansson,
Minister of the Environment, Sweden*

INTERNATIONAL SOCIETY
FOR ECOLOGICAL ECONOMICS

ISLAND PRESS

Washington, D.C. ❑ Covelo, California

Library of Congress Cataloging-in-Publication Data

Investing in natural capital : the ecological economics approach to
    sustainability / [edited by AnnMari Jansson ... et al. : technical
    editor Sandra Koskoff : foreword by Olof Johansson].
        p. cm.
    Contributions to a workshop held in August 1992 near Stockholm,
Sweden.
    Includes bibliographical references and index.
     ISBN 1-55963-316-6 (pbk. : alk paper)
    1. Sustainable development--Congresses. 2. Natural resources-
-Management--Congresses. 3. Economic development--Environmental
aspects--Congresses. 4. Environmental economics--Congresses.
I. Jansson, A.-M.
HD75.6.I59 1994
333.7–dc20   94-7528
                                                                    CIP

Printed on recycled, acid-free paper

Manufactured in the United States of America

10 9 8 7 6 5 4 3 2 1

CONTENTS

PART TWO: ECOLOGICAL ECONOMIC METHODS AND CASE-STUDIES ON THE SIGNIFICANCE OF NATURAL CAPITAL

CONTRIBUTORS

N. Adger   Center for Social and Economics Research on the Global Environment, University of East Anglia; University College

Edward Barbier   Dept. of Environmental Economics and Environmental Management, University of York

Fikret Berkes   Natural Resources Institute, University of Manitoba

James K. Boyce   Universidad Nacional de Costa Rica, San Jose, Costa Rica

Colin Clark   Institute of Applied Mathematics, University of BC, Canada

Cutler Cleveland   Center for Energy/Environment Studies, Boston University

Robert Costanza   Maryland International Institute for Ecological Economics

Herman Daly   School of Public Affairs, University of Maryland

Ralph d'Arge   Dept. of Economics, University of Wyoming

P. Doktor   Environmental Appraisal Group, University of East Anglia

Rudolph de Groot   Center for Environment/Climate Studies, Agricultural University Wageningen

Faye Duchin   Institute for Economics Analysis, New York University

Paul Ehrlich   Center for Conservation Biology, Stanford University

Carl Folke   Beijer International Institute of Ecological Economics

Martha Gilliland   Dept. of Hydrology and Water Resources, University of AZ

Frank Golley   Institute of Ecology, Dept. of Zoology, University of GA

Ing-Marie Gren   Beijer International Institute of Ecological Economics

Monica Hammer   Dept. of Systems Ecology, Stockholm University

C. S. Holling   Department of Zoology, University of Florida, Gainesville

Bengt-Owe Jansson   Dept. of Systems Ecology, Stockholm University

AnnMari Jansson   Dept. of Systems Ecology, Stockholm University

I. Kaganovich   Institute of Economics, Estonian Academy of Sciences

J. P. Kash   Dept. of Political Science, University of AZ

Dennis King   Maryland International Institute for Ecological Economics

Valentina Krysanova   Institute of Economics, Estonian Academy of Sciences

Glenn-Marie Lange   Institute for Economics Analysis, New York University

K. G. Mäler   Beijer International Institute of Ecological Economics

Gordon Munro   Department of Economics, University of BC, Canada

Howard. T. Odum   Center for Environmental Policy, University of Florida

Charles Perrings   Dept. of Environmental Economics, University of York

William Rees   University of BC, School of Community and Regional Planning

Olman Segura   Universidad Nacional de Costa Rica, San Jose, Costa Rica

Kerry Turner   Center for Social and Economics Research on the Global Environment, University of East Anglia; University College

Stephen Viederman   Jessie Smith Noyes Foundation, New York

M. Wackernagel   University of BC, School of Community and Regional Planning

James Zucchetto   National Academy of Sciences, Washington, DC

Tomasz Zylicz   Economics Department, Warsaw University

FOREWORD

To effectively deal with sustainability, more efforts have to be directed to understanding and managing the interface between ecology and economics. This book illustrates that a constructive dialogue can be, and in fact has been, firmly established within the scientific community. There are also signs that such a process is taking place in the policy arena. Ministries of various labels, and in various countries, have started to collaborate, and the international community has agreed on action reflected in *Agenda 21* and the conventions signed in Rio in 1992.

The work presented in this book is of great significance for finding new pathways to redirect development towards sustainability. We need to develop policies that effectively integrate and adapt human societies to our ecological basis; we need to learn how to invest in natural capital. This is an extremely urgent and challenging task. Investing in natural capital—the life-supporting environment upon which our economic, socio-political, and cultural systems depend—is indeed a prerequisite for the economy and for human welfare.

Olof Johansson,
Minister of the Environment,
Sweden 1993

PREFACE

This book is the product of a workshop held August 6–7, 1992, at Grand Hotel Saltsjöbaden, outside Stockholm in Sweden. The workshop immediately followed the second biannual conference of the International Society for Ecological Economics, held August 3–6, 1992, at Stockholm University. The theme of the conference was *Investing in Natural Capital—A Prerequisite for Sustainability*. Over 450 participants from 37 countries participated in the four-day ISEE conference, and the invited plenary speakers attended the subsequent two-day Wallenberg workshop. The institutions involved in the organization of the conference were the Department of Systems Ecology and the Natural Resources and Environmental Research Center at Stockholm University; the Beijer International Institute of Ecological Economics; the Royal Swedish Academy of Sciences; the International Society for Ecological Economics; and the University of Maryland International Institute for Ecological Economics.

In 1982, the first Wallenberg workshop on the Integration of Ecology and Economics with a similar group of participants was held at the same location in Saltsjöbaden. Ten years ago, that meeting was the catalyst and the starting point for subsequent work that addressed the interface of ecology and economics. That workshop ultimately led to several notable events, including the formation of the International Society for Ecological Economics, (ISEE), the creation of the Beijer International Institute of Ecological Economics at the Royal Swedish Academy of Sciences in Stockholm, and the founding of kindred institutions in other nations and much diverse work in numerous countries and disciplines.

The goals of the 1992 Wallenberg workshop were to focus on the directions in which the field of *ecological economics* has developed and to consolidate a core of economists, ecologists, and other scientists into a nucleus of thinkers about *ecological economics*, especially as it applies to natural capital. In particular, we wanted to synthesize the achievements of this transdisciplinary field of study and develop a working agenda for research, education, and policy on how to invest in natural capital for sustainability. The consensus of the group on these matters is presented in the introductory chapter of the book. The remaining chapters provide detailed elaboration on these themes from a number of different perspectives. The papers have benefited from the discussions during the two-day workshop. Each paper was formally reviewed by at least two reviewers, and not all of the invited plenary papers were included in the book.

This book is intended to serve as an academic text primarily for researchers in the interface of the social and natural sciences. Since ecological economics is inherently multidisciplinary, the potential academic audience is broad, including economists, ecologists, conservation biologists, public policy professionals, anthropologists, sociologists, and others. The book will also be useful as a textbook or "sourcebook" of cogent readings in ecological economics that can serve as a basis for graduate courses.

ACKNOWLEDGMENTS

Major funding for the conference and workshop was provided by the Swedish Council for Planning and Coordination of Research (FRN) and the Marcus Wallenberg Foundation for International Cooperation in Science. Additional funding was provided by the Swedish International Development Authority (SIDA), the Beijer International Institute of Ecological Economics, the Royal Swedish Academy of Sciences, Stockholm University, the City of Stockholm, Stockholm County Council, the Swedish Council for Forestry and Agricultural Research (SJFR), Vattenfall AB, Alba Langenskiöld Foundation, the National Swedish Environmental Protection Board (SNV), and the Natural Resources and Environmental Research Center at Stockholm University.

We would like to thank the following persons for their valuable contributions to this conference and workshop: Ms. Margareta Wiberg Roland (general conference secretary), Mona Björnsson, Olimpia Garcia, Florence Eberhart, Reine Elfsö, Janet Barnes, Miguel Rodriguez, Marianne Rylander, and the enthusiastic crew of some twenty students from Stockholm University, University of Maryland, and Boston University; in particular, Laura Cornwell, Jim Nilsson, Julie Sweitzer, and Lisa Waigner, who assisted at the Wallenberg workshop.

Stockholm, September 1993

AnnMari Jansson
Monica Hammer
Carl Folke
Robert Costanza

Investing in
Natural Capital

1 INVESTING IN NATURAL CAPITAL—WHY, WHAT, AND HOW?

Carl Folke
The Beijer International Institute of Ecological Economics
The Royal Swedish Academy of Sciences
Box 50005
S-104 05 Stockholm, Sweden

Department of Systems Ecology
Stockholm University
S-106 91 Stockholm, Sweden

Monica Hammer
Department of Systems Ecology
University of Stockholm
S-106 91 Stockholm, Sweden

Robert Costanza
Maryland International Institute for Ecological Economics
Center for Environmental and Estuarine Studies
University of Maryland
Box 38, Solomons, MD 20688-0038

AnnMari Jansson
Department of Systems Ecology
University of Stockholm
S-106 91 Stockholm, Sweden

ABSTRACT

This introductory chapter is a synthesis and overview of the major ideas contained in the other chapters of the book. It starts with a summary of some of the major characteristics of the rapidly evolving field of ecological economics, with particular emphasis on the concept of natural capital, and its maintenance and enhancement. In the next section a few essential features of the ecological economic approach are described. Thereafter, we present what we believe has evolved as important ecological economic research topics related to the issue of sustainability, and finally, what policy recommendations should be given, considering the state of current knowledge.

This synthesis chapter and the other chapters in the book were developed in part at a workshop which followed the second international conference of the International Society for Ecological Economics (ISEE) held in Stockholm, Sweden in August, 1992. Embedded in the chapter are views and consensus statements developed at the workshop. Basic points of consensus at the workshop included the fact that (1) the economic system is a subsystem of the global ecosystem, (2) fundamental uncertainty is large and irreducible and certain processes are irreversible, and (3) there are limits to biophysical throughput through the economic system. Therefore we need to conserve natural capital, or at least maintain adequate stocks, keep our options open to avoid irreversibilities and create opportunities, and include a broader range of values such as ethics, equity, and intergenerational concerns. Management should be proactive rather than reactive and should result in simple, implementable policy recommendations based on sophisticated understanding of the underlying systems.

The book is organized into three major sections. The first section on "Conceptual Underpinnings" consists of seven chapters that lay the groundwork and develop the "pre-analytic vision" for an ecological economics approach to sustainability. The second section consists of twelve chapters on "Methods and Analysis" that develop some specific methodological approaches, analytical results, and case studies based on this conceptual groundwork. The final section consists of six chapters that deal with "Policy and Institutions" to implement the results of the first two sections.

WHAT IS ECOLOGICAL ECONOMICS AND WHY DO WE NEED IT TO ACHIEVE SUSTAINABILITY?

To achieve sustainability, the global community must deal with new types of problems threatening the future well-being and existence of humanity. These problems are fundamentally cross-scale, transcultural and transdisciplinary, calling for new innovative research approaches and new social institutions (Holling this volume; Berkes and Folke this volume; d'Arge this volume). This research should be integrated rather than divorced from the policy and management process (Golley this volume; Viederman this volume).

The international recognition of sustainable development as a long term goal for human society at the Earth Summit in Rio de Janeiro in 1992, emphasizes the crucial importance of these issues within the human community. Critically important research is now needed to facilitate the transition to sustainable production and consumption systems. Innovative research aimed at articulating the methods and mechanisms by which human populations can strike a dynamic balance between economic development and ecological constraints they face constitutes the foundation on which the future will be built.

Ecological economics is a transdisciplinary field of study that addresses the relationships between ecosystems and economic systems in the broadest sense, in order to develop a deep understanding of the entire system of humans and nature as a basis for effective policies for sustainability. It takes a holistic systems approach that goes beyond the normal narrow boundaries of academic disciplines. This does not imply that disciplinary approaches are rejected, or that the purpose is to create a new discipline. Ecological economics is interdisciplinary in the sense that scholars from various disciplines collaborate side-by-side using their own tools and techniques, and transdisciplinary in the sense that new theory, tools, and techniques are developed to effectively deal with sustainability. It focuses more directly on the problems facing humanity and the life-supporting ecosystems on which we depend. These problems involve: (1) assessing and insuring that the scale of human activities are ecologically sustainable; (2) distributing resources and property rights fairly, both within the current generation of humans and between this and future generations, and between humans and other species; and (3) efficiently allocating resources as constrained and defined by 1 and 2 above, and including both marketed and non-marketed resources.

Humans have a special role to play in the system because we are responsible for understanding our own role in the larger system and managing it for sustainability (Costanza et al. 1991). This responsibility is not only an ethical and a moral issue. It has to do with the fact that saving the environment actually means saving ourselves, including future generations, since we, as a biological species, are dependent on healthy ecosystems for survival. Thus, ecological economics is an anthropocentric field of study in the sense that it cares about the survival and welfare of human beings on this planet.

It differs, however, from many other anthropocentric perspectives because it is embedded more in an ecocentric than an egocentric worldview (Rapport 1993). Ecological economics views the socioeconomic system as a part of the overall ecosphere, emphasizing carrying capacity and scale issues in relation to the growth of the human population and its activities, and the development of fair systems of property rights and wealth distribution. The belief of many that we can continue on the same path of expansion, that technological progress will solve all energy, resources, and environmental limits, and that there is infinite substitutability between human-made and natural capital is considered to be a dangerous one, given the huge uncertainty. This blind faith in technology may be similar to the situation of the man who fell from a ten-story building, and when passing the second story on his way down, concluded "so far so good, so why not continue?"

Instead, it is recognized that uncertainty is fundamental, large, and irreducible, and that particular processes in nature are essentially irreversible (Costanza this volume; Clark and Munro this volume). Instead of locking ourselves in development paths that may ultimately lead to destruction and despair, we need to conserve and invest in natural capital, in the sense of keeping life support ecosystems and interrelated socioeconomic systems resilient to change (Hammer et al. 1993; Holling this volume; Jansson and Jansson this volume; Perrings this volume). Hence, ecological economics has an explicit concern for future generations and long-term sustainability, and works with a broader range of values than the limited perceptions of the current generation of humans (although these perceptions are certainly not ignored). Ethics and equity issues are explored, as well as differences and similarities between worldviews and cultures (Turner et al. this volume; Berkes and Folke this volume).

As an open, dynamic subsystem of the overall, finite global ecosphere, the human population and its socioeconomy is an integral part of the life-supporting environment, physically interconnected by the flows of energy and matter at various scales in time and space (Daly this volume; Ehrlich this volume; d'Arge this volume). Therefore, human exploitation of natural resources and disposal of wastes are not separate activities, but take place in the same environment, and both activities impact on the life support functions provided by natural ecosystems (see Figure 2.1 in Daly this volume). The life support environment is the basis, and healthy ecosystems are a precondition for human welfare. Hence, the ecological economic world view treats humans as *a part of* and not apart from the processes and functions of nature (Costanza 1991). After all, we humans did not create the globe, but evolved from it and with it. Viewing the economic subsystem in its proper perspective relative to the entire system is crucial for achieving a sustainable relationship with the environment, and assuring our own species' continued survival on the planet.

Natural Capital

Ecological economists speak of natural capital, human capital (and/or cultural capital), and manufactured capital when categorizing the different kinds of stocks that produce the range of ecological and economic goods and services used by the human economy (Daly this volume; Berkes and Folke this volume). The latter two are sometimes referred to together as human-made capital (Costanza and Daly 1992). These three forms of capital are interdependent and to a large extent complementary (Daly this volume). As a part of nature, humans with our skills and manufactured tools not only adapt to but modify natural capital, just like any other species in self-organizing ecosystems (Holling this volume; Ehrlich this volume; Jansson and Jansson this volume).

Natural capital consists of two major subtypes: non-renewable resources such as oil, coal, and minerals; and renewable resources such as ecosystems. Environmental or ecological services, which describes a wide range of ecosystem processes and functions, such as maintenance of the composition of the atmosphere, amelioration of climate, operation of the hydrological cycle including flood control and drinking water supply, waste assimilation, recycling of nutrients, generation of soils, pollination of crops, provision of food from the sea, maintenance of species, a vast genetic library and also the scenery of the landscape, and recreational sites, in addition to aesthetic and amenity values are the flows that result from natural capital (Ehrlich 1989 this volume; Folke 1991; de Groot 1992 this volume).

Ecological economists argue that natural capital and human-made capital are largely complements (rather than substitutes), and that natural capital is increasingly becoming the limiting factor for further development (Costanza and Daly 1992, Daly this volume). Therefore, in order to sustain a stream of income, the natural capital stock must be maintained. This does not mean an unchanged physical stock, but rather an undiminished potential to support present and future human generations. A minimum safe condition for sustainability (given the huge uncertainty) is to maintain the total natural capital stock at or above the current level (Turner et al. this volume). An operational definition of this condition for sustainability means that the physical human scale must be limited within the carrying capacity of the remaining natural capital; technological progress should be efficiency-increasing rather than throughput-increasing; harvesting rates of renewable natural resources should not exceed regeneration rates; waste emissions should not exceed the assimilative capacity of the environment; and non-renewable resources should be exploited, but at a rate equal to the creation of renewable substitutes (Barbier 1987; Costanza and Daly 1992).

A working definition of this statement points to maximizing the net benefits of economic development, subject to maintaining the goods, services and quality of the natural environment over time (Pearce et al. 1988). This requirement ensures that we observe the bounds set by the functioning of the natural environment in its role of support system for the economy (Pearce and Turner 1990; d'Arge this volume). The bounds refer to the carrying capacity of the environment to support human activities (Daily and Ehrlich 1992; Ehrlich this volume; Rees and Wackernagel this volume), at various scales. This capacity is dependent on the resilience of ecosystem and the behavior of the economy-environment system as a whole (Common and Perrings 1992; Perrings this volume; Holling 1992 this volume; Jansson and Jansson this volume; d'Arge this volume; Costanza et al. 1993).

Substitution and Complementarity

Solar energy drives the generation of all renewable resources and ecological services. Industrial energy is used through the processes within the industrial economy to sustain socioeconomic activities and to upgrade natural resources into consumable commodities (Zucchetto this volume). The upgrading process is called "production" in economic terminology. This means that economic production of any commodity needs natural resources, and the transformation of natural resources from discovery, extraction, and refinement into useful raw materials and eventually into human-produced goods and services. Economic activity thus requires both the use of human capital (Hall et al. 1986; Cleveland this volume) as well as the support by ecosystems driven by solar energy (i.e., natural capital) (de Groot 1992 this volume; Jansson and Jansson this volume; Odum this volume).

Therefore, from the perspective of ecological economics, it is not possible to have perfect and unlimited substitution between natural and human-made capital. Human-made capital cannot be created and sustained without energy and natural resources. Hence, there will always be a minimum or critical amount of natural capital needed to sustain any individual of the human species, and there will always be a minimum amount of natural capital needed to produce anything in the human economy (Turner et al. this volume; d'Arge this volume). Production of goods and services cannot be decoupled from its biophysical reality. It is not possible to fully substitute human-made capital for natural capital, since the former is in itself made out of the latter (Costanza and Daly 1992). The substitution in monetary terms between natural capital and human-made capital is, in biophysical terms, often a substitution between renewable resources and non-renewable resources (Cleveland this volume; Zucchetto this volume). A frequently observed pattern, particularly in the modern world, is that ever increasing quantities of human-made capital substituting for natural resources means that ever increasing quantities of natural resources are being used elsewhere in the economy to produce that human capital (Regier and Baskerville 1986; Folke and Kautsky 1992). At some point the quantity of energy/matter used to produce the human-made capital will exceed the energy/matter saved by the substitution, after which further substitution of human-made capital will actually increase overall energy/matter inputs (Cleveland 1991).

Life-Support Systems, Biodiversity, and Primary and Secondary Values

The life-supporting environment is that part of the earth that provides the biophysical necessities of life, namely food and other energies, mineral nutrients, air, and water. The "life-supporting ecosystem" is the functional term for the environment, organisms, processes, and resources interacting to provide these physical necessities (Odum 1989). Renewable natural capital is generated by the continuous interactions between organisms, populations, communities, and their physical environment. Species are part of the ecosystem contributing to the production and sustenance of renewable natural capital (Jansson and Jansson this volume; Perrings this volume; d'Arge this volume). For any type of renewable natural capital to be sustained, a minimum amount of species are required to develop the cyclic relations between producers, consumers, and decomposers. These cyclic relations in synergy with the environmental conditions at hand will continuously develop and evolve the structure of the ecosystem. The structure and processes of the ecosystem have to be intact and functioning in order for it to qualify as renewable natural capital. This, the ecosystem life-support, is the primary value of the environment, and species are crucial parts of it. Renewable resources and ecological services are secondary values, and by definition, would not be there without the primary values of the ecosystems' basic existence (Turner et al. this volume). Since we humans and our societies are subsystems of the ecosphere, we are fundamentally dependent on the primary value of the environment. However, valuation of the environment has dealt with secondary values (at least so far), and only a minor part of these values (d'Arge this volume). In addition, the secondary values are often taken out of their context in the ecological economic system. (Gren et al. 1994).

In the context of biological conservation and human welfare, the major challenge from this perspective is to maintain the amount of biodiversity that will ensure the resilience of ecosystems, and thereby the flow of crucial renewable resources and ecological services to human societies (Perrings et al. 1991; Perrings this volume). This does not mean that neither ethical and moral concerns for biodiversity conservation (Norton 1986) nor the preference value of humans for particular species without information as to their role in the system (d'Arge this volume) is of no importance. On the contrary, this hierarchy of values has to be explicitly stressed in discussions of biodiversity conservation and sustainable development.

Development Versus Growth

In addition to being concerned with growth in overall activity we also need to consider the current behavior of humans. Since the growth of the scale of human activities not only affects conditions of the natural environment, but also the possible output and composition of ecologically produced goods and services, the whole human subsystem will become limited by the impacts of its own behavior on the life-supporting environment. If this interdependence is not recognized, the potential range for socioeconomic activity will diminish, and the possibilities to develop sustainable systems will become more constrained (Folke 1991; Barbier this volume).

An appropriate distinction of significance for sustainability is the one between growth and development. *Growth* refers to the quantitative increase in the scale of the physical dimension of the economy, the rate of flow of matter and energy through the economy, and the stock of human bodies and artifacts, while *development* refers to the qualitative improvement in the structure, design, and composition of physical stocks and flows, that result from greater knowledge, both of technique and of purpose (Daly 1987). There is a potential for a more efficient use of natural resources, recycling, and reduction of waste and pollutants. This means that there is a potential for economic progress based on development (qualitative improvement) rather than growth (quantitative improvement)—an economic progress that is not at the expense of the environment, but on the contrary, that tries to fit economic activity and human skills into biogeochemical cycles and adjust the economic system within the framework of the overall finite global life-supporting environment (Gilliland and Kash this volume; Viederman this volume).

Scale, Population Growth, and Appropriation of Natural Capital

A well-known operating principle for researchers at the interface of ecology and economics is that the scale and rate of throughput of energy and matter passing through the economic system is subject to an entropy constraint. Intervention is required because the market by itself is unable to accurately reflect this constraint (Pearce and Turner 1990; Kneese, Ayres, and d'Arge 1970). As stated by Daly (1984), "there is no more reason to expect the market to find the optimum scale than there is to expect it to find the optimum income distribution. Just as we impose ethical constraints on income distribution and let the market adjust, so must we be willing to impose ecological constraints on the scale of throughput, and let the market adjust." Daly's analogy with the "plimsoll line" on a boat clearly illustrates this. Suppose we are economists and want to maximize the load that a boat carries. If we place all the weight in one corner of the boat, it will quickly sink or capsize. We need to spread the weight out evenly, and to do this, we invent a price system. The higher the waterline in any corner of the boat, the higher the price for putting another kilogram in that corner, and the lower the waterline, the lower the price. This is the internal optimizing rule for allocating space (resources) among weights (alternative uses). This pricing rule is an allocative mechanism only. In the real world with lack of information and true uncertainty, it sees no reason not to keep on adding weight and distributing it equally until the optimally loaded boat sinks to the bottom of the sea. What is lacking is a limit (albeit dynamic) on scale, a rule that says "stop when total weight is one ton, or when the waterline reaches the red mark" (Daly 1984).

Figure 1.1, 1.2, and 1.3 illustrate this concept in relationship to two other prevalent visions of the future. Figure 1.1 shows the "conventional" economic optimistic view of ever-continuing growth (in terms of the above definitions) of the human-made components of capital at the expense of natural capital. Environmentally minded individuals within the conventional camp argue that this growth can be used to fund preservation of some of the remaining natural capital, but only as a luxury, since natural capital is not necessary to the operation of the economy and could be driven to zero without causing collapse of the economy. Figure 1.2 illustrates a more realistic (but pessimistic) view that shows over-expansion of the human economy causing collapse of the ecological life-support system and ultimately collapse of the economy which depends on it. The collapse may be more or less severe and allow for recovery afterward, but this is still not a very desirable vision of the future. The third vision, (Figure 1.3), indicates the distinction between growth and development, the ecological "plimsoll line" (including uncertainty) and the possibility for continued development (in terms of the above definitions), if the physical dimensions of the economy are maintained below the planet's carrying capacity. This vision of the future encapsulates the essential characteristics of ecological economics.

Ecological economists argue that the biophysical framework for human actions has to be stressed at all levels of society. Otherwise, we cannot talk about a sustainable use of resources and the environment. Few people, at least in the developed world, have yet started to act as if human societies actually are dependent on the life-support ecosystems for their well-being. As analysts, we cannot simply accept human preferences as given and unchanging. We must instead recognize and try to internalize the environmental costs of economic activities (Barbier this volume; King this volume), and to incorporate such a recognition in collective and individual behavior (Robinson et al. 1990). In economics, this has high relevance for estimating opportunity costs. As we are interested in efficiency the present value criterion itself is not irrelevant, but the objective should be to maximize present value of the uses of all scarce resources. Natural capital is rapidly becoming more and more scarce. It is inappropriate to calculate the net benefits of a project or policy alternative by comparing it with unsustainable options. The economic allocation rule for attaining a goal efficiently, (maximize present value) cannot be allowed to subvert the very goal of sustainability that it is supposed to be serving (Daly and Cobb 1989). In the context of project appraisal involving development versus conservation conflicts, it would seem appropriate to require that cost-benefit analysis be used to choose between alternatives only within a choice set bounded by sustainability (ecosystem stability and resilience) constraints (Common and Perrings 1992; Turner et al. this volume).

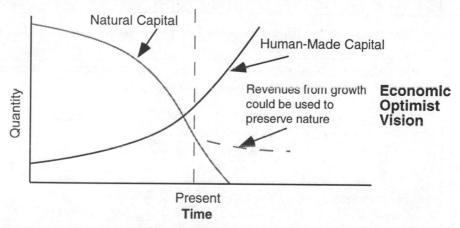

Figure 1.1.   The economic optimist vision of the future with unlimited growth and the environment as a luxury good

Ecological economists stress that markets are not "free," in the sense of being independent. They always act within an institutional or societal framework. It is this framework that determines and directs the behavior of producers and consumers. Markets will contribute to sustainable development only if the societal signals are founded on such a perspective. Hence, the importance of the underlying world view in society.

During the early stages of industrial development, it was not so harmful to neglect the impacts on the ecosystem, because the size of the human population, and the production and consumption patterns of resource use were small relative to the natural environment, and did not significantly interfere with ecosystem functions, including their assimilative capacity for human wastes. Today, the situation is different. The world economy and human population have reached magnitudes at which the effects of its various activities can no longer be absorbed by ecosystems without significant changes and better adjustments of its material flows to the biogeochemical cycles of the biosphere (Ehrlich this volume).

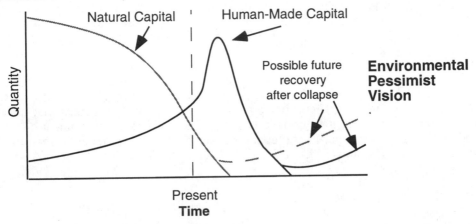

Figure 1.2. The environmental pessimist vision of the future with ultimate collapse of the ecosystem causing collapse of the economy.

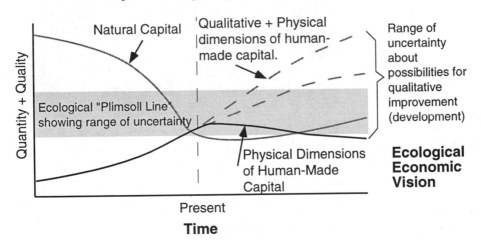

Figure 1.3. The ecological economic vision of the future emphasizing limiting the physical dimensions of the economy below the ecological carrying capacity while encouraging development (qualitative improvement) and acknowledging uncertainty.

Attempts to estimate where we are in relation to carrying capacity limits have been made on a global scale by Vitousek et al. (1986) and Rees and Wackernagel (this volume). They have estimated that as much as 20%–40% of the global net primary production of natural terrestrial ecosystems are diverted to human activities and that human material demand now exceeds the long-term carrying capacity of Earth.

Uncertainty, Irreversibility, and the Precautionary Principle

One of the primary reasons for the problems with current methods of environmental management is the issue of scientific uncertainty. Not just its existence, but the radically different expectations and modes of operation that science and policy have developed to deal with it. If we are to solve this problem, we must understand and expose these differences about the nature of uncertainty and design better methods to incorporate it into the policy- making and management process (Costanza and Cornwell 1992; Costanza this volume).

To understand the scope of the problem, it is necessary to distinguish between *risk* (which is an event with a *known* probability, sometimes referred to as "statistical uncertainty") and *true uncertainty* (which is an event with an *unknown* probability, sometimes referred to as "indeterminacy").

There are risks and uncertainties for reducing natural capital because of our very imperfect understanding of the life-support functions, the inherently limited predictability of ecosystems and social systems (Holling this volume), and our limited capability to invent technical substitutes for natural functions. Complex systems like ecological and economic systems are fundamentally evolutionary and non-linear in causation and of limited predictability (Costanza et al. 1993; Holling this volume). Therefore, policies that rely exclusively on social or economic adaptation to smoothly changing and reversible conditions lead to reduced options, limited potential, and perpetual surprise.

We also generally forget that substituting for natural capital in one place requires natural capital from elsewhere, and that losses of life-support functions are often irreversible. In the face of uncertainty and irreversibility, conserving what there is of natural capital and designing management instruments that are adaptive, flexible, and acknowledge uncertainty ought to be a sound risk-averse strategy (Pearce and Turner 1990; Ludwig et al. 1993; Costanza this volume; Duchin and Lange this volume).

RESEARCH ISSUES FOR ECOLOGICAL ECONOMICS

Much of what is accorded high status as academic research is largely irrelevant to solving the problems of designing a sustainable world. This irrelevance is reinforced because the academic disciplines determine their own research agendas with insufficient communication with other disciplines or with the needs of society. They fall prey to what has been called "the tyranny of illusory precision" (Holdren 1982), "crackpot rigor" (Ehrlich this volume), or the "partial quantification trap" (Costanza and Daly 1987) in attempting to equate "good science" with high precision. They often end up doing in the best possible way something that probably should never be done at all (Ehrlich and Daily 1992; Ehrlich this volume). "Good science" and good academic research needs to be redefined as relevant problem solving in the face of whatever level of precision is possible.

With this in mind and while recognizing that we need to take action now on the basis of imperfect information, and also recognizing that information about the structure and behavior of ecological and economic systems will *always* be imperfect, there are a range of critical issues that would greatly benefit from additional academic research of the more relevant variety. This research should acknowledge and attempt to strengthen the linkages between science and policy in order to develop adaptive and flexible understanding, consensus, and action that can actually achieve sustainability. A partial listing of some important research issues in ecological economics follows.

Sustainability: Maintaining Our Life-Support System

A principal task of ecological economics is to seek ways to keep the scale of human society within sustainable bounds. We need to identify the present status of how large the economic subsystem is compared to the larger sustaining ecosystem and approach the issue of how large the economy can become without irreversibly damaging our future existence. The uncertainty of where society is in relation to the boundaries set by the functioning of the natural environment and the fact that most environmental problems suffer from uncertainty calls for a new approach to environmental research dealing more explicitly with uncertainty (Costanza this volume; Ehrlich this volume; Holling this volume; Clark and Munro this volume).

The boundaries refer to the carrying capacity of the environment to support human activities at various scales. This capacity is dependent on the resilience of ecosystem support, and the behavior of the economy-environment system as a whole. Ehrlich (this volume) and Daily and Ehrlich (1992) distinguish between biophysical and social carrying capacity where social carrying capacity takes the standard of living into account. Appropriated carrying capacity (how much land is required to produce the resource flows or consumption currently enjoyed by that region's human population) has been suggested by Rees and Wackernagel (this volume) and is related to the concept of ghost acreage (Borgström 1967; Odum 1989)

A more complete conceptualization of the economy and the environment thus requires attention to social/cultural/political systems as well. Cultural capital (Berkes and Folke this volume) described as the interface between natural and human-made capital determines how society uses natural capital to create human-made capital. We need to explore the interdependencies between natural, human-made, and cultural capital, as well as the large bank of traditional ecological knowledge that exist in numerous communities around the world, sustaining their resource use (Gadgil et al. 1993). For example, Homer-Dixon et al. (1993) provide evidence that generally deteriorating ecological conditions and growing scarcities of renewable resources will likely contribute to social instability and precipitate civil and international strife.

Natural Resource Valuation

The economic value of ecosystems is connected to their physical, chemical and biophysical role in the long-term, global system—whether the present generation fully recognizes this role or not (Costanza et al. 1991). To achieve sustainability, we need to incorporate ecosystem goods and services into economic accounting. The overriding research issue is to find the most sensible ways of assigning value to natural resources and natural capital (King this volume). Related are the role and value of biodiversity and how to protect the opportunities of choice for future generations (Barbier this volume; d'Arge this volume, Golley this volume; Perrings this volume).

One approach to valuation of environmental functions according to four different categories, regulation functions, carrier functions, production functions and information functions has been suggested by de Groot (1992 and this volume).

Successful attempts to integrate ecological and economic research requires that ecological systems be viewed as sets of processes rather than a collection of resources and that we focus on ecosystem behavior and discontinuities (Holling this volume). The points of discontinuity in ecosystems occur around a set of system thresholds which mark the limits of system resilience. A challenge for ecological economics is to incorporate the dynamic components of ecological systems in economic analysis (Mäler et al. this volume; Perrings this volume).

In valuing biodiversity, the major challenge is to maintain that level of biodiversity which will ensure the resilience of ecosystems. To date, we have tended to invest too much in preserving individual species and not enough in broad categories of ecosystem components. From a policy perspective three important questions need to be analyzed.

- What is the significance and value of biodiversity?

- What are the social and economic forces driving the loss of biodiversity?

- What can be done to reduce or even to reverse the current rate of biodiversity loss?

- What are the priorities within biological conservation? (Perrings et al. 1992; Perrings this volume)

Income and Asset Accounting

Conventional measures of economic performance do not factor resource depletion or environmental degradation into economic trends, thus presenting an incomplete and skewed picture of economic welfare. How can we create better accounting systems that integrate resource depletion or environmental degradation into national and international economic performance?

National income accounting, which has been in use only for the last 50 years, has become the major measure of a country's economic performance. However, natural resources have not been adequately included (Mäler 1991). Several examples show radically different pictures from that depicted by conventional methods, when resource depletion and environmental degradation are included. Daly and Cobb (1989) estimated that while GNP in the USA rose during the 1956–86 period, their "index of sustainable economic welfare" remained unchanged since about 1970. Similar, if less dramatic results, have been obtained in a study of forest, fisheries, and soil depletion in Costa Rica by the World Resources Institute and the Tropical Science Center (Repetto 1992).

A major research issue is to develop consistent sets of sustainability indicators on all levels identifying the resource use of non-renewable and renewable resources, and ecological services, taking into consideration the overall well-being, or health of ecological systems.

Safe minimum standards championed by a few economists may be relevant to the protection of critical levels of natural capital. Under what circumstances is the use of safe minimum standards advisable? What are the optimal forms of safe minimum standards for habitat protection, and biodiversity?

There are a number of additional promising approaches to accounting for ecosystem services and natural capital being developed (de Groot this volume; Mäler et al. this volume; Odum this volume) and "green accounting" is being investigated by several governmental institutions in many countries.

Ecological Economic Modeling

Preserving and protecting threatened ecosystems requires an understanding of the direct and indirect effects of human activities on large geographical areas over time (Duchin and Lange this volume; Krysanova and Kaganovich this volume; Jansson and Jansson this volume). How do we create an integrated, multiscale, pluralistic approach to quantitative ecological economic modeling while developing new ways to effectively deal with the inherent uncertainty involved in modeling complex systems (Costanza et al. 1993)?

Some major researchable questions and opportunities in modeling complex ecological and economic systems are listed below:

1.   *Application of the "evolutionary paradigm" to modeling ecological economic systems.*

    The evolutionary paradigm provides a general framework for complex ecological economic system dynamics. It incorporates the elements of uncertainty, surprise, learning, path-dependence, multiple equilibria, sub-optimal performance, lock-in, and thermodynamic constraints. In applying the evolutionary paradigm, a key feature is the choice of the measure (or multiple measures) of performance on which the system's selection process will work. Several such measures have been proposed and partially tested, but additional research and testing in this area may have a high pay-off. An important research question is the range of applicability of non-equilibrium thermodynamic principles and their appropriate use in modeling ecological economic systems. Key methods include adaptive computer simulation models that incorporate geographic information and evolutionary game theory.

2. *Scale and hierarchy considerations in modeling ecological economic systems.*

The key questions seem to involve exactly how hierarchical levels interact with each other. These questions feed into the larger question of scaling for application to complex, regional, ecological economic systems (Krysanova and Kaganovich this volume).

Additional questions concern the range of applicability of fractals and chaotic systems dynamics to the practical problems of modeling ecological economic systems. In particular, what is the influence of scale, resolution, and hierarchy on the mix of behaviors one observes in systems. This is a key question for extrapolating from small scale experiments or simple theoretical models to practical applied models of ecological economic systems at regional and global scales.

3. *The nature and limits of predictability in modeling ecological economic systems.*

The significant effects of nonlinearities raises some interesting questions about the influence of resolution (including spatial, temporal, and component) on the performance of models, in particular their predictability. There may be limits to the predictability of natural phenomenon at particular resolutions, and "fractal-like" rules that determine how both "data" and "model" predictability change with resolution. To test this we need better measures of model correspondence with reality, and better measures of long-term aggregate system performance.

Institutions for Sustainable Governance

The practical and institutional barriers to sustainability need to be identified and institutions with the flexibility necessary to deal with ecologically sustainable development need to be designed and implemented (Barbier this volume; Gilliland and Kash this volume; Berkes and Folke this volume; Segura and Boyce this volume; Zucchetto this volume; d'Arge this volume). The uncertainties faced by ecological economists on the ecological side are compounded by the uncertainties on the economic and social sides and vice-versa, and over the next few critical decades those uncertainties are unlikely to be significantly reduced. Therefore, two related major challenges for ecological economists are (1) to help society identify and evaluate the "premiums" worth paying for environmental insurance, and (2) to develop strategies for moving toward sustainability that are relatively insensitive to uncertainties—"robust" strategies that are helpful regardless of how uncertainties are ultimately resolved (Ehrlich this volume; Costanza this volume). This will most likely involve systems of governance and decision-making (1) that can involve stakeholders (including nature and future generations) in a more effective way, (2) that distribute property rights more effectively, and (3) that are adaptive and flexible (Golley this volume; d'Arge this volume; Viederman this volume).

Other major questions about institutional reform that can benefit from additional research are listed below.

- How can governmental and other institutions be modified to better account for and respond to the environmental impacts of economic development?

- Can we develop better, more resilient systems of governance that adequately empower the full range of interests, including nature?

- What sociological, political, ethical, or other factors have limited the acceptance of particular instruments, and can these factors be addressed?

- How can we develop experimental economics in order to predict behavioral responses to new management instruments? What role might computer modeling play in this development?

- How do we equitably limit world population without oppressive programs?

- How do we develop mechanisms to lengthen the time horizons of institutions at all levels?

- What institutions are most effective at preserving the pool of genetic information, preserving the ecological knowledge of indigenous people, and facilitating cultural adaptations and/or technological change?

- What international institutions are available or necessary to assure local and global sustainability?

- What are the conditions by which international trade may be both economically equitable and environmentally sustainable for all parties?

- How can intergenerational distribution be addressed analytically as well as ethically?

Hence, ecological economics is seeking an appropriate "political economy" for sustainability. Ecological economics argues that unfettered markets and "economic growth-dominated" policy strategies are definitely unsustainable and carry with them significant environmental degradation and damage cost burdens. Thus, market intervention via some system of regulations (beyond the mere protection of property rights) is a fundamental prerequisite for a practicable sustainable development strategy.

POLICY RECOMMENDATIONS

The Wallenberg workshop aimed, in particular, at discussing and reaching consensus on the issues related to investments in natural capital for sustainability. The following represents a limited set of policy recommendations on which the workshop participants reached general consensus. It is not prioritized, nor is it comprehensive, nor does it imply that all the participants were in complete agreement.

Increasing the Understanding of Ecological Functions for Welfare

Maintaining and investing in natural capital are necessary prerequisites for sustainability. We need to move from abstract notions of environmental problems and interrelations between the functions of nature and the economy towards concrete concepts and participation (de Groot this volume). Incorporating ecological functions is especially important in inter-regional relationships and trade (Rees and Wackernagel this volume). Central and eastern European governments should be encouraged not only to decontaminate polluted environments but to invest in natural capital. A more proactive approach is needed to stimulate these investments (Zylicz this volume). A broader range of values should be promoted, including ethics, equity and intergenerational issues (d'Arge this volume; Perrings this volume; Segura, and Boyce this volume; Golley this volume).

Improving the Use of Policy Instruments

The use of a broad range of policy instruments should be promoted based on criteria of equity, scientific validity, consensus, and environmental effectiveness. We should institute regulatory reforms to promote appropriate use of financial, legal, and social incentives, and implement pollution taxes, taxes on natural capital depletion, regulatory targets, and make use of tradable permits (Clark and Munro this volume; King this volume). Within the baseline regulatory framework (ambient quality standards and targets, etc.) economic incentive instruments have a vital "cost-effectiveness" role to play. Regulations and incentives should be designed and implemented as entirely complementary, mutually supportive devices. The implementation of this set of sustainability-enabling policy measures can both serve to correct for current market and policy failures, as well as being impeded by such failures and the social and political rigidities that underpin them. Both ecological, economic, and other information will be needed to find the optimal trade-offs between more information for a better understanding and the costs for not acting, thereby making sure that information is not wasted and uncertainty reduced optimally. We need to promote the employment of positive sustainability incentives based on the precautionary principle, and the polluter pays principle. (Costanza this volume; Zylicz this volume). In decisions of scale, individual freedom of choice must yield to democratic collective decision making by the relevant community.

Institutional Changes and Property Rights

The largest barriers to sustainability need to be identified in order to redirect institutional structures and patterns (Gilliland and Kash this volume). Presently, institutions with the flexibility necessary to deal with ecologically sustainable development are lacking. At the same time as global connections are becoming more evident, local, and regional adaptive capabilities are eroding. Much environmental knowledge held by indigenous people is being lost. Establishing rules at one level without rules at other levels will create incomplete institutional systems that may not endure in the long run (Ostrom 1991; Berkes and Folke this volume). Public participation and the involvement of local citizens is needed for creating consistent institutional systems capable of handling sustainability (Golley this volume).

Ecological Economics Education

Transforming society towards sustainability requires transdisciplinary, problem-oriented knowledge. Broadly trained environmental professionals are needed, in both private and public institutions to incorporate environmental issues and bridge the gap between disciplines. Capacity building is needed particularly in developing countries and central and eastern Europe (Zylicz this volume; Segura and Boyce this volume; Barbier this volume). We need to encourage interdisciplinary case studies all the way from ecological·understanding to economics, policy, and management.

We should promote transdisciplinary, problem-oriented education on all levels. An ecological economics curriculum is needed that adequately integrates biological, chemical, and physical sciences with social and economic sciences, where the fundamental understanding of the interdependencies between the environment, our life-support systems, and human economic, social, and cultural activities are emphasized.

ACKNOWLEDGMENTS

This chapter benefited from the broad-ranging discussions between the contributing authors of this book, held at the 2 1/2-day workshop at Saltjobaden, Sweden, August 7–9, 1992. We have tried to summarize those discussions, and the ideas presented at the 2nd ISEE International Conference held in Stockholm, Sweden August 3–6, 1992. The workshop was generously funded by the Wallenberg Foundation. We also thank Joy Bartholomew and Lisa Wainger for helpful reviews on earlier drafts.

REFERENCES

Barbier, E. B. 1987. The concept of sustainable economic development. *Environmental Conservation* 14: 101–10.
Borgström, G. 1967. The Hungry Planet. New York: Macmillan.
Common, M., and C. Perrings. 1992. Towards an ecological economics of sustainability. *Ecological Economics* 6: 7–34.

Costanza, R., ed. 1991. Ecological Economics: The Science and Management of Sustainability. New York: Columbia Univ. Press.

Costanza, R., and L. Cornwell. 1992. The 4P approach to dealing with scientific uncertainty. *Environment* 34: 12–20.

Costanza, R., and H. E. Daly. 1987. Toward an ecological economics. *Ecological Modelling* 38: 1–7.

———. 1992. Natural capital and sustainable development. *Conservation Biology* 6: 37–46.

Costanza, R., H. E. Daly, and J. A. Bartholomew. 1991. Goals, agenda and policy recommendations for ecological economics. In Ecological Economics: The Science and Management of Sustainability, ed. R. Costanza. New York: Columbia Univ. Press.

Costanza, R., L. Wainger, C. Folke, and K-G Mäler. 1993. Modeling complex ecological economic systems: toward an evolutionary, dynamic understanding of people and nature. *BioScience* 43: 545–555.

Cleveland, C. J. 1991. Natural resource scarcity and economic growth revisited: economic and biophysical perspectives. In Ecological Economics: The Science and Management of Sustainability, ed. R. Costanza. New York: Columbia Univ. Press.

Daly, H. E. 1984. Alternative strategies for integrating economics and ecology. In Integration of Ecology and Economics: An Outlook for the Eighties, ed. AM. Jansson. Department of Systems Ecology. Stockholm: Stockholm University. Reprinted in Daly, H. E. 1991. Steady-State Economics. Covelo, CA: Island Press.

———. 1987. The economic growth debate: what some economists have learned but others have not. *Journal of Environmental Economics and Management* 14: 323–36.

Daly, H. E., and J. B. Cobb. 1989. For the Common Good: Redirecting the Economy Toward Community, the Environment and a Sustainable Future. Boston: Beacon Press.

Daily, G., and P. R. Ehrlich. 1992. Population, sustainability, and Earth's carrying capacity. *BioScience* 42: 761–71.

de Groot, R. S. 1992. Functions of Nature: Evaluation of Nature in Environmental Planning, Management and Decision-Making. Groningen, The Netherlands: Wolters-Noordhoff BV.

Ehrlich, P. R. 1989. The limits to substitution: meta-resource depletion and a new economic-ecological paradigm. *Ecological Economics* 1: 9–16.

Folke, C. 1991. Socioeconomic dependence on the life-supporting environment. In Linking the Natural Environment and the Economy: Essays from the Eco-Eco Group, eds. C. Folke and T. Kåberger. Dordrecht: Kluwer Academic Publishers.

Folke, C., and N. Kautsky. 1992. Aquaculture with its environment: prospects for sustainability. *Ocean and Coastal Management* 17: 5–24.

Gadgil, M., F. Berkes, and C. Folke. 1993. Indigenous knowledge for biodiversity conservation. *Ambio* 22: 151–6.

Gren, I-M., C. Folke, R. K. Turner, and I. Bateman. 1994. In press. Primary and secondary values of wetland ecosystems. *Environmental and Resource Economics*.

Hall, C. A. S., C. J. Cleveland, and R. Kaufmann. 1986. Energy and Resource Quality: The Ecology of the Economic Process. New York: Wiley.

Hammer, M., AM. Jansson, and B-O. Jansson. 1993. Diversity change and sustainability: implications for fisheries. *Ambio* 22: 97–105.

Holdren, J. P. 1982. Energy risks: what to measure, what to compare. *Technology Review* (April): 32–8.

Holling, C. S. 1992. Cross-scale morphology, geometry and dynamics of ecosystems. *Ecological Monographs* 62: 447–502.

Homer-Dixon, T., J. Boutwell, and G. Rathjens. 1993. Environmental change and violent conflict. *Scientific American* (February): 38–45.

2 OPERATIONALIZING SUSTAINABLE DEVELOPMENT BY INVESTING IN NATURAL CAPITAL

Herman E. Daly
School of Public Affairs
University of Maryland[1]
College Park, MD 20742–8311

ABSTRACT

This chapter argues that we have moved from an "empty world" to a "full world,"—a world relatively full of human beings and their artifacts. The economic strategy appropriate to the new "full world" era is sustainable development. Operationalizing sustainable development requires investment in natural capital. The first part of this chapter consists of seven propositions which, taken together, constitute an argument for shifting the focus and active margin of investment from man-made to natural capital. The second part discusses the meaning of "investment in natural capital"—how to invest (directly or indirectly, actively or passively) in something that, by definition, we cannot make.

WHY INVEST IN NATURAL CAPITAL?—SEVEN PROPOSITIONS

1. The pre-analytic vision at the foundation of standard economics is that of an isolated circular flow of exchange value between firms and households. Nothing enters from the environment nor exits to it. The physical environment is completely abstracted from (see Figure 2.1).

    By contrast, the pre-analytic vision of ecological economics is that the economy, in its physical dimensions, is an open subsystem of a finite, nongrowing, and materially closed ecosystem (see Figure 2.2).

2. The economic subsystem has grown relative to the containing ecosystem to the extent that remaining natural capital has become scarce relative to man-made capital, reversing the previous pattern of scarcity (see Figure 2.3). This forces into practical attention three neglected macroeconomic questions: how big *is* the economic subsystem relative to the containing ecosystem?; how big *can* it be without destroying the larger sustaining system?; and how big *should* it be in order

---

1. The views here presented are those of the author and should in no way be attributed to the University of Maryland, School of Public Affairs.

to optimize life enjoyment? Life enjoyment can be interpreted in an anthropocentric way (for human beings, recognizing only instrumental value of other
species); or in a biocentric way (recognizing intrinsic as well as instrumental
value of other species).

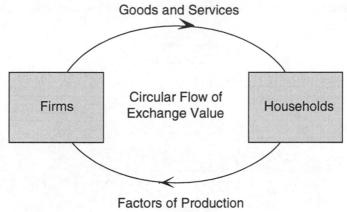

Figure 2.1. Economy as an isolated system.

3. Although the environment has been abstracted from by standard economics, the
    concept of sustainability has been recognized and incorporated into the very defi-
    nition of income as "the maximum amount that a community can consume over
    some time period and still be as well off at the end of the period as at the begin-
    ning" (J. R. Hicks 1946). Being as well off means having the same capacity to
    produce the same income in the next year—i.e., *maintaining capital intact*.[2] The

---

2. *Maintaining capital intact* when technology is constant is the same as maintaining
    physical capital intact. When technology increases the productivity of capital, then it is
    not so clear what "keeping capital intact" means. The maximum amount that the
    community can sustainably consume has increased. To count that increase as income, we
    must continue to maintain the same physical capital intact. This would be the prudent
    course. We could, however, opt to maintain the old smaller income and take the benefits of
    the technological improvement in the form of a one-time increase in consumption (of
    capital), while maintaining only the capital needed to produce the former income stream.
    Capital includes both natural and man-made. We should avoid the error, common in the past
    and even now, of consuming the benefits of increased productivity of man-made capital by
    running down natural capital. As will be argued here, the two forms of capital are
    complementary. Furthermore, the main reason for the historical increase in the
    productivity of manmade capital has been that it has had increasing amounts of natural
    resources to work with, resulting in a decline in the productivity of natural resources (and
    natural capital). The historical increase in manmade capital productivity has been partly at
    the expense of reduced natural capital productivity resulting from its extravagant use—as if
    it were free. Therefore, the historical increase in man-made capital productivity cannot be

criterion of sustainability is thus explicit in this Hicksian definition of income. But the condition of maintaining capital intact has applied only to man-made capital, since in the past natural capital was abstracted from because it was not scarce. The Hicksian definition of income must in the future apply to total scarce capital, which now includes natural capital as well as man-made.

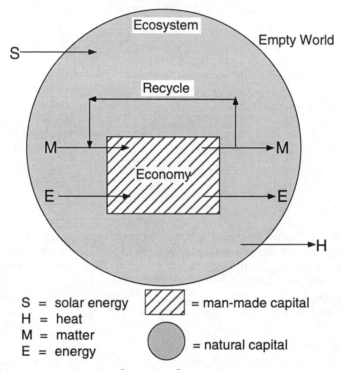

Figure 2.2. Economy as an open subsystem of an ecosystem.

4. There are two ways to maintain total capital intact: (1) the *sum* of man-made and natural capital can be maintained constant in some aggregate value sense; or (2) *each* component can be maintained intact separately, again in some aggregate value sense, but this time there is aggregation only within the two categories and not across them. The first way is reasonable if one believes that man-made and natural capital are substitutes. This view holds that it is totally acceptable to divest natural capital as long as one creates by investment an equivalent value in man-made capital. The second way is reasonable if one believes that man-made and natural capital are complements. The complements must each be maintained

taken as evidence for an increase in natural capital productivity, or as a reason for optimistic expectations in that regard.

intact (separately or jointly in fixed proportion), because the productivity of one depends on the availability of the other. The first case is called *weak sustainability,* and the second case is called *strong sustainability.*

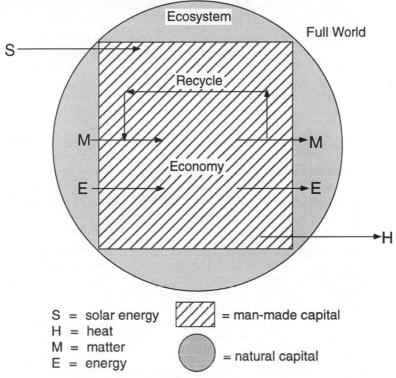

Figure 2.3. Reversing pervious pattern of scarcity.

5. Man-made and natural capital are fundamentally complements and only marginally substitutes. Therefore, strong sustainability is ultimately the relevant concept, although even weak sustainability would be an improvement over current practice. Since this will likely be the most contested of the seven propositions, it is worth taking time to consider the three basic reasons behind it.

    a. One way to make an argument is to assume the opposite and show that it is absurd. If man-made capital were a near perfect substitute for natural capital, then natural capital would be a near perfect substitute for man-made capital. But if so, there would have been no reason to accumulate man-made capital in the first place, since we were endowed by nature with a near perfect substitute. But historically we did accumulate man-made capital—precisely because it is complementary to natural capital.

b. Man-made capital is itself a physical transformation of natural resources which come from natural capital. Therefore, producing more of the alleged substitute (man-made capital), physically requires more of the very thing being substituted for (natural capital)—the defining condition of complementarity!

c. Man-made capital (along with labor) is an agent of transformation of the resource flow from raw material inputs into product outputs. The natural resource flow (and the natural capital stock that generates it) are the *material cause* of production; the capital stock that transforms raw material inputs into product outputs is the *efficient cause* of production. One cannot substitute efficient cause for material cause—one cannot build the same wooden house with half the timber no matter how many saws and carpenters one tries to substitute. Also, to process more timber into more wooden houses, in the same time period, requires more saws, and carpenters. Clearly the basic relation of man-made and natural capital is one of complementarity, not substitutability. Of course, one could substitute bricks for timber, but that is the substitution of one resource input for another, not the substitution of capital for resources.[3] In making a brick house one would face the analogous inability of trowels and masons to substitute for bricks.

The complementarity of man-made and natural capital is made obvious at a concrete and common sense level by asking: what good is a saw-mill without a forest; a fishing boat without populations of fish, a refinery without petroleum deposits, an irrigated farm without an aquifer or river? We have long recognized the complementarity between public infrastructure and private capital—what good is a car or truck without roads to drive on? Following Lotka and Georgescu-Roegen, we can take the concept of natural capital even further and distinguish between *endosomatic* (within-skin) and *exosomatic* (outside-skin) natural capital. We can then ask, what good is the private endosomatic capital of our lungs and respiratory system without the public exosomatic capital of green plants that take up our carbon dioxide in the short run, while, in the long run replenishing the enormous atmospheric stock of oxygen and keeping the atmosphere at the proper mix of gases—i.e., the mix to which our respiratory system is adapted and therefore complementary.

---

3. Regarding the house example I am frequently told that insulation (capital) is a substitute for resources (energy for space heating). If the house is considered the final product, then capital (agent of production, efficient cause) cannot end up as a part (material cause) of the house, whether as wood, brick, or insulating material. The insulating material is a resource like wood or brick, not capital. If the final product is not taken as the house but the service of the house in providing warmth, then the entire house, not only insulating material, is capital. In this case, more or better capital (a well-insulated house) does reduce wasted energy. Increasing the efficiency with which a resource is used is certainly a good substitute for more of the resource. But these kinds of waste-reducing efficiency measures (recycling prompt scrap, sweeping up sawdust and using it for fuel or particle board, or reducing heat loss from a house) are all rather marginal, limited substitutions.

If natural and man-made capital are obviously complements, how is it that economists have overwhelmingly treated them as substitutes? First, not all economists have—Leontief's input-output economics with its assumption of fixed factor proportions treats all factors as complements. Second, the formal, mathematical definitions of complementarity and substitutability are such that in the two-factor case, the factors must be substitutes.[4] Since most textbooks are written on two-dimensional paper, this case receives most attention. Third, mathematical convenience continues to dominate reality in the general reliance on Cobb-Douglas and other constant elasticity of substitution production functions, in which there is nearly infinite substitutability of factors, in particular of capital for resources. Thankfully, some economists have begun to constrain this substitution by the law of conservation of mass! Fourth, exclusive attention to the margin results in marginal substitution obscuring overall relations of complementarity. For example, private expenditure on extra car maintenance may substitute for reduced public expenditure on roads. But this marginal element of substitution should not obscure the fact that private cars and roads are basically complementary forms of capital.[5] Fifth, there may well be substitution of capital for resources in aggregate production functions reflecting a change in product mix from resource-intensive to capital-intensive products. But this is an artifact of changing product aggregation, not factor substitution along a given product isoquant. Also, a new product may be designed that gives the same service with less resource use—e.g., light bulbs that give more lumens per watt. This is technical progress—a qualitative improvement in the state of the art—not the substitution of a quantity of capital for a quantity of resources in the production of a given quantity of a specific product.

---

4. The usual definition of complementarity requires that for a given constant output, a rise in the price of one factor would reduce the quantity of both factors. In the two-factor case, both factors means all factors, and it is impossible to keep output constant while reducing the input of all factors. But complementarity might be defined back into existence in the two-factor case by avoiding the constant output condition. For example, two factors could be considered complements if an increase in one factor will not increase output, but an increase in the other factor will,—and *perfect* complements if an increase in neither factor alone will increase output, but an increase in both will. It is not sufficient to treat complementarity as if it were nothing more than "limited substitutability." The latter means that we could get along well enough with only one factor and less well with only the other, but that we do not need both. Complementarity means we need both, and that the one in shortest supply is limiting.
5. At the margin a right glove can substitute for a left glove by turning it inside out. Socks can substitute for shoes by wearing an extra pair to compensate for thinning soles. But in spite of this marginal substitution, shoes and socks, or right and left gloves are overwhelmingly complements. The same is basically true for man-made and natural capital. Picture their isoquants as L-shaped, having a 90° angle. Erase the angle and draw in a tiny 90° arc connecting the two legs of the L. This seems close to reality. However, this very marginal range of substitution has been so overextrapolated that even Nobel Laureate economist Robert Solow has gravely opined that, thanks to substitution, "...the world can, in effect, get along without natural resources."

No one denies the reality of technical progress, but to call such changes the substitution of capital for resources (or of man-made for natural capital) is confusing. It seems that some economists are counting all improvements in knowledge, technology, and managerial skills—in short, anything that would increase the efficiency with which resources are used—as "capital." If this is the usage, then "capital" and resources would by definition be substitutes in the same sense that more efficient use of a resource is a good substitute for having more of the resource. But to formally define capital as efficiency would make a mockery of the neoclassical theory of production, where efficiency is a ratio of output to input, and capital is a quantity of input.

6. If we accept that natural and man-made capital are complements rather than substitutes, then what follows? *If factors are complements, then the one in shortest supply will be the limiting factor.* If factors are substitutes, then neither can be a limiting factor since the productivity of one does not depend much on availability of the other. The notion of a limiting factor is familiar to ecologists in Leibig's Law of the Minimum. The idea that either natural or man-made capital could be a limiting factor simply cannot arise if the factors are thought to be substitutes. Once we see that they are complements, then we must ask which one is the limiting factor—i.e., which is in shortest supply?

This proposition gives rise to the following thesis: *that the world is moving from an era in which man-made capital was the limiting factor into an era in which remaining natural capital is the limiting factor.* The production of caught fish is currently limited by remaining fish populations, not by number of fishing boats; timber production is limited by remaining forests, not by sawmills; barrels of pumped crude oil is limited by petroleum deposits (or perhaps more stringently by the capacity of the atmosphere to absorb CO_2) , not by pumping capacity; and agricultural production is frequently limited by water availability, not by tractors, harvesters, or even land area. We have moved from a world relatively full of natural capital and empty of man-made capital (and people) to a world relatively full of the latter and empty of the former (see Figure 2.2 and Figure 2.3).

7. Economic logic requires that we maximize the productivity of the limiting factor in the short-run, and invest in increasing its supply in the long-run. When the limiting factor changes, then behavior that used to be economic becomes uneconomic. Economic logic remains the same, but the pattern of scarcity in the world changes, with the result that behavior must change if it is to remain economic. Instead of maximizing returns to and investing in man-made capital (as was appropriate in an empty world), we must now maximize returns to and invest in natural capital (as is appropriate in a full world). This is not "new economics," but new behavior consistent with "old economics" in a world with a new pattern of scarcity.

In conclusion, since natural capital has replaced man-made capital as the limiting factor, we should adopt policies that maximize its present productivity and increase its future supply. This conclusion is far from being trivial or irrelevant, because it means that current policies of maximizing the productivity and accumulation of man-made capital are no longer "economic," even in the most traditional sense. In addition, the Hicksian definition of income imposes the condition that capital be maintained intact. If natural capital is the limiting factor, then the proper measurement of income requires that natural capital maintenance take priority.

But how could the pattern of scarcity have changed so dramatically without economists noticing it? Several factors account for this development. First, exponential growth is deceptive. The bottle goes from half-full to totally full in the same time it took to go from 1% to 2% full. Second, economists have considered man-made and natural capital to be substitutes, when they are basically complements. If factors are substitutes, then a shortage of one does not limit the productivity of the other. Neither factor can be limiting if they are good substitutes. So even as the world moves from 40% to 80% full in the next roughly 35-year doubling time[6] (Vitousek et al. 1986), economists are counting on man-made capital to restore the conditions of relative emptiness by substituting for natural capital. Third, if we subconsciously realize that production growth cannot continue, then the only way to cure poverty is to confront both sharing and population control. Since these are considered impossible by political "realists," it is gratuitously concluded that whatever argument gave rise to this conclusion must be wrong. These three biases may have kept us from seeing the obvious—namely, that man-made and natural capital are complements, and that natural capital has become the limiting factor. More man-made capital, far from substituting for natural capital, just puts greater complementary demands on it, running it down faster to temporarily support the value of man-made capital, making it all the more limiting in the future.

HOW TO INVEST IN NATURAL CAPITAL

Even if one is convinced by the previous argument that the focus of investment should shift from man-made to natural capital, a problem remains. Since natural capital is by definition not man-made, it is not immediately obvious what is meant by "investing" in it. Yet the term "investment" applies because the concept involves the classical notion of "waiting" or refraining from current consumption as the way to invest in natural capital. Before investigating further the meaning of investment in natural capital, we should examine the concept of natural capital itself.

---

6. The figures are mainly suggestive, but do have an empirical basis in the estimate that humans currently preempt 40% of the net primary product of photosynthesis for land-based ecosystems. In other words, 40% of the solar energy potentially capturable by plants and available to other living things passes through the human economy, or is in some way subject to human purposes. This seems a reasonable index of how full the world is of humans and their possessions.

Natural capital is the stock that yields the flow of natural resources—the population of fish in the ocean that regenerates the flow of caught fish that go to market; the standing forest that regenerates the flow of cut timber; the petroleum deposits in the ground whose liquidation yields the flow of pumped crude oil. The natural income yielded by natural capital consists of natural services as well as natural resources. Natural capital is divided into two kinds, as represented in the examples given: renewable (fish, trees); and non-renewable (petroleum). Man-made capital is used here in Irving Fisher's sense to include stocks of both producer and consumer goods.

Several difficulties with these definitions should be noticed. First, capital has traditionally been defined as "produced (man-made) means of production," yet natural capital was not and cannot be produced by man. A more functional definition of capital is "a stock that yields a flow of useful goods or services into the future," and natural capital fits this concept very well, as do durable consumer goods. And, of course, renewable resources can be exploited to extinction and rendered non-renewable—while non-renewable resources can be renewed if we are prepared to wait indefinitely. Subject to these caveats, the terms are well-defined and are in current use (Costanza and Daly 1992).

Also, there is an important category that overlaps natural and man-made capital—such things as plantation forests, fish ponds, herds of cattle bred for certain characteristics, etc., are not really man-made, but are significantly modified from their natural state by human action. We will refer to these things as "cultivated natural capital." This is a broad category, including agriculture, aquaculture, and plantation forestry. How does it affect our distinction between natural and man-made capital, and the claim that they are complements? One can analyze cultivated natural capital into its components of man-made and natural capital proper. For example, a plantation forest has a natural capital component of sunlight, rainfall, and soil nutrients plus a man-made capital component of management services such as planting, spacing, culling, and control of diseases. In general, there seems to be a strong complementary relation between the natural and man-made components of cultivated natural capital. Nevertheless, cultivated natural capital does substitute for natural capital proper in certain functions—those for which it is cultivated, such as timber production, but not wildlife habitat or biodiversity in the case of a plantation forest.[7]

---

7. From the familiar biological yield curve that follows, it is clear that a sustainable harvest of H will be yielded either at a stock of P1 or P2.

Investment in renewable natural capital will now be considered, followed by the more problematic case of non-renewable natural capital.

For renewable resource management, "waiting" investment simply means constraining the annual offtake. Keeping the annual offtake equal to the annual growth increment (sustainable yield) is equivalent to maintenance investment—i.e., the avoidance of running down the productive stock, equivalent to the Hicksian condition that capital remain intact. Net investment in renewables requires additional waiting—allowing all or a part of the growth increment to be added to the productive stock each year rather than be consumed. Investment in natural capital, both maintenance and net investment, is fundamentally passive with respect to natural capital which is simply left alone and allowed to regenerate. Cultivated natural capital investment also involves waiting, except that it is never really left alone. Even during the waiting period, some tending and supervision is required.

By definition investment in renewable natural capital must be only passive. But more active investment is possible in cultivated natural capital. How far can we go with this type of investment? Can we cultivate an entire biosphere? Does Biosphere II in Arizona stand a chance of working? Beyond its un-

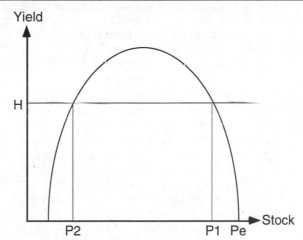

In general, P1 is the natural capital mode of exploitation of a wild population. P2 is the cultivated capital mode of exploitation of a bred population. At P1 we have a large population taking up a lot of ecological space, but providing, in addition to a yield of H, other natural services, as well as maintaining a larger amount of biodiversity. Costs are basically harvest cost of the wild population. At P2 we have a much smaller stock giving the same yield of H, requiring much less ecological space, but requiring greater maintenance, breeding, feeding, and confinement costs if viewed as cultivated natural capital. The appeal of cultivated natural capital is to get H from a low P2, making ecological room for other exploited (or wild) populations. But management costs are high. The appeal of P1 and the mode of natural capital proper is that management service is free, and the biodiversity of larger stocks is greater.

doubted experimental value on a small scale, is a larger version of such active investment in cultivated natural capital at the ecosystem and biospheric scale likely to be a good bet in the future? Can the whole ocean become a catfish pond? Even if it could, would there not still be complementarity with the more basic natural capital of sunlight, chlorophyll, and decomposers? I do not know how far we can rely on cultivated natural capital. However, we are currently so far away from the requisite understanding of ecosystems that their large scale redesign should be ruled out as at best quantitatively insignificant, and at worst qualitatively dangerous. As Paul Ehrlich has reminded us, ecological economics is a discipline with a time limit. We do not have time to learn how to create a cultivated "Biosphere II." We must save the remnants of Biosphere I and allow them to regenerate by the passive investment of waiting. The term *"laissez faire"* thus acquires a new and deeper meaning for ecological economists.

Non-renewable natural capital cannot be increased either actively or passively. It can only be diminished. We can only divest non-renewable natural capital itself, even though we invest in the man-made capital equipment that hastens its rate of extraction and divestment. Non-renewable natural capital is like an inventory of already-produced goods, rather than a productive machine or a reproducing population. For non-renewable natural capital the question is not how to invest, but how to best liquidate the inventory, and what to do with the net wealth realized from that liquidation. Currently, we are counting this liquidated wealth as income (included in both GNP and NNP), which is clearly wrong, because it is not a permanent or sustainable source of consumption.

A better alternative would be to dedicate all or part of the net receipts of non-renewable resource liquidation to finance waiting investments in renewable natural capital—i.e., to allow reduction of the offtake of renewables in order to build up renewable stocks to larger levels producing larger sustainable yields which represent true income. The basic idea is to convert non-renewable natural capital into a renewable substitute, to the extent possible. The general rule would be to deplete non-renewables at a rate equal to the rate of development of renewable substitutes. Thus, extractive projects based on non-renewables must be paired in some way with a project that develops the renewable substitute. Net receipts of non-renewable exploitation are divided into two components (an income component and a capital set-aside), such that the capital set-aside, when invested in a renewable substitute each year, will, by the time the non-renewable is depleted, have grown to a stock size whose sustainable yield is equal to the income component that was being consumed all along. The capital set-side will be greater the lower the growth rate of the renewable substitute and the shorter the lifetime of the non-renewable reserves (i.e., the reserve stock divided by annual depletion). The logic and calculations have been worked out by El Serafy (1988) in the context of national income accounting, but they apply with equal relevance to accounting at the project level. The true rate of return on the project pair would be calculated on the basis of the income component only as net revenue. This differs from the usual cost/benefit evaluation of projects in explicitly costing sustainability by using income (sustainable by definition) rather than cash flow, and by requiring actual rather than hypothetical replacement investment. It does *not* require that the project itself be immortal, nor that non-renewables should forever remain in the ground, never to benefit anyone.

Difficulties remain in the question of defining "substitute"—whether narrowly or broadly. Probably a broad definition would be indicated initially—at least broad enough to encompass improvements in energy efficiency as a renewable substitute for petroleum depletion, and improvements in recycling as a renewable substitute for copper depletion.

In the case of divestment of a *renewable* resource stock, capital consumption is treated as depreciation of a productive asset (the sacrificed base population that was producing a permanent yield). Depreciation should be deducted from gross income to get net income. In the case of non-renewables, the reduction of stocks is treated as a liquidation of existing inventories rather than as depletion of capacity for future production, and consequently, should not even be a part of gross income, as El Serafy rightly insists.

A difficulty in the application of the rule can easily be imagined. Suppose alcohol is the nearest renewable substitute for gasoline. If one tried to invest all of the capital component of current petroleum net revenues in natural capital or cultivated natural capital (wood or sugar cane alcohol), the price of alcohol would have to rise enormously to allow large amounts of land to be bid away from food to plant sugar cane. (This is an indication of how far away we are from sustainability.) The solution would be to begin pricing current petroleum energy at the cost of sugar cane energy—i.c., at the cost of its long-run renewable substitute. Then, as the price of both petroleum and alcohol energy increases, the rate of depletion of petroleum will be slowed. As depletion is slowed, the life expectancy of petroleum reserves will increase, and the percentage of net petroleum rents representing the capital set-aside will diminish. At some higher price, the amount of petroleum rents to be invested in alcohol will fall. In effect, the difficulty of investing the petroleum capital component in a renewable substitute will provide the effective limit on the rate of depletion of petroleum. This remains true even if fusion energy rather than alcohol is used as the long-run permanent substitute.

This rule is similar to that of economist John Ise (1925) who argued that non-renewables should be priced at the cost of their nearest renewable substitute. In the previous example, alcohol was treated as the nearest renewable substitute, but other renewable near substitutes may be cheaper (e.g., technical improvements in energy efficiency) and should be the effective near substitute as long as it remains the cheaper alternative. This last point can be generalized: *although we cannot invest in non-renewables, we can manage their liquidation in such a way as to increase direct passive investment in renewables and indirect active investment in measures to increase throughput productivity that make waiting (throughput reduction) easier.*

Any investment that enables us to reduce the volume of throughput needed to maintain a given level of welfare can be considered an indirect investment in natural capital. There are two classes of investment for reducing the need for throughput. The most obvious is investment in reducing population growth—first stopping growth and then gradually reducing numbers of people by reducing birth rates. Investments in female literacy and social security systems, along with contraceptives and delivery systems, offer investment opportunities of this kind. The second class of investment, (and the focus of the remainder of this discussion), is in increasing the efficiency of throughput use. More generally, this means increasing the efficiency with which capital, both natural and man-made, is used to provide life-support and life-enhancing services.

Think of the world initially consisting of only natural capital—our initial dowry. We convert some of it into man-made capital in order better to serve our wants. The extent to which we should continue this conversion is economically limited. The efficiency with which we use the world to satisfy our wants depends on two things: (1) the amount of service we get per unit of man-made capital, and the amount of service we sacrifice per unit of natural capital lost as a result of its conversion into man-made capital. This overall ecological economic efficiency can be stated as the ratio :

$$\frac{\text{MMK services gained}}{\text{NK services sacrificed}}$$

where MMK is man-made capital and NK is natural capital. In an empty world (or locality), there would be no noticeable sacrifice of NK services required by increases in MMK, so the denominator would be irrelevant. In a full world any increase in MMK would come at a noticeable reduction in NK and its services.

This efficiency ratio can be "unfolded" into four components by means of the following identity (Daly 1991). Each term of the identity represents a dimension of efficiency that might be improved by increased investment in knowledge or technique.

The meaning of each of the four factors of overall efficiency (or efficiency in the ecological economic sense) will be discussed next, along with ways to invest in increasing them.

Ratio 1 is the *service efficiency* of the man-made capital stock. It depends on (1) the technical design efficiency of the product itself, (2) the economic efficiency of resource allocation among the different product uses in conformity with individual preferences and ability to pay, and (3) the distributive efficiency among individuals. The first two are straightforward and conform with standard economics, but the third requires explanation. Usually distribution is carefully separated from efficiency by the Pareto condition that utility cannot be compared across individuals. Of course, in practice we do compare utility across individuals, and it does make sense to believe that total social utility

$$\frac{\text{MMK services gained}}{\text{NK services sacrificed}} = \frac{\text{MMK services gained}}{\text{MMK stock}} \times \frac{\text{MMK stock}}{\text{thruput}} \times \frac{\text{thruput}}{\text{NK stock}} \times \frac{\text{NK stock}}{\text{NK services sacrificed}}$$

Ratio 1   Ratio 2   Ratio 3   Ratio 4

Figure 2.4. Four components of efficiency ratio. Ratio 1 = service efficiency, Ratio 2 = maintenance efficiency, Ratio 3 = growth efficiency, and Ratio 4 = ecosystem service efficiency.

is increased when resources are redistributed from the low marginal utility uses of the rich to the high marginal uses of the poor. One can reject the total egalitarianism implicit in carrying this idea to its logical extreme, while at the same time agreeing with Joan Robinson that it is possible to allow too much of the good juice of utility to evaporate from commodities by allowing them to be too unequally distributed. Investments in distributive efficiency might reasonably be ruled out-of-bounds in an empty world that offered the easier alternative of growth. In a full world, growth is limited, and all improvement must come from efficiency improvement, so we can no longer neglect the possibility of efficiency increase through redistribution. Economists have studied these aspects of service efficiency, especially allocative efficiency via the price mechanism, in great detail. Further refinements from deeper study of Ratio 1 will probably be less productive than the study of the other three ratios, except for the redistributive possibility.

Ratio 2 reflects the *maintenance efficiency* or durability of the man-made capital stock. While Ratio 1 measures the service intensity per unit of time of the man-made stock, Ratio 2 measures the number of units of time over which the stock yields that service. Ratio 2 is the durability of the stock, or the "residence time" of a unit of resource throughput as a part of the man-made capital stock. A slower rate of throughput, *ceteris paribus*, means reduced depletion and pollution. Maintenance efficiency is increased by designing commodities to be durable, repairable, and recyclable, or by designing patterns of living that make certain commodities less necessary to begin with.

Ratio 3 is the *growth efficiency* of natural capital in yielding an increment available for offtake as throughput. Basically, it is determined by the intrinsic biological growth rate of the exploited population in its supporting ecosystem. For example, pine trees grow faster than mahogany, so in uses where either will do, pine is more efficient. Nature generally presents a menu of different species growing at different rates. To the extent that we are able to design our technologies and consumption patterns to depend on the faster growing species, that will be more efficient, *ceteris paribus*.

With the advent of genetic engineering, there will be more attempts to speed up growth rates of exploited species (e.g., the bovine growth hormone). The green revolution involves an attempt to speed up growth rates of wheat and rice—or at least the consumable portions of these plants. Since an increase in biological growth rate frequently comes at the expense of stability, resilience, resistance to disease or predators, it may be that attempts to speed up reproductive rates will usually end up costing more than they are worth. It is for now certainly better for us to slow down our own biological growth rate than to attempt to speed up the growth rates of all the species we depend upon. Nevertheless, we can to some degree adapt our pattern of consumption to depend more on naturally faster-growing species, where possible.

For sustained-yield exploitation, Ratio 3 will vary with the size of the population maintained, according to the familiar inverted U-shaped function. For any chosen combination of population size and yield, the ratio would remain constant over time under sustained yield management. Maximum sustained yield would, of course, maximize this dimension of efficiency over the long run (if harvesting costs are constant). In the short run, this ratio can be driven very high by the nonsustainable practice of exceeding renewable rates of harvest and thereby converting permanent stock into one-time throughput. There is a strong tendency to cheat on this dimension.

Ratio 4 measures the amount of natural capital stock that can be exploited for throughput (either as source or sink), per unit of other natural services sacrificed. For example, if we exploit a forest to get maximum sustainable yield of timber (or maximum absorption of CO_2), then we will, to some degree, sacrifice other natural services of the forest such as wildlife habitat, erosion control, and water catchment. We want to minimize the loss of other ecosystem services per unit of natural capital managed with the objective of yielding a single service—usually that of generating raw material throughput. Ratio 4 might be called *ecosystem service efficiency*, reflecting the minimization of loss of other ecosystem services when a population or ecosystem is exploited primarily for throughput.

The world is complex, and no simple identity can capture everything. But these four dimensions of ecological economic efficiency may be helpful to ecological economists in devising ways to invest indirectly in natural capital. As NK is converted into MMK (as we go from the empty world of Figure 2.2 to the full world of Figure 2.3), we want at each step to maximize the service from the increment of MMK and to minimize the loss of ecosystem service from the decrement of NK. But at some point, even if carried out efficiently, this process of conversion of NK into MMK will itself reach an economic limit, an optimal scale of the economic subsystem beyond which further expansion would increase costs faster than benefits. This optimal scale is defined by the usual economic criterion of equating marginal costs and benefits. This criterion assumes that marginal benefits decline and that marginal costs increase, both in a continuous fashion. It is reasonable to think that marginal benefits decline because humans are sufficiently rational to satisfy their most pressing wants first. But the assumption that marginal costs (sacrificed ecosystem services) increase in a continuous fashion is problematic. As the human niche has expanded, the stresses on the ecosystem have increased, but there has been no rational ordering by human or providential intelligence to ensure that the least important ecosystem services are always sacrificed first. We appear to be sacrificing some vital services rather early. This is another way of saying that Ratio 4, ecological service efficiency, has been ignored. If we begin to pay attention to that dimension of efficiency, then we may expect human rationality to begin to order ecosystem service sacrifices from least to worst, and thus justify the economist's usual assumption of gradually rising marginal costs. That would make the optimal scale of the human niche more definable.

The present lack of rational sequencing of ecosystem costs is due both to nonrecognition of the problem and to ignorance of ecosystem functioning. Prudence in the face of large uncertainties about ecosystem costs should lead us to be very conservative about risking any further expansion. But even with complete certainty and a least-cost sequence of environmental costs, there is still an optimal scale beyond which it is anti-economic to grow. And of course this notion of optimal scale is purely anthropocentric, counting all other species only for their instrumental value to human welfare. If we attribute intrinsic value in some degree to other sentient creatures, then the optimal scale of the human niche would be smaller than if only human sentience counted. Investment in natural capital would then have the additional benefit of increasing life support services to nonhuman species whose enjoyment of life would no longer be counted as zero, though it certainly should not be counted as equal to human life enjoyment. To recognize that a sparrow's intrinsic value is greater than zero, does not negate the fact that a human is worth many sparrows. But not even theology can say how many sparrows are worth a human, much less economics.

REFERENCES

Costanza, R., and H. Daly. 1992. Natural capital and sustainable development. *Conservation Biology* 6 (March): 37–46.

Daly, H. 1991. Steady State Economics. 2d ed. Washington, DC: Island.

El Serafy, S. 1988. The proper calculation of income from depletable natural resources. In Environmental Accounting for Sustainable Development, eds. Y. Ahmad, S. El Serafy and E. Lutz. Washington, DC: The World Bank.

Hicks, J. R. 1946. Value and Capital. 2d ed. Oxford: Oxford Univ. Press.

Ise, J. 1925. The theory of value as applied to natural resources. *American Economic Review* 15 (June): 284–91.

Solow, R. 1974. The economics of resources or the resources of economics. *American Economic Review* (May): 11.

Vitousek, P. et al. 1986. Human appropriation of the products of photosynthesis. *BioScience* 34 (May): 368–73.

3 ECOLOGICAL ECONOMICS AND THE CARRYING CAPACITY OF EARTH

Paul R. Ehrlich
Center for Conservation Biology
Department of Biological Sciences
Stanford University
Stanford, CA 94305

Beijer Institute of Ecological Economics
Royal Swedish Academy of Sciences
Stockholm S–10405, Sweden

ABSTRACT

The central problem of ecological economics is developing ways to modify the world's economic system so that it can both sense and adjust its own scale to remain within the capacity of the biophysical systems that support it. The immediate challenge is helping to reduce the already excessive negative impacts of human enterprise. Novel approaches are required both to take advantage of market efficiency where possible and to respond appropriately to changes in the quality and capacity of ecosystem services and in natural resource stocks and flows.

This chapter addresses critical dimensions of human carrying capacity and recommends ways of correcting the overgrowth of the global economy ("macroeconomic overshoot"). Ecological economists are urged to avoid the temptation of academics to seek overelegant solutions ("crackpot rigor"). They should concentrate instead on the more intellectually challenging task of finding fast, practical answers to the major problems related to carrying capacity, which must be overcome on a time-scale of decades. Ecological economists must take account of the inevitability of bad policy formulation and implementation (the "fog of politics") and of the persistence of massive uncertainties. Both require a conservative approach to resetting the scale of the global economy—one that includes insurance against environmental disaster.

INTRODUCTION

Ecological economics, like conservation biology (the science of preserving biotic diversity), should be viewed as a transdiscipline with a time limit of decades. The depletion of biotic capital (biodiversity) is a bellwether of the general deterioration of natural capital and the ecosystemic supports of the human economy. The various dimensions of the human predicament and the urgency of solving it are clear to many natural scientists (Costanza 1991a, b; Ehrlich and Ehrlich 1970, 1981, 1990, 1991, 1992; Ehrlich et al. 1977; Fremlin 1964; Holdren 1975, 1982, 1991a, b; Jones and Wigley 1989; Myers 1984, 1989, 1991; Schneider 1990; Schneider and Mesirow 1976; Terborgh 1989; Wilson 1988) and increasingly to economists (Boulding 1966; Brown 1992; Cline 1992; Costanza et al. 1991; Daly and Cobb 1989; Dasgupta 1993; Georgescu-Roegen 1973; Goodland and Ledec 1987; Goulder 1990; Perrings 1987). Plenty is known about the directions in which humanity should be moving to establish a "sustainable" society. The world should move toward population decline rather than growth and toward efficiency rather than wasteful use of energy and other resources. It should be moving towards economic equity rather than increasing the gap between rich and poor. Humanity should be struggling to empower rather than to marginalize women and to pursue religious, ethnic, and racial tolerance rather than strife. All nations should be working hard to preserve and restore biodiversity rather than to destroy it.

THE CHALLENGES TO ECOLOGICAL ECONOMICS

The discipline of ecological economics faces two rather distinct challenges. The first is an educational problem which will continue for the next decade or so. The public and policy makers must become convinced that an economic system cannot exist without the support of functioning ecosystems. They must realize that "sustainable growth" of the physical economy is an oxymoron, and that the most severe problem with the global economic system is its inability to sense that its scale has grown too large for its support systems (i.e., that society is in a state of "macroeconomic overshoot"). In short, everyone must come to recognize the imperative to reduce Homo Sapiens' global impact so that the human enterprise becomes sustainable in the long run.

The second task facing ecological economics is to help solve a wide variety of intellectually challenging long-term problems that humanity will face as it attempts to establish a sustainable society (on the optimistic assumption that it will make the attempt). The long list includes (1) adapting the standard theory of resource allocation so that it properly accounts for the physical and biological inputs to and outputs of the economic system (Mäler 1974; Dasgupta and Heal 1979; Dasgupta 1982; Brown and Roughgarden 1992); (2) developing better ways to incorporate the value of ecosystem services into economic calculations (Goodland and Ledec 1987); (3) weaving concepts of equity and energy—and materials-use efficiency—into a series of additional indicators of economic "efficiency" (along with the standard Pareto or allocative efficiency); and (4) in the same vein continuing to work on new approaches to national accounting that give appropriate weight to environmental factors (Norgaard 1989b; Repetto et al. 1987).

Today some economists and politicians seem happy promoting allocative efficiency or growth in the GNP, ignoring the very limited utility of those metrics. Many economists recognize, however, that a Pareto-optimal allocation may be socially extremely "inefficient" (i.e., unstable and dangerous) if it leads to revolution. Similarly, an economy may be maximizing satisfaction given its economic resources, inputs, and technology, as well as expanding its GNP, and still be destroying its biophysical underpinnings because of its size and/or the physical inefficiency of its operations. On the other hand, ecological economists must be alert to the tendency of other economists and politicians (as occurs in many developing nations and in the agricultural sectors of developed economies) to waste precious resources in the name of equity but to the ultimate detriment of the poor.

Achieving appropriate balances between equity and allocative efficiency remains a key problem for all interested in economic and ethical issues.

The list of tasks we face is almost endless:

- Redesigning international trade so that its environmentally damaging aspects (including the perpetuation and exacerbation of economic inequities) are removed, while retaining those features that expand global carrying capacity.

- Improving approaches to intergenerational equity (especially methods for determining suitable rates of depletion of non-renewable resources) and removing hidden assumptions of continued growth (or the "intergenerational invisible hand" of Barnett and Morse) from considerations of discounting.

- Discovering how to take fullest advantage of the allocative efficiency of markets in reaching extrinsically determined goals (e.g., sustainable scale, equity), while recognizing market imperfections (e.g., monopoly, lack of information, wage and price stickiness, and discrimination) and constraining them to prevent environmental and social damage.

- Making advocates of unrestrained capitalism much more aware of potential market failures, missing markets, and the complexities of private property and externalities (i.e., clarifying the limitations of the applicability of the Coase theorem, as in Bromley).

- Emphasizing that externalities are "external" to the bookkeeping of households or firms but that, in fact, internalizing environmental and population externalities (Ehrlich et al. 1992) may be the central problem of economics. Therefore, economists should get serious about social (or economic) profitability (Monke and Pearson 1989; Pagiola 1991) so that its calculation not only considers policy distortions but depletion of resources and damage to life-support systems as well (Daily and Ehrlich 1992).

- Most importantly, allocating and constraining growth so that the scale of the economic system, including its "externalities," remains within supportable bounds.

One challenge that I think ecological economists should worry about less is converting more mainstream economists to an understanding of how the world actually works. To put it bluntly, many of those who are alertable to the nonsensical environmental assumptions of economic theory are alerted by now, and major efforts to inform the remaining economists about the facts of life (Daily et al. 1991; Daly and Cobb 1989; Ehrlich 1989; Ehrlich et al. 1992; Klaassen and Opschoo 1991; Perrings 1987; Victor 1991) seem unlikely to be very cost-effective. Those who are unalertable will likely continue to consider externalities a minor problem, believe in perpetual motion machines, 100% recycling, and eternal growth, and otherwise act as if the laws of nature did not exist (Beckerman 1992). They will continue to assume, as did the Brundtland report (1987, p. 4), that global economic activity can be safely multiplied five- to ten-fold (or even more), and to enhance the general reputation of their discipline as lacking predictive power (*The Economist* 1992). Paradigm shifts are rarely accepted by the middle-aged rank and file in any discipline (Kuhn 1962), and new ideas normally take over only as older scholars retire.

Sadly, since a change in direction is required in the coming decades, there is not enough time to wait for the paradigm shift now under way within economics to be accepted by all, or even a majority—especially since droves of young economists are still being trained under the old paradigm (Colander and Klamer 1987). The economic dimensions of the human predicament are much too important to be left to the old school, so ecological economists must take their results directly to policy makers and the general public—groups that lack a narrow disciplinary orientation. In the process, they can recruit the better minds among professional economists, young and old.

ECOLOGICAL ECONOMICS CANNOT BE VALUE-FREE

Approaching many of the problems described above will require not only brand new insights but also making more value judgments than either ecologists or economists are accustomed to recognizing in their professional lives (of course, no science is ever value-free). But both ecologists and economists are accustomed to making such judgments openly in their private lives—in the professional domain, it is simply necessary to make them as explicit as possible. Some of the value judgments incorporated in this chapter (and shared by many ecologists and economists) are:

1. A world with 2 billion people who have a high standard of living (quality of life) would be superior to one with 10 billion people all living at a subsistence level.

2. Grossly inequitable distributions of wealth are highly undesirable, as are racism, sexism, religious prejudice, and chauvinism.

3. Since knowledge of how environmental systems will react under stress is very imperfect, humanity should be very conservative in imposing those stresses— society should carry a great deal of environmental insurance.

THE DIMENSIONS OF CARRYING CAPACITY

Since the principal task of ecological economics is to seek ways to keep the scale of the human enterprise within supportable bounds, the concept of carrying capacity is central to the discipline. Carrying capacity is defined by ecologists as "the maximum population size of a given species that an area can support without reducing its ability to support the same species in the future." It is usually denoted as "K" (Roughgarden 1979).

K is particularly difficult to define for populations of Homo Sapiens (Daily and Ehrlich 1992). K is a function of both the population's resource base (in the broadest sense) and the characteristics of the organism. But human beings are able to modify both their resource base and their behavior, and to transport resources to permit populations to exceed local K's—abilities that introduce great uncertainty into calculations of global K. For this reason, it is of heuristic value to distinguish between two kinds of K's for human beings. The first kind is biophysical K's, or the maximum number of people that can be supported at given levels of technology; the second is social K's, or the maximum number that can be supported at a given level of technology within a given social organization, including patterns of consumption and trade (Daily and Ehrlich 1992; Holdren et al. 1993). Thus the social K would be higher, *ceteris paribus*, for a global society of vegetarian saints than for a global society with the consumption preferences of rich Americans. Biophysical K will be higher if new knowledge leads to technologies that expand the number of people who can be supported (e.g., high-yielding crops) than if new knowledge reveals previously undetected constraints (e.g., global warming, ozone depletion). Since K is inextricably related to social systems (which generate the crucial knowledge and technologies), I will hereafter use K for human beings as synonymous with social K.

Human beings are long-lived and corporately "plan" to have a civilization that persists for thousands of years. Thus, while discussions of K for animal species usually cover time spans of a few generations (next season, next year, next decade) in which major secular changes ordinarily do not need to be considered, discussions of K for Homo sapiens often must have time horizons of centuries or more. Even the popular press is now discussing population projections, climatic change, and sea-level rises through the end of the next century and sometimes beyond. One must limit one's time horizon for sensible planning; here, I will consider some tens of human generations—hundreds of years to a millennium—following Daily and Ehrlich (1992).

Despite all the difficulties, it is critical that estimates of K be made, especially since human numbers are now above K (Ehrlich and Ehrlich 1990, 1991). Humanity is currently supporting 5.5 billion people, but only by a rapid destruction and dispersal of vital natural capital, especially rich agricultural soils (Brown and Wolfe 1984; WRI 1992), ice-age groundwater (WRI 1992), and biodiversity (Ehrlich and Wilson 1991). Furthermore, we are simultaneously consuming, co-opting, or eliminating some 40% of the basic energy supply for all terrestrial animals (Vitousek et al. 1986). The scale of the physical economy is now clearly too large for the capacity of life-support systems to maintain it over the long term.

Perhaps the most important aspect of the human resource base—the most valuable natural capital that humanity possesses—is the capacity of the environment to absorb abuse and continue to deliver the ecosystem services that are essential to civilization (Holdren and Ehrlich 1974; Ehrlich and Ehrlich 1981). The magnitude of the human impact on that capacity can be described by the equation $I = P \times A \times T$, where I = impact, P = population size, A = per capita affluence (measured by consumption), and T = an index of the environmental damage:human impact on done by the technologies selected to supply each unit of consumption.

There are substantial ecological uncertainties in estimates of carrying capacity. Among them is a cluster of issues relating to the concept of maximum sustainable abuse (MSA) (Daily and Ehrlich 1992). One of the most important is the degree to which the now entrained massive losses of biodiversity will degrade ecosystem services. Could a "weedy" world (one populated largely by organisms adapted to human-caused disturbance), with only a small fraction of today's population or species diversity, maintain those services satisfactorily? Would it be able to satisfactorily control the gaseous quality of the atmosphere, generate and maintain soils, dispose of wastes, recycle nutrients, control potential pests, and carry on the many other critical functions of natural ecosystems (Ehrlich 1993; Ehrlich and Daily 1992)? The world is becoming increasingly weedy, so that question is likely to be answered "experimentally."

A related key uncertainty (Nordhaus 1990; Daily et al. 1991; Schelling 1992; Rubin et al. 1992) concerns the impacts of global change, especially climatic changecaused by global warming, upon agricultural production. At the moment, it is not clear how rapid climate change will be, what directions it will take in key regions such as the U.S. grain belt, or whether carbon dioxide enrichment will significantly offset negative impacts (Daily and Ehrlich 1990; Ehrlich et al. 1993; Schneider 1990). Computer models of the climate lack sufficient resolution to answer the first two questions, and the responses of agricultural ecosystems to changing CO_2 levels have been the subject of little more than speculation so far. What is known provides scant basis for optimism (Ehrlich et al. 1993). The speed with which crops can be adapted genetically to changed climatic regimes without significant losses in productivity (or while increasing yields) is also uncertain.

Similarly, computer models are also inadequate to accurately predict the amount of ozone depletion per unit of chlorofluorocarbon released; put another way, what is the MSA for the ozone shield? While many crops would be adversely affected by significant increases in UV-B flux (Jones and Wigley 1989), there is a great deal of uncertainty about the degree to which UV-B resistant strains of important crops can be developed, and at what cost in productivity. And, of course, one can only speculate on the potential synergisms among climatic change, increased UV-B radiation, low-altitude air pollution, the gradual toxification of the planet, and the loss of biodiversity (including the genetic diversity of crops).

Many resources that at first glance would seem to be crucial aspects of carrying capacity are not, at least at present. For instance, fossil fuels are abundant for now, and there is no real immediate prospect of humanity "running out of energy." Indeed, because of the environmentally damaging aspects of profligate energy use, humanity's energy problem is best described as "too much, too soon," rather than "too little, too late" (Holdren 1975). But there remain many uncertainties about the environmental risks and benefits of any mix of energy technologies—fossil fuel, solar, biomass, fission, and fusion. Furthermore, the maximum sustainable uses (MSUs) (Daily and Ehrlich 1992) of many resources are as difficult to calculate as the MSAs of many components of the global ecosystem.

These ecological uncertainties concerning humanity's current state of macroeconomic overshoot have their counterparts in economic factors influencing K. How much money is it reasonable to spend in preserving biodiversity or in attempting to slow climatic change, and how should those funds be allocated? How does one calculate the costs and benefits of different approaches? Are market-based or command and control mechanisms more likely to provide the most efficient abatement of various kinds of "pollution"?

Food supply is almost always the most critical dimension of K for any animal population. For humanity, that dimension is primarily agricultural, and agriculture is an area fraught with economic problems (as well as ecological uncertainties). Can land reform actually be achieved in poor nations (especially in Latin America) without the disruption of revolutions, and what would be its effect on agricultural productivity? How would the education and liberation of women affect food production in Africa (where much of the farm labor is done by women)? Can (and should) international trade in food be adjusted to promote production of more food for local consumption in poor nations? Should the principle of comparative advantage take precedence, so that global carrying capacity might be maximized at possible costs such as greatly increased environmental disruption (Goodland and Ledec 1987) or famines generated if food-short poor nations cannot afford imports when market prices get too high? Can farm credit, farm-to-market roads, availability of needed inputs, and other key factors required to increase the productivity of the Earth's poorest farmers be supplied? Can ecologically sound technologies be developed to raise yields without adding to environmental deterioration? Will governments remove the economic inefficiencies in agricultural systems, including artificially depressed food prices in third world cities, that can safely (from an ecological and social viewpoint) be removed?

Will other types of inefficiencies be considered? Large-scale cash cropping for export in poor nations may be highly efficient under the standard economic definition, but disastrously inefficient both environmentally and politically. Only if the textbook definition of efficiency could be supplemented by another that included the interests of future generations (or current generations in the future), and if those interests were defined to include socially desirable patterns of distribution, could the market-mediated result of food distribution be embraced because it provided "efficient" allocation. Finally, and most basically, will appropriate attention be paid to agriculture in all nations?

While ecologists can say with certainty that population growth, *ceteris paribus*, exacerbates environmental problems and ultimately will cause the latter to end the former (Ehrlich and Holdren 1971; Holdren and Ehrlich 1974; Ehrlich et al. 1993), there remains great uncertainty about what the eventual consequences of macroeconomic overshoot will be. Will the biggest difficulties eventually be starvation? Or will HIV or a successor virus cause a population crash? Will some form of social breakdown, wars over water, or some totally unanticipated "natural" disaster cause a gigantic rise in death rates? Or will humanity soon see the perils of its present overshoot and make the changes required for a transition to a sustainable society? Finally, the best ways of allocating relatively scarce resources in order to lower P, A, and T respectively remain a central source of uncertainty for economists, and finding these solutions will be a major task for ecological economists.

THE RIGOR TRAP

In approaching their tasks, it is essential that ecological economists avoid *the rigor trap* (sometimes referred to as "crackpot rigor," the "tyranny of illusory precision" (Holdren 1982), or the "partial quantification trap," (Costanza and Daly 1987). The rigor trap is a widespread phenomenon in academia, including both ecology and economics. It is seen whenever some difficult parameter is guesstimated, and then the guesstimate is used in calculations carried out to many decimal places, or when a trivial (or fundamentally intractable) problem is subjected to detailed mathematical modeling. A great deal of intellectual effort by ecologists, for example, is put into sophisticated studies of limited aspects of the ecology of obscure organisms—research that will probably never have much influence on the field because it cannot be placed in any important context.

It is easy to fall into the rigor trap. For many years, my research group has studied the biology of Euphydryas butterflies as a sample system for testing various kinds of theory in ecology and evolution. One of the main tools we have used in examining the dynamics of populations is a censusing technique known as mark-release-recapture (MRR). It requires intensive, time-consuming effort and careful statistical analysis to generate dependable estimates of population size. As our group and others began to be interested in protecting endangered butterfly populations and species, we quite naturally turned to MRR to look for trends in population size. Only gradually did it dawn on us that the effort was largely wasted—at best it could document the disappearance of a population. We were concentrating on method rather than results.

For butterflies, as our long-term work showed (Ehrlich 1992a), determining actual population sizes at a given moment is generally not critical for establishing conservation status or designing nature reserves to protect them. We now believe that the key to the conservation of most butterflies (and perhaps most invertebrates and many small vertebrates) is preserving an array of suitable habitats with somewhat different attributes as sites for populations. This spreads the risk of global species extinction (analogous to the economic concept of risk diversification in an investment portfolio). Details of habitat quality are much more important than details of population size. What is most needed in invertebrate conservation biology is not MRR studies or the equivalent, but rapid evaluation of critical environmental factors (such as interactions between macroclimate and topography) and assessments of the required extent of habitats to meet the needs of the organisms (Ehrlich 1992a). The techniques used for rapid habitat assessment often lack the rigor of MRR, but can produce the required results in time.

Economists have made similar mistakes. In the area of industrial organization and antitrust, for example, elegance has often overwhelmed practicality. Very bright economists have devoted impressive intellectual energies to refining and modifying sophisticated theoretical models of how the interactions of firms in a market are related to the number of firms, the information available to the firms, and the possibility of new entry. These models are of little help to the policy maker, however. Depending on the parameter values inserted, the same model can generate either monopoly pricing or competitive pricing. As MIT's Franklin Fisher put it (1989): "There is a strong tendency for even the best practitioners to concentrate on the analytically interesting questions rather than on the ones that really matter for the study of real-life industries. The result is often a perfectly fascinating piece of analysis. But so long as that tendency continues, these analyses will remain merely games economists play."

The tendency of economists to create sophisticated models of little or no practical applicability and to focus on methods rather than results has been somewhat of a joke both within and outside the field since at least the time of Pareto (1935). Indeed, the basic assumptions of neoclassical theory are all too rarely examined carefully, and the actual applicability of that theory to the real world is often questionable (Christensen 1989; Norgaard 1989a; Victor 1991).

A classic example of falling into the rigor trap from ecological economics itself has resulted from economists and others being misled about biodiversity by ecologists. They have gotten the notion, probably from the overemphasis on endangered species in the conservation literature, that a key task is optimally allocating resources to the preservation of particular species in order to maximize some cladistically based measure of genetic diversity (Solow et al. 1993). Detailed computer analyses involving the assignment of wild estimates of parameters such as extinction probabilities, are classic cases of the rigor trap (Daily 1992). Such efforts are also excellent examples of "suboptimization," i.e., doing in the best possible way something that should never be done at all (Ehrlich and Daily 1992; Daily 1992).

Ecological economics can ill afford the luxury of its practitioners engaging in such pursuits, and many criteria for work well done (such as "elegance") that are beloved in the scholastic realm must be considered secondary goals in ecological economics relative to practicality, especially salability. The most sophisticated ecological economic model is of little or no value unless it can be made understandable and usable by decision makers and, to a degree, the general public. The work of Brown and Roughgarden (1992), offering an ecological economy with three state variables (labor, capital, and a natural resource) as a replacement for the two state variables of neoclassical growth models (labor and capital), exemplifies the sorts of small but sound models that are being developed. Brown and Roughgarden explain their results so that the implications for policy makers are clear.

THE FOG OF POLITICS

Care must also be taken to consider, wherever possible, what I would like to call the "fog of politics" when doing analyses and making prescriptions. Military theorists have long recognized the problem of the "fog of war"— that is, in the heat of battle, communications will be imperfect; some leaders will prove incompetent, cowardly, traitorous, or all three; troops will panic when it is least expected; equipment will not perform as advertised; preparations will have been made badly; the enemy will use unanticipated tactics; and overall luck will play a major role in events. Smart military planners attempt to take the fog of war into account. Similarly, in a "fog of politics" one can generally count on policies being badly formulated and/or badly executed (Andersson 1991; Keyfitz 1991). Politicians and bureaucrats often prove incompetent, cowardly, or venal (Jacobs 1991); citizens will not respond as anticipated; penalties or rewards will not produce the expected results; and, here too, luck will play a major role in events. Scottish poet Robert Burns summed up human "fogs" in his classic lines, "the best laid schemes o' mice an' men gang aft a-gley, an' lea'e us nought but grief an' pain for promised joy."

A classic example of bad policy prescription that ignores the fog of politics is based on the often heard statement that there is no population problem, for if people distributed food equitably and consumed less meat, there would be enough for everyone to eat. Global policy, it is claimed, should therefore be aimed only at improving distribution, changing dietary habits, or abolishing poverty and not at population control. The distribution/dietary change part of the statement is true (Ehrlich et al. 1993), but does not subtract from the urgency of developing policies designed to achieve control over human numbers.

Food is not, and as far as anyone knows never has been, distributed equally among nations, groups within nations, or even within families. Furthermore, dietary habits are among the most difficult of all human habits to alter.

Ecological economists should make policy recommendations that do not depend on flawless execution or saint-like behavior. Rather they should help to plan a world in which everyone can be adequately nourished, even in the face of inadequate distribution of food, as development economists are already doing (Dixon et al. 1989; Conway and Barbier 1990).

While there is abundant need for original ecological economic research on topics related to agriculture, there is also much that specialists know about both food and the fog of politics, but which the public or politicians are unable to appreciate. Examples are the impact of food aid programs on food production in recipient nations and the generally procyclical nature of the aid (Clay 1991; Falcon 1991). Donating ood to nations with shortages tends to depress food prices and thus may hurt the agricultural sector of the economy—the sector that often has already been badly neglected. In the long run, food aid unaccompanied by measures to strengthen the economic position of farmers can be counterproductive. Food aid, when given, also surely should be countercyclical, so that it buffers the impacts of rising food prices in recipient nations. But above all, it must always be recognized that sociopolitical systems may not react to policy recommendations based on such knowledge at all, let alone develop coordinated, well-administered programs.

Many strategies can be adopted to help lift the fog of politics, of course. Perhaps the best of these is to maximize the participation of affected parties in the process of developing policies. Such work was pioneered decades ago by C. S. Holling, using computer models of development schemes. Since then, improvements in computer technology, including the development of the sophisticated Geographic Information System (GIS) technology, have made these approaches much more practical. The dramatic simultaneous presentations of many different variables possible with a GIS should make cutting through the fog of politics easier. Since numerous people can view and manipulate the same data, public participation in decision making could be greatly expanded. Eventually, "virtual reality" capabilities should be added to standard GIS systems so that people could more or less "experience" the results of alternative policy choices.

DEALING WITH UNCERTAINTY

When the Boeing 747 was designed, four independent hydraulic systems, each capable of operating the entire aircraft, were included. Such fail-safe design is accepted practice among engineers and military planners, and is viewed as especially important in potentially high-risk areas such as the aviation and the nuclear power industries. When it comes to Spaceship Earth, however, the prevailing philosophy often appears to be that evolution has incorporated fail-safe negative feedbacks into our life-support systems, essentially assuming that if one part of the system "breaks," some back-up mechanism will keep the system functioning. Thus, many would keep dumping greenhouse gases into the atmosphere until the last glimmer of uncertainty about the results can be removed. Similarly, many would count on science to save us from the consequences of the population explosion. "Don't try to reduce family sizes now" is their view—surely many more people can be accommodated with future miracles of technology. In both cases, uncertainty (about how the atmosphere works or future K) is used as an excuse for inaction. Any of those people riding on a 747 where three of four hydraulic systems have failed would demand not inaction but an immediate landing, even if the pilot claimed that his first officer might come up with an idea that would put one failed system back in service.

Just as individual human beings and firms take out insurance to buffer against uncertainties (e.g., unexpected death, injury, fire loss, or lawsuit), so society as a whole must plan for a future that almost certainly will contain unanticipated and unpleasant discontinuities. It must also plan, by adjustments of discount rates and other mechanisms, for a future that is materially more constrained, not "richer." In short, we must establish safe minimum standards(Ciriacy-Wantrup 1963) and precautionary principles (Perrings 1991) for protecting Earth's life-support systems in the face of virtually inevitable unpleasant surprises.

The uncertainties faced by ecological economists on the ecological side are compounded by the uncertainties on the economic side and vice-versa, and over the next few critical decades those uncertainties are unlikely to be significantly reduced. Therefore, two related major challenges for ecological economists are (1) to help society identify and evaluate the "premiums" worth paying to achieve environmental insurance, and (2) to develop strategies for moving toward sustainability that are relatively insensitive to uncertainties; that is, "robust" strategies—helpful regardless of how uncertainties are ultimately resolved.

On the critical issue of limiting the scale of the human enterprise partly through birth control, there are obvious candidates for robust strategies. Substantial evidence suggests that empowering women—giving them education and some economic and social autonomy, and providing them and their young children with basic health care (including contraceptives)—will reduce total fertility rates (Mauldin and Berelson 1978; Ehrlich and Ehrlich 1990). But whether these measures alone would ultimately reduce those rates to the roughly 1.8 children per family required is very uncertain. Nonetheless, ecological economists can, in my view, recommend improvements in the condition of women on many other grounds (including ethical ones), making that policy choice relatively insensitive to the uncertainties. Indeed, one would guess that the costs of any measures would be well compensated by an increase in the productivity of women and of their children when they mature, setting aside the evident benefits of lower birth rates.

Similarly, it is not clear exactly how much the rate of climate change can be slowed by limiting the flow of greenhouse gases into the atmosphere. In addition, there is considerable uncertainty about the degree to which a given increase in the cost of gasoline (however achieved) will reduce emissions of carbon dioxide. But any reduction in the use of gasoline in vehicles would carry many other benefits—lower rates of deterioration of roads and bridges, less pressure for destroying natural ecosystems to make room for roads and parking lots, less smog damage to human lungs and crops, less acid rain, and so on. Here the costs (lost jobs at service stations and fewer automobiles produced) might even be offset by the benefits of producing new generations of fuel-efficient cars and constructing and staffing mass transit systems, not to mention lower medical costs, increased agricultural, forestry, and fisheries production. So ecological economists can suggest steps to raise gasoline prices in good conscience, regardless of the substantial uncertainties surrounding the possibly catastrophic consequences of global warming. Population limitation qualifies as a "no-regrets" strategy here as well.

In considering any strategies for dealing with uncertainty, one must be careful to analyze the risks of both action and inaction. Generally, with respect to major environmental problems, the latter risks are far greater since deleterious environmental effects (e.g., extinctions, soil erosion, climate change, and sea-level rise) are often irreversible on time scales of interest. Even if only "no-regrets" policies to slow global warming were undertaken, there would be some social cost. But if global warming is not slowed, and the sorts of negative effects that the most knowledgeable of the scientific community predict actually materialize, the costs will almost certainly be considerably higher (Cline 1992). The asymmetry of risks generally argues for action now, although the action must be informed by the best possible cost estimates to ensure that too high an insurance premium is not paid.

Finally, a word on the uncertainties associated with innovation: it is often thought that ecologists (and ecological economists) give insufficient attention to technological changes that could expand K and push back the "limits to growth." This is incorrect. For example, the "Holdren Scenario" (Holdren 1991a; Ehrlich and Ehrlich 1992) is the most optimistic plan yet devised (based on a realistic understanding of technological possibilities) for limiting the scale of the human enterprise while closing the rich-poor gap and supporting a peak population of 10 billion. It envisions moving from a world economy using 13.1 terawatts (10^{12} watts) of energy today to one using 30 terawatts in a century. To avoid ecological catastrophe, such an enormous increase in energy use would require development and deployment of a diverse array of relatively environmentally benign technologies. In such a transition, innovation will be essential just to limit the damage caused by the current overshoot of our life-support systems; the notion that innovation can be counted on to permit infinite expansion of the global economy is without foundation.

The strategies suggested above seem so obvious that one might ask whether there is any real role for deep technical analysis. The answer is that there is plenty. As an example, even though the costs and benefits of a carbon tax of a certain level may be uncertain, ecologists and economists can at the very least try to enumerate, evaluate, and publicize those costs and benefits and recommend ways to minimize the former and maximize the latter. Environmentalists tend to underestimate the costs of ameliorative measures; economists tend to underestimate the social costs of not taking them. Ecological economists can produce the best possible catalogue of both and suggest ways to minimize them. For instance, analyses by Lawrence Goulder (1992a, b) indicate that much of the economic cost of a carbon tax can be eliminated when revenues from the tax are used to finance reductions in existing "distortionary" taxes, such as the regressive FICA tax, the personal income tax, or the corporate income tax (see also Wirth and Heinz 1991).

A major technical task for ecological economists is to examine carefully the costs and benefits to nations and/or the global community of various levels of environmental insurance, as some writers already implicitly recognize (Ciriacy-Wantrup 1963; Giampetro and Pimentel, 1991 p. 142). This issue was front-page news in the spring of 1992 because of the determined resistance of the Bush administration to taking even modest no-regrets steps with regard to global warming. Global warming is far from the only case. The momentum of population growth makes it evident that even relatively large investments in support of population control measures worldwide are likely to be cost-effective in the best case and would insure against total disaster in the worst. Similarly, the irreversibility of the decay of biodiversity argues for an insurance or fail-safe approach to its conservation (Wilson 1989). Ecologists are not certain how much biodiversity must be preserved to avoid large regional and even global collapses of ecosystem services (Ehrlich 1993). We are only likely to find out by running the test, a test roughly akin to seeing how many rivets can be pried from the wing of an aircraft in which we are flying before it will fail in flight (Ehrlich and Ehrlich 1981). Careful calculation of reasonable premiums (and then paying them) seems a vastly superior strategy than letting the experiment run its course.

CONCLUSIONS

Clearly there is a great need for ecologists and economists to work closely to solve the human predicament. Fine economists have already wasted their time by taking a biologically silly approach to the problem of preserving biodiversity, because they have not been in contact with experienced ecologists. Ecologists (including myself) have been equally misled by lack of contact with economists. For instance, I was puzzled as to why the whaling industry was exterminating whales rather than harvesting them sustainably until a Japanese economist explained the industry's behavior to me (Clark 1973, 1991; Ehrlich 1989). Both biological and economic knowledge are essential to understanding ecosystem/economic system relationships and rationalizing them so that the human enterprise can prosper.

The criteria for excellence in research in ecological economics must include relevance to the development of policy prescriptions that have the highest possible probabilities of being accessible to and acted upon by policy makers. The policies must be (1) understandable to the general public, (2) relatively resistant to failure caused by the fog of politics, and (3) at the minimum provide sensible ecological insurance in the face of enormous uncertainties (and the near certainty of future surprises and discontinuities).

Developing such policies is a difficult technical challenge, but their implementation could contribute a great deal to reducing the scale of the human enterprise so that it will again be within the long-term human carrying capacity of the Earth.

ACKNOWLEDGMENTS

I'd like to thank Robert Costanza (Center for Environmental and Estuarine Studies, University of Maryland), Herman Daly (World Bank), Gretchen C. Daily, Anne H. Ehrlich, Jonathan Roughgarden, and Stephen H. Schneider (Department of Biological Sciences, Stanford), Lisa M. Daniel and Timothy Daniel (Bureau of Economics, Federal Trade Commission), Partha Dasgupta (Department of Economics, Cambridge), Walter P. Falcon and Rosamond Naylor (Institute for International Studies, Stanford), Lawrence Goulder (Department of Economics, Stanford), John P. Holdren (Energy and Resources Group, Berkeley), Karl-Göran Mäler (Beijer Institute), and Charles Perrings (Department of Economics, University of California, Riverside) for helpful comments on the manuscript. Needless to say, not all of their advice was taken, and the views expressed are solely my own. This chapter is dedicated to my friend Herman Daly, both because of his always generous intellectual aid and because of his unstinting efforts to return the discipline of economics to rational foundations.

REFERENCES

Andersson, T. 1991. Government failure—the cause of global environmental mismanagement. *Ecological Economics* 4: 215–36.

Barnett, H. J., and C. Morse. 1963. Scarcity and Growth. Baltimore: Johns Hopkins Univ. Press.

Beckerman, W. 1992. The environment as a commodity. *Nature* 357: 371–72.

Boulding, K. E. 1966. The economics of the coming Spaceship Earth. In Environmental Quality in a Growing Economy, ed. H. Jarrett. Baltimore: Johns Hopkins Univ. Press.

Bromley, D. 1991. Environment and Economy. Oxford: Blackwell.

Brown, G., and J. Roughgarden. 1992. An Ecological Economy: Notes on Harvest and Growth. Paper prepared for Beijer Institute of Ecological Economics.

Brown, L. 1992. State of the World, 1992. New York: W. W. Norton.

Brown, L. R., and E. C. Wolf. 1984. Soil Erosion: Quiet Crisis in the World Economy. Worldwatch Paper 60. Washington, DC: Worldwatch Institute.

Brundtland, G. H. 1987. Our Common Future. New York: Oxford Univ. Press.

Christensen, P. P. 1989. Historical roots for ecological economics—biophysical versus allocative approaches. *Ecological Economics* 1: 17–36.

Ciriacy-Wantrup, S. V. 1963. Resource Conservation: Economics and Policies. Berkeley: Univ.. of California.

Clark, C. W. 1973. The economics of overexploitation. *Science* 181: 630–34.

———. 1991. Economic biases against sustainable development. In Ecological Economics: the Science and Management of Sustainability, ed. R. Costanza. New York: Columbia Univ. Press.

Clay, E. 1991. Food aid, development, and food security. In Agriculture and the State, ed. C. P. Timmer. Ithaca: Cornell Univ. Press.

Cline, W. 1992. The Economics of Global Warming. Washington, DC: Institute for International Economics.

Colander, D., and A. Klamer. 1987. The making of an economist. *Economic Perspectives* 1: 95–111.

Conway, G. R., and E. B. Barbier. 1990. Sustainable Agriculture for Development. London: Earthscan.

Costanza, R., ed. 1991a. Ecological Economics: the Science and Management of Sustainability. New York: Columbia Univ. Press.

———. 1991b. Assuring sustainability of ecological economic systems. In Ecological Economics: the Science and Management of Sustainability, ed. R. Costanza. New York: Columbia Univ. Press.

Costanza, R., and H. E. Daly. 1987. Toward an ecological economics. *Ecological Modellng* 38: 1–7.

Costanza, R., H. E. Daly, and J. A. Bartholomew. 1991. Goals, agenda and policy recommendations for ecological economics. In Ecological Economics: the Science and Management of Sustainability, ed. R. Costanza. New York: Columbia Univ. Press.

Costanza, R., and C. Perrings. 1990. A flexible assurance bonding system for improved environmental management. *Ecological Economics* 2: 57–75

Culbertson, J. M. 1991. United States "free trade" with Mexico: progress or self-destruction? *The Social Contract* 2: 7–11.

Daily, G. C. 1992. Accounting for Uncertainty in Natural Systems. Paper presented at second meeting of International Society for Ecological Economics, August 3–6, Stockholm.

Daily, G. C., and P. R. Ehrlich. 1990. An exploratory model of the impact of rapid climate change on the world food situation. *Proc. R. Soc. Lond. B*. 241: 232–44.

———. 1992. Population, sustainability, and Earth's carrying capacity. *BioScience* 42: 761–71.

Daily, G. C., and P. R. Ehrlich, H. A. Mooney, and A. H. Ehrlich. 1991. Greenhouse economics: learn before you leap. *Ecological Economics* 4: 1–10.

Daly, H. E., and J. B. Cobb, Jr. 1989. For the Common Good. Boston: Beacon Press.

Dasgupta, P. 1982. The Control of Resources. Oxford: Blackwell.

———. 1993. An Inquiry into Well-Being and Destitution. Oxford: Clarendon Press.

Dasgupta, P., and G. Heal. 1979. Economic Theory and Exhaustible Resources. Cambridge: Cambridge Univ. Press.

Dixon, J. A., D. E. James, and P. B. Sherman. 1989. The Economics of Dryland Management. London: Earthscan.

Ehrlich, P. R. 1989. The limits to substitution: metaresource depletion and a new economic-ecological paradigm. *Ecological Economics* 1: 9–16.

———. 1992. Population biology of checkerspot butterflies and the preservation of global biodiversity. *Oikos* 63: 6–12.

———. 1993. In press. Biodiversity and ecosystem function: need we know more? In Biodiversity and Ecosystem Function, eds. D. Schulze and H. Mooney. Heidelberg: Springer-Verlag.

Ehrlich, P. R., and G. C. Daily. 1993. In press. Population extinction and the biodiversity crisis. Proceedings of Beijer Institute Workshop on Biodiversity.

Ehrlich, P. R., G. C. Daily, and L. Goulder. 1992a. Population growth, economic growth, and market economies. *Contention* 2: 17–35.

Ehrlich, P. R., and A. H. Ehrlich. 1970. Population, Resources, Environment. San Francisco: Freeman.

————. 1981. Extinction: The Causes and Consequences of the Disappearance of Species. New York: Random House.

————. 1989. How the rich can save the poor and themselves: lessons from the global warming. In Proceedings of the International Conference on Global Warming and Climate Change: Perspectives from Developing Countries, eds. S. Gupta and R. Pachauri. Tata Energy Research Institute, New Delhi, February 21–23.

————. 1990. The Population Explosion. New York: Simon &Schuster.

————. 1991. Healing the Planet. New York: Addison-Wesley.

————. 1992. The value of biodiversity. *Ambio* 21: 219–26.

Ehrlich, P. R., A. H. Ehrlich, and G. C. Daily. 1993. Food security, population and environment. *Population and Development Review* 19 (1): 1–32.

Ehrlich, P. R., A. H. Ehrlich, and J. P. Holdren. 1977. Ecoscience: Population, Resources, Environment. San Francisco: Freeman.

Ehrlich, P. R., and J. P. Holdren. 1971. Impact of population growth. *Science* 171: 1212–17.

Ehrlich, P. R., and E. O. Wilson. 1991. Biodiversity studies: science and policy. *Science* 253: 758–62.

Falcon, W. P. 1991. Whither food aid? A comment. In Agriculture and the State, ed. C. P. Timmer. Ithaca: Cornell Univ. Press.

Fisher, F. M. 1989. Games economists play: a noncooperative view. *The RAND Journal of Economics* (Spring): 113–24.

Fremlin, T. 1964. How many people can the world support? *New Scientist* (October 29): 285–7.

Georgescu-Roegen, N. 1973. The entropy law and the economic problem. In Toward a Steady-State Economy, ed. H. Daly. San Francisco: Freeman.

Giampietro, M., and D. Pimentel. 1991. Energy efficiency: assessing the interaction between humans and their environment. *Ecological Economics* 4: 117–44.

Goodland, R., and G. Ledec. 1987. Neoclassical economics and principles of sustainable development. *Ecological Modelling* 38: 19–46.

Goulder, L. 1990. Using Carbon Charges to Combat Global Climate Change. CEPR Publication No. 226, Center for Economic Policy Research, Stanford.

————. 1992a. Do the Costs of a Carbon Tax Vanish When Interactions with Other Taxes Are Accounted for? Working Paper No. 4061. Cambridge, MA: National Bureau of Economic Research.

————. 1992b. Carbon tax design and U.S. industry performance. In Tax Policy and the Economy 6, ed. J. Poterba. Cambridge, MA: MIT Press.

Holdren, J. P. 1975. Too much energy, too soon. *New York Times*, Op-Ed page, July 23.

————. 1982. Energy risks: what to measure, what to compare. *Technology Review* (April): 32–38.

————. 1991a. Population and the energy problem. *Population and Environment* 12 (3): 231–55.

————. 1991b. Testimony before the Committee on Science, Space, and Technology. (Hearing on Technologies and Strategies for Addressing Global Climate Change), U.S. House of Representatives, July 17.

Holdren, J. P., and P. R. Ehrlich. 1974. Human population and the global environment. *American Scientist* 62: 282–92.

Holdren, J. P., P. R. Ehrlich, and G. C. Daily. 1993. In press. Physical and biological sustainability.

Jacobs, L. 1991. Waste of the West: Public Lands Ranching. Tucson: Jacobs.

Jones, R. R., and T. Wigley. 1989. Ozone Depletion: Health and Environmental Consequences. New York: Wiley.

Keyfitz, N. 1991. Population and development within the ecosphere: one view of the literature. *Population Index* 57: 5–22.

Klaassen, G. A. J., and J. B. Opschoor. 1991. Economics of sustainability and the sustainability of economics: different paradigms. *Ecological Economics* 4: 93–115.

Kuhn, T. 1962. The Structure of Scientific Revolutions. Chicago: Univ. of Chicago Press.

Mäler, K-G. 1974. Environmental Economics: A Theoretical Enquiry. Baltimore: Johns Hopkins Univ. Press.

Mauldin, P., and B. Berelson. 1978. Conditions of fertility decline in developing countries, 1965–1975. *Studies in Family Planning* 9: 104.

Monke, E., and S. Pearson. 1989. The Policy Analysis Matrix for Agricultural Development. Ithica: Cornell Univ. Press.

Myers, N. 1984. Gaia: an Atlas of Planet Management. New York: Anchor Press.

―――― 1989. Tropical deforestation and climate change. In Climate and Geo-Sciences: A Challenge for Science and Society in the 21st Century, eds. A. Berger, S. Schneider and J. C. Duplessy. Dordrecht: Kluwer.

――――. 1991. Population, Resources and the Environment: The Critical Challenges. London: UNFPA.

Nordhaus, W. D. 1990. Greenhouse economics: count before you leap. *The Economist* (July 7): 21.

Norgaard, R. B. 1989a. The case for methodological pluralism. *Ecological Economics* 1: 37–57.

――――. 1989b. Three dilemmas of environmental accounting. *Ecological Economics* 1: 303–14.

Pagiola, S. 1991. The Use of Cost-Benefit Analysis and the Policy Analysis Matrix to Examine Environmental and Natural Resource Problems. Stanford: Food Research Institute.

Pareto, W. 1935. The Mind and Society. New York: Harcourt.

Perrings, C. 1987. Economy and Environment: A Theoretical Essay on the Interdependence of Economic and Environmental Systems. Cambridge: Cambridge Univ. Press.

――――. 1989. Environmental bonds and environmental research in innovative activities. *Ecological Economics* 1: 95–115.

――――. 1991. Reserved rationality and the precautionary principle: technological change, time and uncertainty in environmental decision making. In Ecological Economics: the Science and Management of Sustainability, ed. R. Costanza. New York: Columbia Univ. Press.

Pick a number. *The Economist* (June 13, 1992): 18.

Repetto, R., M. Wells, C. Beer, and F. Rossini. 1987. Natural resource accounting for Indonesia. Washington, DC: World Resources Institute.

Roughgarden, J. 1979. Theory of Population Genetics and Evolutionary Ecology: An Introduction. New York: Macmillan.

Rubin, E., R. Cooper, R. Frosch, T. Lee, G. Marland, A. Rosenfeld, and D. Stine. 1992. Realistic mitigation options for global warming. *Science* 257: 148–9.

Schelling, T. C. 1992. Some economics of global warming. *American Economic Review* 82: 1–14.

Schneider, S. H. 1990. Global Warming. New York: Random House.

Schneider, S. H., and L. E. Mesirow. 1976. The Genesis Strategy: Climate and Global Survival. New York: Plenum.

Solow, A., S. Polasky, and J. Broadus. 1993. In press. On the measurement of biological diversity. *Journal of Environmental Economics and Management* 24: 60–8.

Terborgh, J. 1989. Where Have All the Birds Gone? Princeton: Princeton Univ. Press.

Victor, P. A. 1991. Indicators of sustainable development: some lessons from capital theory. *Ecological Economics* 4: 191–213.

Vitousek, P. M., P. R. Ehrlich, A. H. Ehrlich, and P. A. Matson. 1986. Human appropriation of the products of photosynthesis. *BioScience* 36 (6): 368–73.

Wilson, E. O. 1988. Biodiversity. Washington, DC: National Academy Press.

———. 1989. The value of biodiversity. *Scientific American* (September): 108–16.

Wirth, T., and J. Heinz. 1991. Project 88—Round II. Incentives for Action: Designing Market-Based Environmental Strategies, Washington, DC.

World Resources Institute. 1992. World Resources 1992–93. Oxford: Oxford Univ. Press.

4 NEW SCIENCE AND NEW INVESTMENTS FOR A SUSTAINABLE BIOSPHERE[1]

C. S. Holling
Arthur R. Marshall Jr. Laboratory of Ecological Sciences
Department of Zoology
University of Florida
Gainesville, FL 32611

ABSTRACT

Global increases in greenhouse gases, reduction of stratospheric ozone and new diseases like AIDS represent a new class of problems that are challenging the ability to achieve sustainable development.

- They are more and more frequently caused by local human influences on air, land, and oceans that slowly accumulate to trigger sudden changes directly affecting the health of people, the productivity of renewable resources, and the vitality of societies.

- There is an increasing globalization of biophysical phenomena, combined with globalization of trade and with large scale movements of people that intensify the spatial span of connections. The problems are now fundamentally cross-scale in space as well as in time.

- The problems are ones that emerge suddenly in several places, rather than ones that emerge locally at a speed rapid enough to be remarked, but slow enough to permit considered response. That is, they are fundamentally nonlinear in causation and discontinuous in both their spatial structure and temporal behavior.

- The problems and their potential responses move both societies and natural systems into such novel and unfamiliar territory that aspects of the future are inherently unpredictable. That is, both the ecological and social components of these problems have an evolutionary character.

The consequence of these transformations is a world of confusion, where humanity is revealed as a planetary force, and perhaps one that is out of control. The problems are therefore not amenable to solutions based on knowledge of small parts of the whole, or to

---

1. Parts of this paper were prepared for the Committee on Environmental Research, U.S. National Academy of Sciences.

assumptions of constancy or stability of fundamental relationships—ecological, economic or social. Such assumptions produce policies and science that contribute to a pathology of rigid and unseeing institutions, increasingly fragile natural systems, and public dependencies.

Sustaining the biosphere is not an ecological problem, a social problem, or an economic problem—it is an integrated combination of all three. Effective investments in a sustainable biosphere are, therefore, ones that simultaneously retain and encourage the adaptive capabilities of people, business enterprises, and nature. Those adaptive capacities depend on those processes that permit renewal in society, economies, and ecosystems. For nature, it is biosphere structure; for businesses and people, it is usable knowledge; and for society as a whole, it is trust.

Recent advances in theories of nonlinear change and complexity, in methods of perceiving and analyzing natural patterns, and in regional experience in partnerships of scientists, businessmen, and citizens are leading to the understanding needed for such investments in sustainable development in a world of surprises.

INTRODUCTION

Resource and environmental policies and management are in a decision gridlock in many regions of North America, Europe, and Australia. Issues are polarized. It is a time of deep frustration, when conflicts are extreme, mutual suspicions dominate, and cooperation seems the road to personal defeat. Identifying enemies and utterly destroying them seems more important than finding win/win solutions. The result is ecosystem deterioration, economic stagnation, and public mistrust.

At the same time that these local and regional adaptive capabilities are eroding, intensifying global connections are becoming more evident. The resulting surprises have the flavor of almost archetypal unknowns. AIDS, the ozone hole, species extinction, and possible climate change are occurring because of human transformations of local landscapes or of the atmosphere, transformations that spread and accumulate to become global in consequence.

The processes that make them problems are all fundamentally ecological, environmental, and evolutionary. AIDS, for example, seems likely to have mutated and moved from animal populations to human, perhaps as a consequence of transformation of land uses and human population increases. That, at least, was the story of malaria in Africa (Desowitz 1991). Intensified movement of people around the planet then turned a local disease that was potentially self-extinguished into a threatening global catastrophe.

The consequences of these transformations reveal humanity as a planetary force—perhaps one that is out of control.

So how does a politician react in those circumstances? How does the head of a regional resource management agency react? In the United States, few politicians or administrators are now demanding immediate solutions and practical actions as they did in the 1970s. The world is too fundamentally confusing. At best (or worst) they are asking for more and more precise data in order to be invulnerable in a court room! This is science for understanding, not understanding for policy design, but data for litigation!

The issue should not be lack of certainty and precision of data or predictions. There is simply a fundamental loss of certitude—lack of conviction that any of the ground rules work anymore. Any action seems to be full of costs and absent of benefits. The only comfort is a retreat to unsupported ideology and beliefs.

There are two possible responses. One is to seek a spurious certitude by increasing control over information and action. The U.S.S.R. learned the price of that strategy! The other is to seek understanding.

SEEKING UNDERSTANDING

A critical minority of politicians and inquiring public are now not so much driven by fear of prophecies of doom but by the need for understanding. To whom can they turn for understanding? Science is not helping, in large measure, because there are not only conflicting voices, but conflicting modes of enquiry and conflicting criteria for establishing the credibility of a line of argument.

In particular, the philosophies of two streams of science are often in conflict. The tension between those two are particularly evident in biology. One is brilliantly represented by the advances in molecular biology and genetic engineering. That stream of science promises to lead to health and economic benefits through biotechnology, but also to a journey on an uncertain sea of changing social values and consequences. It is a stream of biology that is essentially experimental, reductionist, and narrowly disciplined in character.

The other stream is represented within biology by evolutionary biology and by systems approaches that extend to include the analysis of populations, ecosystems, landscape structures and dynamics, and more recently, further extend to include biotic and human interactions with planetary dynamics. The applied form of this stream has emerged regionally in new forms of resource and environmental management, where uncertainty and surprises become an integral part of an anticipated set of adaptive responses (Holling 1978; Walters 1986). It is fundamentally interdisciplinary and combines historical, comparative, and experimental approaches at scales appropriate to the issues. It is this combination that provides the necessary foundations for any kind of global science, if for no other reason because we have but one planet to live on, for the present at least, and cannot experimentally manipulate lost pasts. It is a stream of enquiry that is fundamentally concerned with integrative modes of enquiry and multiple sources of evidence.

It is this stream that has the most natural connection to related ones in the social sciences that are historical, analytical, and integrative. It is also the stream that is most relevant for the needs of policy and politics.

The first stream is a science of parts (e.g., analysis of specific biophysical processes that affect survival, growth, and dispersal of target variables). It emerges from traditions of experimental science where a narrow enough focus is chosen in order to pose hypotheses, collect data, and design critical tests for rejecting invalid hypotheses. The goal is to narrow uncertainty to the point where acceptance of an argument among scientific peers is essentially unanimous. It is appropriately conservative and unambiguous, but it achieves that by being incomplete and fragmentary.

The other is a science of the integration of parts. It uses the results and technologies of the first, but identifies gaps, develops alternative hypotheses and multi-variate models, and evaluates the integrated consequence of each alternative by using information from planned and unplanned interventions in the whole system that occur or are implemented in nature. Typically, the goal is to reveal the simple causation that often underlies the complexity of time and space behavior of complex systems. Often, there is more concern that a useful hypothesis might be rejected than a false one accepted—"don't throw out the baby with the bath water." Since uncertainty is high, the analysis of uncertainty becomes a topic in itself.

The premise of this second stream is that knowledge of the system we deal with is always incomplete. Surprise is inevitable. Not only is the science incomplete, the system itself is a moving target, evolving because of the impacts of management and the progressive expansion of the scale of human influences on the planet.

In principle, therefore, there is an inherent "unknowability," as well as unpredictability, concerning evolving managed ecosystems and the societies with which they are linked. There is, consequently, an inherent unknowability and unpredictably to sustainable development. Therein lies the paradox addressed in this chapter. Most importantly, evolving systems require policies and actions that not only satisfy social objectives but, at the same time, achieve continually modified understanding of the evolving conditions and provide flexibility for adaptation to surprises.

This is the essence of active regional experimentation by management at the scales appropriate to the question, i.e., Adaptive Environmental Management (Holling 1978; Walters 1986). Otherwise, the pathologies of exploitive development are inevitable—increasingly brittle ecosystems, rigid management, and dependent societies leading to crises.

Faced with the partial understanding we have of the problems and with the conflicting voices of science, it is no wonder, therefore, that public concern and mistrust is great, and that public understanding is disturbingly bad. Political responses have a weak foundation for confident action that will not make the cure worse than the disease.

THE COMPETITION OF BELIEFS

So much presently seems uncertain or unknown that many of the calls for action or inaction, however well-supported by technical argument, are largely determined by beliefs and opinions. Because each belief is partially relevant, impressive and convincing technical arguments can be mobilized for each one regardless of conflicting resulting calls for action or inaction.

Four belief systems and an emerging fifth one are driving present debate and public confusion. Each reflects different implicit assumptions concerning stability and change, as I have suggested elsewhere (Holling 1987). Alternatively they can be labeled as follows (albeit unfairly) by a caricature of their causal assumptions.

- The first is a view of smooth exponential growth where resources are never scarce because human ingenuity always invents substitutes. This was the basic view of Hermann Kahn and Julian Simon (1984). It assumes that humans have an infinite capacity to innovate and that nature changes gradually—fast enough to be detected yet slow enough to be managed.

- The second is a hyperbolic view where increase is inevitably followed by decrease. It is a view of fundamental instability, when persistence is only possible in a decentralized system where there are minimal demands on nature. It is the view of Schumacher (1973) or of the proponents of so-called "deep ecology." If the previous view assumes that infinitely ingenious humans do not need to learn anything different, this view assumes that humans are *incapable* of learning how to deal with the technology they unleash.

- The third is a view of logistic growth where the issue is how to navigate a looming and turbulent transition—demographic, economic, social, and environmental—to a sustained plateau. This is the view of several institutions (e.g., the Brundtland Commission and the World Resources Institute) with a mandate for reforming global resource and environmental policy. Many individuals are also contributing skillful scholarship and policy innovation. They are among some of the most effective forces for change, but they function within an essentially conservative and static view of nature and society.

- The fourth is a view of nested cycles organized by fundamentally discontinuous events and processes. That is, there are periods of exponential change, periods of growing stasis and rigidity, periods of readjustment or collapse, and periods of reorganization for renewal. Instabilities organize the behaviors as much as do stabilities. That was the view of Schumpeter's economics (1950), and it has more recently been the focus of fruitful scholarship in a wide range of fields—ecology, technology, economics, and social science, as well as the body of my own ecological research for the past 20 years. There are striking similarities in Harvey Brook's view of technology as being cycles of innovation, leading slowly to a technological "monoculture," then to a participatory paralysis that can suddenly unlock a new wave of innovation. Brian Arthur's recent studies of the economics of innovation and competition (1989) provides a rigorous theoretical and empirical foundation for that view. Similarly, in cultural anthropology, Mary Douglas (1978) and Mike Thompson (1983) describe the discontinuous change that can be generated through the interaction among entrepreneurs, bureaucracies, sectarian ideologues, and the general public. And historians have long perceived abrupt events as ones that interrupt long periods of accumulating experience and rigidities with suddenly released opportunities for reorganization and renewal (Barbara Tuchman 1978; William McNeill 1979).

- The emerging fifth view is evolutionary and adaptive. It has been given recent impetus by the paradoxes that have emerged in successfully applying the previous, more limited views. Complex systems behavior, discontinuous change, chaos and order, self-organization, nonlinear system behavior, and evolving systems all characterize the more recent activities. It is leading to integrative studies that combine insights and people from developmental biology and genetics, evolutionary biology, physics, economics, ecology, and computer science. The Santa Fe Institute and its ebullient enthusiasts from physics, economics, biology, and computer science provide an interesting experiment in applying collaborative approaches to exploring insights and opportunities opened by an evolutionary paradigm (Waldrop 1992).

The point is not that one of these beliefs is correct and the others wrong—each contain partial truths. Because we are only now beginning to understand the changing reality, there is consequently no limit to the ability of a good scientists to invent compelling lines of causal explanation that inexorably support their particular beliefs. How can even the best-intentioned politician possibly be expected to deal with that? How can even the most reflective of the public? With every issue having supporting evidence and explanation, and denying evidence and counter explanation—all legitimate—the issues seem to be ones in which there is no independent reality of nature, only moral issues that can be dealt with by social debate. Can we ever separate belief from fact?

FOUNDATIONS FOR INTEGRATION—SCIENCE, UNDERSTANDING AND POLICY

The previous argument explains my basic unease in calls for action that are dominated exclusively by prophesies of crisis. Certainly it is appropriate to cite clear examples of the critical new class of problems, particularly those that clarify the need for action (e.g., AIDS, the ozone hole, and carbon dioxide increase). Those are so clear, so unambiguous, so growing, so global, and so novel that action can be taken—actions we would want to take, in any case, for other reasons of efficiency, health, and economic sustainability.

Maybe I have been in the game too long to be sympathetic to "Chicken Little." In 1969 *Time* ran an article, "The New Jeremias," and featured six scientists who were prophesizing doom—an environmental doom that was perhaps novel then, but is familiar now. They included Paul Ehrlich, Barry Commoner, Ken Watt and—me! Now, 23 years later, I find the articles, projects, and proposals that repeat the same litany of doom to be not necessarily wrong, but tiresome, unconvincing, and weak. Certainly continuing growth of world human population and of material consumption is fundamentally unsustainable. However spasmodically, humans and their institutions do learn.

There have been remarkable advances and learning in the intervening years. Opportunities for conversation among and actions by previously polarized individuals have increased both understanding and the ability to develop and apply integrated and adaptive policies. The present problems and topics revolve around five interrelated themes: (1) regional resource management and development, (2) ecosystem restoration, (3) sustainable development, (4) global change, and (5) biodiversity. Population growth and exploitive technology drive them all.

The last 20 years have seen a stunning advance in understanding how the planet has evolved and how its physical aspects function. The reconstruction of the composition of our atmosphere over the last 160,000 years using bubbles trapped in the Vostok ice core, and its correlation with climate using proxy biological and chemical signals can be read as much as an engrossing detective story, or as history, as a tour-de-force of international science. It is also useful for politicians. It tells them that the present concentration of CO_2 in our atmosphere is higher than it has been for the last 160,000 years.

The detection of the "ozone hole" in the Antarctic came as a complete surprise to existing "gradualist" theories of the atmosphere, and the demonstration of its reality and of the role of industrial chloroflourocarbon (CFC) emissions on atmospheric chemistry has been an example of the passionate application of the best kind of cooperative, and at times combative, science in a complex new area—one useful for politicians again, as countries now move to ban CFCs, as an act of international cooperation.

However narrow the mainstream of molecular biology might be, it has yielded techniques that are now transforming the evolutionary, ecological and conservation sciences. Is it true we can trace all human mitochondrial DNA back to an "Eve" in Africa (Vigiland et al. 1991)? Biologists can now certainly unravel affinities in related groups of species and individuals, and join the geophysicists in compelling reconstructions of our past that at the least place our present problems within a perspective—from the role of past extinctions to present declines in biodiversity.

The understanding needed for the changes we now experience or anticipate draws on more than this knowledge from geophysics, atmospheric science, and techniques of cellular and molecular biology. The core to understand such changes lies in *integrating* ecosystem and community ecology with the more physically-based environmental sciences.

But recognize what that means, and the challenge it presents. The relevant biophysical processes operate over an enormous range of scales, potentially from soil processes operating with time constants of hours or days in meter square patches, to ecosystem successional processes of decades to centuries covering tens to thousands of square kilometers, to global biotic processes involved in the regulation and isolation of elements like carbon that have time lags of millennia and are global in their impact. That is why satellite imagery, remote sensing, and geographic information systems now routinely available to analyze patterns are of such major importance. Computer advances, both towards the portable but powerful and the large and parallel, have opened ways to visualize complexity in both space and time. It is a picture of discontinuous behavior, of multiple stable states, of the interaction between slow forces that accumulate environmental capital, and fast processes that slowly exploit, and suddenly release and renew the capital.

That is as far a cry from public perceptions of fragile, stable, and equilibrium nature as could be imagined. And that knowledge, too, is useful and used. It is the foundation for the regional experiments in adaptive policy design and management that are as much examples of institutional learning as they are of using science for public policy.

THE PARADOX OF SUSTAINABLE DEVELOPMENT

Sustainable development is something of a paradox. The phrase says something must change, but something must remain constant. The paradox appears in a number of forms, and its resolution can provide the direction for investments that could sustain development.

I encountered the paradox in the form of two puzzles, when I reviewed some 23 examples of managed ecosystems (Holling 1986). Those examples fell into four classes—forest insect, forest fire, savanna grazing, and aquatic harvesting. One puzzle seemed to be a paradox of the organization of ecosystems. The other seemed to be a paradox of the management of ecosystems. Both have turned out to be the consequence of the natural workings of any complex, evolving system.

The first paradox suggested that the great diversity of life in ecosystems is traceable to the function of a small set of variables, each operating at a qualitatively different speed from the others. The second suggested that any attempt to manage ecological variables inexorably led to more brittle ecosystems, more rigid management institutions and more dependent societies. A discussion of each paradox follows.

The Ecosystem Organization Puzzle

How could the great diversity within ecosystems possibly be traced to the function of a small number of variables? The models developed and tested for these examples certainly generated complex behavior in space and time. Moreover, those complexities could be traced to the actions and interactions of only 3 to 4 sets of variables and associated processes, each of which operated at distinctly different speeds. The speeds were therefore discontinuously distributed and differed from their neighbors often by as much as an order of magnitude. A summary of the critical structuring variables and their speeds are presented in Table 4.1. For the models at least, this structure organizes the time and space behavior of variables into a small number of cycles, presumably abstracted from a larger set that continue at smaller and larger scales than the range selected.

Table 4.1. Key Variables and Speeds in Four Groups of Managed Ecosystems

The System	Variables			Key reference
	Fast	Intermediate	Slow	
Forest insect	insect, needles	foliage crown	trees	McNamee et al. 1981; Holling 1991
Forest fire	intensity	fuel	trees	Holling 1980
Savanna	annual grasses	perenn. grasses	shrubs	Walker et al. 1969
Aquatic	phyto-plankton	zooplankton	fish	Steele 1985

But are those features simply the consequence of the way modellers make decisions rather than the results of ecosystem organization? This uneasy feeling that such conclusions reflect the way we think, rather than the way ecosystems function, led to a series of tests using field data to challenge the hypothesis that ecosystem dynamics are organized around the operation of a small number of nested cycles, each driven by a few dominant variables (Holling 1992).

The critical argument is that if there are only a few structuring processes, their imprint should be expressed on most variables (i.e., time series data for fires, seeding intensity, insect numbers, bird and mammal populations, water flow—indeed any variable for which there are long-term, yearly records—should show periodicities that cluster around a few dominant ones). In the case of the eastern maritime boreal forest of North America, those periodicities were predicted to be 3–5 years, 10–15 years, 35–40 years, and over 80 years (Holling 1992). Similarly, there should be a few dominant spatial "footprint" sizes, each associated with one of the disturbance/renewal cycles in the nested set of such cycles. Finally, the animals living in specific landscapes should demonstrate the existence of this lumpy architecture by showing gaps in the distribution of their sizes and gaps in the scales at which decisions are made for location of region, foraging area, habitat, nests, protection, and food.

All the evidence so far confirms those hypotheses for boreal forests, boreal region prairies, pelagic ecosystems (Holling 1992) and the Everglades of Florida (Gunderson 1992). A variety of alternative hypotheses based on developmental, historical, or trophic arguments were disproved in the fine traditions of Popperian science, leaving only the "world-is-lumpy" hypothesis as resisting disproof.

Thus there is strong evidence for the following conclusions:

1. A small number of plant, animal, and abiotic processes structure biomes over scales from days and centimeters to millennia and thousands of kilometers. Individual plant and biogeochemical processes dominate at fine, fast scales; animal and abiotic processes of meso-scale disturbance dominate at intermediate scales; and geomorphological ones dominate at coarse, slow scales.

2. These structuring processes produce a landscape that has lumpy geometry and temporal frequencies or periodicities. That is, the physical architecture and the speed of variables are organized into distinct clusters or quanta, each of which is controlled by one small set of structuring processes. These processes organize behavior as a nested hierarchy of cycles of slow production and growth alternating with fast disturbance and renewal.

3. Each quantum is contained to a particular range of scales in space and time and has its own distinct architecture of object sizes, interobject distances, and fractal dimension within that range.

4. All of the many remaining variables, other than those involved in the structuring processes, become entrained by the critical structuring variables, so that the great diversity of species in ecosystems can be traced to the function of a small set of variables and the niches they provide. The structuring processes are the ones that both form structure and are affected by that structure. Therefore, when investing to protect biodiversity, priority should be placed on these structuring variables.

5. The discontinuities that produce the lumpy structure of vegetated landscapes impose discontinuities on the behavior and morphology of animals. For example, there are gaps in body mass distributions of resident species of animals that correlate with scale-dependent discontinuities in the geometry of vegetated landscapes. Thus these gaps, and the body mass clumps they define, become a way to develop a rapid bioassay of ecosystem structures and of human impacts on that structure. This discovery therefore opens the way to develop a comparative ecology across scales that might provide the same power for generalization that came when physiology became comparative rather than species specific.

6. Conversely, changes in landscape structure at defined scale ranges caused by land use practice or by climate change will have predictable impacts on animal community structure (e.g., animals of some body masses can disappear if an ecosystem structure at a predictable scale range is changed). Therefore, predicted (using models) or observed (using remote imagery) impacts of changing climate or land use on vegetation can also be used to infer the impacts on the diversity of animal communities.

The lessons for both sustainable development and biodiversity are clear—focus on the structuring variables that control the lumpy geometry and lumpy time dynamics. They are the ones that set the stage upon which other variables play out their own dramas. It is the physical and temporal infrastructure of biomes *at all scales* that sustains the theater. Given that, the actors will look after themselves!

The Ecosystem Management Puzzle

There was an even more surprising and puzzling feature that emerged from comparing the 23 examples. All were associated with management of a resource where the very success of management seemed to set the condition for collapse. Is there some general property of unsustainability represented in these examples, or does the observation simply result from cases selected because of an unconscious attraction to catastrophic visions? Again, some independent tests are necessary.

Each of the examples represented policies of management whose goal was to control a target variable in order to achieve social objectives, typically maintaining or stimulating employment and economic activity. In the case of management of eastern North American spruce/fir forests, the target was an anticipated outbreak of a defoliating insect—the spruce-budworm (Clark et al. 1979); for the forests of the Sierra Nevada Mountains, the target was forest fires (Holling 1980); for the savannas of South Africa, the target was the grazing of cattle (Walker 1969); for the salmon of the Pacific Northwest Coast, the target was salmon populations (Walters 1986).

In each case, the goal was to control the variability of the target—insects and fire at low levels, cattle grazing at intermediate stocking densities, and salmon at high populations. The level desired was different in each situation, but the common feature was to reduce variability of a target whose normal fluctuations imposed problems and periodic crises for pulp mill employment, recreation, farming incomes, or fishermen's catch.

That is the typical response to threats of fire or pestilence, flood or drought! Narrow the purpose, focus on it exclusively, and solve that problem as defined. Modern engineering, and technological, economic, and administrative experience can deal well with such narrowly defined problems. And in each example the goal was successfully achieved—insects were controlled with insecticide; fire frequency and extent was reduced with fire detection and suppression techniques; cattle grazing was managed with modern rangeland practice; and salmon populations were augmented with hatchery production.

At the same time, however, elements of the system were slowly changing. First, reducing the variability of the ecological target produced a slow change in the spatial heterogeneity of the ecosystem. Forest architecture became more contiguous over landscape scales, so that if fire or defoliating insects were released, the outbreaks could cover larger areas with more intensive impacts than before management. Rangeland gradually lost drought-resistant grasses because of a shift in competition that favored more productive but less drought-resilient grasses. If drought occurred, the consequences were therefore more extensive, more extreme, and more persistent—grasslands turned irreversibly into shrub-dominated semi-deserts. Wild populations of salmon in the many streams along the coast gradually became extinct because fishing pressure increased in response to the increased populations achieved by enhancement. That left the fishing industry precariously dependent on a few hatcheries whose productivity declines with time.

In short, the success in controlling an ecological variable that normally fluctuated led to more spatially homogenized ecosystems over landscape scales. It led to systems more likely to flip into a persistent degraded state, triggered by disturbances that previously could be absorbed.

Those changes in the ecosystems could be managed if it were not for concomitant changes in two other elements of the interrelationships—in the management institution(s) and in the society that reaped the benefits or endured the costs. Because of the initial success, in each case the management agencies shifted their objectives from the original social and ecological ones to the laudable objective of improving operational efficiency of the agency itself—spraying insects, fighting fires, producing beef, or releasing hatchery fish with as much efficiency and with the least cost possible. Efforts to monitor the ecosystem for surprises rather than for product therefore withered in competition with internal organizational needs, and research funds were shifted to more operational purposes. Why monitor or study a success? Thus, the gradual reduction of resilience of the ecosystems was unseen by any but maverick and suspect academics!

Success brought changes in the society as well. Dependencies developed for continuing sustained flow of the food or fiber that no longer fluctuated as it once had. More investments therefore logically flowed to expanding pulp mills, recreational facilities, cattle ranches, and fishing technology. That is the development side of the equation, and its expansion can be rightly applauded. Improving efficiency of agencies should also be applauded. But if, at the same time, the ecosystem from which resources are garnered is becoming more and more brittle (i.e., more and more sensitive to large scale transformation triggered by events that earlier could be absorbed), then the efficient but myopic agency and the productive but dependent industry simply becomes part of the source of crisis and decision gridlock.

So here is the paradox: success in managing a target variable for sustained production of food or fiber apparently leads to an ultimate pathology of (1) more brittle and vulnerable ecosystems, (2) more rigid and unresponsive management agencies, and (3) more dependent societies. That seems to confirm one opinion that sustainable development is an oxymoron.

But there seems to be something inherently wrong with that conclusion, implying, as it does, that the only solution is a radical return of humanity to being "children of nature."

The argument is based on two critical points. One is that reduced variability of ecosystems inevitably leads to reduced resilience and increased vulnerability. The second is that there is, in principle, no different way for agencies and people to manage and benefit from resource development.

Again, some independent evidence is needed. Are there counter examples? Oddly, nature itself provides counter examples of tightly regulated yet viable systems in the many examples of physiological homeostasis. Consider temperature regulation of endotherms ("warm-blooded" animals), for example. That represents a system where internal body temperature is not only tightly regulated within a narrow band, but at an average temperature perilously close to lethal. Moreover, the costs of achieving that regulation requires ten times the energy for metabolism than is required by an ectotherm ("cold-blooded" animals). That would seem to be a recipe not only for disaster, but a very inefficient one at that. And yet evolution somehow led to the extraordinary success of the animals having such an adaptation (i.e., birds and mammals).

In order to test the generality of the variability-loss/resilience-loss hypothesis, I have been collecting data from physiological literature on the viable temperature range of the internal body of organisms exposed to different classes of variability. The data is organized into three groups ranging from terrestrial ectotherms exposed to the greatest variability of temperature from unbuffered ambient conditions, to aquatic endotherms exposed to an intermediate level of variability because of the moderating attributes of water, to endotherms that regulate temperature within a narrow band. As predicted, the viable range of internal body temperature decreases from about 40°C for the most variable group, to about 30°C for the intermediate, to 20°C for the tightly regulated endotherms. That is, resilience, in this case the range of temperatures that separates life from death, clearly does contract as experience with variability is reduced. Therefore, reduction of variability of living systems from organisms to ecosystems to institutions inevitably leads to loss of resilience.

What remains is an even starker paradox that says that successful control inevitably leads to collapse. But in fact endothermy does persist. It therefore serves as a revealing metaphor for sustainable development. This metaphor contains two features that were not evident in my earlier descriptions of examples of resource management.

First, the type of regulation is different. Five different mechanisms, from evaporative cooling to metabolic heat generation, control the temperature of endotherms. Each mechanism is not notably efficient by itself. Each operates over a somewhat different range of conditions and with different efficiencies of response. It is this overlapping "soft" redundancy that seems to characterize biological regulation of all kinds. It is not notably efficient or elegant in the engineering sense. But it is robust and continually sensitive to changes in internal body temperature. That is quite unlike the examples of regulation by management where goals of operational efficiency gradually isolated the regulating agency from the things it was regulating. The result was not only a lack of response, it was a blindness to the need for response.

Second, endothermy is a true innovation that explosively released opportunity for the organisms evolving the ability. Maintaining high body temperature, short of death, allows the greatest range of external activity for an animal. Speed, stamina increase, and activity can be maintained at both high and low external temperatures. A range of habitats forbidden to an endotherm is open to an ectotherm. The evolutionary consequence of temperature regulation was to suddenly open opportunity for dramatic organizational change and the adaptive radiation of new life forms. Variability is therefore not eliminated. It is reduced and transferred from the animal's internal to its external environment as a consequence of the continual probes by the whole animal for opportunity and change. Hence, the price of reducing internal resilience and maintaining high metabolic levels is more than offset by that creation of evolutionary opportunity.

That surely is at the heart of sustainable development—the release of human opportunity! It requires flexible, diverse, and redundant regulation, monitoring that leads to corrective responses and experimental probing of the continually changing reality of the external world. Those are the features of Adaptive Environmental and Resource Management (Holling 1978; Walters 1986). And those are the features which are missing in my earlier description of traditional, piecemeal, exploitive resource management that leads to an ultimate pathology.

In fact, such a switch to an adaptive strategy is what eventually happened in at least one of the examples quoted. In New Brunswick, the intensifying gridlock in forest management, combined with slowly accumulated and communicated understanding, led to an abrupt transformation of policy whose attributes became much like those just described for sustainable systems . It is a policy that functions for a whole region by transforming and monitoring the smaller scale stand architecture of the landscape and by focusing the productive capacities of industry on opportunities that can be created and yet sustained. ecosystem management

A PROFILE FOR THE NEW SURPRISES

Citizens and politicians are frustrated because they are not hearing simple and consistent answers to the following key questions concerning present environmental and renewable resource issues:

- What is going to happen under what conditions?
- When will it happen?
- Where will it happen?
- Who will be affected?
- How certain are we?

The answers are not simple or consistent because the concepts, technology, and methods to deal with the generic problems are just being developed. Those generic features can be described as follows:

- The problems are essentially systems problems, where aspects of behavior are complex and unpredictable and where causes, while at times simple (when finally understood), are always multiple.

Therefore interdisciplinary and integrated modes of enquiry are needed for under-standing. And understanding (not complete explanation) is needed to form poli-cies.

- They are fundamentally nonlinear in causation. They demonstrate multi-stable states and discontinuous behavior in both time and space.

Therefore, the concepts that are useful come from nonlinear dynamics and theo-ries of complex systems. Policies that rely exclusively on social or economic adaptation to smoothly changing and reversible conditions lead to reduced op-tions, limited potential, and perpetual surprise.

- They are increasingly caused by slow changes reflecting decadal accumulations of human influences on air and oceans and decadal transformations of landscapes. Those slow changes cause sudden changes in fast environmental variables that di-rectly affect the health of people, productivity of renewable resources, and vital-ity of societies.

Therefore, analysis should focus on the interactions between slow phenomena and fast ones, and monitoring should focus on long-term, slow changes in struc-tural variables. The political window that drives quick fixes for quick solutions simply leads to more unforgiving conditions for decisions, more fragile natural systems, and more dependent and distrustful citizens.

- The spatial span of connections is intensifying so that the problems are now fundamentally cross-scale in space as well as time. Local, regional, and national environmental problems can now more and more frequently have their source both at home and half-a-world away—witness greenhouse gas accumulations, the ozone hole, AIDS, and the deterioration of biodiversity. Natural planetary pro-cesses mediating these issues are combining with the human, economic, and trade linkages that have evolved among nations since World War II.

Therefore, the science needed is not only interdisciplinary, but cross-scale. And yet the very best of environmental and ecological research and models have achieved their success by being either scale-independent or constrained to a nar-row range of scales. Hierarchical theory, spatial dynamics, event models, satellite imagery, and parallel processing perhaps open new ways to violate, successfully, the hard-won experience of the best ecosystem modellers (i.e., never include more than two orders of magnitude, or the models will be smothered by detail).

- Both the ecological and social components of these problems have an evolution-ary character. That is why the phrase "sustainable development" is not an oxy-moron. The problems are therefore not amenable to solutions based on knowl-edge of small parts of the whole or to assumptions of constancy or stability of fundamental relationships ecological, economic, or social. Assumptions that such constancy is the rule might give a comfortable sense of certainty, but it is spurious. Such assumptions produce policies and science that contribute to a pathology of rigid and unseeing institutions, increasingly brittle natural systems, and public dependencies.

Therefore, the focus best suited for the natural science components is evolutionary; for economics and organizational theory, it is education and innovation; and for policies, it is actively adaptive designs that yield understanding as much as they do products.

Sustaining the biosphere is not an ecological problem, nor a social problem, nor an economic problem. It is an integrated combination of all three. Effective investments in a sustainable biosphere are therefore ones that simultaneously retain and encourage the adaptive capabilities of people, of business enterprises, and of nature. It is the effectiveness of those adaptive capabilities that can turn the same unexpected event (e.g., drought, price change, market shifts) into an opportunity for one system, or a crisis for another. Those adaptive capacities depend on those processes that permit renewal in society, economies, and ecosystems. For nature it is biosphere structure; for businesses and people it is usable knowledge; and for society as a whole it is trust.

This chapter concludes that investments to increase productivity can only be sustained if all these sources of renewal capacity are maintained or enhanced. Temporary erosion of any one might be bearable as long as recovery occurs within the critical time unit of a human generation (note relation to intergenerational equity and freedoms of choice). But continued erosion of even one ultimately reaches the point where it cannot be reversed by normal, internal recovery. That state is the condition defined as poverty—a condition of inability to cope. It leads to an unsustainable biosphere.

REFERENCES

Arthur, B. W. 1989. Competing technologies, increasing returns and lock-in by historical events. *The Economic Journal* 99: 116–31.

Clark, W. C., D. D. Jones, and C. S. Holling. 1979. Lessons for ecological policy design: a case study of ecosystem management. *Ecological Modelling* 7: 1–53.

Desowitz, J. W. 1991. The Malaria Capers. New York: W. W. Norton.

Douglas, M. 1978. Cultural Bias. Paper for the Royal Anthropological Institute, No. 35. London: Royal Anthropological Institute.

Gunderson, L. H. 1992. Spatial and Temporal Hierarchies in the Everglades Ecosystem with Implications for Water Management. Ph.D. dissertation, Univ. of Florida, Gainesville, Florida.

Holling, C. S. 1973. Resilience and stability of ecological systems. *Annual Review of Ecology and Systematics* 4: 1–23.

————., ed. 1978. Adaptive Environmental Assessment and Management. London: Wiley.

————. 1980. Forest insects, forest fires, and resilience. In Fire Regimes and Ecosystem Properties, eds. H. Mooney, J. M. Bonnicksen, N. L. Christensen, J. E. Lotan and W. A. Reiners. USDA Forest Service General Technical Report WO–26. Washington, DC: 445–64.

————. 1986. Resilience of ecosystems: local surprise and global change. In Sustainable Development of the Biosphere, eds. W. C. Clark and R. E. Munn. Cambridge: Cambridge Univ. Press.

————. 1987. Simplifying the complex: the paradigms of ecological function and structure. *European Journal of Operational Research* 30: 139–46.

————. 1991. The role of forest insects in structuring the boreal landscape. In A Systems Analysis of the Global Boreal Forest, eds. H. H. Shugart, R. Leemans and G. B. Bonan. Cambridge: Cambridge Univ. Press.

————. 1992. Cross-scale morphology, geometry and dynamics of ecosystems. *Ecological Monographs* 62 (4): 447–502.

Kahn, H., and J. Simon. 1984. The Resourceful Earth: A Response to Global 2000. Oxford: Blackwell.

McNamee, P. J., J. M. McLeod, and C. S. Holling. 1981. The structure and behavior of defoliating insect/forest systems. *Researches on Population Ecology* 23: 280–98.

McNeill, W. H. 1979. The Human Condition: An Ecological and Historical View. Princeton: Princeton Univ. Press.

Schumacher, E. F. 1973. Small is Beautiful: Economics as if People Mattered. New York: Harper.

Schumpeter, J. A. 1950. Capitalism, Socialism and Democracy. New York: Harper.

Steele, J. H. 1985. A comparison of terrestrial and marine Systems. *Nature* 313: 355–8.

Thompson, M. 1983. A cultural bias for comparison. In Risk Analysis and Decision Processes, eds. H. C. Kunreuther and J. Linnerooth. Berlin: Springer-Verlag.

Tuchman, B. W. 1978. A Distant Mirror. New York: Ballantine.

Vigiland, L., K. Hawkes, and A. C. Wilson. 1991. African populations and the evolution of human mitochondrial DNA. *Science* 253: 1503–7.

Waldrop, M. M. 1992. Complexity. New York: Simon & Schuster.

Walker, B. H., D. Ludwig, C. S. Holling, and R. M. Peterman. 1969. Stability of semi-arid savanna grazing systems. *Ecology* 69: 473–98.

Walters, C. J. 1986. Adaptive Management of Renewable Resources. New York: McGraw Hill.

5 ECOSYSTEM PROPERTIES AS A BASIS FOR SUSTAINABILITY

AnnMari Jansson and Bengt-Owe Jansson
Department of Systems Ecology
Stockholm University
S-106 91 Stockholm

ABSTRACT

The somewhat late realization of the finite character of the natural resources have led to a healthy reevaluation of their basic role in the economics of mankind. Because the natural systems form the basis for economic development and sustainability, their short- and long-term dynamics and structure have to be recognized and fully understood. The evolution of ecological economics as an extended "ecological regime" is both qualitatively and quantitatively dependent on an adequate understanding of the behavior of living systems. The stimulating participation in this evolution has for us repeatedly highlighted basic, but apparently less well-known properties of the ecosystem, which, when neglected, often cause adverse and unexpected results of environmental management practices. This chapter highlights a series of ecologically established concepts, and their implications for the choice of investment strategies, such as gross and net production of ecosystems and their maintenance costs; biodiversity and its energetic background; fragmentation, patchiness, and metapopulation concepts; resistance and resilience; stability domains and equilibrium points; self-maintenance and succession; and life-support systems and life-support area.

INTRODUCTION

For a long time now ecologists have tried to explain how natural systems work and why it is necessary to preserve their species, structure, and functions for the benefit of mankind. Economists in general, on the other hand, tend to regard the services from functioning ecosystems—like producing food, fresh air, clean water—as self-evident, seemingly free, unlimited, and lasting forever. Evolution during millions of years has developed systems with those properties, still evolving but susceptible to "new" disturbances like man's fossil fuel-based activities. It is imperative for the human society to develop behavior patterns in tune with the properties of ecosystems and the carrying capacity of the earth's ecosystems to support economic growth.

In a natural ecosystem, each species performs work like fixing solar energy, filtering water for food particles, or decomposing fallen leaves to get fuel for its own metabolism. But each species is also dependent on the rest of the ecosystem for its maintenance. It has a life-support system, which must be kept intact. The actual base of a species is thus not the mere pool of its potential food, but the amount of solar energy fixed by its ecosystem. The human species is not exempt from this general rule even if an abundant supply of fossil energy has made it believe so for some time.

Knowledge of the natural systems and how they work is required for an ecological-economic governance of our natural resources. There is a great need for fundamental research and critical examination of principles for ecological-economic interactions. In that work we need to use the same language. Calibrating the terminology may sound trivial, but as some conceptual names have deteriorated in the everyday vocabulary, it is crucial for avoiding serious misunderstandings. In this chapter we present and discuss some fundamental notions in their proper ecological context. The cited publications have been selected more for their basic treatment of the respective concepts than in an attempt to review the most recent literature.

SUSTAINABILITY REQUIREMENTS

As one of today's buzzwords, "sustainability," means for most people sustainability of the economic activities regardless of how large they may grow. However, what only a few decades ago was taken for granted—an infinite and stable environment for man to use without restrictions, inexhaustible both as a resource producer and a waste dump—has shown to change at an increasing rate, and has reached the limits for providing society with fresh air, clean water, and fertile soils. The growth of the human population and resulting environmental impacts have become so extensive that they are threatening to lead to significant global changes inimical to human survival. In its basic ecological sense, sustainability is actually the main property of the ecosystem, long known under other names such as *persistence*, which is attained through the complex structures and functions of natural systems, some of which will be addressed here.

Before turning to the separate properties, let us look at one example of an ecosystem (see Figure 5.1) and its contributions to the human society to stress that the separate properties and services are working in concert and not individually. The illustration shows a forest ecosystem, using solar energy, rain, soil, water, and minerals to produce trees, a diverse bird life, mushrooms, berries, and other products for the pleasure and use of the human community. The forest ecosystem also maintains a stable groundwater reservoir of low mineral content. Man transfers the trees to timber and the groundwater to drinking water. Through overexploitation like clearcutting, however, the groundwater table is lowered, the soil is eroded, and the essential minerals are leached from the soils. The spreading of toxic substances for controlling outbreaks of noxious insects poisons birds and other wildlife species. The atmospheric deposition of pollutants from the burning of fossil fuels acidifies the soil water and wipes out the fish populations of the lakes in areas lacking sufficient lime.

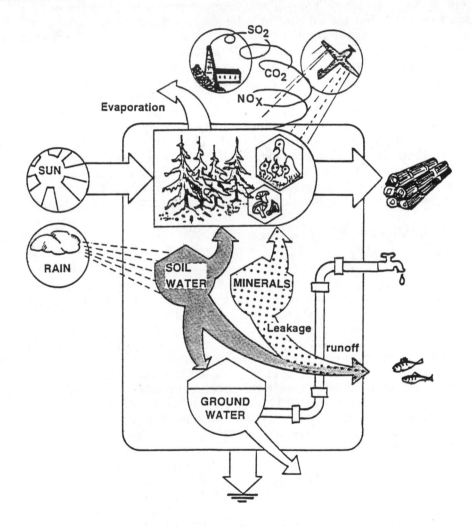

Figure 5.1. A forest ecosystem acting as a source and sink for human society. Based on solar energy, soil water, minerals, and carbon dioxide, the forest produces timber, wildlife, mushrooms, and berries, and provides various environmental services including the production of clean groundwater.

Obviously several of the essential properties of ecosystems have been violated in this single example. Figure 5.2 shows these major properties and how they relate to each other. Greatly simplified, an ecosystem consists of a *production base* of green plants, constituting the resource base for a complex *network of consumers*—animals arranged in *trophic levels* according to their diet. Each level displays different degrees of *species diversity*, and the number of levels corresponds to the *functional diversity*. The total *community metabolism* of this assembly of organisms is maintained through a *recycling* of critical elements via dead *organic matter,* which is broken down by *decomposers* where bacteria and fungi play an important role. As ecosystems are very complex, the theoretical meaning of many concepts are still subject to scientific debate. The view presented here is personal, based on long experience from work in the field.

Production Base

The establishment of an energy base is a first precondition for sustainability (see Figure 5.2). In a virgin area this base is established by the invasion of plants, fixing solar energy by the cell's cycling receptor of chlorophyll. This synthesis is the only real production of organic matter that, constituting food for a consumer network of animals, forms the ultimate base for all living organisms on earth, including humans. The term has been misinterpreted, however, through the use of the concepts "secondary and tertiary" production, when describing the energy flow in a food chain. These processes should rather be termed *transformations* as no "new" organic matter is produced—only the already synthesized organic matter is involved. In a pure ecological context most "industrial production" is therefore a form of consumption, because the manufacturing of goods implies the transformation of given natural resources, requires large inputs of extra energy, and produces large amounts of wastes capital to the environment.

It is important to discriminate between two measures of production: *gross production* and *net production*. This is seldom done, however. The difference can be explained by studying a plant enclosed in a jar during a day and night cycle. During the day plants synthesize organic matter from carbon dioxide, nutrients and water. Oxygen is produced in equivalent amounts in this process. By measuring the oxygen concentration in the jar for 24 hours, we can observe how the plant produces oxygen during the day, but consumes oxygen during night. The night consumption constitutes the *respiration* of the plant and is a measure of the metabolic work, the "cost" necessary for maintaining the life processes of the organism. The respiration process also takes place during the day, in fact even more intensively, but it is masked in the measurements by the much larger production at the presence of light. The production value during the day corresponds to the net production and adding the respiration gives a value of gross production. Figure 5.3 shows real values from measurements of community metabolism of a seaweed community during two diurnal cycles. A similar dynamic pattern is exhibited during

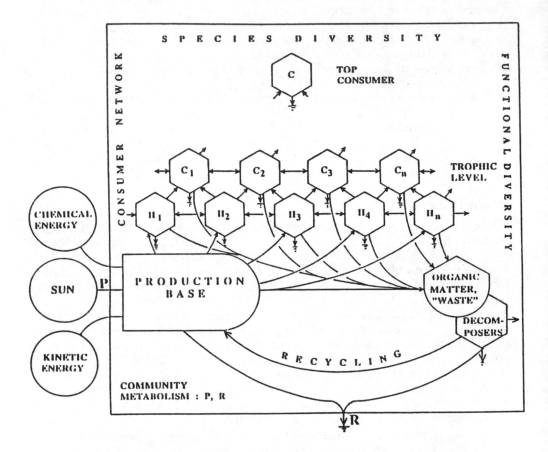

Figure 5.2. The main structure of an ecosystem. The green plants of the production base
create living organic material by fixing solar energy (P) and using carbon
dioxide and nutrients. The consumer network starts with many species of
plant-eating animals (H), sources for different levels of carnivore species (C).
The waste from the plants and animals, including the dead organisms, is
broken down by decomposers, especially bacteria and fungi. Resulting
inorganic matter like phosphorus and nitrogen compounds are recycled to the
plants and the biogeochemical cycle starts again. The community metabolism
incorporates the total production of the system (one measure is the total
amount of fixed solar energy, P) and the total energy loss from all the
processes, mainly as heat (R; for organisms this is the respiration). R is a
crude measure of the total costs of running the system.

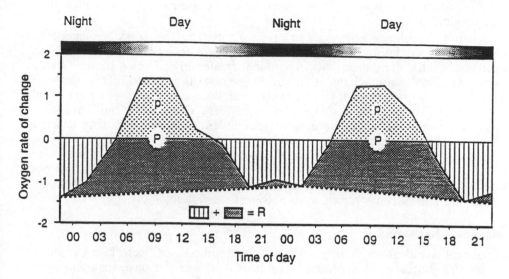

Figure 5.3. Production and consumption (respiration) of a marine seaweed system, enclosed in a transparent plastic bag and measured as diurnal oxygen changes. The whole area below the 0-base line to the horizontal dashed line is the respiration or maintenance costs (R) during the two diurnal cycles. The lightly stippled area above the 0-base is the net production (p), which together with the dark stippled area constitutes the gross production (P).

the year with a productive pulse during the summer and a respiration phase during the dark winter period.

Compared to the annual economy of a tourist industry, which in temperate latitudes also has its main activity during the light part of year, the respiration costs would be analogous to the cost of maintaining the business: keeping the stock, human labor, advertising (in terrestrial systems comparable to exhibiting the bright color of flowers to attract insects for the necessary pollination). The net income during summer would have to pay for the costs during the rest of the year. When the term *production* is used in everyday conversation, it usually means net production, probably because the physical result we see in the form of, for example, the green grass of the lawn or the manufactured cars in the factory are the mere products. But they do not show all the energy-consuming processes that made their production possible. The careless use of the word production actually has far-reaching consequences. It makes us forget that persistence requires much potential energy for pure maintenance. This is probably easier to recognize when it involves personal labor like the necessary watering of potted plants or the cleaning of the personal car. When we admire the effectiveness of genetically improved organisms or technically sophisticated racing cars, we tend to disregard that these technical wonders actually need a more expensive maintenance than their original prototypes.

Because the *energy efficiency* of green plants (i.e., the gross production as percentage of incoming solar insolation), is only a few per cent, it has often been wrongfully compared with the much greater efficiencies of man-made machines. The comparison is invalid because the green plant through respiration pays not only for running its metabolism but also for repair and propagation. To be fully comparable, a car, for example, would not only repair itself but also generate new specimens. When comparing energy efficiencies, it is important to consider both solar and fossil energy inputs during the whole life of a given product. Industrial agriculture may, for example, seem highly productive but compared to natural ecosystems that are not fossil fuel-subsidized, it actually consumes more energy than it delivers and should therefore be regarded as a consumer system.

Consumer Network

In the real world no organism, whether plant, animal, or bacterium can carry out its life processes in isolation. The producers are the base in a *foodweb* containing consumers of various kinds, linked in *foodchains,* and aggregated into *trophic levels* (Figure 5.4). There are seldom any straight foodchains in nature since most animals eat from more than one trophic level. But for the sake of clarity, we can depict an unramified flow of energy and matter from the producer level through the successive consumer levels. Summarizing several empirical attempts to quantify *the food transformation efficiency* showing variable but generally low efficiencies, we find an average of 90% reduction at each step in the foodchain. This is due not only to the basic respiration of the consumers and the varying quality of their food items but also to the costs for obtaining the food by running, swimming, and digging. Theoretically, therefore, the producers may put a limit to the number of possible trophic levels in the foodchain, which seldom exceed five. In most instances, however, disturbances rather than the basic energy flow seem to determine food chain length (Pimm 1988).

According to the "Emergy" concept developed by H. T. Odum (1988), there is an increase of the *energy quality* with each step of the food chain, which means that more solar energy is needed to generate and maintain one gram of fish than one gram of its food items like the water fleas. Also, the number of individuals usually decreases at higher trophic position. In an aquatic system the number of microscopic algae per unit area is much higher than the number of individual water fleas that feed on them and in turn are food for a much smaller number of fish. This means that the size of territory per individual is larger the higher the trophic position. Food has to be collected over a much larger space—*the life-support area* is large (Odum 1988). As this area has to comprise the entire lifespan of an organism, it can

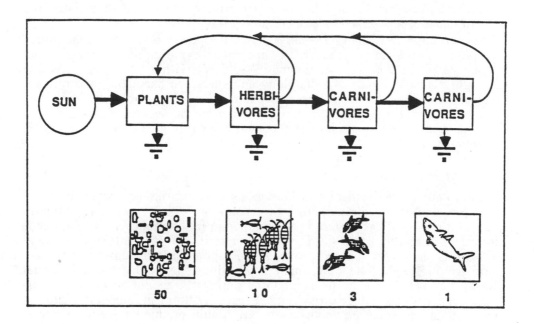

Figure 5.4. This simplified foodchain shows increasing life support areas and thus decreasing number of organisms per unit area higher up in the food chain. The control functions of the higher trophic levels on the lower ones are indicated by feedback loops (upper diagram).

be quite large and difficult to estimate for migrating species like a salmon, an Arctic tern, or a reindeer.

The life-support area concept is also valid for the human species. Some 40% of the net terrestrial productivity on earth is presently channeled to human usage (Vitousek et al. 1989). In addition, about 8% of the productivity of the sea surface area is exploited by world fisheries. These studies imply that even if the built up area of a world population of 5.3 billion people occupies only 2 million km^2, the human support area is in the order of 100 million km^2. Like an ecosystem, the economic system consists of a network of sectors, each sector requiring inputs of energy, goods, and services from other sectors. The total energy costs (*embodied energy*) of the final products from a sector that is further removed from the initial energy sources (solar as well as fossil) are usually very high, but this fact is seldom reflected in their market value.

Another important aspect of Odum's emergy theory is that the *control functions* exerted by the species are stronger the higher up they are in the food chain. There are numerous examples of such "invisible control wires" between species, (e.g., pollination of flowers by insects, birds eating fruits and dispersing their seeds, and humans performing work that stimulates the development of a diverse landscape pattern).

As top consumer, man has really shown the power of affecting the whole biosphere. For example, the greatly increased exploitation of intermediate trophic levels by world fisheries since World War II has been made possible through increased inputs of technology and energy in harvesting and by near extinction of the competing top consumers of marine mammals. This may lead to large indirect effects in lower trophic levels, such as increased algal blooms and changes in the biogeochemical cycling (Folke et al. 1989; Hammer et al. 1993).

Community Metabolism

The community metabolism of whole systems is the summed production and respiration of all the participating organisms during their work (see Figure 5.2), and tells us if the system is self-supported by its producers or if it needs import of already synthesized organic material. Coral reefs and tropical rainforests show very low net production. Respiration (maintenance) costs are almost equal to the rate of primary production. Apparently much potential energy has been invested in adaptations during the evolution of these self-organizing systems. Other ecosystems, especially those in temperate climates like saltmarshes and hardwood forests, are highly productive during summer but are "dormant" during winter. Man is changing these large-scale patterns. The coastal areas may be driven heterotrophic through the increased inputs of organic matter and nutrients from various human activities. Instead of serving as carbon sinks taking up carbon dioxide from the atmosphere, they are now switching to become sources of carbon dioxide and other gases speeding up the global warming process (Smith and Mackenzie 1987).

Recycling

Recycling is necessary for ensuring the continuous flow of critical elements like nutrients and trace metals to the producers. In a balanced ecosystem a large part of the material flows is recycled. In the upper layers of the open sea carbon and nutrients in the organic matter are recycled many times through microbial activity, and only low calorific fecal pellets from the animals are sinking out of the pelagic system to settle on the bottoms. The coral reef and the tropical rain forest are systems with a very efficient recycling. The seawater flushing the reef and the soil supporting the rainforest contain only small amounts of nutrients—they are all in circulation in the organisms. The organic material produced in the reef is consumed there by the myriads of organisms. Similarly, the nutrients in the old leaves continuously shed in the tropical rainforests are resorbed before they fall to the ground. But these are rather rare examples. Usually ecosystems accumulate materials in storages of long duration, (e.g., tree trunks or peat layers). Fossil fuels, coal, oil, and gas, are the transformed products of organic matter deposited outside the active flows in ancient ecological systems and stored in the ground or at the bottom of the sea for millions of years.

Lately, recycling has become a goal of the industrial society for managing the impending problem of waste generation. In these efforts it should be remembered that speeding up recycling requires inputs of extra energy. An alternative strategy to decrease the use of materials would probably afford a better solution. In any case the flows should be compatible with the biogeochemical cycles of the biosphere and to the capacity of natural ecosystems to store and process wastes.

Self-organization

Although industrial man is the most powerful species in reorganizing natural structures and redirecting flows of energy and matter to his own advantage, other organisms do not just passively adjust to their environment either. In their instinctive struggle to sustain and disperse their species, they change and control the energy flow through what might be described as numerous feedback loops. Each species has been forced by natural selection to "pay" for its existence in the system by feeding back potential energy that amplifies the network it uses at least as much as it drains (Odum 1971, 1983). This interchange between the living and nonliving compartments gives the ecosystem its *self-organizing* ability. For example, by modifying the waterflow, wetland plants cause the sedimentation to increase giving their roots better attachment. Tree crowns create aerodynamic conditions that make pollen, seeds, and other airborne particles accumulate at the fringe of the wood. Predators control the grazing of primary production through eating the grazers, and grazers stimulate the plant production by the fertilizing effect of their excretion. There is much to learn from studying such self-organizing mechanisms in nature about how to develop better feedbacks between human societies and the life-supporting ecosystem.

The self-organizing ability makes it possible for devastated areas to redevelop persistent ecosystems given a set of climatic and edaphic variables. Stripped of its organisms by some disturbance of the environmental conditions—such as a forest fire or an oil spill—the system is open for colonization by invading organisms, starting a *succession* process. Given enough time for exploitation of the released resources, a conservation or climax phase will be reached of largely the same structure as the one that existed before. The climax phase will last until the next "creative destruction" starts a renewal process again (Holling 1986). The strength of this process is very strong. If we for example want to exchange the native grass of our lawn for some foreign, greener species, we have to work against the natural succession by watering, fertilizing, and cutting the lawn. Left unattended, the foreign species will otherwise be outcompeted by native ones because it has other requirements. Falk (1976) estimated that the energy costs for keeping a suburban lawn free from being overgrown with weeds corresponded to more than half the net primary production of that system. In the same way the maintenance of our agrarian monocultures have to be paid for by large inputs of human labor and fossil energy.

Diversity

Conserving biodiversity is regarded as one of the major issues for maintaining the natural capital of the life-supporting ecosystems. At least 5 million species (Wilson 1988), in size classes from microns up to tens of meters and lifetimes from hours to hundreds of years, are involved in the living machinery that is driven by solar insolation and helps turn the biogeochemical cycles. The accelerating loss of species and populations of animals and plants has created an increased interest in the concept of biodiversity and what it means for sustainability. *Biodiversity* (Rosen, in Wilson 1988) is a well-known term that has not been clearly defined, but seems in most cases to refer to species diversity. However, biodiversity rightly encompasses many different levels in time and space that are all important to consider.

Species diversity is a property of the population level dealing with the numbers and distribution of species and populations. Because it is relatively easy to get data on species diversity by merely counting the different animals (and plants, which is seldom done, however) in an area, it has often been improperly used as an indicator of ecosystem complexity and organization.

Functional diversity is focused on what the different species do (their "jobs") and is a property of the ecosystem level (O'Neill et al. 1986). It is strongly related to ecosystem stability. In an ecosystem with a high diversity of species, there are many different species performing the same life-support functions like solar energy fixation, nitrogen fixation, filterfeeding, or decomposition (see Figure 5.2). Through the elasticity of the total gene pool in each functional group or *guild,* the persistence of the system is secured even during adverse conditions.

With a low species diversity the buffering capacity of an ecosystem decreases even if the species involved are rather tolerant. A comparison between the North Sea and the Baltic Sea is a good example. The former high-salinity system has at least 1000 macroanimal species, while the brackish Baltic has fewer than one tenth of that number to perform the same ecosystem functions. The northernmost part of the Baltic lacks macroscopic filterfeeders—clams and mussels—resulting in a less effective processing of organic material (Elmgren et al. 1984). This might be one of the reasons for the very low primary productivity of that area.

Spatial diversity or spatial heterogeneitys is of basic importance for both the structure and performance of an ecosystem. This can be exemplified in the vertical distribution of production and respiration (decomposition) of different systems due to the gradual extinction of sunlight (see Figure 5.5). In the open sea the main production takes place in the surface layer, while the primary and long-term decomposition goes on several thousand meters down.

The settling of organic material and the up-transport of inorganic nutrients occur over long distances. The recycling of nutrients necessary for the primary production therefore takes place mostly within the water column through the bacteria and microscopic animals and through transport by vertically migrating crustaceans and fish. In a forest the distance between the primary production of the green leaves in the tree crown and the respiration that goes on in the root zone is much less, and the transports take place in the circulatory canals in the tree trunk. In the coral reef the two processes are intimately connected thanks to the small distance between the producing, microscopic green algae and the embedding, consuming coral animals. This might be one of the explanations for the extremely efficient recycling in a coral reef. If a similar principle for spatial organization were applied to human societies, there should be a much closer association between the producers and the consumers than in a typical industrial society. Decreasing the distance between the production in agricultural areas and the consumption in urban centers by developing more diverse settlement patterns is probably a feasible policy for improving recycling and energy efficiency.

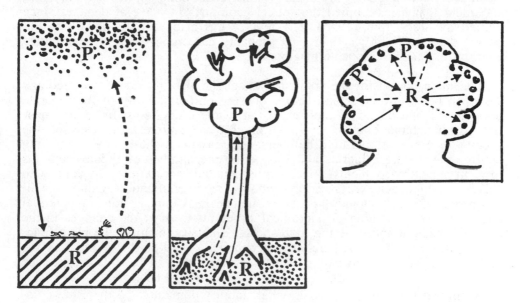

Figure 5.5. The distribution of main centers of production (P) and respiration-decomposition (R) in three different ecosystems: the open ocean (A) , a tree (B) and a coral head (C), small circles indicate symbiotic green algae. The solid arrows indicate transport of synthesized organic material, the dashed ones indicate recycling of regenerated inorganic nutrients (adapted from Odum 1971).

The diversity in space is supplemented by a *temporal diversity*. Pulses of energy and matter, like the daily and seasonal variations in sunlight are, important "clocks" for the ecosystem and its individual organisms. The temporal diversity of species at the producer and consumer levels provide a solid energy basis for top consumers like the human species. In economics, however, one seeks to level out as much as possible the natural fluctuations to secure a steady delivery of certain goods or functions preferred by the consumers.

Stability

This is a property of ecological systems that is a rather vague concept in spite of efforts to amalgamate its ecological meaning with terms from physics (Orians 1975). Still, accurately defined in space and time, the stability concept gives valuable information on ecosystem dynamics. We prefer here a broad definition: the ability of a system to stay in a steady-state condition or oscillating equilibrium, rapidly recovering its balance after perturbation with an intact structure. "Diversity begets stability," meaning species diversity, is a statement that has occupied theoretical ecologists for several decades since coined by MacArthur (1955). The reverse has also been found—"stability begets diversity"—(May 1973) and probably has higher relevance, provided energy flow has a reasonable magnitude.

The main reason for the confusion about the stability concept in ecology seems to be that some scientists are focusing on the stability of the separate populations, and others on the stability of whole ecosystems. In environmentally stable habitats like the coral reef, the tropical rainforest, the deep sea, the species diversity attains high numbers, whereas in stochastically fluctuating environments, like estuaries and rockpools, the number of species remains low. In such environments, species must expend more energy to cope with survival, while less excess energy remains for diversification. Odum (1983) has provided a convincing explanation that there is competition for potential energy between creating physiological structures to meet the changes in the external medium and evolving high species diversity to fulfill the main functions of the ecosystem. A parallel in economics would be the choice that has to be made between investing in new infrastructure to be able to produce more of one particular good or to start diversifying the business.

Nonlinear systems may have several stability domains, and there are common examples of large-scale systems switching to a new steady state when forced beyond a certain threshold level. Overgrazing of the African savannah has turned grasslands to deserts. Overfishing in the Great Lakes has turned the former community, dominated by large species into a community characterized by small species. Such *surprises* (Holling 1986) create a feeling of great uncertainty about the future development of the global ecosystems, considering the ongoing large-scale changes occurring in the terrestrial, aquatic, and atmospheric environments.

A fruitful way of looking at ecosystems from a stability point of view is the *mosaic-cycle concept* (Aubreville 1938; Remmert 1991). It mainly stresses the spatial heterogeneity of the ecosystem, where the different patches undergo a similar succession but in a desynchronized fashion (see Figure 5.6). In a tropical rainforest, the dense cover of vegetation is occasionally broken up by a falling tree, letting in sunlight and starting a succession, ultimately leading to the regrowth of another large tree. Because of the desynchronized pulsing of patches or subsystems, the separate minima and maxima even out, giving stability to the total system. The cycles are clearly coupled to the species diversity, which shows a maximum in areas undergoing a fast succession, but a minimum in the mature stages. This stresses both the importance of stochastic events like storms and proper recognition of the time and space scales. A deeper insight into this problem area seems decisive for achieving a sustainable relationship between ecological and economic systems.

Resilience

The ability of a system to keep structure and function after disturbance has been labeled *resilience* (Holling 1973). Contrary to stability, which emphasizes equilibrium, low variability, and resistance to and absorption of change, *resilience* is characterized by events far from equilibrium. It therefore stresses the boundaries of the stability domain, high adaptation, and high variability. Temperate systems exposed to changes in climate usually have high resilience and low stability. Systems evolved during stable conditions like coral reefs and tropical rainforests show high stability and low resilience probably because in the stable environment there has been no need for establishing structures to meet perturbations. Consequently, these systems, as well as the deep sea, are easily disturbed by man's activities like eutrophication, overfishing, forestry, or exploitation of manganese nodules.

Hierarchy

The large span of both space and time scales in natural systems offers numerous examples of hierarchical organizations that can be subject to successful analysis using hierarchy theory (Webster 1979). A horizontal organization is seen in the patchy distribution of structures in the space dimension. Structures of similar quality and time characteristics have been called *"holons"* (Koestler 1967), and can be exemplified by the different species within the vertical levels of a wood—herbs, shrubs, and trees (see Figure 5.6). A patchiness may start with a small nucleus resulting from some heterogeneity in the substrate or from an uneven dispersal of propagules and successively grow to larger areas. The larger the patch, the stronger its influence on the local environment. A patch also needs to attain a critical size to persist using the normal resource flows of the surrounding environment. Single trees, groves, and forests may have the same species assembly of inhabiting species but as entities at different time scales. This *vertical hierarchial structure* (see Figure 5.6) is created by the difference in metabolic rates. Small scale structures like individual cells show higher turnover rates than populations, that in turn have faster dynamics than ecosystems. From the point of a view of an observer at a lower level, the dynamics at the upper level seem nearly constant. Seen from an upper level, the high process frequencies of the lower levels appear as a sampled statistical behavior. What looks like an unexpected catastrophic event from a lower level may just be a normal pulse in the larger time scale of the higher level. Disturbances at the lower levels of a multilevel system might be easily incorporated and repaired like a forest, which continues its normal behavior in spite of the cutting of individual trees.

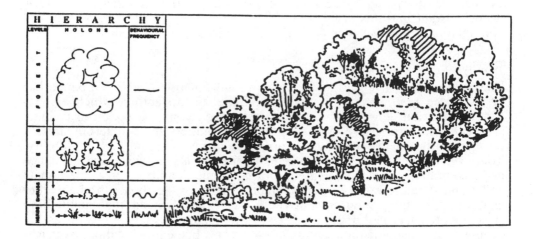

Figure 5.6. Basic ecological concepts demonstrated for a woodland. Two previous
treefalls or clearcuttings of recent (A) and older date (B) have created clearings
where herbs and shrubs invade, constituting *successions* of different stages.
The hierarchial structure of the system comprises a *horizontal division* in
structures of similar quality and time characteristics—holons—and a *vertical
division*, which is based on the frequency of the behavior of the structures.
The interactions between the holons are usually stronger than between the
vertical levels.

The growth of the human society with increasing cultivation and urbanization of land has led to a successive *fragmentation* of the landscape with great implications for the dynamics of the natural systems. The plant and animal populations have been cut up into *metapopulations,* small "satellites" that are totally dependent on immigration from large "cores" (Hanski 1982). This increasing reduction of patches to undercritical sizes is a large threat for the proper functioning of our life support system. In many regions of the world the fast expansion of the traffic system is one of the major factors disintegrating the terrestrial landscape, which at the same time impairs the capacity of the ecosystems to assimilate and process the exhausts from the car traffic.

TOWARDS A SUCCESSFUL MAINTENANCE OF OUR LIFE-SUPPORT SYSTEM

In the course of evolution the size of flows and storages of the systems were strongly related to the influx of solar energy, and the speed of the biogeochemical cycles was set by the naturally aging systems. When man learned to use fossil fuels, additional, auxiliary, and high quality energy was put into processes that could replace, outcompete, or amplify the natural flows of energy and matter (Odum 1971). For example, the use of artificial fertilizers has increased the primary production both on land and in the sea. But the decomposition rate of dead organic matter from the increased biomasses of plants and animals has not increased by any comparable stimulation. Well known results of this mismatch are the eutrophication problems with inland and coastal waters, and the garbage problems in densely populated areas. Aware of the demand for recycling, how do we cope with the waste problems? Does efficient sorting and reuse of glass, plastics, paper, and metals really solve the problem? Eugene Odum (1989) argues for *input management*, that is, to reduce the use of fossil energy and raw materials instead of focusing primarily on the polluting outputs, as a major strategy for minimizing waste. The only way to ecological sustainability is to keep our natural life-support systems working. We cannot withdraw from maintaining the structures we want to use. For a successful herring fishery, the spawning grounds, although far from the fishing areas, have to be kept intact. There is no sense in spraying with pesticides to get a higher production of fruits when the bees performing the pollination are killed off at the same time.

The dynamic behavior of the ecosystems has to be respected as a basic rule in human affairs. While other species are exposed to and regulated by nature's pulsing patterns at many levels, humans, due to the extensive use of fossil energy, not only show an exponential population growth but also a steady increase in the per capita use of natural resources. If we are not able to change this trend, we will soon exceed the carrying capacity of our natural resource base with accompanying social disruptions and economic crises. Equally important is an increased attention to the life-support areas required for various human activities. Just as an oyster reef with a restricted substrate area can have many individuals sitting in several stories on top of one another when the waterflow is good, so can humans build skyscrapers as the space in a city grows scarce if there is enough energy to support these structures. The life-support area does not decrease because people live on top of each other. Instead, the maintenance costs often rise rapidly as car parks and maintenance areas have to be added proportionally to maintain the flow of workers and materials. Conversely, if we restrict the waterflow for the oyster reef by constructing embankments, we kill off most of the oysters as their life-support area becomes subcritical.

Extensive natural areas have to be saved around human settlements to perform the necessary life-support functions (Odum and Odum 1972). A virgin archipelago may be a tempting subject for a developer, who sees it as still unused. But it is actually continuously working producing oxygen, taking up carbon dioxide, filtering the water, recycling nutrients. The exploited areas further away would be worse off without the life-support from fully functioning coastal systems that generate clean air, process wastes, help reduce erosion, and act as barriers against the oceans. To replace these functions by technical substitutes would be very expensive, if at all possible. Instead we should try to maximize the services from natural ecosystems by investing in technologies that take advantage of their capabilities for self-organization, productivity, and recycling. The global ecosystems have done well without human beings, especially without their overuse of fossil fuels. In order to survive in the post-fossil era, humans have to learn how exponential growth can be switched into a pulsing stability without destroying human cultural achievements. This is perhaps the most urgent and difficult challenge of ecological economics.

ACKNOWLEDGMENTS

The authors want to thank Frank Golley, Rudolph de Groot, and James Zucchetto for constructive criticism of an early version of this manuscript, and Jon Norberg for help with a figure. We are much obliged to the Wallenberg Foundation for the confidence and support given to us for arranging for a second time an international workshop ten years after the first meeting on Integrating Ecology and Economics.

REFERENCES

Aubreville, A. 1936. La foret coloniale: les forets de l´Afrique occidentale francaise. *Annales de L'Academie des Sciences Coloniales* 9: 1–245.

Elmgren, R., R. Rosenberg, A-B. Andersin, S. Evans, P. Kangas, J. Lassig, E. Leppäkoski, and R. Varmo. 1984. Benthic macro- and meiofauna in the Gulf of Bothnia (Northern Baltic). *Finnish Marine Resources* 250: 3–18.

Falk, J. H. 1976. Energetics of a suburban lawn ecosystem. *Ecology* 57: 141–50.

Folke, C., M. Hammer, and AM. Jansson. 1991. Life-support value of ecosystems: a case study of the Baltic Sea Region. *Ecological Economics* 3: 123–37.

Hammer, M., AM. Jansson, and B-O. Jansson. 1993. Diversity, change, and sustainability: implications for fisheries. *Ambio* 22: 97–105.

Hanski, I. 1982. Dynamics of regional distribution: the core and satellite species hypothesis. *Oikos* 38: 210–21.

Holling, C. S. 1973. Resilience and stability of ecological systems. *Annual Revue of Ecological Systems* 4: 1–23.

———. 1986. Resilience of terrestrial ecosystems: local surprise and global change. In Sustainable Development of the Biosphere, eds. W. C. Clark and R. E. Munn. Cambridge: Cambridge Univ. Press.

Koestler, A. 1967. The Ghost in the Machine. New York: Macmillan.

May, R. M. 1973. Stability in randomly fluctuating versus deterministic environments. *American Naturalist* 107: 621–50.

McArthur, R. H. 1955. Fluctuations of animal populations, and a measure of community stability. *Ecology* 36: 533–6.

Odum, E. P. 1989. Global Stress on Life-Support Ecosystems Mandates Input Management of Production Systems. Crafoord Lectures. Stockholm: The Royal Swedish Academy of Sciences.

Odum, E. P., and H. T. Odum. 1972. Natural Areas as Necessary Components of Man's Total Environment. Proceeding 37, North American Wildlife and Natural Resources Conference. Washington: Wildlife Management Institute.

Odum, H. T. 1971. Environment, Power and Society. New York: Wiley.

———. 1988. Self-organization, transformity, and information. *Science* 242: 1132–9.

O'Neill, R. V., D. L. DeAngelis, J. B. Waide, and T. F. H. Allen. 1989. A Hierarchial Concept of Ecosystems. Princeton: Princeton Univ. Press.

Orians, G. H. 1975. Diversity, stability and maturity in natural ecosystems. In Unifying Concepts in Ecology, eds. W. H. van Dobben and R. H. Lowe McConnell. The Hague: Dr. W. Junk B. V. Publishers.

Pimm, S. L. 1988. Energy flow and trophic structure. In Concepts of Ecosystem Ecology, eds. L. R. Pomeroy and J. J. Alberts. Berlin: Springer-Verlag.

Remmert, H. 1991. The Mosaic-Cycle Concept of Ecosystems. Berlin: Springer-Verlag.

Smith, S. V., and F. C. McKenzie. 1987. The ocean as a net heterotrophic system: implications from the carbon biogeochemical cycle. *Global Biogeochemical Cycles* 1: 187–98.

Vitousek, P., P. R. Ehrlich, A. H. Ehrlich, and P. M. Matson. 1986. Human appropriation of the products of photosynthesis. *BioScience* 36: 368–73.

Webster, J. R. 1979. Hierarchial organization of ecosystems. In Theoretical Systems Ecology, ed. E. Halfon. New York: Academic Press.

Wilson, E. O. 1988. The current state of biological diversity. In Biodiversity, ed. E. O. Wilson. Washington: National Academy Press.

6 BIOTIC DIVERSITY, SUSTAINABLE DEVELOPMENT, AND NATURAL CAPITAL

Charles Perrings[1]
Department of Environmental Economics and Environmental Management
University of York
York YO1 5DD

ABSTRACT

Ecological and economic concepts of sustainability yield different insights into the development process. However, both imply that a necessary condition for development is that the opportunities associated with the stock of natural capital should be widening over time. If the opportunity set is constant over time, there is a stationary state, and although this may be sustainable, it is not development. Since the value of capital can be shown to be a function of the opportunities associated with it, then a necessary condition for development is that the value of natural capital should be increasing over time. At issue is whether or not this is consistent with the loss of biological diversity. This chapter maintains that despite the role of technological change in widening the opportunities associated with some subset of the stock of natural capital, there are few greater threats to the sustainability of development than the depletion of the aggregate stock of natural capital through loss of biological diversity.

INTRODUCTION

What criteria should govern collective decisions which, by changing the mix of biotic resources, change the opportunities available to all future generations? Such decisions are characterized by fundamental uncertainty about the long-run ecological implications of biodiversity change. They also raise questions about the ethics of depleting a common resource that contains the genetic blueprint for the stock of natural capital available to all future generations. The future effects of species extinction are neither known in a probabilistic sense, nor are free of ethical judgments. What may be known are the net benefits to the present generation from exploitation of existing biota un-

---

1. I am indebted to Ralph d'Arge, Herman Daly, and Prasanta Pattanaik for their comments on an earlier draft of this chapter. The usual disclaimer applies.

der current preferences. But the social opportunity cost of the existing biota exploitation includes the opportunities forgone or gained by future generations from the resulting change in a future biotic resource base, under technologies and preferences yet to be determined. It is a problem of decision-making under ignorance about both the physical and the economic implications of current actions.

Part of the problem lies in the irreversibility of change. Errors in the management of global biotic resources cannot, in general, be undone. But irreversibility is not always a source of difficulty. We can no more recapture the services of ecosystems lost through the "development" of Manhattan in the 17th century than we can the genetic uniqueness of *didus ineptus* (the dodo) hunted to extinction in the course of the "development" of Mauritius in the same period. The loss of biodiversity involved in both instances is just as irreversible. But the regret now felt at the loss will be very different in the two cases. Biodiversity loss does not always serve the future ill. Indeed, the enormous world-wide growth in material production over the last three centuries is largely based on the specialization of natural resource use, and thus on the loss of biodiversity. While it can be argued that many of the future costs of past specialization in the use of biotic resources were consistently ignored and are only now beginning to be understood, it is also very clear that much biodiversity loss has improved the human condition.

The net result is that we cannot be sure what the current global assault on biodiversity implies for the well-being of future generations. Although we are in the midst of an extinction "spasm" that has prompted massive demands for the preservation of species, it is not at all clear that this is what we should be doing. Indeed, a decision to preserve the biotic status quo may very well condemn future generations to progressive impoverishment, especially in the light of the continuing expansion of the global human population. Although growth at the extensive margin in tropical megadiversity zones is the single greatest current cause of species extinction, it may very well be that growth is a necessary condition for the protection of biodiversity in the future. Growth based on extension of the agricultural frontier in Amazonia, Central America, or Central Africa is destroying at an unprecedented rate both habitats and the species those habitats support. But without growth in the personal incomes of the poorest two-thirds of the world's inhabitants, and without a demographic transition that depends on growth in personal incomes, there is little hope of halting the march into the world's remaining forested areas.

Many of the choices to be made in these circumstances are social choices. That is, society cannot abdicate responsibility for the future impact of present decisions by leaving the problem to the mercy of private decisions based on current market signals. The pervasiveness of intertemporal environmental externalities and the equity implications of the intergenerational redistribution of assets it implies, obliges society to make collective decisions about the loss of biotic diversity. Reliance on an incomplete set of current markets will not do. But if society is to make choices that affect future generations, it is important that those choices be sensitive both to the limitations on our understanding of the dynamic behavior of the biosphere, and the ethical dimensions of those choices.

The spirit of the Brundtland Report (WCED 1987) defines in the broadest sense the ethic that guides the decision process in this case to be described. This is not the only ethic that links environmental care with the well-being of future generations, but it is one that has sufficiently deep roots in the history of social philosophy and a wide enough measure of popular support that it is at least credible. In what follows, therefore, the touchstone for any principle of social choice is whether it satisfies Brundtland in the sense that it is consistent with both meeting the needs of present generations and preserving the opportunities available to future generations. That is, the touchstone for any principle of social choice is whether it is sustainable in the sense of Brundtland.

This chapter reduces the collective decision process to its simplest components, and constructs corresponding principles of collective choice that may be used to test for the sustainability of development in the face of biodiversity loss. The decision problem breaks down, as I have observed elsewhere, into aspects over which the current generation exercises some control, and aspects over which it does not (Perrings 1991a, b). Confronted with a given set of biotic resources, a given technology and a given preference ordering over the set of all alternatives, the current generation can choose the "best" outcome in terms of the set of foreseeable opportunity costs. But the opportunities lost by future generations of resource users as a result of a change in the mix of biotic resources depend on preferences and technologies currently unavailable. The present generation can opt to limit its forward vision, either by truncating the time horizon over which it evaluates the opportunity costs of the options before, or by discounting those costs at a high enough rate. Because present actions do change the opportunities available to all future generations, the responsibilities for decisions made becomes apparent. This, it turns out, is very closely related to an intratemporal problem discussed in the social choice literature—the role of freedom of choice in the ranking of sets of opportunities under incomplete information about preferences (Pattanaik and Xu 1990; Sen 1991).

The chapter contains five sections. In approaching the decision problem, the next section discusses the requirements of a sustainable development process, and the implications of a change in biodiversity for that process. A third section then considers the links between biodiversity, the set of opportunities open to society, and the value of natural and produced capital, as it states the principles of social choice that seem to be consistent with sustainability in the sense of Brundtland. A fourth section derives sufficient conditions for the sustainability of development on the basis of these principles, and a final section offers some concluding remarks.

DEVELOPMENT, BIODIVERSITY, AND OPPORTUNITY

As a first approximation we may take development to be given by any welfare-improving change in the set of opportunities open to society, where the set of opportunities is given by the potential uses to which the available assets may be put. Although development is often approximated by the change in per capita GDP, it is now conventional to ask for a more general index which takes into account measures of the "quality of life"— educational attainment, nutritional status, access to basic freedoms, and spiritual welfare. Pearce and Turner (1990), for example, take development to be given by positive change in any of the components of a vector of the attributes that society values. This would include increases in per capita GDP, but it would also depend on improvements in health and nutritional status, educational achievement, access to resources, distribution of income, and basic freedoms. Such approaches recognize that the market value of consumption may not be a good measure of the welfare—one involving an increase in the value to society of a given set of opportunities, and the other involving an expansion in the set of opportunities. Development may be satisfied by any of the following: (1) an expansion in the set of opportunities open to society without reduction in the instrumental or use value of those opportunities, (2) an increase in the instrumental value of the opportunities available to society without reduction in the range of opportunities, or (3) both.

Since a positive rate of change in per capita GDP may be positively correlated with an expansion in the opportunity set, the traditional proxy for development may not be wholly misleading. However, much of what is currently recorded as economic growth—the accelerated mining of non-renewable resources or increasing defensive expenditures against environmental degradation—is actually evidence of declining future opportunities. Development is clearly not the same as economic growth. Indeed, wherever economic growth has the effect of narrowing the range of opportunities, it may be said to be development-retarding or regressive. Development is accordingly conceptualized as any evolutionary process which extends either or both of the set of opportunities available to society and the instrumental value of those opportunities.

In any given society, and for any given set of opportunities, the alternatives open to society may be ranked according to some preference ordering over those alternatives. There will exist an indirect function which defines the welfare yielded at the best alternative from the set of all alternatives. The relative value of each alternative under this preference ordering lies in the utility or welfare it yields to society (i.e., value is instrumental). Using such criteria, it is also possible for society to rank distinct sets of opportunities—to judge whether the best outcome in the set of opportunities available today, for example, is better or worse than the best outcome of other sets of opportunities that may be available in the future. In other words, using a set of weights of the sort that lies behind Pearce and Turner's vector of attributes, one generation may compare present and future opportunity sets in terms of the best outcome in each.

The limitation of this comparison lies in the fact that one generation's ranking between distinct sets of opportunities is based on its own preference ordering of the alternatives in each opportunity set. No generation can know the preference ordering that will inform the ranking that another as yet unborn generation might make amongst the same opportunity sets. What this means is that opportunities will have value to future generations of producers and consumers in ways that cannot be assessed in instrumental terms by the present generation. If opportunity sets are to be ranked in this case, it will have to be on some non-instrumental (ethical) basis.

Since Brundtland, one basis suggested by the debate is the freedom of choice offered by different opportunity sets, where freedom of choice is equivalent to the range of alternatives in each set. That is, the present generation may discriminate between present and future opportunity sets on the basis of the number of options offered in each set. In the absence of a common preference ordering between generations, it is the range of choice that counts. While the present generation may be certain that development implies more automobiles, tobacco, open heart surgery, theme parks, nuclear power stations, and fewer wetlands, tropical moist forests, and semi-arid savannas, it is not at all clear that future generations will take the same view. Hence, a process which narrows the range of choice open to future generations by locking them into the combination preferred by the present generation might satisfy a ranking between present and future opportunity sets by the preference ordering of the present generation. However, it would violate a ranking in which the range of alternatives in each opportunity set is a factor. That is, while it might satisfy a ranking between present and future opportunity sets on instrumental grounds, it would violate a ranking grounded in the principle of freedom of choice.

Biotic diversity derives its instrumental value to human society in two ways. Specific organisms have properties that make them of *direct* (actual or potential) use in production or consumption activities. In addition, the mix of organisms and their role in a variety of ecological functions makes them of *indirect* (actual or potential) use to producers and consumers who depend in some way on those ecological functions. Actual use gives rise to what is normally referred to as *use value,* while potential use gives rise to what is normally referred to as *option, quasi-option* or *scientific value.* The first of these is the value of the option to make use of the resource in the future (Weisbrod 1964), the second is the value of the future information made available through the preservation of a resource (Arrow and Fisher 1974; Henry 1974; Fisher and Hanneman 1983), and the third is the value of organisms as scientific resources (Krutilla 1967).

For a small number of species the direct value of the chemical or physical properties which make them useful as sources of food, drugs, construction materials, and clothing is closely approximated by the market price. For many other species, however, their direct value is known only to a small group of users, usually indigenous to the area from which such species derive, and is not closely approximated by a market price. The biophysical basis of the indirect value of biodiversity is even less well understood and captured in the market price of species. The basis of the indirect value of species lies in their individual role in food webs (Paine 1980), in solar fixation, and in the regulation of ecosystem processes (Westman 1990), as well as in the collective contribution of the diversity of genes, genotypes, species, and communities to ecosystem functions (Solbrig 1991). This is, however, an area in which little is now known with certainty. There is also no reason to believe that the indirect value of species is reflected in either the current set of market prices, or the private valuation of species for which there exist no markets. The private valuation of a species is given by the value of the goods and services which the consumer is prepared to forgo by committing the species to some particular use (i.e., its opportunity cost to the private user). Since much of the indirect value of a species accrues to agents other than the direct user, and since there exist no markets for the indirect effects of an increase or decrease in most species, that indirect value will be external to the private valuation of the species.

With respect to the non-instrumental value of biodiversity, it is intuitive that the set of opportunities associated with a stock of biotic resources is an increasing function of the diversity of those resources. The genetic diversity of existing biota fixes the range of the properties of existing species, which, though not now valuable, may have value under future preferences and technologies. It also fixes the evolutionary potential of species, and so the potential range of valuable properties of future species. Put another way, the genetic diversity of existing biota is what limits the range of alternative uses of future biota to future generations. An increase in diversity, *ceteris paribus,* implies an increase in the opportunity sets available to future generations. Symmetrically, a decrease in diversity, *ceteris paribus*, implies the reverse.

Environmental economists have devoted considerable effort to uncovering the basis of the non-instrumental value of biotic resources. It has been argued at various times to include bequest value (Krutilla 1967) and the value conferred simply by the existence of a resource (Pearce and Turner 1990). Both notions come close to capturing the value to current generations of the range of opportunities available to future generations as those of bequest and existence value, since neither implies the ranking of opportunities in terms of the preference ordering of the current generation. Weisbrod's option value and Fisher and Hanemann's quasi-option value are both concerned with the potential value of existing biota to existing users, given the current preference ordering. In other words, they are instrumental values.

Mäler (1992) has argued that the instrumental or use values and non-instrumental or non-use values of environmental resources can be accommodated within a single preference relation, providing that it is separable in the two sorts of value. If we are considering some biotic resource, he suggests that if utility can be written as the sum of one function of the resource alone, and a second function which is not additive separable in the same resource and all other goods and services, then the non-use value of the resource is the value associated with the first term, and the use value of the resource is the value associated with the second term. In other words, the non-instrumental value attached to some species by an individual is independent of any use they may or may not make of that resource. It is independent, therefore, of any direct value- based preference ordering of the set of all alternatives. This implies that the value of the mix of species to society can likewise be reflected by a separable function, in which the non-instrumental value of biodiversity is independent of the use society makes of it. To be consistent with the definition of development adopted here, the non-instrumental value of biodiversity would be constrained to be an increasing function of the number of species. That is, it would be constrained to be an increasing function of the existence of species. It is interesting that authors taking a quasi-theological view of the problem have something of this sort in mind (Daly and Cobb 1989; Blasi and Zamagni 1991).

BIODIVERSITY, NATURAL CAPITAL, AND SUSTAINABLE DEVELOPMENT

It is well understood that the definition of (Hicksian) income[2] on which all economic analysis of the sustainability problem is based embodies a notion of sustainability. The maximum flow of benefits from the exploitation of some asset base is defined to be income only if it leaves intact the capacity of that asset base to yield a similar flow of benefits in the future. This implies that true income is maximum sustainable income. It is not a priori clear why Hicksian income should be a goal of policy, though there does a exist a social welfare function that generates Hicksian income as an optimal outcome. A Rawlsian or intertemporally egalitarian social welfare function is satisfied only if the level of welfare of successive generations is the same—viewed from the perspective of the first generation (Solow 1974; Hartwick 1977). Most recent treatments of the problem of sustainable development are less restrictive in the form of intertemporal social welfare function assumed, but all require that the welfare successive generations gain from the exploitation of the resource base is at least as great as the welfare of the current generation (Turner 1988; Pearce and Turner 1990; Pearce, Markandya, and Barbier 1989; Pezzey 1989). As one would expect, the two requirements are slightly different in the implications they have for the value of the asset base. Constant welfare requires that the value of the asset base be constant over time, whereas non-declining welfare implies that the value of the asset base be non-declining (Solow 1986).

---

2. Hicksian income is defined as the maximum amount that can be spent on consumption in one period without reducing real consumption expenditure in future periods [Hicks 1946].

To approach the link between development (as it has been defined in this chapter), welfare, and the value of the asset base, it will be useful to clarify these conditions. Let us first define the set of all possible produced and natural assets over the time horizon of interest, T, to be K_T, the number of elements in which, denoted $\#K_T$, is n. That is, there are a maximum of n produced and natural assets available to the economic system over this period. Let us further define the alternative combinations of these n assets by the power set, $\prod(K_T)$. Any element of $\prod(K_T)$, $K_i(t) \in \prod(K_T)$ denotes an opportunity set, the cardinality of which is n at most: i.e., $\#K_i(t) \leq n$. This last means that the opportunity sets between which society chooses includes some opportunity sets for which the range of assets is narrower than in others. Since biodiversity loss implies a reduction in the range of assets available to society, it also implies a reduction in the cardinality of the corresponding opportunity sets. $K_i(t)$ defines the discounted net benefit (welfare) of the alternatives in the opportunity set evaluated at time t. This is generated by a social preference ordering on K_T that is both reflexive and transitive, and that enables society to rank opportunity sets within $\prod(K_T)$ and over T. If $K_j(t+s) \underline{f} K_i(t)$ the jth opportunity set at time t+s may be said to offer at least as high a level of welfare as the ith opportunity set at time t. Hence, the criteria of non-declining welfare and non-declining asset value will be satisfied for a society exercising choice over opportunity set i at time t only if $K_j(t+s) \underline{f} K_j(t)$ for any i and for all $s \in \{1,...,T-t\}$.

My interest lies in the significance of biodiversity in the ranking of opportunity sets implied by this relation. More particularly, my interest is in the conditions for one set of opportunities to offer at least as high a level of welfare as another, and in the role of the stock of biotic resources in satisfying those conditions. To characterize the axioms of social choice for this problem, it may be useful to consider three cases: (1) where the only value of assets to society is instrumental, (2) where the only value of assets to society is non-instrumental, and (3) where the value of assets to society has both instrumental and non-instrumental components. The first might be thought to correspond to the case of the hedonistic but sovereign consumer who asks not what he can do for the future but what the future has done for him. The second might be thought to correspond to the case of the ascetic altruist on his deathbed. The third might be thought to correspond to some improbable combination of the two. In breaking down the problem into its simplest components, we need to consider the principles corresponding to the instrumental and non-instrumental value of opportunities, and these three cases may make it easier to motivate each principle.

The Instrumental Value of Biotic Resources and the Axioms of Social Choice

If all assets have only instrumental value, it has been shown that where the capital stock includes exhaustible or depletable natural resources, a necessary condition for the value of capital to be non-declining is that the rents deriving from resource depletion should be reinvested in reproducible capital to compensate for the user costs of depletion (Hartwick 1977, 1978, 1991). This is the *Hartwick rule*. The intuition behind it is quite clear: if the loss of biotic resources has a social opportunity cost in terms of forgone benefits of future use, this should be compensated by an equivalent gain in produced assets. The net user cost of the depletion of biotic resources should be zero. Sufficiency requires a rather stronger set of conditions, including that there exist substitutes for all biotic resources with instrumental value. That is, the Hartwick rule will be effective in maintaining the use-value of the asset base where certain biotic assets are deleted from the list providing that there exist substitutes capable of delivering equally valued services. Similar conditions have been derived for other models of growth with exhaustible resources (Kemp and Long 1984).

I shall consider the physical implications of biodiversity loss momentarily. What is important about these conditions for the non-declining value of the asset base is that if biotic resources are valued for the services they offer under a given set of preferences and technology, then any set of resources yielding the same services will be as valuable. That is, there exists a focus to each opportunity set which enables the decision maker to rank that opportunity set against others regardless of the particular mix or absolute number of resources in each. The focus is an index of the instrumental value of the opportunity set to society. One way of thinking about the focus of the ith opportunity set at time t, $f(K_i(t))$, is that it is the dominant element in $K_i(t)$ given the social preference ordering of the alternatives it contains. But this is not the only possibility. If the opportunity set comprised the elements of some ecosystem, for example, its focus might be a particular indicator species. More generally, if the opportunity set were characterized by fundamental uncertainty, its focus would be the outcome with the most power to attract the decision-maker's attention—the focus-gain of the opportunity set (Katzner 1986, 1989; Perrings 1991a).[3]

The foci of opportunity sets are a function of social preferences. If there is a change in the composition of the alternatives under some opportunity set that does not affect its focus, that change will be irrelevant to the decision-making process. So, for example, the deletion or addition of a species that has no affect on the instrumental value of the services yielded by some ecological system would be irrelevant to the well-being of the society using that ecosystem. This gives rise to the following principle of social choice.

Axiom of focus:

A1:   For all $K_j(t+s) \neq K_i(t) \in \prod(K_T)$ and for all $s \in \{0, ..., T-t\}$, if $\#K_j(t+s) = \#K_i(t)$, $f(K_j(t+s)) \sim f(K_i(t))$ implies that $K_j(t+s) \sim K_i(t)$.

---

3. The concept of focus-gain derives from Shackle (1955, 1969). It may be interpreted in this case as the least-surprising outcome of a decision process based on an opportunity set about which there is fundamental uncertainty.

This says that if society is indifferent (~) between the foci of any two distinct opportunity sets of equal cardinality, it will be indifferent between the opportunity sets themselves. It implies the irrelevance of social preferences between alternatives that do not affect the focus of the opportunity sets being compared. If society is to rank a world of automobiles and bears against a world of automobiles and bearskin rugs, for example, and if automobiles are the focus of each set of alternatives, it is irrelevant that society might prefer live bears to dead ones.

A second axiom merely extends this idea. It says that adding or deleting alternatives that are irrelevant to the focus of compared opportunity sets does not change the ranking of those opportunity sets. The axiom corresponding to this is an extension of Pattanaik and Xu's (1992) weak independence principle, and is named accordingly.

Axiom of weak independence:

A2i: For all $K_h \in \prod(K_T)\backslash(K_j(t+s) \cup K_i(t))$ such that $f(K_j(t+s)) \succ f(K_h(t+s))$ and $f(K_i(t)) \succ f(K_h(t))$, $K_j(t+s) \succ K_i(t)$ implies that $K_j(t+s) \cup K_h(t+s) \succ K_i(t) \cup K_h(t)$; and

A2ii: For all K_h contained in $(K_j(t+s) \cup K_i(t))$ such that $f(K_j(t+s)) \succ f(K_h(t+s))$ and $f(K_i(t)) \succ f(K_h(t))$, $K_j(t+s) \succ K_i(t)$ implies that $K_j(t+s)\backslash K_h(t+s) \succ K_i(t)\backslash K_h(t)$,

where f implies strict preference. The first of these says that if any set of opportunities, not a subset of the ith opportunity set at time t or the jth opportunity set at time t+s, is added these opportunity sets, and if its focus is dominated by the foci of the opportunity sets to which it is being added, then it will not change the ranking of those sets. The second says that if any set of opportunities which is a subset of the ith opportunity set at time t and the jth opportunity set at time t+s, is deleted from these opportunity sets, and if its focus is dominated by the foci of the opportunity sets from which it has been deleted, then it will not change the ranking of those sets. By this axiom, non-declining welfare and non-declining asset value are both consistent with a change in the number and composition of species.

It follows that there is nothing in an instrumental valuation that privileges the existence of species. At the same time, however, there is nothing that precludes the preservation of species. If there exist no substitutes for some species of indirect value, then the preservation of that species will be required under an instrumental approach. It is the substitutability of species that determines their preservation value. There is certainly a high degree of substitutability between species in terms of their direct instrumental value (think of plants as photosynthesizers, or animals as protein sources). But plants and animals also perform a range of ecological functions of indirect value in which the possibilities for substitution between species is much reduced (think of key-stone or critical-link species). In such cases, a change in the balance between the species in an ecosystem, whether or not this involves species deletion, may induce changes in the dynamics of that system that have far-reaching consequences for the ecological services available to society.

Ecosystems typically evolve in a process of discontinuous change, involving both successional and disruptive processes, and are driven by a complex set of feedbacks. They are, in other words, typically complex, non-linear, feedback systems. The points of discontinuity in such systems occur around a set of system thresholds, which mark the limits of system resilience. Note that an ecosystem is said to have lost resilience whenever a change in its environment disrupts the system to the extent that its internal organization breaks down. Loss of resilience implies the (generally irreversible) loss of a system's ability to maintain functions in the face of exogenous stress and shock (Holling 1973, 1986).[4] It turns out to be important for the instrumental value of future resources for two reasons. First, loss of resilience implies a qualitative change in the structure and functioning of the system, and so in the nature of ecological services it delivers—the source of its instrumental value to society. Second, even if the qualitative changes in ecological functions caused by loss of resilience are not economically significant, the disruption of an ecological system is frequently associated with a reduction in its biological productivity, and so a quantitative reduction in the ecological services it delivers. For both reasons, maintenance of the resilience of ecosystems delivering instrumentally valued services for which there are no ready substitutes is as necessary to the sustainability of the economic system as the Hartwick rule (Perrings 1991b; Common and Perrings 1992).

There is a direct link between resilience and biodiversity. Holling (1986) has argued that resilience is an increasing function of the complexity of ecosystems, where complexity refers both to the number of constituent populations in a system and to the interdependence between them. This implies that the ability of an ecosystem to co-evolve with its environment depends on the breadth of evolutionary options open to it. Its resilience, like the riskiness of any portfolio of assets, rises as the diversity of its constituent populations increases. That is, resilience is an increasing function of the size of the opportunity set. In such cases, even an instrumental valuation of natural capital will require maintenance of the size of the opportunity set—maintenance, that is, of species within the thresholds of system resilience. This is reflected in arguments for the protection of some upper bound on the assimilative capacity of the environment to absorb wastes, and some lower bound on the level of biotic stocks that can support sustainable development (Barbier and Markandya 1990). Indeed, the "multifunctionality" of biotic resources and the irre-

---

4. More precisely, resilience defines the ability of the system to accommodate the stress imposed by its environment through selection of a different operating point along the same thermodynamic path without undergoing some "catastrophic" change in organizational structure.

versibility change associated with loss of ecosystem resilience have been cited as reasons to set such bounds conservatively (Pearce and Turner 1990).

The Non-instrumental Value of Biotic Resources and the Axioms of Social Choice

What of the non-instrumental value of biotic resources: the bequest and existence values referred to earlier? How do these enter the collective decision-making process? The Hicks criterion implies that if the welfare of future generations is assessed on the basis of an intertemporal social welfare function rooted in the present, then future generations should be at least as well off as the present generation by the standards of the latter. If all generations could be assumed to share the preference ordering of the present generation, and in the absence of fundamental uncertainty, this condition would be satisfied if, for all $s \in \{1, ..., T\text{-}t\}$, there exist reachable $K_j(t+s)$ such that $f(K_j(t+s)) \succeq f(K_i(t))$, where $f(K_i(t))$ is the focus of the opportunity set currently subject to choice. Put another way, the non-declining welfare and non-declining value conditions will be satisfied if there is a path between opportunity sets such that, given choice of $K_i(t)$, there exist opportunity sets over the whole time horizon which are attainable, and under which society will be at least as well off in instrumental value terms as at time t.

As we have already seen, the non-instrumental value of biotic resources has a number of facets, but a common feature is that the value assigned to species is unrelated to the use currently made of those species. Since an intertemporal comparison of opportunity sets based on instrumental value takes the current preference ordering of society as the yardstick for comparison, it does depend on the use currently made of biotic resources. The problem considered in this section exists because the social preference ordering of future generations over the universal set of alternatives cannot be known. Hence, it cannot be assumed that satisfaction of the condition $f(K_j(t+s)) \succeq f(K_i(t))$ will ensure that future generations are no worse off than the present generation. What is needed is a criterion that is independent of the specific uses to which biotic resources are now put.

This problem is very similar to the role of freedom of choice addressed in the choice literature by Pattanaik and Xu (1990), Sen (1991a, b), and Bossert, Pattanaik, and Xu (1992). The problem discussed in this literature concerns the ranking of opportunity sets based on considerations of freedom of choice, where freedom of choice is equivalent to the range of options open to the individual, and is given by the number of elements in the opportunity set over which the individual has to choose. Hence, if one opportunity set has more elements (more choice) than another, it is said to involve more freedom than the other. The notion that freedom is given by the range of choice is an intuitive one, but there is a difference in the approach to the problem taken by Sen, who argues that preferences are important in defining the areas over which individuals wish to exercise choice, and Pattanaik et al., who argue that freedom is independent of the preferences of the individual. It is the second approach that is relevant in the context of this chapter.

The justification for regarding freedom as independent of preferences (at least in some cases) is illustrated by Bossert, Pattanaik, and Xu (1992) in the following simple example. Suppose that an individual is confronted by two opportunity sets, X and Y, each consisting of only one element. Suppose that the element in X is preferred to that in Y. Clearly X will dominate Y in terms of the indirect utility of the two alternatives. But X and Y both involve exactly the same range of choice—none at all. The individual may be better off in utility terms with X than with Y, but it is not obvious that they are better off in terms of the freedom they enjoy. The problem considered in this chapter is slightly different, but may also be posed through a similar simple example. Suppose that there are no common elements in two opportunity sets, X and Y, available to two individuals, A and B. Suppose further that X has one element, that Y has two elements, and that individual A is indifferent between the two sets. If B's preferences are unknown to A, but A is charged with ranking the two opportunity sets on behalf of B, which set dominates? In this case it is not at all obvious whether X or Y would leave B better off in utility terms, but it is quite clear that Y offers greater freedom of choice.

In the biodiversity problem, future generations are not in a position to determine which out of all possible opportunity sets in $\prod(K_T)$ they would prefer. They are stuck with an inheritance from the present generation that depends on current use of assets. So, for example, current strategies that involve the widening and deepening of human capital extend the future opportunities. Conversely, current strategies involving the destruction of tropical moist forest narrow future opportunities. The present generation is, in a very important sense, responsible for determining which opportunity sets are available to future generations. Using the social preference ordering of the first generation, it is possible to rank opportunity sets over time in a way that establishes which reachable future opportunity sets have at least as high a (discounted) instrumental value to the present generation as the current opportunity set. But since it cannot know the social preferences of future generations, the present generation cannot determine which of these opportunity sets will yield the greatest instrumental value to future generations. All that it can know is the freedom of choice associated with each opportunity set. Freedom of intergenerational choice accordingly provides a means of ranking opportunity sets that is independent of any ranking that may be obtained from application of the preference orderings of different generations.

Freedom of choice, in this sense, is identical to the cardinality of the opportunity sets, $K_i(t)$. If $\#K_j(t+s) = \#K_i(t)$ then the opportunity sets $K_j(t+s)$ and $K_i(t)$ have the same range of choice. The principle implied by this (analogous to Bossert, Pattanaik, and Xu's simple monotonicity principle) is reflected in the following axiom:

Axiom of choice:

A3:   For all $K_j(t+s) \neq K_i(t) \in \prod(K_T)$, and for all for all t, s $\in$ {0, ..., T}, #$K_j(t+s)$ > #$K_i(t)$ and $f(K_j(t+s)) \sim f(K_i(t))$ implies that $K_j(t+s) \succ K_i(t)$.

If the jth opportunity set at time t+s offers greater freedom of choice than the ith opportunity set at time t, and if the foci of each are equally preferred, then it will also offer a higher level of well-being. In other words, independent of the time at which it is being assessed, if one opportunity set involves a greater range of opportunities than another which is as valuable in instrumental terms, then it offers a higher level of well-being.

Notice that since future opportunities depend on the underlying list of assets, and since the cardinality criterion weights all alternatives associated with a particular list of assets equally, this axiom has a strong preservationist bias in certain cases. If two opportunity sets are compared, each having the same focus, and identical except for the fact that a species which is irrelevant to that focus has been deleted from one, the opportunity set without the deletion will dominate the opportunity set with the deletion. The axiom does not preclude a trade-off between produced and natural capital. But it does imply that any strategy involving the deletion of species will be at least as good as a preservationist strategy only if the number opportunities lost through depletion are more than compensated by the number of opportunities gained through the introduction of produced assets or the expansion of other species. This is very different from the implications of the axiom of focus, by which alternatives that do not affect the focus of an opportunity set are irrelevant to its ranking. By the axiom of focus, the deletion of less valued species and the addition of inferior alternatives have no bearing on the valuation of any opportunity set. By the axiom of choice, they are crucial.

An even more important characteristic of the axiom of choice is that the criterion by which opportunity sets are ranked is independent of time. The cardinality of opportunity sets is not sensitive to the rate of discount. In other words, the axiom of choice insists that a necessary condition for non-declining welfare is a non-declining range of choice. Since the global system is evolutionary, and since there does exist potential for substitution between resources, the axiom of choice does not require the preservation of existing opportunities. But it does require that changes in the set of opportunities should not diminish the freedom of choice open to future generations.

THE PRINCIPLES OF SOCIAL CHOICE AND THE SUFFICIENT CONDITIONS FOR THE SUSTAINABILITY OF DEVELOPMENT

Development was defined earlier as a process by which human welfare is improved over time. The process was argued to have two components: (1) the expansion of the set of opportunities open to society, and (2) the growth of the instrumental value of the asset base. It is now evident that the axioms of social choice discussed above are directly related to these components, and may be used to establish the sufficient conditions for the sustainability of the development process.

Before addressing this, however, consider for a moment a slightly different question. What conditions are sufficient to assure not development but the absence of regression—the stationary state? In other words, what conditions are sufficient to assure that welfare is constant over time: the Hicks criterion. By the axiom of choice, for all $K_j(t+s) \neq K_i(t) \in \prod(K_T)$ and for all $s \in \{0, ..., T-t\}$, if $\#K_j(t+s) = \#K_i(t)$, $f(K_j(t+s)) \sim f(K_i(t))$ implies that $K_j(t+s) \sim K_i(t)$. That is, if distinct opportunity sets offer the same freedom of choice if society is indifferent between the foci of those opportunity sets, then welfare will be constant over time. But what if opportunity sets are not distinct? What if we consider the preservation of the existing set of opportunities? In this case, it turns out that the sufficient conditions include what may be interpreted as a very stringent restriction on the social rate of time preference. Consider the following proposition.

P1:   If there exist $K_i(t) \in \prod(K_T)$ such that $f(K_i(t+s)) \sim f(K_i(t))$ for all $s \in \{1, ..., T-t\}$, and if the relation $\succeq$ satisfies the axiom of focus, then $K_i(t+s) \sim K_i(t)$.

That is, if there exist opportunity sets such that the foci of those sets are equally preferred when evaluated over the whole time horizon, then those opportunity sets will yield a constant level of welfare. Since the cardinality of $K_i(t)$ is independent of time, $\#K_i(t+s) = \#K_i(t)$ for all $s \in \{1, ..., T-t\}$. Let this cardinality be equal to m, and define the opportunity sets $K_j'(t+s) = K_j(t+s)\backslash f(K_j(t+s))$ and $K_i'(t) = K_i(t)\backslash f(K_i(t))$, both of cardinality m-1. If $f(K_j(t+s)) \sim f(K_i(t))$, then by the axiom of focus, $f(K_j(t+s)) \cup K_j'(t+s) \sim f(K_i(t)) \cup K_i'(t)$, and hence $K_i(t+s) \sim K_i(t)$ for all $s \in \{1, ..., T-t\}$ which is what we want. But consider what the restriction $f(K_i(t+s)) \sim f(K_i(t))$ implies. Recall that the function $K_i(t)$ discounts the instrumental value of the opportunity set K_i. If the rate of discount is strictly positive, then $f(K_i(t)) \succ f(K_j(t+s))$: present well-being will be preferred to future well-being. Hence, $f(K_i(t+s)) \sim f(K_i(t))$ only if the future benefits of an opportunity set are discounted at the zero rate. So if simple preservation of the set of opportunities is to be consistent with a constant level of welfare, the rate of discount is constrained to be equal to zero.

Sustainable development requires something more than maintenance of the status quo, whether that is conceived in terms of a set of opportunities or a level of welfare. In this case, we are interested in the conditions for welfare to be strictly increasing over time. Consider the following propositions:

P2:   If there exist reachable $K_j(t+s)$ such that $\#K_j(t+s) = \#K_i(t)$ and $f(K_j(t+s)) \succ f(K_i(t))$, and if the relation $\succeq$ satisfies axiom of focus, then $K_j(t+s) \succ K_i(t)$; and

P3:   If there exist reachable $K_j(t+s)$ such that $\#K_j(t+s) > \#K_i(t)$ and $f(K_j(t+s)) \sim f(K_i(t))$, and if the relation $\succeq$ satisfies the axioms of weak independence and choice, then $K_j(t+s) \succ K_i(t)$.

The first says that if there exist opportunity sets that are reachable from $K_i(t)$ that offer the same freedom of choice but that have a focus of greater value than $K_i(t)$, then there exists a development path along which welfare will be strictly increasing. The second says that if there exist opportunity sets that are reachable from $K_i(t)$, the foci of which are equally preferred, but that offer greater freedom of choice than $K_i(t)$, then there exists a development path along which welfare will be strictly increasing.

Both are quite intuitive, and can be verified very shortly. First, consider the case where $\#K_j(t+s) = \#K_i(t)$ and $f(K_j(t+s)) \succ f(K_i(t))$. Suppose, as before, that the cardinality of $\#K_j(t+s) = \#K_i(t) = m$, and define the opportunity sets $K_j'(t+s) = K_j(t+s)\backslash f(K_j(t+s))$ and $K_i'(t) = K_i(t)\backslash f(K_i(t))$, both of cardinality m-1. Since $f(K_j(t+s)) \succ f(K_i(t))$, by the axiom of focus $f(K_j(t+s))\cup K_j'(t+s) \succ f(K_i(t))\cup K_i'(t)$, and hence, $K_j(t+s) \succ K_i(t)$. The preservation of the range of choice combined with an increase in the value of the focus of successive opportunity sets is sufficient to assure a development path along which welfare will be strictly increasing.

The second case is only a little more complicated. If $\#K_j(t+s) > \#K_i(t)$, and $f(K_j(t+s)) \sim f(K_i(t))$, there exists a set of opportunities, $K_o = K_j(t+s)\backslash K_i(t)$, which does not contain the focus of either opportunity set. It follows from the axiom of choice that $f(K_j(t+s))\cup K_o \succ f(K_i(t))$, and from the axiom of weak independence that $f(K_j(t+s))\cup K_o\cup K_i(t)\backslash\{f(K_i(t)) \succ f(K_i(t))\cup K_i(t)\backslash\{f(K_i(t))$. From P1, $K_j(t+s) \sim \{f(K_j(t+s))\cup K_o\cup K_i t)\backslash\{f(K_i(t))$, hence, by transitivity of $\succ$, $K_j(t+s) \succ K_i(t)$. That is, if there exist reachable opportunity sets that offer strictly greater freedom of choice, and have the same the focus as the set $K_i(t)$, there exists a development path that will yield strictly increasing welfare over time.

These propositions help us to think about the sufficient conditions for sustainable development in the case where $\#K_j(t+s) < \#K_i(t)$ and $f(K_j(t+s)) \succ f(K_i(t))$, or the opposite. Indeed, something of this kind probably better characterizes reality than conditions of the type indicated in P2 and P3. While the instrumental value of the opportunities available to the present generation may be growing (albeit very slowly during the current recession), the range of options open to future generations is narrowing with the destruction of more and more biotic resources. It is not intuitively obvious what conditions of this sort mean for the sustainability of the development process. In such a case the states, $K_j(t+s)$ and $K_i(t)$ are non-comparable under the axioms of focus, weak independence, and choice. Neither instrumental value nor freedom of choice is privileged over the other. Nor is there is any scope for a trade-off between these two things. In this case, a minimal condition for $K_j(t+s)$ to be part of a sustainable development strategy, is for the opportunity set $K_j(t+s)$ to be augmented by some set, $K_h(t+s)$, equally preferred to the opportunities lost from the current opportunity set, such that $\#(K_j(t+s) \cup K_h(t+s)) = \#K_i(t)$. In particular:

P4:   If there exist reachable $K_j(t+s)$ such that $\#K_j(t+s) < \#K_i(t)$ and $f(K_j(t+s)) \succ f(K_i(t))$, if there exist $K_i''(t) \sim K_h(t+s) \in \Pi(K_T)\backslash K_j(t+s)$ such that $\#K_i''(t) = \#K_h(t+s) = (\#K_i(t) - \#K_j(t+s))$, and if the relation $\succ$ satisfies the axioms of focus and weak independence, then $(K_j(t+s) \cup K_h(t+s)) \sim K_i(t)$.

If $\#K_j(t+s) < \#K_i(t)$ and $f(K_j(t+s)) \succ f(K_i(t))$ there exists an opportunity set $K_i'(t)$ such that $\#K_j(t+s) = \#K_i'(t)$, with $f(K_i(t)) \in K_i'(t)$. By the axiom of weak independence $f(K_j(t+s)) \cup K_i'(t)\backslash f(K_i(t)) \succ f(K_i(t)) \cup K_i'(t)\backslash f(K_i(t))$, and by the axiom of focus, $f(K_j(t+s)) \cup K_i'(t)\backslash f(K_i(t)) \sim f(K_j(t+s)) \cup K_j(t+s)\backslash f(K_j(t+s))$. It follows that $f(K_j(t+s)) \cup K_i'(t)\backslash f(K_i(t)) \sim K_j(t+s)$, and by transitivity of $\succ$, $K_j(t+s) \succ f(K_i(t)) \cup K_i'(t)\backslash f(K_i(t))$. Let $K_i''(t) = K_i(t)\backslash K_i'(t)$. If there exists $K_h(t+s) \sim K_i''(t)$, then by the axiom of focus, $K_j(t+s) \cup K_h(t+s) \succ f(K_i(t)) \cup K_i'(t)\backslash f(K_i(t)) \cup K_i''(t)$. Hence, $K_j(t+s) \cup K_h(t+s) \succ (K_i(t))$. The proposition states that a sufficient condition for the sustainability of a reachable opportunity set that does not offer the same freedom of choice as the current opportunity set is for that set to be augmented by a range of choice of equal size and (present) value to that lost from the current opportunity set. If the loss of biotic diversity implies a narrowing of the options open to future generations, even though positive rates of economic growth signal an increase in the instrumental value of opportunities now available, sustainability requires that those options be restored or replaced with options of comparable value.

CONCLUSIONS

The burden of these propositions is that any development path will be sustainable if it offers either an increase in the range of choice, holding the discounted instrumental value of the opportunities open to society constant, or an increase in the discounted instrumental value of the opportunities open to society holding the range of choice constant. Current preference is one criterion for non-declining welfare, the range of choice is another. Aspects of the axiom of choice are present in the treatment of natural capital in some existing models of intertemporal resource allocation. It has already been remarked that Mäler (1992) argues that the existence of use and non-use value implies the separability of individual utility functions. I have elsewhere characterized the social optimization problem, in control terms, as one in which society seeks to maximize welfare over a finite period as the sum of an algebraic function of the terminal value of the state variables (bequest or existence value) and an integral function of the state and the control variables (use value) where the state variables are the set of all assets available to society and the control variables are the resource allocations made by society (Perrings 1991b).[5] Both these and analogous approaches accordingly imply that their

---

5. The specific problem considered in Perrings (1991) is as follows:

exists a set of benefits from natural biotic resources that does not in any way depend on their use in economic processes. But both also retain the notion that current preferences determine the relative non-instrumental value of such resources. Even the most significant recent work on the non-instrumental valuation of collections of different objects requires some structure of preferences to determine the criterion of distance between objects (Weitzman 1991a, b). But the axiom of choice is blind to preferences. Providing that future opportunity sets are no worse than present opportunity sets under the preference ordering of the present generation, and that expansion of opportunities is welfare-improving.

The ethical content of the axiom of choice lies in the implication that expansion in the set of choices available to future generations is a benefit to the present generation. In this sense the axiom captures the spirit of Brundtland very well. It insists that a necessary condition for the sustainability of development is that future generations do not forfeit freedom of choice as a result of the activities of the present generation. Hence, if the present generation is committed to the principle of sustainability, it implies that it should accept the responsibility that is the obverse of the right of choice conferred on future generations. It should be said that it is not a right that is necessarily burdensome to the present generation. Bossert, Pattanaik, and Xu (1992) make the point in respect to the role of freedom in individual choice, that if agents have the power to choose the outcome corresponding to some opportunity set, then they cannot lose by adding an alternative to that opportunity set, even if that alternative is strictly worse than existing alternatives. In terms of the biodiversity problem discussed here, an analogous argument is that if the present generation has the power to determine the outcome corresponding to a given opportunity set, it cannot be made worse off by retaining an alternative (some species of plant or animal) that is strictly worse than all other alternatives under the current preference ordering (i.e., that has no effect on the focus of the opportunity set).

---

maximize$_k$

$$J = W[\mathbf{x}(T),T]e^{-\delta T} + \int_0^T Y[\mathbf{x}(t),\mathbf{u}(t),t]e^{-\delta t}dt$$

subject to

$$\dot{\mathbf{x}}(t) = \mathbf{f}[\mathbf{x}(t),\mathbf{u}(t),t], \qquad\qquad 0 \leq t \leq T,$$
$$\mathbf{g}[\mathbf{x}(T), T] = \mathbf{0}$$
$$\mathbf{x}(0) = \mathbf{x}_0,$$

with

$$\mathbf{u}(t) = \mathbf{u}[\mathbf{k}, \mathbf{p}(t), t] \qquad\qquad 0 \leq t \leq T,$$
$$\mathbf{p}(t) = \mathbf{p}[\mathbf{x}(t),t] \qquad 0 \leq t \leq T,$$

where $\mathbf{x}(t)$ is a vector of state variables (the assets available to society); $\mathbf{u}(t)$ is a vector of control variables (the allocation of resources); $W[\cdot]$ defines the bequest or existence value of assets at the terminal date, and $Y[\cdot]$ defines the use value of those assets during the time horizon of interest. $\mathbf{f}[\cdot]$ defines the equations of motion of the system, and $\mathbf{g}[\cdot]$ a set of terminal boundary conditions (restrictions on the admissible value of assets bequeathed to the next generation). $\mathbf{p}[\cdot]$ is a price vector, and $\mathbf{k}$ is a vector of control parameters.

It is worth noting that while the axiom of choice does have a preservationist bias in some cases, it does not privilege the preservation of existing opportunities. Sustainable development is incompatible with the maintenance of the status quo. In an evolutionary system, the economic and ethical problems converge in the need to maintain that level of biodiversity which will guarantee the resilience of the ecosystems on which human consumption and production depend. Indeed, this is the central goal of a strategy of biodiversity conservation. It requires neither the preservation of all species, nor the maintenance of the environmental status quo. Where economic activity changes the level or composition of biodiversity, it requires that the opportunities foregone by future generations be compensated, and any change in biotic diversity that affects the flow of ecological services on which this and subsequent generations depend is affected by this principle.

This said, the weight of available evidence suggests that the loss of biotic diversity that so exercised the world's attention at UNCED is compromising the interests of future generations precisely because it is narrowing the range of options open to those generations. There is no evidence that any care is being taken to assure that the accelerating destruction of biota by the present generation will not limit the range of choice open to future generations. Indeed, all available evidence indicates just the opposite. Biota are being destroyed to maintain consumption, not to promote investment. Options are being narrowed, not expanded. Perhaps it is unrealistic to expect anything different. Since much of the threat to biodiversity comes from people driven by poverty, one would expect them to be driven by the exigencies of the present. Yet many of the best examples of social protection of the rights of future generations come from societies—the so-called primitive societies—which are poor in terms of material consumption. Admittedly, such societies respect the axiom of choice more through the preservation of options than the generation of alternatives, but the point is that care for the rights of future generations is not the prerogative of the rich. Something very much like the axiom of choice has influenced collective decision making in a wide range of societies over a long period of time. It may have become lost in the consumption-oriented individualism of many existing societies, but that is merely to demarcate the battle ground for those who do worry about the effects of biodiversity loss.

REFERENCES

Arrow, K. J., and A. C. Fisher. 1974. Environmental preservation, uncertainty, and irreversibility. *Quarterly Journal of Economics* 88 (2): 312–19.

Barbier, E. B., and A. Markandya. 1990. The conditions for achieving environmentally sustainable development. *European Economic Review* 34: 659–69.

Blasi, P., and S. Zamagni, eds. 1991. Man-Environment and Development: Towards a Global Approach. Rome: Nova Spes International Foundation Press.

Bossert, W., P. K. Pattanaik, and Y. Xu. 1992. Ranking Opportunity Sets: an Axiomatic Approach. Unpublished paper.

Common, M., and C. Perrings. 1992. Towards an ecological economics of sustainability. *Ecological Economics* 6: 7–34.

Daly, H., and J. B. Cobb. 1989. For the Common Good. Boston: Beacon Press.

Fisher, A. C., and W. M. Hanneman. 1983. Option Value and the Extinction of Species. Working Paper 269. Giannini Foundation of Agricultural Economics. Berkeley: Univ. of California.

Hartwick, J. M. 1977. Intergenerational equity and the investing of rents from exhaustible resources. *American Economic Review* 66: 972–4.

———. 1978. Investing returns from depleting renewable resource stocks and intergenerational equity. *Economics Letters* 1: 85–8.

———. 1991. Economic Depreciation of Mineral Stocks and the Contribution of El Serafy. World Bank Environment Department. Working Paper 4. Washington, DC: The World Bank.

Henry, C. 1974. Investment decisions under uncertainty: the irreversibility effect. *American Economic Review* 64: 1006–12.

Hicks, J. R. 1946. Value and Capital. Oxford: Oxford Univ. Press.

Holling, C. S. 1973. Resilience and stability of ecological systems. *Annual Review of Ecological Systems* 4: 1–24.

———. 1986. The resilience of terrestrial ecosystems: local surprise and global change. In Sustainable Development of the Biosphere, eds. W. C. Clark and R. E. Munn. Cambridge: Cambridge Univ. Press.

Katzner, D. W. 1989. The comparative statics of the Shackle-Vickers approach to decision-making in ignorance. In Studies in the Economics of Uncertainty in honor of Joseph Hadar, eds. T. B. Fomby and T. K. Seo. Berlin: Springer-Verlag.

———. 1986. Potential surprise, potential confirmation and probability. *Journal of Post Keynesian Economics* 9: 58–78.

Kemp, M., and N. V. Long. 1984. Essays in the Economics of Exhaustible Resources. Amsterdam: North Holland.

Krutilla, J. V. 1967. Conservation reconsidered. *American Economic Review* 57 (4): 778–86.

Mäler, K-G. 1992. Multiple Use of Environmental Resources: the Household Production Function Approach. Beijer Discussion Paper 4. Stockholm: Beijer Institute.

Paine, R. T. 1980. Food webs: linkage interaction strength and community infrastructure. *Journal of Animal Ecology* 49: 667–85.

Pattanaik, P. K., and Y. Xu. 1990. On ranking opportunity sets in terms of freedom of choice. *Recherches Economiques de Louvain* 56: 383–90.

Pearce, D. W., and R. K. Turner. 1990. Economics of Natural Resources and the Environment. London: Harvester-Wheatsheaf.

Pearce, D. W., A. Markandya, and E. B. Barbier. 1989. Blueprint for a Green Economy. London: Earthscan.

Perrings, C. 1991a. Reserved rationality and the precautionary principle: technological change, time and uncertainty in environmental decision-making. In Ecological Economics: The Science and Management of Sustainability, ed. R. Costanza. New York: Columbia Univ. Press.

———. 1991b. Ecological Sustainability and Environmental Control, Structural Change and Economic Dynamics 2. Oxford: Oxford Univ. Press.

Pezzey, J. 1989. Economic Analysis of Sustainable Growth and Sustainable Development. World Bank Environment Department. Working Paper No. 15. Washington DC: The World Bank.

Sen, A. K. 1991. Welfare, preference and freedom. *Journal of Econometrics* 50: 15–19.

Solbrig, O. T. 1991. The origin and function of biodiversity. *Environment* 33: 10.

Solow, R. M. 1974. Intergenerational Equity and Exhaustible Resources. Review of Economic Studies, Symposium: 29–46.

———. 1986. On the intertemporal allocation of natural resources. *Scandinavian Journal of Economics* 88 (1): 141–49.

Turner, R. K. 1988. Sustainability, Resource Conservation and Pollution Control: An Overview. In Sustainable Environmental Management: Principles and Practice, ed. R. K. Turner. London: Bellhaven Press.

Weisbrod, B. 1964. Collective consumption services of individual consumption goods. *Quarterly Journal of Economics* 77: 71–7.

Weitzman, M. L. 1991a. On Diversity. Discussion Paper 1553. Cambridge, MA: Harvard Institute of Economic Research.

———. 1991b. A Reduced Form Approach to Maximum Likelihood Estimation of Evolutionary Trees. Unpublished paper. Cambridge, MA: Harvard Institute of Economic Research.

Westman, W. E. 1990. Managing for biodiversity: unresolved science and policy questions. *BioScience* 40: 26–33.

World Commission on Environment and Development 1987. Our Common Future. Oxford: Oxford Univ. Press.

7 SUSTENANCE AND SUSTAINABILITY: HOW CAN WE PRESERVE AND CONSUME WITHOUT MAJOR CONFLICT?[1]

Ralph C. d'Arge
Department of Economics
University of Wyoming
P.O. Box 3985, University Station
Laramie, WY 82071

ABSTRACT

All species have an inherent right to survive. All species also have an inherent capability of not surviving. Human knowledge on rights of survival are well developed while knowledge on capabilities is highly uncertain, limited, and often driven by false beliefs. This fundamental lack of knowledge of environmental processes places decisions on biodiversity outside the realm of normal human decision making. Expected utility makes little sense when nothing is accurately known about expected consequences and little is known about the true utility of a given species or ecosystem. Since normal decision processes cannot capably lead to wise management decisions, what type of choice or ethical process is efficient, reasonable, and likely to be accurate?

This chapter explores alternative ways of thinking about investments in species diversity and other management tools when there is little or no information on a species' proclivity for survival (and where the utility of the species is unknown). In this case, a market economy would quite naturally place little or no value on the species. Values totally external to markets or to human consciousness are likely to be there, but are unknown and unrecognized. In this type of "blackhole" global ecosystem what is a rational strategy toward preservation and sustainability? Two biases are readily apparent. We tend to preserve what we like and what we know a bit about (e.g., cheetahs, black-footed ferrets, and elephants) and do nothing for species we know little about or do not like. We also tend to invest too much in preserving individual species and too little in broad categories of ecosystem components. There is also a bias toward not preserving species that

1. I have benefited significantly from discussions with more than twenty participants at the Second ISEE Meeting in Stockholm. I particularly wish to thank Herman Daly, Charles Perrings, and Tomasz Zylicz without implying that they necessarily agree with any part of this essay.

harm us (e.g., Japanese beetles, "Killer" bees, and rattlesnakes). These inherent biases are likely to seriously reduce biological diversity and thereby harm the future economy and societies that reside within it.

To resolve these dilemmas, the author recommends some substantive institutional changes in the economic system, particularly to vest complete, fee simple ownership of all natural species with the world's children.

INTRODUCTION

Increasing species diversity, in simple terms, means that there are a greater number of species present in the biosphere. If the biosphere is operating at near it's carrying capacity, an increase in species diversity implies a reduction in the mass and/or numbers of at least some other species. Increasing the number of species may also result in fewer varieties or phenotypes within any one species category. From an economic perspective, species diversity may well be increased at the expense of a broader genetic diversity. This fundamental trade-off is induced by overall limits placed by carrying capacities.[2] An example of this trade-off can be observed for the thoroughbred horse. The thoroughbred breed of horse was created by successive patterns of inbreeding of the bloodlines of three Arabian stallions imported into England in the 18th century. Records of the historical breeding, mating, and performance of this breed of horse (capable of great speed over a distance of 1–3 miles) suggest that continuous inbreeding of the species has resulted in a reduction in genetic diversity. While speed over a distance has more or less continuously increased since the 18th century, the breed has also developed some less satisfactory characteristics, including respiratory bleeding, unsoundness, reproductive difficulties, and psychotic behavior with tendencies toward aggression. Thus, natural or anthropogenic manipulation that creates species or subspecies may also reduce genetic diversity. Further, preserving a species that nature is trying to get rid of may reduce genetic diversity because the natural environment in which it resides could have been potentially utilized by another emerging species that is now displaced by the act of preservation. Thus, genetic and species diversity may be reduced through conscious anthropogenic choices attempting to preserve an existing species. In addition, the economic resources devoted to the preservation activity rather than to other conservation practices may well reduce the number of species or genetic diversity because of that investment decision. The economic consequences of a particular choice for preservation may be quite pervasive in impact on genetic diversity and thereby on species diversity.

Several important points and issues emerge from the above discussion. There may be fundamental natural and economic trade-offs between species diversity and genetic diversity. Is species or genetic diversity inherently more

---

2. On the crucial role of the concept of carrying capacities, see G. Hardin, Paramount positions in ecological economics, in Ecological Economics: The Science of Management of Sustainability, ed. R. Constanza. (New York: Oxford University Press, 1991), 47–57.

valuable? Do values that the economy directly or indirectly places on species, including conservation and preservation values, reflect in any way the inherent or real values associated with biological diversity? If these values are unknown for most species, what is the best course of action for managing natural species? Given the common property nature of most natural species, can efficient institutional systems of rights and ownership be designed to protect them and in some cases, preserve them? This chapter will explore some very basic questions of genetic diversity from the perspective of economics. I apologize beforehand to my colleagues in the natural sciences for oversimplifying, inferring, and drawing conclusions that may not be warranted in light of careful scientifically based analytic dissection.

ECONOMIC AND ECOLOGICAL PRICES

Every binding constraint in the economy manifests itself as either an explicit or implicit price. Thus, if the availability of leather or rubber impacts the manufacturing of shoes, it will impact shoe prices and therefore have at least an implicit price. If there are markets for leather and rubber, there will be explicit prices for these commodities as well. Likewise, elk will have an implicit price as the number of elk impact elk hunting. The browse the elk graze on also has an implicit price in terms of the impact of browse availability on elk numbers and, indirectly, hunter success. Thus, the economy's price system contains a large number of tentacles each connected in some way to natural ecosystems. There is hardly any commodity in the economy that does not directly or indirectly have a well-defined connection through prices to at least one natural species in the biosphere, although some of the tentacles are obscure and/or very long. Thus, a complete economic price system would contain all explicit market prices as well as implicit prices for all binding constraints (i.e., the value of all pieces of the ecosystem to the economy). This leads to the conclusion that parts of the natural ecosystem are valued and thereby are priced in the economy. However, there are substantially fewer markets than binding ecosystem constraints in the economy, so market prices are unlikely to reflect all implicit prices derived from the connection between the economy and the natural ecosystem.

Presume that the natural ecosystem contains it's own relevant set of prices. If ecosystems taken together form a convex and compact set and something is maximized (or minimized) over that set, then the ecosystem or biosphere will "naturally" generate a price for each binding constraint of that set.[3] What natural ecosystems tend to maximize or minimize, or even if they do, has yet

---

3. It should be noted that the set of prices are generated by each species acting in its self interest tries to survive and sustain itself. In so doing, it is "acting" identically with producers and consumers in a market economy. The relevant prices may be the outcome of species maximizing something very different than profits or "utility." For a discussion of this process in the context of compact and convex sets, see G. Debreu (1959) or K. Arrow and G. Debreu (1954).

to be proven, although there are some interesting hypotheses. Note that all natural ecosystem prices are implicit ones in that organized or any other kind of markets for exchange are not present. We can think of the natural ecosystem as having a large number of tentacles between it's set of implicit prices and various commodities in the economy.

How do natural ecosystems generate prices? Quite simply, each price is an implicit one for the constraint on the ecosystem itself. Thus, if browse limits elk herd expansion, then browse has a positive value to elk. This positive value is in units of browse productivity to elk and elk's productivity in augmenting whatever objective the natural ecosystem strives for (i.e., maximizing species stock, net energy, stored energy, etc.). Another way of stating this is to ask if a natural ecosystem would generate implicit prices even without the presence of humans in the system? The answer is a relatively straightforward "yes." If the actual ecosystem had one or more binding constraints and if there was an overall driving force or objective of the system, then implicit prices would manifest themselves in directing changes in the natural ecosystem. Thus, under very general assumptions natural ecosystems have an inherent pricing system. However, this system appears to be devoid of traditional economic concepts such as supply, demand, equilibrium, or tatonnement processes of market movements toward an equilibrium. It is interesting to note that economists have never had a realistic process of how buyers and sellers interact to reach an equilibrium price (recontracting). Alternatively, understanding species interactions and thereby implicit price changes (even if they are not called that) is one of the achievements of modern ecology.

Economic and ecological systems can be characterized as two distinct processes where one is continuously exploiting the other, as opposed to economic trade, where the terms of trade are totally dictated by the exploiting economies. Traditional models of international trade presume that the terms of trade are known by both economies, and that exploitation is ruled out by national sovereignty and adherence to international law. In this world of both implicit and explicit prices, the natural ecosystem cannot evaluate or respond to the economy's price signals. It is as if two planets were negotiating with each other without a common language (two sets of prices), where one planet was entirely passive and accepted any possible trade proposed by the other.

Since Kneese, Ayres, and d'Arge linked production and consumption with mass balance to a simple general equilibrium (welfare model of the economy), there has been a growing set of even more complete models of the economy and environment. (Kneese et al. 1970). Each of these models has emphasized more general (Mäler 1974) or specific environmental interaction relationships (Tietenberg 1972; Baumol and Oats 1988). However, the basic structure of these models has remained relatively constant over the past two decades. What many of these models have attempted to show is that a perfectly competitive economy, with sufficient adjustments in pricing for natural ecosystems, will be Pareto efficient. That is, the economy will be operating efficiently when it *fully* takes into account all impacts on itself from its interaction with the natural ecosystem. Each general equilibrium model proposed generates a set of prices for the various attributes and constraints evolving from the natural environment. It is then proposed by the model developers that if such prices were actually paid in the economy, that efficiency would result. In these models, each manifestation of the natural ecosystem is represented by a binding constraint, and an implicit price is identified for it. The set of explicit and implicit prices taken together form an efficient price system. What alarmed Kneese, Ayres, and d'Arge in their first effort was both the magnitude in number of implicit prices to be obtained, and the sheer complexity in their construction. For each commodity there is a set of implicit prices which requires mass balance for each type of environmental impact. Thus, production and consumption along with price formation associated with one type of commodity would entail the derivation of implicit prices for all types of water pollution, air pollution, solid waste, hazardous waste, and species impact, along with mass balance both within and across all economic sectors.

Mäler, in his model, proposes a singular government agency to compute and implement all of these environmental prices (Mäler 1974). However, conceptual difficulties arise. The agency would literally need to compute and charge millions of different implicit prices. The level of each market price is partially determined by the level of all other explicit and implicit prices and vice versa. So it would be impossible to compute each in isolation of the other. This observation has led some observers to believe that it is impossible to practically measure all of the implicit prices for connections between economic commodities and the natural ecosystem. Therefore, we have "pervasive externalities." It is unlikely that any single agency could both compute and implement such a grand pricing scheme. There appear to be two rather distinct views on this problem. First, pervasive externalities are so encompassing and the general equilibrium computations so complex that it is futile to ever presume a set of efficient, economy-wide prices can be obtained. Thus, we should either ignore the problem of efficient prices or look for other paradigms that would be more meaningful for environmental policy. Second, and somewhat distinctly, while such a grand scheme is unfeasible, actions by many autonomous, local, regional, and national environmental agencies through standards, fines, and subsidies have caused the price system in general to move towards the efficient one, and thus pervasive externalities have been at least partially accounted for. However, from the theory of second best, we know that such a partial correction may have harmed more than it helped. (Davis and Whinston 1965; Buchanan 1967). Third, one could just accept the integration of general equilibrium and the natural environment as an intellectual "curiosum" and ignore it.

To recapitulate the discussion thus far, we can propose three distinct vectors of explicit and implicit prices. The first vector P_1 denotes existing prices in the economy with no adjustment for environmental interactions or impacts. Such a vector contains positive or zero market prices, and also contains zero's for all implicit prices. This vector would be appropriate where humans were little more than just another slightly advantaged member of the food chain. Next let P_2 denote the complete set of both explicit and implicit prices necessary to achieve Pareto efficiency. This vector would contain much fewer zero's than the previous price vector, P_1. Finally let P_3 represent the price vector where some environmental problems have been solved and where some natural ecosystems have been priced by environmental agencies. Note P_3 will have many more zeros than P_2 but fewer than P_1. Models in environmental economics have examined P_1 and P_3 and have proved that such price vectors exist conceptually. Proofs on their uniqueness have not been as complete. Further, it can be agreed on, but not proved in the general case, that P_3 will yield a better state of the economy than P_1 but less than P_2. It can be proved that P_3 will always be at least as good as P_2.

If ecological systems contain their own consistent pricing processes, inclusive of man as a traditional hunter and gatherer, an implicit price system exists for it. Without being capricious, let us propose that a price vector exists for the ecosystem with it's own units represented by Z, where for convenience a single numeraire is used to evaluate all units. Instead of money, the numeraire is likely to be BTUs of stored energy, kilograms of mass, or some other such natural unit. Then let Z denote this implicit price vector as it exists today. Since each unique species member confronts approximately the same number and type of constraints within that species, it is probably true that there are many more binding constraints, and therefore, implicit prices for the ecosystem than for the economy. That is, the vector Z is dimensionally much larger than the vector P_2. Another way of stating this premise is that the natural environment is likely to contain many more prices and hence, be more complex than any economy. The social contrivance of money has led to a single numeraire for most economies, whereas it cannot be expected that an equally simplifying biological contrivance has occurred. Following the identification of price vectors in economics, let Z_1 and Z_2 represent implicit price vectors for the natural environment when Pareto efficient prices and environmentally adjusted prices exist in the economy.

There are some rather striking comparisons between the two price systems just proposed. The first is that the Pareto-efficient price vector P_1 influences and is influenced by the implicit price vector for ecosystems of Z. However, since Z is unknown dimensionally or in terms of the extent of it's domain, P_1 cannot be truly efficient. It will only be relatively efficient subject to the extent that binding constraints in natural ecosystems are represented in the economy's pricing system. If most are not binding, then it is highly unlikely that the P_1 system will indeed be efficient. Rather, it will only be efficient over the subset of ecosystems which by design or default, are priced in the economy. The rather simple conclusion to be drawn from these remarks is that a Pareto-efficient set of prices in the economy depends on the relevant set of implicit prices in ecosystems, and likewise, these ecosystem prices depend on the level of Pareto-efficient prices in the economy. With little or no knowedge of these ecosystem prices, it is impossible to discover the magnitude of Pareto-efficient prices.

INVESTMENT IN NATURAL CAPITAL

How much of the output of economic systems should be devoted to preserving and sustaining the natural environment as we know it and how much should be channeled toward human comfort and sustenance? A sub-category of this investment is how much we should invest in protecting and sustaining the existing level of biological diversity. The answers to these questions must begin with a few observations from modern growth theory directed toward investment decisions on biological diversity. Imagine we are in a world where humans consume, but to consume, they must eliminate a species, and to continue to consume more, they must eliminate more species. Thus, the more they consume, the more species they consume and eliminate. Even if they consume at some minimal sustenance level, species are eliminated and not replaced by some "manna from heaven" process. In such a world the central question is simply how much to consume or how fast should species be harvested and eliminated?[4] The more that are immediately consumed, the more species that are irreversibly removed and not available for future consumption or benefit. If species diversity is valueless, the optimal rate of consumption is a simple variant of the Hotelling model of mining an exhaustible resource. More is consumed now than in the future because discounted future consumption is less valuable. If we make the future equal to the present, then consumption will remain constant on an optimal path at the level that equates marginal utilities of consumption over each time interval until all species except humans are devoured. Now let us assume in this scenario that species *inherently* have value to humans. They, through gene pools, breeding stocks, or just enjoyment, yield utility or positive values to the economic system. The more species there are, the better off humans are, but this process is subject to diminishing returns (i.e., diminishing marginal utility of additional species or increasing marginal disutility of diminishing numbers of species). The optimal strategy is then to eliminate species at a much slower rate since the value of delaying extinction will more than offset the value of immediate consumption, at least at the margin. Thus, in this simple "devour now or later process," it pays to delay. In such a world, since species themselves tend to be common property resources with no well-defined market and external values, they will tend to be exploited too fast. Thus, a divergence between actual consumption and optimal consumption of species arises from two primary causes: (1) the omitted value of the species in the future as a stock resource, and (2) the omitted value of the species because of its common property resource

---

4. The model utilized here was first developed for examining the optimal path for consumption when there is a finite environment to pollute. See d'Arge (1971) and d'Arge and Kogiku (1973). For similar results utilizing a different model structure, see Schimmelpfennig (1991).

characteristics. These two causes reinforce each other in the same direction to severely undervalue the current exploitation of the species. The net result, even in this most simple paradigm of species exploitation, is their rapid decimation within a market-oriented economy. If the social rate of discount is high enough, the divergence may not occur between optimal and actual paths of consumption. But if it is high, the future has no value so that future consumption of species or their preservation becomes valueless.

Let me restate the conclusions of this most simple paradigm. In a model where species extinction is inevitable, it still pays to delay extinction unless there is no value to both future consumption and preservation. A divergence between optimal and actual rates of extinction occurs from two sources—the common property character of the natural species, *and* it's future value. Note that we have omitted considering the fundamental interdependencies between species and determining how these interactions would influence harvest or decimation rates. In many instances, the elimination of one may stress or even cause the elimination of a second, which would increase the current value of preserving any one species. This becomes a third factor contributing to too rapid an extinction rate for a single species. Also, if one adds the idea that some natural species are inherently or economically more valuable than others, it is clear from the theory of mining exhaustible resources of differential qualities, that the less valuable species should be devoured first (Schulze 1974). This is a rather simple application of the principle that less valuable species can be sacrificed at less opportunity cost to the economy than those of greater value.

NATURAL SPECIES AND UTILITY FUNCTIONS

A fundamental preposition in economics is that humans have well-defined preferences over bundles of commodities such that they can rationally select the best commodity bundle for themselves. Children, to a much lesser degree, have been acknowledged to have such preferences, although most microeconomic-based empirical studies try to identify a dictatorial "head of household" with one set of preferences for the entire household. What should be recognized is that if *all* adults have well-defined preferences, so do children to a greater or lesser degree. And, if all children have such preferences, all other species have such preference relations to a greater or lesser degree.[5] It can be documented that at least some animals exhibit more consistency and completeness in their preferences than some children or adults. It has been shown that some animals exhibit behavior which is completely consistent with well-

---

5. Boundaries between human and other animals are becoming less pronounced. The major idea is that if animals suffer pain, then that pain should be considered along with human pain. See the excellent examination of this idea in T. Page (1991). See also Van DeVeer and Pierce (1986) and Stone (1974).

defined preference orderings.[6] Thus, at least some species have well-defined utility functions. In modern welfare economics, animal or natural species utility functions are given zero weights except when they enter an adult person's utility function who is designated as the "head of household." This is as absurd as the 19th century dictum of the basic inferiority of minorities or women and lack of basic rights of children. Do animals and plants really have economically specified utility functions? The answer is obviously "yes" if it is accepted that humans, particularly adult humans, have them. What worries economists and policy makers is that if some or all natural species have well-defined preferences, is it ethically correct to quite arbitrarily exclude these preferences? I would argue that it is not, anymore than excluding women's or children's preferences was in the past. But, if such preferences are actually there, how can they be adequately represented? Should we opt for a one animal/one vote system of elections? How should the preferences of trillions of micro-organisms be tabulated, or even discovered. It suggests to me that political bodies need to develop "representatives" for natural species that can adequately defend and elucidate natural species preferences. It is somewhat ironic that the two groups with potentially the most to gain or lose in global climate change—future human generations and future natural species—have absolutely no direct representation at international tribunals dealing with their welfare.

How should we account for natural species' voting of their rights and preferences? One organism/one vote might be the fairest, but through the tyranny of the majority, humans would have little or no control over their destiny or livelihood, given the almost uncountable number of organisms residing both within and outside the human body. Realistically, we must make sure that natural species preferences are represented and that the spectrum of these preferences is accounted for. For example, two natural species with very different preferences or utility functions should have different representation (i.e., hawks and rabbits, polar bears, and sea otters). While one organism/one vote will not work, it is likely that a weighted body of representatives, where weights reflect biomass extent of critical habitat and intensity of species preferences, could be designed. There would be no democratic voting processes as we know them today, but mechanisms to recognize preferences could clearly be developed, at least for some species. Should animals have rights? If children, women, minorities, poor, or any other formerly disenfranchised segment of our population has them, animals should also.

---

6. The pioneering work of Kagel, Battalio, and others in examining the consistency of animal behavior with hypotheses derived from consumer demand theory is well known. If animals exhibit behavior totally consistent with humans in terms of preference orderings, consumer choices, demand theory, income/leisure tradeoffs, and income distribution based on effort, it seems absolutely incomprehensible to anyone they do not have well-defined utility functions or that such value relationships should not matter. See Kagel, Battalio, Rachlin, and Green (1981); Kagel et al. (1975); Battalio et al. (1977); Battalio et al. (1981).

If economists can discover, through scientific investigations of choices, preference relations (and demand curves) for humans, they certainly can discover them for natural species. The only economic technique that cannot be used to infer implicit prices and choices is the contingent valuation method, which depends upon knowledge and capability of using human language and analysis. The travel cost, time allocation, opportunity cost, and other implicit methods to discover relevant prices and values are directly applicable to at least some natural species. Even the contingent valuation method may have limited application where animals (particularly large mammals) become somewhat conversant in human language, signals, or symbols.

Economists, to date, have concentrated on perfecting techniques to discover human preferences on non-market commodities, some of which are associated with the environment and species. Efforts should also be devoted to discovering natural species preferences over both market and non-market goods, perhaps even including humans.

HUMAN MANIPULATION OF NATURAL SPECIES

Just as some groups of humans have prejudice, contempt, aggression, or disdain for some humans afflicted with minor differences in our species, they associate similar negative beliefs with some classes of animals. Natural species also exhibit similar traits toward each other. Thoroughbreds discriminate against quarter horses when they are maintained in a single herd, even though their genetic differences are very slight. Thus, humans and natural species discriminate, exhibit aggression, and demonstrate other bad social characteristics, values, and traits. Given these general tendencies of both human and natural species, can the set and number of natural species be sustained without some form of external monitoring and even control? Probably not. The first virus humans would most likely eliminate if given a choice is the AIDS virus. Yet, if eliminated, future species would lose their genetic pool, and genetic diversity would be reduced. Humans would also eliminate rattlesnakes, "killer" bees, all disease-carrying vectors, and any other species that harmed humans without being cuddly or sweet. If humans were able to eliminate *all* natural species that harmed them, it almost goes without saying that we would immediately be plagued by a variant of the Malthus starvation model—explosive net population growth would rapidly obliterate food supplies and cause substantial deaths from malnutrition and aggression caused by malnutrition. Thus, the human species, at least in its present form, needs the stability it receives from human predator species.

How successful has our existing socio-economic system been in manipulating and preserving species? A few success stories are obvious. Buffalo are being maintained in slightly increasing numbers on the Great Plains, although mostly in artificial environments or even feedlots. To some degree, the African cheetah, elephant, and lion are being preserved along with several related species. But the number of controlled species (species thought to be subject to endangerment) exceeds 50,000 (see *Controlled Wildlife*, Vol. 2). And the number of actual endangered species probably exceeds 1,500 just in the United States. Most of these species are not cute, cuddly, or majestic, so there is human motivation towards preservation.

I would like to evaluate how well humans have done at manipulating or preserving a few species. The first is a breed of equinus, popularly known as the thoroughbred horse. The English thoroughbred horse evolved from mixing the bloodlines of three Arabian stallions in the early 18th century. These stallions were crossed with native English bloodstock, which were themselves developed by crossing with imported Arabian and barb stock earlier. The thoroughbred breed is relatively unique for large mammals, since there are very accurate breeding records for more than 15 generations, and there are also substantive records of performance.[7] The overriding objective of the more than 250 years of outbreeding and inbreeding was to produce a horse with the speed and stamina to run over a prescribed distance faster than any other horse. In several centuries, the prescribed distance has been reduced by nearly 75% per race, and the number of heats per race has been likewise reduced. Speed has more or less steadily increased while stamina has been significantly reduced. The breed has emphasized speed to the exclusion of almost every other characteristic including: stamina, soundness, respiratory ailments (including "bleeding"), and temperament or disposition. The genetic diversity of the breed has been severely reduced by continuous inbreeding. It is exceedingly rare to find two thoroughbreds without at least one common ancestor within 5 or 6 generations of their pedigrees. The initial genetic stock has not been preserved, so it would be almost impossible to reclaim or restore it.

While genetic diversity has been reduced by prolonged and selective inbreeding, it also has been reduced by economic events and social calamities. The World Wars reduced breeding stock by at least 50% in Western Europe and recent foal crops in the United States have decreased by 40% due to industry market conditions. Thus, genetic diversity can be significantly reduced by market factors that have nothing to do with natural species but are determined by changes in human preferences.

The second species I wish to examine is the black-footed ferret. A small weasel-looking mammal, the black-footed ferret was thought to be extinct in the 1940s or 1950s in the western United States. A small colony (18) was discovered near Meteetse, Wyoming in 1984. The black-footed ferret, while delightful to watch, was a relatively inefficient hunter, incapable of sustaining itself in its largely natural niche. An aggressive scientific program was installed to preserve and sustain the ferret. The original 18 were mated selectively to maximize outcrossing over this small species. Several millions of dollars were

---

7. There is a vast literature on the evolution and breeding of thoroughbred horses and on their genetic diversity and defects. See Lambton and Offen (1987), Longrigg (1972), Cook (1904), Vamplew (1976), Robertson (1964), and Bongianni (1984).

spent on preserving the ferret. However, when it was returned to its natural habitat, the yearly mortality rate was very high, exceeding 90%. Unless substantive innovations are discovered on maintaining the black-footed ferret in a natural environment, it will cost at least $1 million per year to sustain a population of 100–200 ferrets in artificial conditions. Is the black- footed ferret worth it? Preliminary studies of individuals' willingness to pay to preserve the black-footed ferret indicate a value of at $1.00 per month per household, or $12 per year. With approximately 104 million U.S. households, the total willingness-to-pay exceeds $1.2 billion per year, yielding a benefit-cost ratio exceeding 1,000.[8] However, such a comparison is valid only if the black-footed ferret is the only endangered species relevant for social choice. If one gives a list of 1,500 species to the average respondent, the black-footed ferret is typically ignored and overwhelmed by cuter, more cuddly, species, such as baby deer, moose, or rocky mountain sheep. The black-footed ferret should not be preserved if all other endangered species and the federal funds (about $20 million in the United States) available for their preservation are weighed. In the black-footed ferret case, human intervention has maintained genetic diversity associated with this ferret. It may well have reduced genetic diversity overall because the public funds utilized to sustain it were not available for other endangered species.

NATURAL SPECIES MANAGEMENT: A MODEST PROPOSAL

From previous sections, it should be clear that slight tinkering with the economic price structure to restore Pareto-efficient prices is not going to work. More fundamental institutional changes need to be derived to adequately protect species and genetic diversity.

It is a common pronouncement in economics that real property owned by no one in the economy is mismanaged or not efficiently allocated. Another way of saying this is that common property resources or "green access" resources are exploited and overutilized by all who can use them. If there is a dominant tenet of natural resource economics, it is probably this one. Natural species numbers and genetic diversity are examples of common property resources where no one, including specialized natural resource management agencies, manage them properly. With few exceptions, such as hunting season on selected species, there is little or no protection offered to many species. No one requires you to register or buy a permit to eliminate a pesky, common house fly in your home, or a rat, mouse, mosquito, spider, squirrel, moth, or

---

8. Whether the contingent valuation method is yielding a reliable, replicative, and accurate measure of individual preferences for natural species is a matter of much debate in the U.S. Commodities such as preservation of the black-footed ferret seem to be subject to a number of severe problems using a traditional CVM approach. Part of the problem stems from embedding where individuals tend to bid on a concept rather than a commodity. Also, the context of "buying" such a commodity may inherently cause problems. See d'Arge (1988), Rowe, d'Arge, and Brookshire (1980), and Kahneman and Knetsch (1991).

butterfly. The only species that tend to be regulated are those that are larger than humans and that produce immediate financial gain by slaughter.

It seems clear that species cannot regulate themselves to maintain genetic diversity and the preservation of natural species. One economic solution is to define representation for natural species and genetic diversity. This can be done by giving entitlement to the natural species to one group in society. But to whom and on what basis? What characteristics should the group have? In the abstract, we should like the group to have a long-term outlook on both the economy and the natural environment (i.e., the group should be able to not only be aware of short-term economic growth and gain, but also of the preferences of natural species). The group should be decisive, multi-national, and easily identified. The group should be in awe of nature, appreciate biological diversity, and have special benevolence for animals and natural beauty. We might think of a young and uncorrupted conservationist such as John Muir, John Wesley Powell, or Teddy Roosevelt. The group should have a natural affinity for plants and animals and champion weaker species. The group should not search for a monetary numeraire for every environmental decision and every event. Finally, the group should be able to represent the preferences of the human and animal future. Is there such an ordained group within our society who could successfully manage and own all of the natural species?[9] I think there is. (The group is not academics.) It is the world's children. They find unpaid joy in zoos and nature. Typically, they are the only explorers of local ecosystems. Our economic system preys on these interests, ranging from picture books to animal cartoons. They are also natural representatives of the future. Thus, my simple proposal to the problems of preservation of natural species and biological diversity is to give fee simple and complete ownership rights of all natural species to all children, collectively and without reservation.

They collectively would own all undomesticated animals and plants inclusive of genetic pools. They could and would manage them for their own benefit, and hopefully ours. They may deem to develop both local and global institutions for insuring proper management. Or, they may find such adult institutions a waste of time. They may elect to save or eliminate a species by whatever mechanisms they choose. Their institutions could be operated from funding derived from hunting and fishing licenses, access fees, or royalties on genetic structure. They might provide subsidies for preserving certain species as well as penalties for overexploitation. Would it work? I have some faith that it would. The right values would be there and if efficient institutions can be designed to sustain their efforts, then natural species preferences would count and be counted.

---

9. Farber suggests that the failure of sustaining ecosystems arises from "(1) short time horizon, (2) failures in property rights, (3) concentration of economic and political power, (4) immeasureability, and (5) institutional and scientific uncertainty." The correct choice of a represented group should eliminate problems 1, 2, and 3. See Farber (1991).

MAJOR OBSERVATIONS AND CONCLUSIONS

1. If species diversity is maintained, genetic diversity may be reduced. This is due to carrying capacity and other physical limits in numbers and types of species. The biosphere is therefore likely to be subject to the same characteristics of scarcity similar to economies.

2. If the natural environment can be characterized as a compact and convex set and if there is an overriding objective in this system such as maximizing stored energy, then the natural environment (all ecosystems taken together) will have its own price system. That is, it will generate a set of implicit prices that may or may not be consistent with market prices. Pareto efficiency is possible only when both price systems are completely consistent with each other.

3. Since many animals have demonstrated behavior consistent with microeconomic predictions of human behavior, there is almost as much evidence that they have as well-defined utility functions and preferences as humans. Yet these preferences are not considered when decisions are made about them. They are relegated to the status of women, minorities, and children in the 17th and 18th centuries. Their preferences should count and can be discovered utilizing many of the techniques developed recently to value non-market goods, including travel cost, hedonic pricing, and in some instances, even contingent valuation.

4. The problem of maintenance of genetic diversity and natural species is a classic one of inefficient management of common property measures. Since no one owns them, they are misused and overexploited. Is there a group in society who (1) has a special interest in them, (2) has values that encompass future generations' values, (3) is particularly sensitive to non-market values, (4) is naturally compassionate toward the weak and small, (5) can be decisive, and (6) would consider preferences of natural species in making decisions? Is there a group who naturally has the values and ethics to represent natural species interests? There is, and this group is children collectively. Thus, children as a group should be given complete entitlement to all natural species including all genetic resources. Institutions should be designed to allow children to efficiently and fairly manage this extremely important common property resource.

REFERENCES

Arrow, K. J., and G. Debreu. 1954. Equilibrium for a competitive economy. *Econometric* 22: 265–90.

Battalio, R., J. Kagel, and M. Reynolds. 1977. Income distribution in two experimental economics. *Journal of Political Economy* 85 (6): 1259–71.

Battalio, R., L. Green, and J. Kagel. 1981. Income—leisure trade-offs of animal workers. *American Economic Review* 71 (September): 621–32.

Baumol, W. J., and W. E. Oats. 1988. The Theory of Environmental Policy. 2d ed. Cambridge: Cambridge Univ. Press.

Bongianni, M. 1984. Champion Horses: An Illustrated History of Flat Racing, Steeplechasing, and Trotting Races. New York: Bonanza Books.

d'Arge, R. C. 1971. Essay on economic growth and environmental quality. *Swedish Journal of Economics* 73 (March): 27–43.

———. 1990. A Practical Guide to Economic Valuation of the Natural Environment. Rocky Mountain Mineral Law Institute. Proceedings at the 35th Annual Institute, May 1–20. New York: Matthew Bender.

d'Arge, R. C., and K. C. Kogiku. 1973. Economic growth and the environment. *Review of Economic Studies* 40 (January): 61–77.

d'Arge, R. C., R. Rowe, and D. Brookshire. 1980. An experiment in the economic valuation of visibility. *Journal of Environmental Economics and Management* 7 (March): 1–19.

Davis, O., and A. Whinston. 1965. Welfare economics and the theory of second best. *Review of Economic Studies* 22 (January): 113.

Debreu, G. 1959. Theory of Value: An Axiocratic Analysis of Economic Equilibrium. Cowles Foundation. New York: Wiley.

Estes, C., and K. W. Sessions. 1983. Federally Controlled Species. Vol. 2. Lawrence, Kansas: Association of Systematic Collections.

Farber, S. 1991. Local and global incentives for sustainability: failures in economic systems. In Ecological Economics: The Science and Management of Sustainability, ed. R. Constanza. New York: Columbia Univ. Press.

Kagel J. H., R. C. Battalio, L. Green, H. Rachlin, R. Basmaun, and W. Klein. 1975. Experimental studies of consumer demand behavior using laboratory animals. *Economic Inquiry* 13: 22–38.

Kagel, J. H., R. C. Battalio, H. Rachlin, and L. Green. 1981. Demand curves for animal consumers. *Quarterly Journal of Economics* 96 (February): 1–15.

Kahneman D., and J. Knetsch. 1992. Valuing public goods: the purchase of moral satisfaction. *Journal of Environmental Economics and Management* 22: 57–70.

Kneese, A. V., R. U. Ayres, and R. C. d'Arge. 1970. Economics of the Environment: A Materials Balance Approach. Baltimore: John Hopkins Press.

Lambton, A., and K. Offen. 1987. Thoroughbred Style. Topsfield, MA: Salem House.

Longrigg, R. 1972. The History of Horse Racing. London: MacMillian.

Mäler, K-G. 1974. Environmental Economics: A Theoretical Inquiry. Baltimore: Johns Hopkins Press.

Odum, H. T. 1983. Systems Ecology: An Introduction. New York: Wiley.

Page, T. 1991. Sustainability and the problem of valuation. In Ecological Economics: The Science and Management of Sustainability, ed. Robert Costanza. New York: Columbia Univ. Press.

Rico, J. 1879. History of the British Turf: From the Earliest Times to the Present Day. London: Low, Marston, Scarle and Rivington.

Robertson, W. H. P. 1964. History of Thoroughbred Racing in America. New York: Prentice Hall.

Schimmelpfnnig, D. 1991. Long Run Equilibrium in an Economy with a Greenhouse Effect. Econometrics and Economic Theory Paper No. 9067. East Lansing: Michigan State Univ.

Schulze, W. D. 1974. The optimal use of non-renewable resources: the theory of extinction. *Journal of Environmental Economics and Management* 1 (May): 53–72.

Stone, C. 1974. Should Trees Have Standing? Toward Legal Rights For Natural Objects. Las Altos, CA: William Kaufman.

Vamplew, W. 1976. The Turf: A Social and Economic History of Horse Racing. London: Allen Lane.

Van deVerr, D., and C. Pierce, eds. 1986. People, Penguins, and Plastic Trees. Belmont, CA: Wadsworth.

8 INVESTING IN CULTURAL CAPITAL FOR SUSTAINABLE USE OF NATURAL CAPITAL

Fikret Berkes[1] and Carl Folke[2]
[1]*Natural Resources Institute*
University of Manitoba
Winnipeg, Manitoba
R3T 2N2 Canada

[2]*The Beijer International Institute of Ecological Economics*
The Royal Swedish Academy of Sciences
Box 50005
S-104 05 Stockholm
Sweden

[2]*Department of Systems Ecology*
Stockholm University
S-106 91 Stockholm
Sweden

ABSTRACT

The importance of natural capital and the relationships between natural capital and human capital are of fundamental interest in ecological economics. But consideration of these two kinds of capital exclusively falls short of providing the essential elements for the analysis of sustainability. A more complete conceptualization of the interdependency of the economy and the environment requires attention to social, cultural, and political systems as well. We use the term "cultural capital" to refer to factors that provide human societies with the means and adaptations to deal with the natural environment. Cultural capital, as used here, includes factors such as social/political institutions, environmental ethics (world view), and traditional ecological knowledge in a society. The three types of capital are closely interrelated. Natural capital is the basis for cultural capital. Human-made capital is generated by an interaction between natural and cultural capital. Cultural capital will determine how a society uses natural capital to create human-made capital. Aspects of cultural capital, such as institutions involved in the governance of resource use and the environmental world view, are crucial for the potential of a society to develop sustainable relations with its natural environment.

INTRODUCTION

There has been considerable conceptual progress towards achieving ecological economics by combining conventional economics with conventional ecology (Costanza et al. 1991). For example, the fundamental importance of life-support functions of the natural environment to economic development and sustainability has gained recognition in economics as well as in ecology. Ecological economists have distinguished between *natural capital* and *human-made capital* and have come to regard human-made capital and natural capital as fundamentally complementary. Natural capital and its derived goods and services have been considered the preconditions or the basis for economic development. Ecological economists recognize that it is not possible for human ingenuity to create human-made capital without support from natural capital (Daly 1990).

However, it is not possible to analyze sustainability by focusing only on natural capital and human-made capital. For a more complete conceptualization of economy-environment relations, a third dimension is needed, here referred to as *cultural capital*. From a systems perspective, the three types of capital are strongly interrelated and need to be considered together as the essential elements for the analysis of sustainability.

The three kinds of capital, their interrelations, and the systems view of some of these relationships are described initially. Then we explore how self-regulation of social systems (as described in the literature of common property resources) is achieved, and how common action may be developed and sustained. We compare the fields of common property and ecological economics to demonstrate that self-regulatory systems are of central concern in both. Finally, we touch upon a variety of themes currently discussed in a number of fields, in search of ways of enhancing cultural capital towards sustainability.

THREE TYPES OF CAPITAL

Human-made capital is capital generated through economic activity, through human ingenuity and technological change—the produced means of production. This is a common definition of capital in economic textbooks.

Natural capital consists of three major components (1) non-renewable resources, such as oil and minerals, that are extracted from ecosystems; (2) renewable resources, such as fish, wood, and drinking water that are produced and maintained by the processes and functions of ecosystems; and (3) environmental services, such as maintenance of the quality of the atmosphere, climate, operation of the hydrological cycle including flood controls and drinking water supply, waste assimilation, recycling of nutrients, generation of soils, pollination of crops, provision of food from the sea, and the maintenance of a vast genetic library. These crucial services are generated and sustained by the work of ecosystems (Odum 1975; Folke 1991).

Ecological economists argue that a minimum condition for sustainability is to maintain the total natural capital stock at or above the current level (Daly 1990). An operational definition of this condition for sustainability means that:

- the human scale must be limited within the carrying capacity of the remaining natural capital,

- technological progress should be efficiency-increasing rather than throughput-increasing,

- harvesting rates of renewable natural resources should not exceed regeneration rates,

- waste emissions should not exceed the assimilative capacity of the environment, and

- non-renewable resources should be exploited, but at a rate equal to the creation of renewable substitutes.

The multifunctionality of natural capital needs to be acknowledged in this respect, including its role as integrated life-support systems. Only through maintenance of an integrated, functional ecosystem can each environmental good and service be assured: such goods and services cannot be managed one-by-one as independent commodities.

Humans reduce natural capital because of their capability to invent technical substitutes for those functions, generally forgetting that such substitutes require environmental goods and services from other ecosystems (substituting natural capital in one place requires natural capital from elsewhere) and because of imperfect understanding of the life-support functions on which society depends. It is the significance of this understanding that we discuss in terms of cultural capital.

Cultural capital refers to factors that provide human societies with the means and adaptations to deal with the natural environment and to actively modify it. We use the term "culture" in the general anthropological sense of a set of rules for a society, recognizing the existence of many distinct societies (and also of many different definitions of culture among anthropologists). Culture implies commonality, providing a basis for collective action within that group. But, as documented by anthropologists, different societies have developed a variety of ways to deal with the environment (Bennett 1976); the concept of nature, for example, is culture-specific (Hjort af Ornäs and Svedin 1992). The diversity of ways to deal with the environment is a significant part of cultural capital, and perhaps as important to conserve as biological diversity (Gadgil 1987).

Also included in the concept of cultural capital are people's views of the natural world and the universe, and the *source* of these values or cosmology (Skolimowski 1981); environmental philosophy, values, and ethics, including religion (Leopold 1949; Naess 1989); and local and personal knowledge of the environment, including traditional ecological knowledge (Johannes 1989). An important dimension is the organization of human societies by the evolution of various kinds of resource management institutions, defined as "the conventions that societies establish to define their members' relationships to resources, translate interests in resources into claims, and claims into property rights" (Gibbs and Bromley 1989, p. 22). These institutions, both formal and informal, governmental and non-governmental, dealing with the use of resources or any aspect of the natural environment, are part of cultural capital (Ostrom 1990).

We have used "cultural capital" for the lack of a better term and at the risk of criticism for combining complex concepts. The "capital" label in both cultural capital and natural capital should be thought of not as a reduction of these fields into economic terminology, but as a short-hand to allow the exploration of a systems approach with three fundamentally different but nevertheless interrelated clusters of variables.

As an alternative term, we considered using "adaptive capital" to emphasize that we are referring to all of these factors important to ecological economics from an evolutionary—mainly culturally evolutionary—sense. But the term would have been inadequate to capture the systems perspective that we present here, in which organisms not only adapt to, but also actively modify their environment. This concept, referred to as *autopoiesis* (Varela et al. 1974), is a key to a systems perspective, because it emphasizes a cyclic relationship necessary for the analysis of sustainability. In this cyclic relationship the system boundaries and the components necessary for the development of the system are the result of the system's own actions, in which the components mutually shape each other. Recursive processes of this type are necessary for the evolution of all ordered systems (Jantsch 1980; Günther and Folke 1993).

Various authors have grappled with the challenge of finding a term that would capture a range of social-cultural variables similar to those that concern us here. Coleman (1990, 300–21) used "social capital" to refer to features of social organization, such as trust, norms, and networks, which can facilitate coordinated action. The term "social capital" is used by many social scientists to refer to richness of social organization (Ostrom 1990, 190). Ostrom also uses the term "institutional capital" to mean the supply of organizational ability and structures, literally the "capital" of institutions that a society has at its disposal (Ostrom 1990, 190, 211). Presumably, institutional capital would be a subset of social capital.

Other authors have explored other terms in search of the missing variable in ecological economics. Bormann and Kellert (1991), for example, have focused on ethics as the relevant variable, and considered "ecology, economics and ethics as parts of a whole, of an interconnected circle." Daly (1980) has used the term "moral" to refer to the important social dimension. Daly and Cobb (1989) have distinguished between moral capital and physical capital in reference to ethics and community. But ethics or "moral capital" as terms are too limited to cover some of the other essential social considerations such as institutions. Economists generally use the term "human capital" for the stock of education, skills, culture, and knowledge stored in human beings (see Becker 1993 for a review; see also Cleveland this volume). "Social-overhead capital" has been used as a concept in economic analysis, and includes natural capital, social infrastructure, and institutional capital (Uzawa 1974, 1992). Costanza and Daly (1992) used the labels natural capital, human capital, and manufactured capital to correspond roughly to the traditional economic factors of production of land, labor, and capital, and lumped human capital and manufactured capital together as human-made capital.

No doubt all of the above terms, including cultural capital, are inadequate. But it is difficult, if not impossible, to find a term that would sufficiently cover all aspects of the human societal/ethical/political dimension. These areas fall into a number of different fields of social sciences and humanities (including philosophy, theology, anthropology, sociology, geography, and political science); there is no common technical literature that binds them. Yet from a systems point of view, they are clearly related as they all pertain to adaptations dealing with natural systems of which human systems are a part. Together they shape the way society interacts with its environment, and defines and uses natural capital.

There exists a fundamental interrelation between natural capital, human-made capital, and cultural capital. In simplest terms, cultural capital is the *interface* between natural capital and human-made capital (Figure 8.1). Our world view, values, knowledge, and institutions shape the way in which we treat the environment. If ecological economics "addresses the relationships between ecosystems and economic systems in the broadest sense" (Costanza 1989, 1), then this interface is part of the proper field of study of ecological economics. A more complex view of the interrelationships among the three kinds of capital is provided in Figure 8.2. Natural capital is the basis, the precondition, for cultural capital. Human-made capital is generated by an interaction between natural and cultural capital. Human-made capital, in turn, may cause an alteration of cultural capital. Technologies (tools, skills, and know-how) which mask the society's dependence on natural capital encourage people to think that they are above nature. The more extensive this change, the more of similar type of technologies will be developed and the more impacts on natural capital there will be. Positive feedbacks between cultural capital, and human-made capital are established which enhance this trend. There will be resource depletion and environmental degradation to feed an industrial society that requires ever-increasing amounts of raw materials, and that generates ever-increasing amounts of waste. Therefore, cultural capital plays an important role in how we use natural capital to "create" human-made capital. Thus, human-made capital is never value-neutral, but a product of evolving cultural values and norms. Technologies that we develop are not simply tools

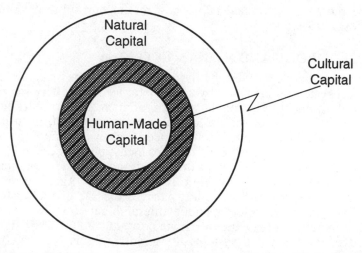

Figure 8.1. Cultural capital is the interface between natural capital and human-made capital.

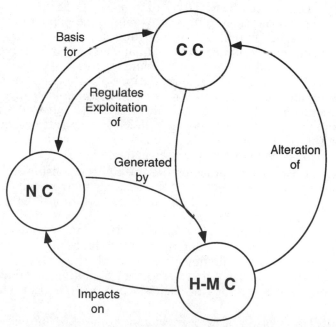

Figure 8.2. First-order interrelationships among natural capital (NC), human-made capital (H-MC) and cultural capital (CC).

that can be put to good or bad use—they reflect our cultural values, world view, and institutions.

SYSTEMS ECOLOGY AND HUMAN ECOLOGY

First, consider the systems view of the environment. The structure and function of the ecosystem is sustained by synergistic feedbacks between organisms and their environment. The physical environment puts constraints on the growth and development of the biological subsystem which, in turn, actively modifies its physical environment to enhance its chance of survival. The focus of the evolutionary perspective is on the incessant process by which organisms adapt to and co-evolve with their environment. The ecological system as a whole is seen to be in a dynamic process of self-organization and self-maintenance (homeostasis). Solar energy drives the use of matter for self-organization, and complex, interdependent hierarchical structures evolve. It is this self-organizing ability, the resilience, organization, and vigor of the ecosystem that generates and sustains the goods and services which form the necessary material basis for human societies.

Now consider the systems view of the human/environmental relationship. The structure and function of the ecosystem is sustained by synergistic feedbacks between human societies and their environment. The physical and biological environment places basic physical constraints on the growth and development of the human subsystem. For example, the population growth in a certain area would be limited by the carrying capacity of the environment (see also Ehrlich this volume). The human subsystem, in turn, actively modifies its physical and biological environment: carrying capacity of an area may be decreased through the degradation of life-support systems, or increased by organizing differently or using new technology that works with the environment (Mitsch and Jörgensen 1989). The self-organizing ability and homeostasis of the ecosystem is paralleled by the self-organizing ability and homeostasis of the human subsystem. These adaptations, in turn, shape the way in which society defines and uses natural capital.

The systems view of the human/environmental relationship in human ecology is not nearly as well-developed as the systems view in ecology, but there are some notable works (Jantsch 1980, chapter 9; Moran 1990). One of the more helpful examples for our purposes of systems view applications in human ecology is the concept of *co-evolution*. Human/environmental interactions may be viewed as a co-evolutionary interrelationship in which the two sides change one another continuously by mutual feedback. This is the logical extension into the human subsystem of an evolutionary concept that has been in common use in ecology at least since the 1960s (Ehrlich and Raven 1964).

Historically, the world can be seen as consisting of a "mosaic of co-evolving social and ecological systems" (Norgaard 1987). In each part of the mosaic, the human subsystem selected for species that fulfilled its needs, and itself evolved under the selective pressure of having to use natural capital sustainably. "Co-evolution is a local process," Norgaard (1987) pointed out, "specific to local cultural knowledge, technology, and social organization." Thus, these local human subsystems are a significant starting point for a discussion of coupled biological/cultural evolution in ecological economics.

A major impediment to the development of a full systems view of human/environmental relationships is our heritage of a reductionistic science world view which excludes humans from the system to be studied (Clark 1989). Many ecologists have been reluctant to extend the notion of mutually interactive relations to the study of human ecology, concentrating instead on other species. Other environmental scientists merely study the impacts of humans on the environment, a one-way relationship, effectively treating human systems as exogenous to the ecosystem. Examples of studies of two-way relationships consistent with the systems view may be found in the emerging field of common property resources.

COMMON PROPERTY SYSTEMS AND INSTITUTIONS

The literature on common property rights is of special interest to ecological economics because it deals with the success or failure of self-regulatory systems in the sustainable use of resources across many different cultures and geographic areas (McCay and Acheson 1987; Berkes 1989; Ostrom 1990; Bromley 1992). The current literature builds, on the one hand, on the observation by Hardin (1968) and others of the divergence between individual and collective interests, and the tendency of individuals to free-ride and to behave opportunistically. On the other hand, it builds on the insight of social scientists (mainly anthropologists) such as Geertz (1963), Ingold (1980), and Netting (1981) that many societies evolve institutions to reconcile individual interests with the collective interest. Much of the literature has appeared since 1985—the International Association for the Study of Common Property (IASCP) was established in 1989, with annual meetings after 1990. The newsletter of the IASCP network, the *Common Property Resource Digest*, had a circulation of some 4,000 as of 1992. What accounts for the phenomenal development of interest in common property resources?

Like the International Society for Ecological Economics (ISEE), the IASCP is interdisciplinary and represents new alliances amongst old players from established disciplines. Concentration on property rights, resource management institutions, and social self-regulating mechanisms has allowed specialists in fisheries, water resources, forestry, and land resources to discover new approaches across old disciplinary boundaries. It has provided a coherent, practical approach to problem solving. The parallels between ISEE and IASCP do not end there. These two international, interdisciplinary alliances also show convergent development towards the objective of sustainability.

Figure 8.3 outlines the development of the fields of common property and ecological economics. The starting points are quite different but not mutually

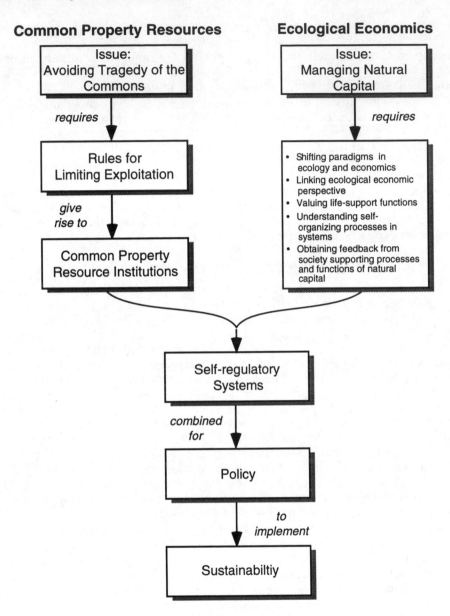

Figure 8.3.   Convergent developments in the fields of Common Property Resources and Ecological Economics.

exclusive. In the case of common property, the emphasis is on avoiding or solving the "tragedy of the commons" (Hardin 1968). In the case of ecological economics, the emphasis is on conserving and valuing natural capital. The central issue in the case of common property, requires the development of rules (and their enforcement) for limiting exploitation to levels that are sustainable. This in turn requires attention to resource management institutions, both governmental and other. In the case of ecological economics, the central issue requires a paradigm shift in discipline boundaries to allow for the development of a linked ecology-economics perspective where social systems are regarded as subsystems of the overall ecosphere. Costanza et al. (1991) have provided a detailed comparison of ecological economics with conventional ecology and economics.

The common property literature deals directly with the local resource base. The usual emphasis is on the local community of resource users and their informal institutions. A major finding of the common property literature is that much of effective management occurs at the local level and tends to be community-based, both for rule making and rule enforcement. By contrast, the ecological economics literature is explicitly not concerned with community-level processes. These processes are often masked by activities at the national, regional, and international levels, such as economic growth, development projects, and pollution.

Whatever the differences in emphasis and scale, common property and ecological economics fields converge because they are both concerned with self-organization and self-regulatory systems. In the common property area, negative feedback to keep resource exploitation within the limits of sustainability is provided by communal institutions, government institutions, or private property regimes (or co-management regimes representing various combinations of the three). In the case of ecological economics, the study of self-regulation of ecosystems, especially with regard to life-support functions, is a priority area. The study of market processes and incentives in the aid of sustainability is another major area. Self-regulatory systems in the use of natural capital do not seem to receive as much attention. In terms of scale, the emphasis appears (implicitly) at the national and international levels, and on specific natural capital problems such as those involved in acid rain, eutrophication, and particular ecosystems such as wetlands.

Perhaps the main lesson from the common property literature is that, given a resource management problem, a group of people often organize themselves to deal with it in a manner similar to the formation of a "bucket brigade" to put out a fire in a rural neighborhood. Communal management of resources also has the advantage of reducing transaction costs. The evolution of rules and self-regulatory mechanisms within the group has adaptive significance for sustainability and survival, and can arise over a time period of as little as ten years (Berkes 1986), and may endure over centuries (Ostrom 1990). Conceptually, it is not surprising to find self-organizing capabilities in social systems, similar to those in ecosystems. In both ecological economics and common property frameworks, human systems are subsystems of ecosystems, and if our premises are correct, should follow the same laws of general systems theory (von Bertalanffy 1968).

Many of the earlier studies of common property systems involved isolated communities. More recent work has focused largely on common property use in contemporary settings, such as irrigation water use in India and coastal fisheries in Turkey (Bromley 1992). Of particular interest from a sustainability viewpoint are ultra-stable systems. A number of long-enduring, self-organized, and self-governed common property institutions have been analyzed by Ostrom (1990). Examples include communal land tenure in high mountain meadows and forests in Törbel, Switzerland; common land management in Hirano and area villages in Japan; and the *huerta* irrigation system in the Valencia area and elsewhere in Spain. From these and other cases, Ostrom (1990) has derived a set of principles important for sustainable resource management outcomes.

Many simpler common property systems involve on the order of a hundred users. More complex systems, organized hierarchically, function with thousands. For example, the number of irrigators in Spanish *huertas* may be some 13,000. Such systems seem to start with small numbers of local resource users which later federate into larger units. In the case of irrigation systems, users may be organized hierarchically from the smallest canals to the main branch of the river, so that the system that eventually evolves may be four layers deep (Ostrom 1990, 1989). Can more layers be added? What are the prospects for a bottom-up hierarchical organization that regulates resource use all the way to the global level?

Common property institutions run into jurisdictional problems, user-group conflicts, and the barrier of national laws and regulations that often contradict them. Thus, for practical purposes, local rules are not likely to extend to international or even national levels. What is more promising, however, is the possibility of making consistent rules simultaneously at various levels of organization. The main impediment is the difficulty of developing common action involving different cultural groups and nation states in a world in which cultural differences are often the cause of civil strife.

Much of the recent commons literature focuses on local commons. Global commons issues have nevertheless started to receive attention (Bromley 1991; Keohane et al. 1993). The experience from commons research at the local level seems to have direct implications for regional and international resource and environmental management issues (Ostrom 1990). In particular, design principles for successful local institutions for managing the commons have turned out to be very similar to those for the management of international commons. This finding has stimulated comparative work among two groups of specialists—those in the area of local commons and those in international relations/global commons fields (Keohane et al. 1993).

ENHANCING CULTURAL CAPITAL

It is likely that approaching sustainability at local, regional, or global levels will require attention to factors that provide societies with the means and adaptations to develop common action. That is, it will be necessary to "invest in cultural capital" as well as in natural capital. Figure 8.4 summarizes some of the themes in the current literature which we consider relevant for conserving and enhancing cultural capital.

Cultural Diversity

Diverse cultures hold the key not only to diverse adaptations to the environment, but also to a diversity of world views that underpin these adaptations. The "dominion over nature" world view, emerging in part from the values and perspectives of the Industrial Revolution, is best geared for the efficient exploitation of resources as if they were boundless, but not for the sustainable use of natural capital (Gadgil and Berkes 1991). With only a limited number of dominant world views, the chances of finding sustainable patterns will be diminished. Thus, as Gadgil (1987) observed, human cultural diversity and biological diversity go hand-in-hand as prerequisites for long-term sustainability. Cultural diversity may be considered a pool of social system adaptations spanning many millenia, a "library" from which a new science of sustainable resource management can borrow. Cultural diversity is no doubt important for its own sake as well, but our emphasis here is on cultural diversity as fundamental to adaptations and knowledge to enable the sustainable use of the environment.

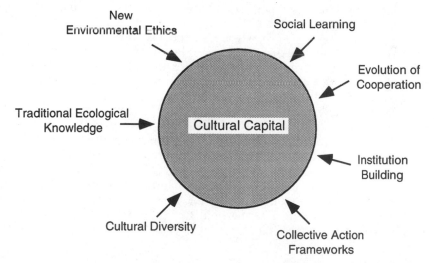

Figure 8.4. Conserving and enhancing cultural capital towards self-organization for sustainability.

Traditional Ecological Knowledge

Indigenous knowledge has received a great deal of attention in the fields of agriculture, pharmacology, and ethnobotany. More recently, the importance of indigenous ecological knowledge has been recognized for the management of tropical forest, arctic and subarctic, marine coastal, mountain, and dryland ecosystems (Johannes 1989; Posey and Balee 1989; IUCN/UNEP/WWF 1991). Traditional systems of communal resource use have been the main ways in which resources have been managed in human history. If these traditional systems had not been, for the most part, sustainable, we would not have any resources left today to speak of. Indigenous knowledge has proved useful for supplementing scientific information, assisting development planning and impact assessment studies, and helping design resource management systems. For example, the aquaculture and integrated farming systems of ancient China, Egypt, and Hawaii show remarkable diversity and ecological sophistication (Ruddle and Zhong 1988; Costa-Pierce 1987). The experience with integrated farming and aquaculture in ancient cultures provide clues for the identification and management of feedbacks for sustainable modern aquaculture (Folke and Kautsky 1992).

Institution Building

One of the reasons for the degradation of natural resources is the degradation of institutions that once provided for their use. For example, the destruction of Hawaii's traditional land tenure system by colonialists in 1848 precipitated the decline of the traditional ecosystem-based land use system (*ahupua'a*), with its integrated farming and aquaculture systems. Decline of fishpond complexes was such that some 98%–99% of the earlier production was eventually lost (Costa-Pierce 1987). How can the lost production be recovered? Traditional production systems may be replaced by new production systems and traditional institutions by the building of new institutions. Or alternatively, institution building may use elements of traditional institutions if these are still relevant for the solution of resource management problems in hand, as in the case of wildlife co-management in James Bay (Berkes et al. 1991). A current institution-building issue, and a very major one, concerns Eastern European countries. Central state mismanagement of resources, coupled with the virtual elimination of private property and communal property regimes, have left these countries with an unenviable task of institution building. In general, institution building is relevant to natural capital conservation from the local to the international scale, and includes such challenges as building international institutions to deal effectively with global atmospheric change.

Collective Action

When a number of individuals (or countries) have a common or collective interest, unorganized individual action will often be insufficient to advance that interest. The problem is one of organizing—how to get the independent actors to adopt coordinated strategies to obtain higher benefits for all. This is the key problem behind the theory of collective action (Olson 1965; Sandler 1992), and it also underpins a variety of social traps (circumstances in which rational individual choices are inconsistent with long-term or collective interests), including the "tragedy of the commons" (Costanza 1987). Users of common property resources such as fish, forests, and water are interdependent on one another, as are countries fighting international pollution problems. Much of the observed self-organization among common property users may be explained in terms of collective action frameworks (Wade 1987). The concept of *interdependency*, with all the feedback loops that it implies, has been in widespread ecological use for decades. It is equally applicable in human ecology and is the basis for the development of common property institutions at all levels from the local to the international.

Evolution of Cooperation

The dilemma in the "tragedy of the commons" is that there is no mechanism in Hardin's (1968) version of the theory to reconcile rational individual interest with the collective interest. In reality, the recent common property literature indicates the prevalence of social self-regulatory mechanisms in constraining individual behavior (Feeny et al. 1990; Bromley 1992). There are sophisticated and ingenious social controls, enforced by mutual consent of community members, on interdependent resource users such as farmers along an irrigation network; these have been documented in great detail by such scholars as Wade (1987). Cooperation confers selective advantage, at the ultimate (or evolutionary) level of causation, to a society of interdependent individuals. But how does cooperation evolve in the first place? Various mechanisms are potentially available, and a particularly promising model is Axelrod's (1984) game-theory approach, based on reciprocity and a repeated Prisoner's Dilemma game; it is used by both common property and international relations/global commons theorists (Keohane et al. 1993). In addition, the Folk-theorem of game theory (Friedman 1986) has unraveled how individuals *in their own interest* may evolve a variety of social norms that will support sustainable uses of resources. These arrangements are fragile, however, and their robustness very much depends on the condition under which these systems evolve (Dasgupta and Mäler 1991).

FINDING A PROSPEROUS WAY DOWN TO SUSTAINABILITY

The present high levels of seemingly limitless natural capital use is a historically unique phenomenon that is a product of the Industrial Revolution. In the context of the history of the human species, it is a very thin slice of time. For most of their existence, human societies have lived with biophysical limits. A generalized picture of natural capital use over recent human history may indicate more-or-less sustainable levels of utilization of various resources until the Industrial Revolution, followed by sharply increasing use, peaking at the global scale perhaps sometime in the 20th or 21st century. Most ecologists and many ecological economists would probably argue that, on the whole, we have overshot the sustainability levels of our resource base (Figure 8.5).

Possibilities for resource substitutability and technological change notwithstanding, most would agree that physical growth cannot continue indefinitely. If so, we are nevertheless uncertain as to (1) whether we have reached the peak of the curve, (2) where the limits really are, and (3) whether the peak will be followed by a collapse, and (4) how far such a collapse might extend. Since not too many people would like to revert to a pre-Industrial Revolution way of life, there is obvious motivation in finding "a prosperous way down to sustainability" (Odum 1988)—that is, along the lines of A instead of B or C in Figure 8.5.

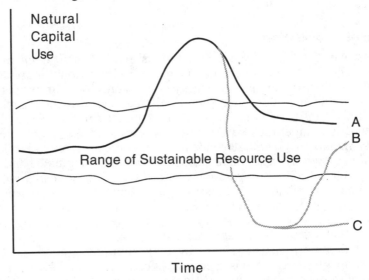

Figure 8.5. Natural capital: depletion or sustainable use? What level of sustainability?

Through discoveries of new energy sources (mainly fossil fuels) throughout the industrial era, which made it possible to exploit minerals and other non-renewable resources, the expansion of the human sub-system accelerated. The cultural capital components of accumulated knowledge and human ingenuity found these new sources of natural capital which released the direct constraints set by the natural environment. Throughput production and consumption behaviors were made possible during this rapid and creative expansion phase because of this additional (but non-renewable) natural capital.

Many systems developed during the industrial era are, at the current level of human population and activity, unsustainable and inefficient from a biophysical perspective. They are throughput systems which require large amounts of resource inputs often imported from distant areas. At the same time they generate waste and pollutants concentrated in time and space. Throughput systems lack feedbacks for a sustainable use of resources and the environment. They have not been developed for integration with the processes of ecosystems on which they ultimately depend. In fact, the way these "extra" resources are used masks the degradation of natural capital, both in biophysical reality and in human perception.

Intensive salmon farming is an example of such systems. The salmon are densely reared in cages, which requires huge imports of feed pellets continuously "pumped" into the system, as well as inputs and support from an industrial infrastructure. The salmon culture would have no chance of surviving without these inputs. In addition, large ecosystem areas are needed to produce the salmon food. In fact, that ecosystem area is about 50,000 times larger than the surface area of the cages, or 1 km² per tonne of salmon produced. The concentration of large quantities of imported resources in a limited area generates much waste, which creates severe environmental impacts that threaten the survival of the industry itself (Folke and Kautsky 1989). If such an industry expands (and this is equally true for other industries behaving in a similar manner), the limits to growth will appear that much earlier.

A necessary condition for sustainability is that the functions of natural capital be supported, rather than disrupted, by feedbacks from society. In the case of salmon farming, creating a more diverse system by managing feedbacks between the farm and its surrounding ecosystem would turn waste into new resources, such as seaweeds and mussels (Folke and Kautsky 1992). Such "new" resources would serve as valuable products, both at the salmon farm and in society. Recycling of waste into new resources would also minimize impacts on the environment, and considerably reduce the system's dependence on inputs from large ecosystem areas. The culturing system would be much more self-sufficient. Such approaches, referred to as ecological engineering (Mitsch and Jörgensen 1989), would build the capacity for the management system to work in synergy with ecosystem processes. As well, reducing the spatial scale of the operation makes it more suitable for communal control and management.

In a natural system, feedbacks from the environment direct the system towards restructuring resource use from a throughput behavior towards a more complex and sustainable use pattern. Quantitative growth is replaced by qualitative improvements, simply because of necessity. We would argue that many of the self-regulatory social and behavioral patterns that exist in local communities have evolved in a similar way. In addition, the increased attention to sustainable development at the national and international levels is a reflection of environmental feedbacks as limits are approached. The question is whether humanity will be able to respond to these feedbacks rapidly enough to find and adopt sustainable development pathways. Successful transition requires creating conditions conducive for such pathways, under which we will deal with three topics.

Ethics of Sustainability

The way in which resources and environment are managed depends on our values. As it is clear enough that we have not been managing our natural capital sustainably over the last two centuries or so, some of our values need to be modified. The current Western world view of the environment has a complex background rooted in our religious, scientific, and industrial history (Clark 1989; Skolimowski 1981). We need to restore a semblance of the "community of beings" world view of ancient pantheistic traditions; this may be useful in curing widespread alienation from nature. Everenden (1985) has suggested that humans are akin to aliens from space because they, in most Western cultures, have a self-identity distinct from the world around them. The root of this aberration, according to Everenden, is our cultural emphasis on objectivity. Skolimowski (1981) has argued further that our cosmology is based far too much on empiricism and scientism. It is too mechanistic and analytic; it is not sufficiently based on humanistic notions and morality toward nature (Skolimowski 1981). Developing an ethic of sustainability, taking into account these considerations and re-integrating people with nature, is a challenge for using our cultural capital creatively.

Hierarchy of Systems

The issue of the speed of change in values leading to sustainability is obviously crucial. Some authors have pointed out that natural systems may be seen as consisting of a hierarchy of nested subsystems with a two-way flow between levels (Norton 1990; Günther and Folke 1993). The discrete functional components of the ecosystem may be operating at different scales. The higher levels may be operating at a time scale that is slower than that of the lower levels. Norton (1990) analyzed Leopold's (1949) land ethic and interpreted his *Thinking Like a Mountain* as a metaphor for thinking in larger time frames: "He recognized that human individuals consider their actions in short frames of time whereas the mountain—the larger ecological community—must 'think' in the longer frames of ecological time" (Norton 1990).

Little is known about effective time scales in social-ethical change in this context and how they relate to time scales in natural systems. Social systems, like natural systems, *are* organized hierarchically. However, the potential for change at the higher levels (i.e., national and international) is *not* necessarily slower than that at the lower levels (i.e., community). Given the global media coverage and electronic communications of our age, there is great potential for rapid change at the highest levels. (Effective political action at the global environmental level, however, is yet another story). The image of the "global electronic village" has been with us for the last few decades. However, this term conveys the sense of a mass society alienated from nature (*gesellschaft* rather than *gemeinschaft*). The image we would prefer is that of a "community of communities" (Daly and Cobb 1989).

Social Learning

A concept that is difficult to define precisely, *social learning* carries the sense of a "self-educating community" (Milbrath 1989, 1988). It refers to learning that takes place at the level of the population or society, rather than at the individual level. It is a learning process at an altogether different time scale; it carries great promise as a way in which cultural capital can be put to work for sustainability. The classical cases of social learning in sociobiology concerned the spread of learned adaptive behavior. The general finding was that the adaptive behavior at first spread very slowly, gathering momentum as it spread to a certain critical number of individuals. When that critical threshold was reached, however, the behavior was learned very quickly— explosively—by the rest of the population.

Applied to human societies, analogies may be found with regard to smoking and the avoidance of cholesterol-rich food. The spread of individual behaviors consistent with sustainability principles (e.g., family planning, recycling waste, sorting of household toxic substances, avoidance of pesticides in the garden, bicycling to work) may follow similar pathways. But these individual behaviors may not be sufficient in themselves to bring about sustainable society. Community and corporate-level behavioral change may also be necessary. There are national-level "behaviors" as well, such as assessing the potential impact of development projects before they are built, encouraging the development of energy policies based on renewable resources, and natural capital accounting. These also seem to be spreading but have not yet reached the "take-off" stage internationally. It may be suggested that incentives for social learning towards sustainability may be provided by episodic events such as the Chernobyl accident, the Rhine chemical spill, the death of seals in the North Sea—shocks to the value system to stimulate change. Furthermore, the behavior of critical nodes in the social system such as media, international banks, multinational enterprises and others, serve an important role in the process of social learning.

CONCLUSIONS

There are no pristine environments on earth untouched by human influence. Even Antarctica is affected by pollutant fallout and ozone-layer thinning. The positive approach for ecological economics is not to bemoan the human influence on the natural environment, but to adopt a systems view of human-environment relationships to trigger measures that would stimulate social, as well as ecological, self-regulatory patterns for sustainability. Policies of sustainability need to recognize the importance of *institutions*, at the various levels in the "community of communities," and the lessons available from the common property literature which deals with the organization of sustainability across various cultures, resource types, and geographical areas.

Institution-building, collective action, cooperation, and social learning towards a new environmental ethic are some of the ways in which social self-organization may help us adapt rapidly enough to meet the constraints of sustainability. We need institutional structures which are adaptable. Cultural diversity and traditional ecological knowledge are part of the cultural capital into which society needs to invest to provide the raw material for the process of sustainable development. Successful transition to sustainability requires building the capacity for the environmental/socio-political system to change. Sustainable development is a continuous process and not a state; it is never achieved once and for all, but only approached. As such, it can only be reinforced, not attained. Sustainable development calls for maintenance of the dynamic capacity to respond adaptively, which is a property of all successful species and societies. It is not meaningful to measure the absolute sustainability of a society at any point in time. "Our concern should be more with basic natural and social processes, than with the particular forms those processes take at any time" (Robinson et al. 1990).

We need to learn to live with uncertainties and surprises, and be prepared for them. We can use the great creative activity of the current energy-rich world and the pervasive information network that we have developed to find "a prosperous way down" to sustainable steady-state societies (Odum 1988). To rethink and reconstruct a new science that is better adapted to deal with the interdependencies among natural capital, human-made capital, and cultural capital, the systems perspective will be of great value to the field of ecological economics.

ACKNOWLEDGMENTS

This chapter is an expanded version of a commentary published in *Ecological Economics*. We thank Garrett Hardin, D.C. Lee and an anonymous referee for useful comments on the commentary. We are indebted to many colleagues who commented on the ISEE Conference version of the article, in particular to Anders Hjort af Ornäs and an anonymous reviewer who acted as referees for the chapter. Berkes' work was supported by the Social Sciences and Humanities Research Council of Canada, and Folke's, in part, by the Swedish Council for Forestry and Agricultural Research (SJFR).

REFERENCES

Axelrod, R. 1984. The Evolution of Cooperation. New York: Blackwell.

Becker, G. S. 1993. The Economic Way of Looking at Life. Les Prix Nobel. The Nobel Prizes 1992. Stockholm: The Nobel Foundation and the Royal Swedish Academy of Sciences.

Bennett, J. W. 1976. The Ecological Transition: Cultural Anthropology and Human Adaptation. Oxford: Pergamon.

Berkes, F. 1986. Local-level management and the commons problem: a comparative study of Turkish coastal fisheries. *Marine Policy* 10: 215–29.

———, ed. 1989. Common Property Resources: Ecology and Community-Based Sustainable Development. London: Belhaven.

Berkes, F., P. George, and R. J. Preston. 1991. Co-management: the evolution in theory and practice of the joint administration of living resources. *Alternatives* 18 (2): 12–18.

Bormann, F. H., and S. R. Kellert. 1991. Ecology, Economics, Ethics: The Broken Circle. New Haven: Yale Univ. Press.

Bromley, D. W. 1991. The Law, Agency and Global Climate Change. Report to the U.S. Agency for International Development. Washington, DC: U.S. Agency for International Development.

———, ed. 1992. Making the Commons Work: Theory, Practice and Policy. San Francisco: Institute for Contemporary Studies.

Clark, M. E. 1989. Ariadne's Thread: The Search for New Modes of Thinking. New York: St. Martin's Press.

Cleveland, C. J. 1994. Re-allocating work between human and natural capital in agriculture. In Investing in Natural Capital: the Ecological Economics Approach to Sustainability, eds. AM. Jansson, M. Hammer, C. Folke and R. Costanza. Washington, DC: Island Press.

Coleman, J. S. 1990. Foundations of Social Theory: Cambridge, MA: Harvard Univ. Press.

Costa-Pierce, B. A. 1987. Aquaculture in ancient Hawaii. *BioScience* 37: 320–31.

Costanza, R. 1987. Social traps and environmental policy. *BioScience* 37: 407–12.

———. 1989. What is ecological economics? *Ecological Economics* 1: 1–7.

Costanza, R., and H. Daly. 1992. Natural capital and sustainable development. *Conservation Biology* 6: 37–46.

Costanza, R., H. Daly, and J. A. Bartholomew. 1991. Goals, agenda, and policy recommendations for ecological economics. In Ecological Economics: The Science and Management of Sustainability, ed. R. Costanza. New York: Columbia Univ. Press.

Daly, H. 1980. The steady-state economy: toward a political economy of biophysical equilibrium and moral growth. In Economics, Ecology, Ethics: Essays Toward a Steady-State Economy, ed. H. Daly. San Francisco: Freeman.

———. 1990. Toward some operational principles of sustainable development. *Ecological Economics* 2: 1–6.

Daly, H., and J. B. Cobb. 1989. For the Common Good: Redirecting the Economy toward Community, the Environment, and a Sustainable Future. Boston: Beacon.

Dasgupta, P., and K-G. Mäler. 1991. The environment and emerging development issues. In Proceedings of the World Bank Annual Conference on Development Economics. The International Bank for Reconstruction and Development. Washington, DC: The World Bank.

Ehrlich, P. R., and P. H. Raven. 1964. Butterflies and plants: a study in coevolution. *Evolution* 18: 586–608.

Everenden, N. 1985. The Natural Alien. Toronto: Univ. of Toronto Press.

Feeny, D., F. Berkes, B. J. McCay, and J. M. Acheson. 1990. The tragedy of the commons: twenty-two years later. *Human Ecology* 18: 1–19.

Folke, C. 1991. Socio-economic dependence on the life-supporting environment. In Linking the Natural Environment and the Economy: Essays from the Eco-Eco Group, eds. C. Folke and T. Kåberger. Dordrecht: Kluwer.

Folke, C., and N. Kautsky. 1989. The role of ecosystems for a sustainable development of aquaculture. *Ambio* 18: 234–43.

———. 1992. Aquaculture with its environment: prospects for sustainability. *Ocean and Coastal Management* 17: 5–24.

Friedman, J. W. 1986. Game Theory with Applications to Economics. Oxford: Oxford Univ. Press.

Gadgil, M. 1987. Diversity: cultural and ecological. *Trends in Ecology and Evolution* 2: 369–73.

Gadgil, M., and F. Berkes. 1991. Traditional resource management systems. *Resource Management and Optimization* 8: 127–41.

Geertz, C. 1963. Agricultural Involution: The Process of Ecological Change in Indonesia. Berkeley: Univ. of California Press.

Gibbs, C. J. N., and D. W. Bromley. 1989. Institutional arrangements for management of rural resources: common-property regimes. In Common Property Resources, ed. F. Berkes. London: Belhaven.

Günther, F., and C. Folke. 1993. In press. Characteristics of nested living systems. *Journal of Biological Systems.*

Hardin, G. 1968. The tragedy of the commons. *Science* 162: 1243–8.

Hjort af Ornäs, A., and U. Svedin. 1992. Cultural variation in concepts of nature. *GeoJournal* 26: 167–72.

Ingold, T. 1980. Hunters, Pastoralists and Ranchers: Reindeer Economies and Their Transformation. Cambridge: Cambridge Univ. Press.

IUCN/UNEP/WWF. 1991. Caring for the Earth: A Strategy for Sustainable Living. Gland: World Conservation Union.

Jantsch, E. 1980. The Self-Organizing Universe: Scientific and Human Implications of the Emerging Paradigm of Evolution. New York: Pergamon.

Johannes, R. E. 1989. Traditional Ecological Knowledge: A Collection of Essays. Gland: World Conservation Union.

Keohane, R., M. McGinnis, and E. Ostrom, eds. 1993. Proceedings of a Conference on Linking Local and Global Commons. Cambridge: Harvard Univ./Bloomington: Indiana Univ.

Leopold, A. 1949. A Sand County Almanac. Oxford: Oxford Univ. Press.

McCay, B. J., and J. M. Acheson, eds. 1987. The Question of the Commons: The Culture and Ecology of Communal Resources. Tucson: Univ. of Arizona Press.

Milbrath, L. W. 1989. Envisioning a Sustainable Society: Learning Our Way Out. Albany: State Univ. of New York Press.

Mitsch, W. J., and S. E. Jörgensen, eds. 1989. Ecological Engineering: An Introduction to Ecotechnology. New York: Wiley.

Moran, E. F., ed. 1990. The Ecosystem Approach in Anthropology: From Concept to Practice. Ann Arbor: Univ. of Michigan Press.

Naess, A. 1989. Ecology, Community and Lifestyle. Cambridge: Cambridge Univ. Press.

Netting, R. McC. 1981. Balancing on an Alp. Cambridge: Cambridge Univ. Press.

Norgaard, R. B. 1987. Economics as mechanics and the demise of biological diversity. *Ecological Modelling* 38:107–12.

Norton, B. G. 1990. Context and hierarchy in Aldo Leopold's Theory of Environmental Management. *Ecological Economics* 2: 119–27.

Odum, E. P. 1975. Ecology: The Link Between the Natural and Social Sciences. 2d. New York: Holt-Saunders.

Odum, H. T. 1988. Living with complexity. In The Crafoord Prize in Biosciences 1987: Crafoord Lectures. Stockholm: The Royal Swedish Academy of Sciences.

Olson, M. 1965. The Logic of Collective Action. Cambridge, MA: Harvard Univ. Press.

Ostrom, E. 1990. Governing the Commons: The Evolution of Institutions for Collective Action. Cambridge: Cambridge Univ. Press.

Posey, D. A., and W. Balee, eds. 1989. Resource management in Amazonia: indigenous and folk strategies. *Advances in Economic Botany* 7 (special issue).

Robinson, J., G. Francis, R. Legge, and S. Lerner. 1990. Defining a sustainable society: values, principles and definitions. *Alternatives* 17 (2): 36–46.

Ruddle, K., and G. Zhong. 1988. Integrated Agriculture-Aquaculture in South China: The Dike-Pond System of the Zhujiang Delta. Cambridge: Cambridge Univ. Press.

Sander, T. 1992. Collective Action: Theory and Application. Hemel Hempstead, U.K.: Harvester/Wheatsheaf.

Skolimowski, H. 1981. Eco-Philosophy. London: Boyars.

Uzawa, H. 1974. Sur la théorie économique du capital collectif social. Cahier du Séminaire d'Économétrie, 103–22. Translated in Preference, Production and Capital: Selected papers by Hirofumi Uzawa, 340–62. 1988. Cambridge: Cambridge Univ. Press.

————. 1992. Towards a General Theory of Social Overhead Capital. Beijer Discussion Paper Series No. 13. The Beijer International Institute of Ecological Economics. Stockholm: The Royal Swedish Academy of Sciences.

Varela, F. J., H. R. Maturana, and R. Uribe. 1974. Autopoiesis: The organization of living systems, its characterization and a model. *Biosystems* 5: 187–96.

von Bertalanffy, L. 1968. General System Theory: Foundations, Development, Applications. New York: George Braziller.

Wade, R. 1987. Village Republics: Economic Conditions for Collective Action in South India. Cambridge: Cambridge Univ. Press.

PART TWO

Ecological Economic
Methods and Case-Studies
on the Significance
of Natural Capital

9 ENVIRONMENTAL FUNCTIONS AND THE ECONOMIC VALUE OF NATURAL ECOSYSTEMS

Rudolf S. de Groot
Coordinator, Center for Environment and Climate Studies
Agricultural University Wageningen,
PO Box 9101
6700 HB Wageningen, The Netherlands

ABSTRACT

Because the importance of nature and a healthy natural environment to human welfare is still not fully reflected in economic planning and decision making, degradation and loss of natural ecosystems by human activities still continues on a large scale. Current methods of evaluation in decision making, such as cost benefit analysis, inadequately reflect the true environmental and socioeconomic value of natural ecosystems and the goods and services they provide.

This chapter (1) presents a checklist of environmental functions provided by natural and semi-natural ecosystems, based on various case studies; (2) describes a comprehensive and systematic method for assessing the full economic value of these functions; (3) discusses possibilities for determining the total socioeconomic and capital value of natural ecosystems and protected areas; (4) shows that, when all factors are considered most protected areas not only have great ecological and intrinsic values but also provide considerable economic benefits; and (5) concludes that better information on the (economic) value of natural areas alone, is insufficient for sound environmental decision making. Unless ecological information is structurally integrated in the planning and decision-making process, solving the environmental problems of today will prove difficult, if not impossible. The last section suggests how the function-concept can be used as a tool in planning and decision making in order to stimulate investments aimed at the maintenance and sustainable utilization of the remaining "natural capital" on earth.

ENVIRONMENTAL FUNCTION EVALUATION: A COMPREHENSIVE—METHOD TO ASSESS THE FULL VALUE OF NATURE TO HUMAN SOCIETY

An important obstacle to the inclusion of environmental concerns in environmental planning and decision making is the translation of ecological data into useful information for planners and decision makers. Current methods of evaluation for decision making, such as cost-benefit analysis and environmental impact assessment, inadequately reflect the true environmental and socioeconomic value of natural resources and ecosystems. To better assess the full value of natural systems to human society, a so-called function evaluation system was developed by the author (de Groot 1992) which integrates four separate assessment procedures (see Figure 9.1).

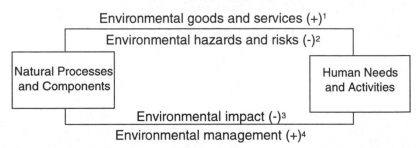

The functional interactions between the natural environment and human society have both positive (+) and negative (-) aspects and can be divided into four types of interactions: (1) Environmental Function Evaluation, (2) Environmental Risk Assessment, (3) Environmental Impact Assessment, (4) Environmental Management Evaluation

Figure. 9.1. Functional interactions between human society and the natural environment (from de Groot 1992).

An important element in this evaluation procedure is the *function-concept*. In de Groot (1992), environmental functions are defined as "the capacity of natural processes and components to provide goods and services that satisfy human needs (directly and/or indirectly)." Four main function categories are distinguished:

1. **Regulation functions**: this group of functions relates to the capacity of natural and semi-natural ecosystems to regulate essential ecological processes and life support systems which, in turn, contributes to the maintenance of a healthy environment by providing clean air, water, and soil;

2. **Carrier functions**: natural and semi-natural ecosystems provide space and a suitable substrate or medium for many human activities such as habitation, cultivation and recreation;

3. **Production functions**: nature provides many resources, ranging from food and raw materials for industrial use to energy resources and genetic material; and

4 **Information functions**: natural ecosystems contribute to the maintenance of mental health by providing opportunities for reflection, spiritual enrichment, cognitive development, and aesthetic experience.

Two steps in the evaluation-procedure are of special importance and are briefly described below.

Step 1: Ecological Assessment of Environmental Functions

The capacity of a given natural or semi-natural ecosystem to provide certain goods and services depends on particular characteristics of the natural processes and components of the area in question. Since the environmental characteristics of most ecosystems vary substantially, the functions of different ecosystems are also quite different. To develop a general checklist of parameters that may be used to assess the contribution of a given ecosystem to certain environmental functions, several case-studies have been carried out by the author on various ecosystem complexes, notably tropical moist forests, coastal wetlands, and the Galapagos National Park.

Based on these (and other) case studies, and additional information from literature, Figure 9.2 shows a checklist of 37 functions that can be attributed to natural ecosystems. Which functions are most relevant for a given natural ecosystem or protected area depends on the ecological characteristics, the cultural and socioeconomic setting, and the management objectives of the area in question.

Many of the functions (goods and services) listed in Figure 9.2 are provided by all natural ecosystems regardless of their management status; i.e., these functions can be attributed to both protected and unprotected areas (provided these areas are in a primarily natural state). This is especially true for the regulation functions. The other functions (carrier, production, and information functions) are more strongly related to specific human needs and activities and, thus, to the management objectives. In protected areas which allow certain forms of human use (such as national parks), these latter functions seem more important for the (direct) economic benefits they provide (e.g., aquaculture, recreation, harvestable resources, and research). Nevertheless, the maintenance of regulation functions is essential to the proper functioning of all natural systems and should always be considered when assessing the (economic) benefits of protected areas.

Step 2: Socioeconomic Valuation of Environmental Functions

Once the many functions provided by natural and semi-natural ecosystems have been identified and described, the contribution of these goods and services to human welfare can be analyzed. Over the years, a variety of methods have been developed for assigning values to nature and natural resources, and there are many titles available on this subject, such as Krutilla and Fisher (1975); Cooper (1981); Hufschmidt et al. (1983); Brown and Goldstein (1984); Johansson (1987); Barrett (1988); McNeely (1988); Pearce and Turner (1990); and Folke and Kaberger (1991). This list of publications on valuing nature and natural resources is by no means complete. The multiplicity of ways and means for assessing environmental values is not surprising, because the benefits provided by the many environmental functions are so diverse that methods to measure the socioeconomic value of one

Regulation Functions	
1.	Protection against harmful cosmic influences
2.	Regulation of the local and global energy balance
3.	Regulation of the chemical composition of the atmosphere
4.	Regulation of the chemical composition of the oceans
5.	Regulation of the local and global climate
6.	Regulation of runoff and flood-prevention (watershed protection)
7.	Watercatchment and groundwater-recharge
8.	Prevention of soil erosion and sediment control
9.	Formation of topsoil and maintenance of soil-fertility
10.	Fixation of solar energy and biomass production
11.	Storage and recycling of organic matter
12.	Storage and recycling of nutrients
13.	Storage and recycling of human waste
14.	Regulation of biological control mechanisms
15.	Maintenance of migration and nursery habitats
16.	Maintenance of biological (and genetic) diversity

Carrier Functions	
	providing space and a suitable substrate for:
1.	Human habitation and (indigenous) settlements
2.	Cultivation (crop growing, animal husbandry, aquaculture)
3.	Energy conversion
4.	Recreation and tourism
5.	Nature protection

Production Functions	
1.	Oxygen
2.	Water (for drinking, irrigation, industry, etc.)
3.	Food and nutritious drinks
4.	Genetic resources
5.	Medicinal resources
6.	Raw materials for clothing and household fabrics
7.	Raw materials for building, construction and industrial use
8.	Biochemicals (other than fuel and medicines)
9.	Fuel and energy
10.	Fodder and fertilizer

Information Functions	
1.	Aesthetic information
2.	Spiritual and religious information
3.	Historic information (heritage value)
4.	Cultural and artistic inspiration
5.	Scientific and educational information

Figure 9.2. Functions of the natural environment.

function may not be appropriate for measuring the value of other functions. For example, the value of a tropical forest as the provider of logs for export of hardwoods is measured in a different way than the value of the forest to the local inhabitants as a provider of their daily living needs; or the value of the forest for tourism or watershed protection.

The major types of values that can be attributed to environmental functions, and the natural ecosystems which provide them, are summarized in Figure 9.3.

	Ecological values(1)		Social values (2)		Economic values (3)		
Environmental Functions	Conserva-tion value	Existence value	Health	Option value	Consumptive use value	Productive use value	Employ-ment
Regulation							
Carrier							
Production							
Information							
TOTAL for ecosystem or natural area							

(1) The ecological value of environmental function scan often only be described in qualitative terms; quantification will usually only be possible in "natural" dimensions (e.g., number of species, amount of runoff prevented).

(2) Social values may be quantified by setting standards for minimum requirements for the availability of a given function (e.g., air-quality or maximum limits to ensure sustainable harvesting of natural resources).

(3) The economic value of environmental function may be expressed in their "natural" dimensions (e.g., quantity of resources harvested), in monetary units (i.e., the value of the resources harvested) or by the number of people employed by activities that depend on a given function.

Figure 9.3. Functions and values of natural ecosystems.

A brief description of the various types of socioeconomic values that can be attributed to environmental functions is given below.

1. **Conservation value**. Many environmental functions do not provide direct economic benefits but are, nevertheless, quite essential to human welfare. The so-called non-use or conservation values are mainly provided by the services (as opposed to the goods) of natural and semi-natural environments, such as the regulation and information functions. The regulation functions maintain and conserve the environmental conditions necessary for most of the other functions that provide more direct economic benefits. The non-use benefits of nature are illustrative examples of environmental functions which are considered to be "free." Because of the problems involved in quantifying the economic and monetary value of these non-use benefits, they are usually not reflected in national income accounts.

2. **Existence value.** This type of value relates to the intangible, intrinsic, and ethical values attributed to nature. Pearce and Markandya (1987) call this type of value *existence value*, stemming from feelings of stewardship on behalf of future generations and non-human populations. The responsibility people feel towards future generations is also called the *bequest value* (Krutilla and Fisher 1975): even if we do not benefit ourselves directly, we do have a responsibility to our children and grandchildren to conserve natural ecosystems and enhance the evolution of biological diversity as much as possible.

3. **Contribution of nature to human health.** Many environmental functions contribute directly or indirectly, to human health. Oxygen, drinking water, and food are essential resources to maintain human life. Natural regulation processes contribute to the maintenance of clean air water and soils. Nature provides a large array of medicinal resources and contributes to mental health by providing a multitude of opportunities for recreation and cognitive development. The socioeconomic importance of this value is evident and the specific contribution of a given ecosystem or function may be expressed in terms of human lives "saved" or in monetary indicators, such as the costs of medical treatment required in the absence of a given function (or the actual loss of life and/or economic damage suffered after a given function is disturbed), for example, the disturbance of the protective function of the stratospheric ozone layer.

4. **Option value.** The option value of natural ecosystems and environmental functions relates to the importance people place on a safe future (i.e., the future availability of a given amenity, good, or service) either within their own lifetime, or for future generations. This value is therefore sometimes also referred to as *bequest value* (see above), or *serendipity value* (Pearsall 1984; Meyers 1984). The underlying concept of option value was expressed well by Hueting (1984), who stated that "... man derives part of the meaning of existence from the company of others, which in any case include his children and grandchildren. The prospect of a safer future is therefore a normal human need, and dimming of this prospect has a negative effect on welfare." Since the future is uncertain, all types of option value can be seen as a means of assigning a value to risk aversion in the face of uncertainty (McNeely 1988). It is a type of life insurance for access to future benefits from natural ecosystems.

5. **Consumptive use value.** The consumptive use value of environmental functions relates to the use of natural products, which are harvested directly from the natural ecosystem. This value, therefore, mainly relates to natural resources in the narrow sense, which are included in the category of production functions. Because these natural products are consumed directly, without passing through a market, consumptive use values seldom appear in national income accounts such as Gross National Product (GNP) or Gross Domestic Product (GDP), although their economic value is often considerable. For example, in Sarawak, Malaysia, a detailed field study showed that wild pigs harvested by hunters had an (estimated) market value of some $100 million per year if they would have been sold on the market (Caldecott 1988).

6. **Productive use value.** The most important part of the traditional economic value of a given good or service is probably still its contribution to the (economic) production process which consists of many different sectors, such as agriculture, energy conversion, transportation, and industry. Since this value can relatively easily be expressed in monetary units, the productive use value of natural resources is usually the only economic value of environmental functions which is reflected in national income accounts.

7. **Contribution of natural ecosystems and protected areas to employment.** In many economic sectors, employment depends directly or indirectly on environmental functions, such as people who are employed in the management of protected areas and the guiding of recreational activities in nature. In addition jobs held by fishermen and farmers depend on a healthy natural environment and many jobs in industry are in one way or another dependent on environmental functions.

THE NATURAL CAPITAL—HOW MUCH IS IT WORTH?

To stimulate investments into the maintenance and sustainable use of the remaining natural capital on earth, better understanding of the full economic value of this natural capital is needed. The previous section demonstrated that natural ecosystems and protected areas fulfill a multitude of functions with many different values to human society. Some of these values can be expressed in monetary terms, and the next section briefly lists a few methods available to calculate monetary values for environmental functions.

Since most natural areas provide various functions simultaneously, the total socioeconomic value of a given ecosystem or protected area consists of the sum of the economic and monetary benefits of the individual functions. The "total value" of a given environmental function, in turn, is determined by a number of values, including non-monetary values, market values, and shadow prices. When calculating the *total monetary value* of a given area or ecosystem, values must not be double-counted.

Another factor to consider is that the benefits from environmental functions should be determined for *sustainable use levels*. When several functions are used simultaneously, this often means that not all functions can be utilized to their maximum potential but that an optimal mixture should be found to ensure the continued integrity of the full range of functions of the area in question.

When all functions and values are properly taken into account, it will often become clear that sustainable use of a combination of functions provides more long-term economic benefits, than non-sustainable use of only a few functions (such as logging in tropical moist forests versus sustainable use of all of its functions).

Furthermore, it must be realized that natural areas can provide functions in perpetuity, when used in a sustainable manner. The calculated annual monetary value of individual functions or of entire ecosystems or protected areas must, therefore, somehow be translated into a capital or net-present value, as presented in upcoming sections.

Monetary Valuation of Environmental Functions

For several types of functions and values, notably those which are of direct economic importance, it is possible to calculate monetary values. Assessing monetary values of environmental functions is a rather complicated, and somewhat controversial procedure. Figure 9.4 shows some of the methods available to calculate monetary values for environmental goods and services, which broadly fall into two categories: market pricing and shadow pricing.

Discussion of these methods would prove too lengthy for this chapter. However, an overview is given in de Groot (1992), while further details can be found in other Wallenberg contributions and in existing literature in this field. A few recent titles are: Dixon and Hufschmidt (1986), Opschoor (1986), Farber (1987), Pearce (1987), Pearce and Markandya (1987), Pearce, Markandya and Barbier (1989), Bojo (1991), Folke and Kaberger (1991), and Howe (1991).

Types of	Monetary valuation methods						
Socio-economic value	Market Price	Shadow price					
		Cost of environmental damage	Mainten-ance costs	Mitigation costs	Willingness to pay/accept	Property pricing	Travel cost
Conservation value		X	X	X	X		
Existence value			(X)#		(X)#		
Health			X		X		
Option value			X		X		
Consumptive use value	(X)*						
Productive use value	X					X	X
Employment	X						

The existence value could be quantified by these techniques, but it is argued that it is principally wrong to put a monetary price on this value.

*   Is usually derived form a surrogate market price.

Figure 9.4. Types of socioeconomic values and methods to determine their monetary value.

The total socioeconomic value of a given natural area or ecosystem is the sum of the different values listed in the vertical axis of Figure 9.4. Within one value-category, the benefits of all functions can be added to arrive at a sum-total for the conservation value, use value, or the contribution to employment of a particular ecosystem or natural area. Since the seven types of values are not comparable, it is impossible to determine one total "end value" for a given function or natural area, and therefore, each type of value must be used independently in the decision-making process.

It should also be stressed here that socioeconomic evaluation of environmental functions does not necessarily mean that the importance of nature and wildlife is entirely reduced to dollar values. As Figures 9.3 and 9.4 show, the total value of environmental goods and services to human society consists of many different values which are described and quantified by different parameters, of which monetary units are but one element. Furthermore, quantification of the socioeconomic benefits of natural areas and wildlife in monetary units must always be seen as an addition to, and not a replacement of, their many intrinsic and intangible values.

Total Annual Monetary Benefits of the Environmental Functions Provided by Three Types of Natural Ecosystems/Protected Areas

Several case studies have been carried out to illustrate the evaluation method presented earlier in this discussion and include assessments of the functions and values of tropical moist forests (the Darien Rainforest, Panama), wetlands (notably the Dutch Wadden Sea), and the Galapagos National Park (Ecuador). For each case study a standardized matrix based on Figure 9.3 was used to assess the functions and associated values in a systematic and comparable way. The three case studies are described in detail in de Groot (1992). A few results arc summarized below.

For those functions for which monetary benefits could be calculated, values have been determined for maximum sustainable use levels based on the (estimated) carrying capacity of the area in question for each function.

It must be stressed here that monetary values and the sum-totals discussed below, are a first attempt for a comprehensive assessment of the functions and values of a particular natural ecosystem or protected area. So far, emphasis has been on developing the methodology and consequently the figures are mainly indicative and give a rough indication of the types and magnitudes of monetary returns and economic benefits to be considered. Much more research is needed to obtain more complete and accurate data for each individual function (which should not be taken as an excuse not to use the information already available for more balanced planning and decision making).

(1) Economic Value of Tropical Moist Forests

According to Myers (1988), a tropical forest tract of 500 square kilometers could, with effective management, produce a self-renewing crop of wildlife with a potential value of at least $10 million per year, or slightly more than $200 per ha/year. The study of Peters, Gentry and Mendelsohn (1989) of the actual and potential benefits of 1 hectare of tropical forest in Peru found that of the total number of trees on the site, 72 species (26.2%) and 350 individuals (41.6%) yielded useful products, mainly fruits, and latex (rubber) with an actual market value of about $700. After deducting collecting and transport costs, the net annual revenues were $422. Fruits and latex represent more than 60% of the total market value of the forest products, yet they are but two of many different products provided by the forest. If it were possible to include revenues from other resources, the relative importance of the non-wood products would increase even further. The economic and monetary value of sustainable use of "minor" forest products is, therefore, probably much higher than $422/ha/year. Simultaneously, the area could generate income from tourism and may perform important regulating functions in the form of watershed protection and local climate regulation. According to the case study discussed in de Groot (1992), the total monetary value of the functions of 1 hectare of tropical moist forest is at least $500/ha/year.

(2) Economic Value of Coastal Wetlands

Tidal areas, and the wide shallows, marshes, and swamps that accompany them perform a multitude of ecological functions, many of which have great socioeconomic value to man. Because of their location between land and water, wetlands are among the most productive of the world's ecosystems, and play a major role in food production. In addition, they provide a wide range of other goods (such as timber and thatch) and services (like flood prevention, shoreline protection, water purification, and possibilities for recreation). Wetlands also have great natural value as breeding, feeding, and resting grounds for great numbers of fish, migratory birds, and other animals. Another important function of estuaries is their role in the continued normal functioning of nutrient cycles. Due to their high mineralization rate, estuaries can recycle large amounts of (organic) human waste without negative side effects.

Coastal marshes, and other shallow water areas such as reefs, export mineral and organic nutrients that support much of the biological production of adjacent estuarine and coastal waters. These estuaries and coastal waters, in turn, serve as important nursery grounds for coastal fish and shellfish. About two-thirds of the fish caught throughout the world were hatched in tidal areas (Wagenaar Hummelinck 1984).

The most important wetland functions for which a monetary value can be calculated are (1) flood prevention, (2) storage and recycling of human waste, (3) nursery value, (4) aquaculture and recreation, (5) food production and (6) education and science uses. Adding these values for the Dutch Wadden Sea amount to a total of over $6,200/ha/year (de Groot 1992). Of course, values for other wetland areas may be different. In comparison to calculations made for other wetlands, this is a rather moderate estimate; Gosselink et al. (1974), for example, calculated a monetary value of $10,000/ha/year for several estuaries along the east coast of the United States. Thibodeau and Ostro (1981) arrived at a figure of $28,000/ha/year for the Charles River Basin in Massachusetts. Another estimate placed the economic value of a hectare of Atlantic Spartina Marsh at over $72,000 per year (Hair 1988).

(3) Economic Value of the Galapagos National Park

The Galapagos Islands are situated on the equator in the Pacific Ocean, approximately 1,000 km west of Ecuador, of which the archipelago is a province. The archipelago consists of 14 major islands and a larger number of smaller islets and rocks with a total surface area of almost 8,000 km². Including the territorial waters, the total surface area is approximately 60,000 km². In 1959, about 90% of the total land area was set aside as a National Park, and in 1984 about 4,300 km² of the marine area was declared a Marine National Park

For the Galapagos National Park, total monetary benefits amount to about $140 million per year, based on maximum sustainable use levels mainly from tourism, harvesting natural resources, and value to scientific research. This amounts to an average value of about $120/ha/year (de Groot 1992).

Capital or Net Present Value of Environmental Functions

When interpreting the total monetary value of a given function, natural area or ecosystem, it must be realized that this value only represents the annual return from the respective functions. Since natural areas can provide many environmental goods and services in perpetuity, if used in a sustainable manner, the total annual value should somehow be transformed into a capital value to reflect the true economic value of the area or ecosystem concerned.

In order to arrive at a capital value, the present worth of future benefits (and costs) must somehow be estimated. A practice applied in business economics is *discounting*. An important difficulty in arriving at an acceptable capital value for environmental functions is the choice of discount rate to be used, which depends on the time horizon applied. Usually the time-horizon in market economics is rather limited—50 years or less—resulting in (market) discount rates of 10% or more. Discount rates for future benefits of environmental functions found in literature usually range between 5% and 6%. However, there are some objections to using discount rates for calculating the depreciation of the monetary return of the benefits derived from environmental functions.

1. A discount rate of 5% effectively means that the value of a given function 30–40 years from now is considered to be near zero today. However, the benefits of the "works of nature" will last in perpetuity when used in a sustainable manner, and the "economic lifetime" of these goods and services can (or should) not be calculated in the same manner as is customary for man-made goods and services which usually lose their economic value after about 20 years. Therefore, placing discount rates on the functions of natural ecosystems ignores the interests of future generations.

2. Human constructions that provide certain goods and services, such as a factory, are replaceable, while natural ecosystems usually are not.

Although much has been written on this subject (Stokoe 1988; Markandya and Pearce 1988; Hueting 1990, 1991), a satisfactory solution is not in sight yet.

In so far as the use of discount rates for calculating the future benefits of environmental functions is unavoidable, it should be calculated as low as possible, preferably in accordance with the time it takes for the ecosystem to reach its climax stage. Succession times differ strongly between various types of ecosystems and may range from a few years for certain aquatic or grassland ecosystems to a thousand years or longer for tropical moist forests and over 10,000 years for a bog ecosystem. For practical purposes, it is proposed here to apply a range of discount rates between 1% and 6 % for environmental functions provided by natural ecosystems, whereby the higher figure applies to pioneer communities and the lower figure to climax communities.

Instead of the use of discount rates, a much better approach to calculate the capital or net present value of environmental functions would be the *interest-on-capital* approach. Since natural environments could provide goods and services indefinitely, when used in a sustainable manner, it would seem more appropriate to consider the annual return in monetary terms as the interest on the capital stock of the natural processes and components that provide these functions. This implies that there is no time limit to the benefits derived, making the value of the capital immeasurable. Although this is essentially correct, for practical purposes it may sometimes be necessary to estimate a capital or net present value for a given good, service or natural area.

Assuming a 5% interest rate, the capital value of the three case study areas presented in this chapter would amount to about $2,400/ha for the Galapagos National Park, $120,000/ha for the Dutch Wadden Sea, and $10,000/ha for tropical moist forests. By contrast, non-sustainable use of tropical moist forest resources has a capitalized value of not more than $4,000 (de Groot 1992). Thus, sustainable use of "minor" forest products is clearly more economical than logging.

Finally, it must be realized that the Capital or Net Present Value (NPV) of the functions of protected areas is only based on the (estimated) annual return of those functions for which a "real" or derived market value can be calculated. Therefore, the values given here must be considered a minimum estimate since for many functions no monetary value could be calculated, although their (potential) contribution to the economy is considerable.

SOME CONCLUSIONS AND RECOMMENDATIONS ON THE APPLICATION OF FUNCTION EVALUATION IN PLANNING AND DECISION MAKING

Human welfare and the quality of life depend directly or indirectly on the availability of environmental goods and services in many ways. As has been shown in the previous sections, natural ecosystems and protected areas, besides their ecological and intrinsic value, represent a considerable socioeconomic and monetary value as well.

Unless ecological information is structurally integrated in the planning and decision-making process, non-sustainable use of environmental functions will continue to prevail over investments aimed at securing long-term benefits of nature and natural resources. The final discussion, therefore, presents some possibilities to use function evaluation (i.e., information on the monetary value and economic benefits of natural ecosystems) in planning and decision making, in order to stimulate investments aimed at conservation and sustainable use of the "natural capital."

Function Evaluation as a Tool in Environmental Planning

In order to make more balanced decisions, ecological data must somehow be translated into useful information for planners and decision makers. The function-evaluation system presented here can be helpful in this difficult process in many ways, notably in Environmental Impact Assessment and Cost-Benefit Analysis.

a.   An important application area is the use in providing *ecological base line information* for development projects. Especially for drafting environmental profiles, country reviews and formulating carrying capacity limits, information on environmental functions and values is essential.

b.   Information on environmental functions and values may also be used to provide more general *guidelines for development assistance* or national government policies. A description of the many functions of tropical forests, for example, was an important element in formulating the Dutch Government Policy concerning Tropical Forests (de Groot 1992).

c.   In order to obtain a clear insight into the environmental trade-offs involved in alternative development projects, *environmental impact assessment* studies should assess both the direct environmental effects of certain human activities or interventions in a given area, and the environmental functions and hazards affected by the activity.

d.   *Cost-benefit analyses* are a much-used instrument to help decide where to invest or sacrifice in natural capital. A major shortcoming of (traditional) cost-benefit analysis is that it is often limited to economic (financial) trade-offs. This last feature is probably one of the main causes of the undervaluation of natural ecosystems and environmental aspects in most economic accounting procedures, since many environmental goods and services are still considered to be "free," and losses of environmental functions are seen as "external effects." Because many functions of natural ecosystems cannot (as yet) be expressed in monetary units, traditional cost-benefit analysis inadequately reflects the true environmental and socioeconomic value of natural resources and ecosystems. Previous discussions demonstrated that the combined economic value of the sustainable use of environmental functions is often considerable and usually far exceeds the (short-term) returns from non-sustainable use of only some of the functions of a given natural ecosystem or protected area. When all factors are taken into account, most natural ecosystems and protected areas have a highly positive benefit-cost ratio. For example, two case studies on economic benefits of protected areas (Cahuita NP in Costa Rica and the Virgin Islands NP, St. John), both show a benefit/cost ratio of about 10 (de Groot 1991).

An important aspect in the planning process is the **time factor**. Often environmental impact assessments or cost-benefit analysis are not carried out properly because the time available for collecting the necessary information is too short. At first site, the function-evaluation system proposed here may also seem rather time-consuming, considering the large number of functions (37) and cells in the evaluation matrices. However, the list of functions must be seen as an ideal "checklist"; for most ecosystems or natural areas a much smaller number of functions is of direct importance, which also depends on the purpose of the evaluation and the nature of the planned intervention(s). With the help of expert judgment and the case studies, a first screening can be accomplished quite rapidly. In addition, it must be realized that most of the information needed for the socioeconomic evaluation is, or should also be, collected for "traditional" EIAs or CBAs.

The proposed function-evaluation procedure is mainly designed to make the planning process more systematic, and the decision-making process more transparent, to ensure that no important factors are left out of the analysis. Experience has shown that 3–5 months should be enough for a first general assessment of the most important functions and values of a particular area or ecosystem.

Function Evaluation and Ecological Pricing

Current economic incentives still mainly favor short-term profits and strongly neglect the need for conservation and sustainable use of environmental functions. The market price of many goods and services does not adequately reflect the increasing scarcity, nor the true socioeconomic value of the natural resource base on which they depend. Also, the (environmental) costs involved in the production and/or extraction of natural resources and the loss of environmental functions are largely neglected (usually referred to as *external effects*). In addition, the contribution of many natural goods and services to the economic production process is insufficiently or never accounted for in the pricing mechanism, since the market is unable to provide realistic "environmental" or "ecological" prices for nature's works (many of which are entirely neglected because they are considered to be "free").

As a result, natural ecosystems, and the goods and services they provide, are usually undervalued in conventional economic accounting procedures leading to over-exploitation, non-sustainable development and environmental disasters.

Information on the true socioeconomic value of environmental functions should therefore be translated into ecological prices for man-made goods and services. Adjustment of the pricing mechanism, in combination with changes in the tax system, is essential in order to develop economic incentives which favor the conservation and sustainable use of the functions of natural ecosystems and protected areas.

Adjustment of Economic Indicators and Accounting Procedures

Because the current pricing system of the market economy is based on incomplete cost-benefit analyses, economic accounting procedures must be reconsidered in order to better account for the benefits of natural goods and services and the costs of the loss of these goods and services (i.e., loss of environmental quality) due to non-sustainable production activities. In present national accounts (such as GNP), man-made assets are valued as productive capital, while natural resources and other environmental functions are not.

If we are to solve the environmental problems of today, the conflict between economic interests (i.e., increasing the economic capital) and conservation goals (i.e. maintaining the natural capital) need to be solved. Much has been written on the need for better integration of ecological principles and environmental considerations in economic assessment and accounting procedures (Dorcey 1984; Knetsch and Freeman 1979; Dixon and Bojo 1988; Pearce, Markandya and Barbier 1989; and Daly 1991). Without the development of a new kind of "environmental economics," most efforts to halt environmental degradation and to bring economic development more in harmony with the carrying capacity of nature will fail. The question now is not so much anymore **whether** economics should be "ecologized," but rather **how** this can be done.

In this respect, the function-concept can be useful as a common paradigm in ecology and economics. Instead of the narrow concept of natural resources, the availability of environmental functions is a much better indicator for measuring both environmental quality and the quality of life. By definition, environmental functions satisfy human needs, which is also an important element of economic theory. The maintenance of environmental functions serves, therefore, both ecological interests (i.e., environmental health) and economic goals (i.e., human welfare). The thought of environmental functions as a unifying concept for ecology and economics is developed in more detail in de Groot (1987).

By including the value of the goods and services provided by the natural capital (i.e., natural ecosystems) more systematically into national accounting systems, it becomes clear that an increase of the man-made economic capital, especially in the market sector, often goes at the expense of the natural capital. The net effect of many so-called development activities on human welfare is, thereby, greatly reduced, and may sometimes even be negative. For example, the "mining" of tropical forests for hardwood converts the natural capital into economic capital which quickly evaporates into paying off national debts. What remains are devastated landscapes, which have lost their productive potential and the many other functions they provided, bringing much hardship to the local communities. Natural ecosystems are more than just a cheap source of resources and land to be used at will for short-term economic gains. They should be seen as a productive natural capital that could provide many goods and services indefinitely if conserved and used in a sustainable manner.

Only when ecological principles become an integral part of economic planning and political decision making is there a chance of achieving sustainable development based on a new kind of ecological economics which integrates conservation objectives and economic interests into one common goal—the maintenance and sustainable use of environmental functions provided by nature and natural ecosystems.

REFERENCES

Barrett, S. 1988. Economic Guidelines for the Conservation of Biological Diversity. Paper prepared for the workshop on The Economics of Sustainable Development during the IUCN General Assembly in San Jose, Costa Rica, February 1–10.

Bojo, J. 1991. Economic analysis of environmental impacts. In Linking the Natural Environment and the Economy: Essays from the Eco-Eco Group, eds. C. Folke and T. Kaberger. Dordrecht: Kluwer Academic Publishers.

Brown, G. M., Jr., and J. H. Goldstein. 1984. A model for valuing endangered species. *Journal of Environmental Economics and Management* 11:303–9.

Caldecott, J. 1988. Hunting and Wildlife Management in Sarawak. Gland: IUCN

Cooper, C. 1981. Economic Evaluation and the Environment. London: Hodder and Stoughton.

Costanza, R., and H. E. Daly. 1990. Natural Capital and Sustainable Development. Paper prepared for Workshop on Natural Capital, March 15–16. Canadian Environmental Assessment Research Council, Vancouver, Canada.

Daly, H. E. 1991. Towards an environmental macroeconomics. *Land Economics* 67: 255–9.

de Groot, R. S. 1987. Environmental functions as a unifying concept for ecology and economics. *The Environmentalist* 7 (2): 105–9.

————. 1991. Functions and Socio-economic Values of Coastal/Marine Protected Areas. Paper prepared for the thematic meeting on "Economical Impact of Protected Areas" of the Mediterranean Protected Areas Network (MEDPAN) in Ajaccio (Corsica), September 27.

————. 1992. Functions of Nature: Evaluation of Nature in Environmental Planning, Management and Decision Making. Groningen, The Netherlands: Wolters Noordhoff BV.

Dixon, J. A., and J. P. Bojo. 1988. Economic Analysis and the Environment. Report to the African Development Bank based on a workshop held at the World Bank, Washington, DC, June 7–9.

Dixon, J. A., and M. M. Hufschmidt, eds. 1986. Economic Valuation Techniques for the Environment: A Case Study Workbook. Baltimore: Johns Hopkins Univ. Press.

Dorcey, A. H. J. 1984. Interdependence between the Economy and the Environment: from Principles to Practice. Paper prepared for the OECD International Conference on Environment and Economics in Paris, France, June 18–21.

Farber, S. C. 1987. The value of coastal wetlands for protection of property against hurricane wind damage. *Journal of Environmental Economics and Management* 14: 143–51.

Folke, C., and T. Kaberger, eds. 1991. Linking the Natural Environment and the Economy: Essays from the Eco-Eco Group. Dordrecht: Kluwer.

Gosselink, J. G., E. P. Odum, and R. M. Pope. 1974. The Value of the Tidal Marsh. Center for Wetland Resources (Publ. No. LSU-SG-74-03). Baton Rouge, LA: Louisiana State Univ.

Hair, J. D. 1988. The Economics of Conserving Wetlands: A Widening Circle. Paper presented at Workshop on Economics, IUCN General Assembly, February 4–5 in Costa Rica.

Howe, C. W. 1991. Frontiers in the Valuation of Non-market Amenities: Progress and Remaining Problems. A paper prepared for "Convegno Internazionale di Studi in Onore di Carlo Forte." Estimo ed Economia Ambientale: Le Nuovo Frontiere nel Campo della Valutazione.

Hueting, R. 1984. Economic Aspects of Environmental Accounting. Paper prepared for the Environmental Accounting Workshop, organized by UNEP and hosted by the World Bank, Washington, DC, November 5–8.

———. 1990. Correcting national income for environmental losses: a practical solution for a theoretical dilemma. In Ecological Economics. The Science of Management of Sustainability, ed. R. Costanza. New York: Columbia Univ. Press.

———. 1991. The use of the discount rate in a cost-benefit analysis for different uses of a humid tropical forest area. *Ecological Economics* 3: 43–57.

Hufschmidt, M. M., D. E. James, A. D. Meister, B. T. Bower, and J. A. Dixon. 1983. Environment, Natural Systems, and Development: An Economic Valuation Guide. Baltimore: Johns Hopkins Univ. Press.

Johansson, P. O. 1987. The Economic Theory and Measurement of Environmental Benefits. London: Cambridge Univ. Press.

Knetsch, J. L., and P. H. Freeman. 1979. Environmental and economic assessments in development project planning. *Journal of Environmental Management* 9: 237–46.

Krutilla, J. V., and A. C. Fisher. 1975. The Economics of Natural Environments: Studies in the Valuation of Commodity and Amenities Resources. Resources for the Future. Baltimore: Johns Hopkins Univ. Press.

Markandya, A., and D. Pearce. 1988. Environmental Considerations and the Choice of the Discount Rate in Developing Countries. Environment Department Working Paper No. 3. Washington, DC: World Bank.

McNeely, J. A. 1988. Economics and Biological Diversity: Developing and Using Economic Incentives to Conserve Biological Resources. Gland: IUCN.

Myers, N. 1984. The Primary Source. New York: Norton.

———. 1988, Tropical forests: much more than stocks of wood. *Journal of Tropical Ecology* 4: 209–21.

Opschoor, J. B. 1986. A Review of Monetary Estimates of Benefits of Environmental Improvements in the Netherlands. Paper prepared for the OECD Workshop on the Benefits of Environmental Policy and Decision Making, Avignon, France, October 8–10.

Pearce, D. W. 1987. Economic Values and the Natural Environment. Univ. College of London Discussion Papers in Economics 87 (8): 1–20.

Pearce, D., and A. Markandya. 1987. The Benefits of Environmental Policy: an Appraisal of the Economic Value of Environmental Improvement and the Economic Cost of Environmental Damage. Paper prepared for the OECD Workshop on the Benefits of Environmental Policy and Decision Making, Avignon, France, October 8–10.

Pearce, D., A. Markandya, and E. B. Barbier. 1989. Blueprint for a Green Economy. London: Earthscan.

Pearce, D., and R. K. Turner. 1990. Economics of Natural Resources and the Environment. New York: Harvester Wheatsheaf.

Pearsall, S. 1984. In Absentia Benefits of Natural Preserves: A Review. *Environmental Conservation* 11(1): 3–10.

Peters, C. M., A. H. Gentry, and R. O. Mendelsohn. 1989. Valuation of an Amazonian rainforest. *Nature* 339 (June 29): 655–56.

Sedjo, R. A. 1983. The Comparative Economics of Plantation Forestry. Washington, DC: Resources for the Future.

Stokoe, P. K. 1988. Integrating Economics and EIA: Institutional Design and Analytical Tools. Background paper prepared for the Canadian Environmental Research Council.

Thibodeau, F. R., and B. D. Ostro. 1981. An Economic Analysis of Wetland Protection. *Journal of Environmental Management* 12: 19–30.

Wagenaar Hummelinck, M. G. 1984. Tidal areas, a blessing in disguise. *Environment Features* 84–3. Strasbourg: Council of Europe.

10 REBUILDING A HUMANE AND ETHICAL DECISION SYSTEM FOR INVESTING IN NATURAL CAPITAL

Frank B. Golley
Institute of Ecology and Department of Zoology
University of Georgia
Athens, GA 30602

ABSTRACT

A case study of modern irrigation in northern Spain is used to demonstrate the positive and negative aspects of investing in natural capital. The project, La Violada, receives water from the Spanish Pyrenees. La Violada project is 5000 hectare in size and is organized into about 700 farms. Production of crops is relatively high, and the economic return is favorable. However, the landscape produces salt pollution which comes from gypsum in the soil. Gypsum is dissolved by the irrigation water and pollutes the rivers receiving the discharge from the project. The case study is an example of discounting environmental and social costs in the development of natural capital. How can the costs be reduced or avoided and the system be made more humane, ethical, and sustainable? The recommended steps toward a human and ethical system involve three points: (1) a form of decision making that involves all stakeholders, (2) recognition of spatial and temporal hierarchies, and (3) an ethical system that recognizes limits and intergenerational justice.

INTRODUCTION

The phrase "investment in natural capital for sustainability" is both deceptively simple and dreadfully complex. It is simple by its reference to nature and the need for humans to relate to and care for nature. Who can disagree with that? It is complex in its transference of the concept of nature, first, into natural resources and then, second, into natural capital, implying that humans can invest in and therefore receive a return from natural capital in perpetuity. I hope that this very brief deconstruction of the phraseology of this book title challenges the reader to examine the assumptions and ideas underlying this concept. What are we really talking about?

The answer to this question is not immediately clear. My contribution to a process of challenge and examination of the concept is to present a case study and think about its implications for sustainable development, evaluating its positive and negative aspects and considering how we could avoid the unacceptable costs that damage the environment and reduce human well-being. The case study comes from northern Spain, from the Autonomous Region of Aragon. It is based on the work of nine Master of Science students at the Instituto Agronomico Mediterraneo de Zaragoza and several technical papers which summarize their studies (Bellot and Golley 1989; Bellot et al. 1989; Golley et al. 1990; Golley and Bellot 1991). The basic data are available in the theses located in the library of the Institute. The case study involves a project called "The La Violada" after the place where it was carried out.

LA VIOLADA PROJECT

The La Violada project is located on a 5,000 hectare area of irrigated land in an arid region of Europe. The project was constructed in the 1940s and 1950s in a basin which originally contained a saline lake. Geologically the substrate is made up sedimentary deposits in horizontal strata. Some of the beds contain calcium sulfate or gypsum, which is water soluble. Farmers were moved from other parts of Spain to this new agricultural development, and currently the area is divided into about 700 farms. Agricultural production is comparatively quite high. Production is subsidized with water, fertilizer, pesticides, and other forms of advanced technology. The project also produces large quantities of salts, which are leached from the soils and leave the irrigated area in the water that flows from the fields to receiving rivers. The quantity of salt pollution almost reaches 11 MT per hectare per year and is largely derived from gypsum. These salts contaminate the Gallego and Ebro rivers and reduce the utility of the river water for further irrigation and other uses (Alberto et al. 1986).

In the La Violada project the Spanish government invested in natural capital. The capital was in the form of rainfall in the Spanish Pyrenees and in the availability of a flat arid landscape with ample sunshine and temperature which could be converted into farmland for year-round crop production. The investment was about 235 million (1940) Spanish pesetas. The investment converted an area that had been largely dry pasture for sheep into a highly productive agricultural system. Traditionally irrigation was confined to the land on the side of rivers, and essentially all of this type of land had been converted to irrigation hundreds of years ago (Maass and Anderson 1978). The irrigated huertas or gardens of Zaragoza were famous in the past. Crop yields on the new land are relatively high and the financial yield from the project has been large (see Table 10.1). The main negative aspect of the project has been the salt pollution, and the pollution only became of broad public concern as more and more of the arid landscape was renovated into agriculture and as urban and industrial demand for river water increased. The question posed in our studies was how could the salt production be decreased and crop production maintained? Our approach was to examine the biogeochemistry of the landscape and determine the controls on the principal flows into

Table 10.1. Economic Return on La Violada Irrigated Landscape, in Spanish Pesetas

Crop	Area	Yield	Value of Yield	Value Per Area
	Ha	MT/Ha	pstas/Ha	1,000,000 pstas
Maize	2,392	12	208,000	498
Barley	718	5	53,250	38
Wheat	206	6	103,500	21
Alfalfa	483	13	202,340	98
Totals	3,799			655

This chapter will focus on the controls on the biogeochemistry rather than the basic data. There are two sets of controls: those that come from the environment outside the system of interest, and those that come from within the system. External controls involve the rainfall which feeds the irrigation system, the economic market that supplies the farms and receives their products, and the system of regulations and laws of the society in which the system is located. Internal controls concern the technology which dictates alternatives and the cultural and social character of the farmers which determines how they act.

Rainfall is variable from year to year (see Figure 10.1) but this variability is controlled through the construction of reservoirs to collect and hold water and canals to transport the water to the projects and fields. Irrigation is by gravity. In this way variability in rain input in space and time can be managed by the irrigation authorities. At this time water is available in excess and the canals flow year-round, so there is always a base flow through the system. The price of this water is fixed at a relatively low rate of about 250 pesetas per 1,000 cubic meters.

The economic environment controls the prices farmers pay for inputs and services and the price of the products they produce. In Spain the index prices received for crops increased 2.5 times from 1976 to 1985, while costs of all requisites to farmers increased 3.1 times (FAO 1986). As the profit margin has shifted, farmers on the La Violada area have compensated by growing more maize. The variability in the economic environment has also been controlled in part by the formation of a cooperative which provides farmers with supplies and services and receives, stores, and markets production. The shifting ratio of costs and benefits has also resulted in more salt pollution because maize requires more water and fertilizer.

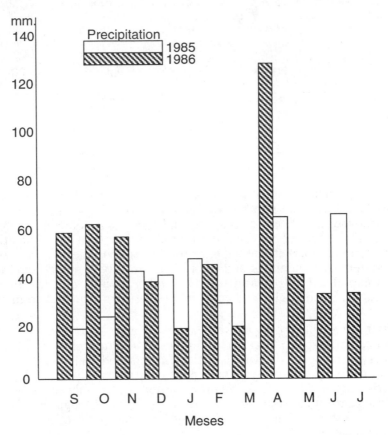

Figure 10.1. Monthly rainfall in mm, at La Violada in 1985 and 1986, data from the thesis of Fermin Cerezo.

In order to evaluate the relationships between crops and water, an energy analysis was conducted along with the economic analysis (a summary of this analysis is shown for maize in Table 10.2). In general, the energy cost of water using the constants of H. T. Odum (1970) was 40% of all energy costs. In contrast, water was only about 2% of the peseta costs. It appears that water is underpriced. Because of this subsidy the farmer has little or no information about the value of water and the relation between water use and crop production. This lack of information means that a market mechanism to reduce salt pollution is not available and the effort to control pollution must be shifted to regulation. However, in this situation the regulatory environment is weak and ineffective.

System Behavior (Kg/ha)

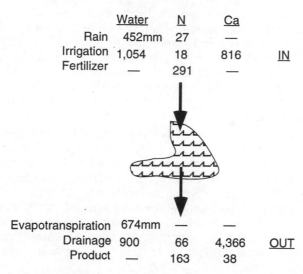

	Water	N	Ca	
Rain	452mm	27	—	
Irrigation	1,054	18	816	IN
Fertilizer	—	291	—	

Evapotranspiration	674mm	—	—	
Drainage	900	66	4,366	OUT
Product	—	163	38	

Figure 10.2. A diagram of flows into and out of La Violada landscape, illustrating the patterns observed for a variety of elements. Water and nitrogen represent subsidies, while calcium represents an earth element subject to solution by the water.

Table 10.2. A comparison of energy and monetary data for the production of a crop of maize on La Violada landscape. Energy is reported in million joules per hectare and the economic expenses and gains in Spanish pesetas per hectare. The data represent one crop period, not a full year.

Inputs	Energy MJ/Ha	Economics Pesetas/Ha
Irrigation Water & Ppt	29,535	3,000
Seeds	405	13,000
Fertilizer	37,722	46,333
Machinery (sequestered)	1,849	
Farming operations	10,018	10,500
Fuel	6,707	7,680
Pesticides	151	4,202
Total	86,387	84,715
Outputs		
Harvested Product	218,252	212,500
Drainage Water	701	
Evapotranspiration	30,472	

Finally, we can consider farmer attitudes. The farmers on this project have migrated to the region from other parts of Spain. The project was conceived as a development project providing land to landless farmers. The consequence is that the farmers do not have long association with place, are less influenced by traditional practices and are more responsive to innovation. They also tend to have less commitment to the farm environment. For example, several retirees have left the project villages and returned to their original home villages. This looseness in place relationship may lead farmers to pay less attention to maintaining the environment for the future and to focus more on short-term economic gain.

Our answer to the initial question of control of salt pollution from this irrigated landscape was to suggest that the price of water to the farmers be altered. Of course, we do not know the real price of water but we do know that water was about 40% of the total energy costs. Except for water and a few other inputs, changes in energy values and monetary values of inputs and outputs in Table 10.2 appear parallel. Water seems to be substantially underpriced. If the price of water is raised, farmers might be motivated to shift toward crops which use less water and more closely allocate water to plant needs. Agronomists and hydrologists in Aragon feel that water use might be reduced 20% by these means (Faci et al 1984).

The problem of sustainability remains. The La Violada project is about 50 years old. The rate of solution of the gypsum in the soil and in the parent material beneath the soil is uncertain. However, solution caverns have already appeared in some fields, which mean that the field will no longer hold irrigation water unless the cavern is filled with soil and rock moved in by mechanical equipment. The solution pipes in the gypsum may threaten the life of the project. Thus, under the current management regime the project may not be sustainable, but how long production can be sustained remains uncertain due to the lack of an inventory of the gypsum and an understanding of the rate of solution of gypsum in this local situation.

A further problem of sustainability involves the mountain watersheds which are the source of water for the project. These watersheds are being developed for tourism, and the dynamics of water flow are being changed. How these developments will influence water yield and ultimately agricultural production has not yet been determined.

THE PROBLEM OF PARTIAL ANALYSIS

In this case study we have an example of an investment in the natural capital of rainfall and arid land to create a humanized landscape highly productive of agricultural product, which also causes salt pollution of the rivers. The 5,000 hectare La Violada project is the locus of flows of water from a mountainous region in northern Aragon, and it influences another region through its salt pollution of the Gallego and Ebro rivers, which enter the Mediterranean Sea. In this case a relatively small area receives large subsidies and has a correspondingly large impact on the surrounding environment. The case is a familiar one in arid regions. For example, participants at a seminar in Sacramento, California commented that the data on biogeochemistry and pollution on La Violada were almost identical to those for California irrigated landscapes.

The source of the problem of salt pollution is relatively clear in this case study. The engineers ignored the fact that gypsum is water-soluble or could not calculate the rate of solution of gypsum and its contribution to runoff from the fields to the receiving rivers and therefore disregarded the problem. Further, there has been no attempt to calculate the costs of the salt pollution to downstream users. Here, we have an example of the familiar problem of partial analysis. Perhaps the problem should have been anticipated and incorporated into the project design.

Partial analysis is both a philosophical and a practical problem. To avoid partial analysis one must be able to analyze a problem deeply and deconstruct it, as well as see the problem broadly and understand it as part of a larger whole. To do this requires disciplined training and a recognition of hierarchies of systems. A possible theoretical resolution of this fundamental problem will be considered in the next section. But if we could solve it, then we are left with a methodological problem. How does one distinguish key factors in a design problem from insignificant factors? How do we deal with the problems of scale? The human brain has a limited capacity to hold information in immediately recoverable form. The human group size for close interaction to achieve an objective is relatively small. The ecological world is organized hierarchically so that the scale of the ecosystem in space and time is important to its function. How can we incorporate the relevant knowledge, the role of culture, religion, and other nonquantitative yet fundamentally important information in problem solving? How do we avoid the problem of power through which humans manipulate culture, ideology, and economics in order to accumulate and maintain personal and organizational status and authority?

In other words, partial analysis has both positive and negative impacts. It can be used to conceal the accumulation of power and resources by specific groups or individuals, and it confuses those objecting to a given action. The responsibility for partial analysis is diffused in a bureaucracy so that personal responsibility is avoided (which is one of the reasons that the professions support and expand bureaucracy). The costs may also be transferred to future generations or distributed over entire populations so that they become almost invisible. The latter situation occurred in the case study where salt pollution costs were distributed over all users of downstream river water, while the benefits accrued to a limited and well-defined population of farmers.

THEORETICAL RESOLUTION

In order to resolve the dilemmas posed by the La Violada case study, we need methods which will help us avoid partial analysis. I have suggested that the problem lies in the social organization that underlies decision making, in the

concepts used to make decisions, and in the methods used to evaluate alternatives and resolve conflicts. My suggestions are not original nor unique, and
they do not form a complete solution. They do, however, highlight three areas
which absolutely require attention.

First, it is essential that decisions be made by the people being effected by
the decision. This is obvious yet it is seldom accomplished in practice. The
large number of people involved, their lack of knowledge and emotional response to alternatives, and their lack of experience in conflict resolution all
make public discussion and decision making difficult. It is often necessary to
have individuals represent the parties involved directly and indirectly in the
decision. How are these stakeholder representatives chosen? This turns out to
be a key question in resolving the partial analysis problem. Frequently, certain
individuals and professions claim that they represent the stakeholders. These
are frequently elected officials, political party members, and experts, including scientists. Such people are motivated to advance their own interests,
through the decision process. While this is not intrinsically bad, it is inefficient
and leads to the problem of partial analysis. Such people need to be around
the table but so do the other stakeholders or their legitimate representatives.
Naturally, this slows down the process. The argument and discussion required
to find compromises that meet the goals of all the users is difficult. Sometimes
there is no solution and the project should be abandoned for that time.
Nevertheless, the stakeholders are those who bear the costs and reap the benefits, and they are the ones with traditional and local knowledge. In the La
Violada case, what would have been the decision if the traditional irrigators of
the region had been included in the decision-making system?

Second, there are two forms of thinking required in these stakeholder
meetings—the capacity to think hierarchically and ethically. Hierarchical
thinking is a way of structuring decisions in time and space. There are as
many hierarchies as we want—hierarchies are merely mental constructs used
as tools to organize thinking. Hierarchical thinking recognizes that phenomena are related. Every event is part of large-sized, long-term, more complex
sets of events and is also composed of smaller, short-term, simpler events.
Hierarchies do not necessarily imply that higher or lower order phenomena
control events—they may be merely a structural way of organizing objects or
events. Any problem which is the object of our attention can be decomposed
into subproblems or subsystems and components and also structured into
suprasystems. Study of these subsystems and suprasystems helps us understand the mechanisms that define the problem. Rudolf de Groot, in this book,
describes a way to organize our thinking about these "functions."

In an evaluation of a complicated environmental system, there is a temptation to convert the functions into a single currency. This appears to make the
discussion less complex, but it can be a trap. The problem is that in converting
all functions into one currency requires that we accept assumptions about
substitutability of functions. In order to make the value of biodiversity, for
example, comparable to the market value of the trees in a forest, it is necessary

to create an imaginary market for biodiversity, construct shadow pricing, or ask consumers how much they value biodiversity. Each of these alternatives has conceptual problems associated with it. Actually, the human mind is fully capable of comparing apples and oranges—we do it every day. In a committee of stakeholders apples and oranges can be compared and a decision reached without conversion problems. Under bureaucratic decision making, however, a common currency is desirable for a variety of positive and negative reasons. For those with no expert knowledge, a decision is simplified because a choice can be made between the alternative with largest or smallest value. Common currency also may be an advantage when an administrator wants to create abstractions to conceal decisions and confuse the stakeholders. We need to accept that everything is not convertible or substitutable and that ordinary people can handle such criteria in their decision making. This does not mean that the arguments about value go away. Actually they become more fierce, but they focus on the impacts to specific ends rather than on abstract valuation.

Finally, decision making needs to be informed by an ethical form of thinking that has at least two parts:

(1) The ethics must include the concept of *limits*. Every dynamic action, from individual human behavior to the behavior of nations, must reflect the limitations of the environment, human potential and capacity, and the human mind. The concept of continual progress and unlimited growth is an artifact of a particular cultural history of colonization and despoiled environments. This way of thinking can no longer be sustained. Progress now reflects other characteristics—more qualitative, downsized, craft-like, skilled, caring, loving, and humane. We need to insist that these softer ways of thinking and feeling assume dominant roles in decision making. It is not enough to ask what will it cost. We must first ask how will it effect people and the environment.

(2) The ethic needs to recognize intergenerational justice. We can no longer maintain systems of exploitation by transferring the costs to the future. This is a fundamentally immoral act. We do not know the future. It is irresponsible to incur costs now on the assumption that they will be paid by someone else. Ethical behavior requires that we live within the limits that exist here and now, without damaging the use of resources in the future.

Obviously our decision-making systems are far from these ideal forms. In order to move from the current state toward the ideal, we need education of the citizens in a social ethical value system that attempts to embrace the world as it is, without proposing romantic utopias or hegemonies of power. Such a system would seek standards of human well-being that could be used to guide preservation and management of nature, reestablish stability of the human population and its demands, and create designs that would build health, creativity, beauty, and integrity into the human environment. This form of education requires more breadth and depth, in the traditional sense of those terms in education. The traditional university no longer seems to fit the task of education. It seldom trains the young to solve problems of the present, because there is little broad social-ethical training given to them—narrow disciplinarity is the norm. The perceptive student recognizes these weaknesses in the curriculum and seeks ways to cross the disciplinary gaps. The scholar recognizes the problems engendered by partial analysis and invents bridges across the gaps. One such bridge is ecological economics. There seems to be a movement, a convergence, throughout the intellectual community toward a new center. Our task is to encourage and foster this convergent movement.

ACKNOWLEDGMENTS

This work was based on the Masters work of Ana Campillo, Fermine Cerezo, Cecilia Esquisabel, Esperanza Amezqueta, Puy Trebol, Maria del Mar Serrano, Jose Carlos Gonzalez, Maria del Mar Torres, and Maite Aguinaco, and I acknowledge with appreciation their contributions to understanding La Violada. The study was conceived and administrated by Juan Bellot, former administrator of IAMZ. The present administrator is Maite Aguinaco. I also acknowledge with thanks their contributions. Finally, the work was supported by the Instituto Agronomico Mediterraneo de Zaragoza, and I appreciate the kind attention of the Institute's director, Dr. Miguel Valls, and the helpful comments of Colin Clark, who reviewed the original paper.

REFERENCES

Alberto, F., J. Machin, and R. Aragues. 1986. *La problematica general de la salinidad en la cuenca del Ebro*. In Sistema Integrado del Ebro. Estudio Interdisciplinar. Madrid: Editorial Hermes.

Bellot, J., and F. B. Golley. 1989. Nutrient input and output of an irrigated agroecosyystem in an arid Mediterranean landscape. *Agriculture, Ecosystems, and Environment* 25: 175–86.

Bellot, J., F. B. Golley, and M. T. Aguinaco. 1989. Environmental consequences of solts exports from an irrigated landscape in the Ebro River basin, Spain. *Agriculture, Ecosystems, and Environment* 27: 131–8.

Campillo, A. R. 1987. Estudio agroecologica del poligono de riego de la Violada: comparacion do los flujos de energia y la productividad de los cultivos trigo y maiz. Masters thesis, IAMZ.

Faci, J., R. Aragues, F. Albert, D. Quilez, J. Machin, and J. L. Arrue. 1984. Water and salt balance in an irrigated area of the Ebro River Basin (Spain). *Irrigation Science* 6: 1–9.

FAO. 1986. FAO Production Yearbook, Vol. 40. Rome: FAOUN

Golley, F., B. A. Campillo Ruiz, and J. Bellot. 1990. Analysis of resource allocation to irrigated maize and wheat in northern Spain. *Agriculture, Ecosystems, and Environment* 31: 313–23.

Golley, F. B., and J. Bellot. Interactions of landscape ecology, planning and design. *Landscape and Urban Planning* 21: 3–11.

Maass, A., and R. L. Anderson. 1978. And the Desert Shall Rejoice: Conflict, Growth and Justice in Arid Environments. Cambridge, MA: MIT Press.

Odum, H. T. 1970. Energy value of water resources. In 19th Southern Water and Pollution Conference Proceedings. Durham, NC: Duke Univ.

11 RE-ALLOCATING WORK BETWEEN HUMAN & NATURAL CAPITAL IN AGRICULTURE: EXAMPLES FROM INDIA & THE UNITED STATES

Cutler J. Cleveland
Center for Energy/Environmental Studies; Dept. of Geography
Boston University
675 Commonwealth Avenue
Boston, MA 02215

ABSTRACT

This chapter explores the interrelation between human capital and natural capital in agroecosystems. The evolution from traditional to industrial agriculture was driven by two substitutions. The first is the substitution of non-renewable natural capital (principally fossil fuels) for animate energy sources (draft animals and human labor). The second is the substitution among forms of human capital, principally manufactured capital for traditional knowledge. From this perspective, the Green Revolution is based on lager and larger subsidies of solar energy with non-renewable natural capital and human capital. Plant breeding has increased the harvest ratio—the percentage of total crop substance in organs that humans demand. Increasing the harvest ratio requires the reallocation of work between natural and human capital in agroecosystems. The result is a higher yield per hectare and per laborer, but also a greater impact on the natural capital basis of agriculture (soil erosion, diminished biodiversity, etc.). Empirical case studies of Indian and U.S. agriculture demonstrate the large reliance on non-renewable natural capital, and the impact of food production on natural capital.

INTRODUCTION

The manipulation and alteration of natural ecosystems to produce food and fiber are the largest agents of environmental change in human history. In the last century alone, more than 2.3 billion hectares of wetlands, forests, and deserts, and other ecosystems—equivalent to about 17% of the land surface—were converted to arable land and pastures (Buringh 1989). Agriculture now accounts for about 20% of terrestrial net primary production, and arable (1.5

billion ha) plus permanent pasture (3.3 billion ha) occupy about 35% of the land surface (Odum 1989). Until the mid-twentieth century the impact of agriculture was due largely to spatial and temporal intensification—the increase in the quantity of land under cultivation and the increase in the number of crops cultivated per year. In the past 50 years, that impact has been accentuated by the technological intensification of agriculture—the increased use of fuel, tractors, fertilizers, chemicals, and water per hectare of cultivated land. Technological intensification increases food production, but it also causes a range of environmental problems—soil erosion, diminished genetic diversity, pesticide runoff, salinization of soils, groundwater pollution, and depletion.

The problems associated with all forms of agricultural intensification are likely to worsen in the coming decades due to the attempt to feed a global population of 5.5 billion people growing at 1.74% each year. There is an upper limit to the spatial intensification, although estimates of the ultimate quantity of land that can be cultivated vary from 2 to 3.4 billion ha (Pimentel et al. 1989). What is certain, however, is that most new land brought under cultivation will be lower quality, implying that technological intensification will accompany any future spatial intensification. That implies an increase in the use of fossil fuels and non-renewable minerals such as phosphate rock. In addition, the degradation of land already under cultivation—much of it high quality land, requires larger inputs of fossil fuels to offset the loss of soil fertility. Rapid intensification is under way in many developing nations.

The current modern version of the population-food-environment nexus clearly demands an ecological economic framework. The field of ecology has traditionally ignored the study of agroecosystems as systems that are dominated by human manipulation, although the growing field of agroecology has begun to bring together agricultural scientists and ecosystem ecologists, academics that until recently were relatively uninterested in each others' research. Agricultural economics has a rich history, but like other branches of standard economics, it ignores the mechanisms of agroecosystems that bind the agricultural sector to the biophysical world. The economic and environmental challenges that confront global food production can be met only with an analytical framework that consciously reflects the nature of agriculture as the co-evolution between society, culture, and the environment (Gliessman 1990).

THE HUMAN AND NATURAL CAPITAL BASIS OF AGROECOSYSTEMS

Food production in agroecosystems depends on the availability of natural capital and human capital. Renewable natural capital (RNC) includes the products and services generated continuously by environmental processes that are powered by solar energy and residual heat energy from the earth's core. Renewable natural capital is self-reproducing and self-maintaining. Examples of RNC are ecosystem goods and services, such as a stand of timber or the operation of the hydrologic cycle. Non-renewable natural capital (NNC) includes products of biogeochemical processes that were powered by solar energy and residual heat energy in the past, and which do not re-generate themselves on a time scale of interest to the humans species. Examples of NNC are fossil fuel and mineral deposits. Human capital (HC) includes manufactured capital (MC) and cultural capital (CC). Manufactured capital includes the factories, buildings, and other physical artifacts usually associated with the term "capital." Cultural capital includes the stock of education, skills, culture, and knowledge stored in humans and their institutions.

Climate is a unique form of renewable natural capital because light, temperature, and water availability are important determinants of the productivity of crop species. Climate is a principal determinant of the crop species, the types of cultivar, and the management practices that are suitable for agriculture in different regions of the world (Hall 1990). In addition, there are important feedbacks between the use of HC that controls the use of NNC, such as fossil fuels, which in turn can modify climate. Any change in global temperature and precipitation patterns that result from human-induced climate change will affect the distribution and productivity of agroecosystems. Indeed, of all the changes that may result from climate change, the impact on agriculture is likely to be one with far-reaching socioeconomic effects.

Soil is another critical form of renewable natural capital in agroecosystems. Soil formation results from the interaction of physical, chemical, and biological processes that produce a fertile top layer of soil, which supports plant growth. The processes that yield fertile land are extremely rare. About 78% of the earth's landscape is too wet, dry, stony, cold, steep, or shallow to support crop production. The area suitable for crop production is only 22%, and less than 3% of the land area is endowed with high quality soil. The biogeochemical cycles that yield fertile land are also extremely slow. Under tropical and temperate agricultural conditions, soil is formed at the rate of 0.3 to 2 t/ha/yr (Pimentel et al. 1987), which means that it takes hundreds or thousands of years in most regions to generate a layer of fertile topsoil.

Biological diversity is another component of the renewable natural capital basis of agriculture. About 90% of our food comes from 15 plant species and 8 animal species, but agricultural productivity also depends on the vast array of services generated by the genetic diversity of other ecosystems. These services include: (1) maintaining the genetic diversity necessary for successful crop breeding; (2) recycling nutrients such as carbon and nitrogen; (3) conserving soil and water; (4) pollinating crops; (5) removing toxic pollutants from water and soil; and (6) buffering the impacts of air pollutants, and sources of certain medicines, pigments, and spices.

Ecosystems generate many other goods and services that sustain plant growth in agroecosystems. Examples are the hydrologic cycle that provides precipitation, runoff, and groundwater, and biogeochemical cycles that provide nutrients such as nitrogen and carbon. These services are provided only by healthy, intact ecosystems. Thus, the structure and diversity of ecosystems is a critical component of the natural capital base of agroecosystems.

Industrial agroecosystems are also powered by large quantities of NNC. Crude oil is extracted and transformed into fuel and chemicals such as pesticides, herbicides, and insecticides. Natural gas is the principal feedstock and energizer of the production of nitrogen fertilizer. Phosphate rock is excavated and refined into phosphate fertilizer.

The HC component of agroecosystems is as large and diverse as the NC component. Manufactured capital includes the tractors, irrigation pumps, storage facilities, and other equipment used in farm operations. Cultural capital refers to factors that provide human societies with the means and adaptations to deal with the natural environment, and to actively modify it (Berkes and Folke 1992). Cultural capital includes the stock of knowledge, education, skills, and culture that is actually stored in human beings themselves. Like natural capital, cultural capital yields a flow of services—the attitudes, morals, ethics, and information that determines a culture's relationship to the natural environment, and especially its resource management strategies. Cultural capital determines which types of natural capital are used by a society, how quickly it is used, the degree of emphasis placed on the acquisition of material goods, the ethical relationship with non-human species, and the degree of concern for future generations. Some important aspects of cultural capital in agriculture are knowledge of the physical environment as it affects crop growth (i.e., soil type, and rainfall patterns) and knowledge about farming practices (i.e., provision of water and nutrients, and control of succession). In industrial societies, farmers must also possess information about the use of sophisticated forms of MC (e.g., tractors, combines).

THE RELATION BETWEEN HUMAN CAPITAL AND NATURAL CAPITAL IN AGRICULTURE

The interdependencies among HC, NC, and climate lie at the heart of ecological economics (Costanza and Daly 1992). Agroecosystems are embedded in a complex web with many other ecosystems, a web that is formed and maintained by the flow of energy, the cycling of materials, and the exchange of ecosystem services. These stocks and flows of NC exist at spatial scales that extend far beyond the field or pasture. The flow of energy and materials (e.g., groundwater) and the provision of ecosystem services (e.g., crop pollination) link the agroecosystem to the rest of the environment by providing the life support necessary for food production. Many of those flows and services are controlled by HC, either purposefully or inadvertently, with both beneficial and deleterious results.

Two substitutions characterize the evolution from traditional to industrial agriculture. The first is the substitution between the principal energizers of production in agroecosystems. For the first 10,000 years of agricultural society, various forms of RNC powered the production of food and fiber. In traditional agriculture, human labor and draft animals provided the energy for land conversion, cultivation, harvest, and pest and weed control. In industrial agriculture, NNC forms of energy are the dominant energy sources. Refined oil products, natural gas, and electricity power machines that now perform tasks done previously by humans and draft animals, or tasks that were once performed for "free" by ecosystem services. Examples are diesel fuel used to power tractors, electricity used to pump groundwater for irrigation, and natural gas used in crop drying. Non-renewable forms of energy are also used indirectly—the natural gas used as a feedstock and energy source in the production of nitrogenous fertilizer, the crude oil used as a feedstock and energy source in the production of pesticides, insecticides, and herbicides, and the fuels used to produce tractors and implements.

The second important substitution occurs among different forms of human capital. The importance of manufactured capital has increased enormously. Farm implements, tractors, vehicles, storage facilities, and other forms of MC are a large and integral part of industrial agriculture. Many forms of MC are energy converters powered by fossil fuels.

Perhaps the most important change in the use of human capital in agroecosystems is the change in the role of cultural capital. Traditional agriculture relied on traditional knowledge—information about agroecosystems that originated locally and naturally, selected through a complex evolutionary process, and passed on from generation to generation. Traditional knowledge in agriculture includes (Altier 1990):

- intimate knowledge about the local physical environment (precipitation, and soil types),

- biological classification systems for local plants and animals, and

- knowledge of farming practices that are suitable for local conditions.

Traditional knowledge has allowed farmers throughout the world to adapt successfully to the environmental constraints specific to their region for long periods of time. These resource management strategies include intercropping, agroforestry, terracing, green manuring, canal irrigation, use of drought and shade tolerant species, and enhancement of natural pest enemies.

Cultural capital in industrial agriculture is very different. Most of the intimate ecological knowledge embodied in traditional agriculture has been lost or is ignored. Farmers today must have sophisticated information about finance, business, engineering, and technology. They must know something about how internal combustion engines work and how to properly apply chemical sprays. Thus, there has been a fundamental shift in the role of cultural capital away from the understanding and synchronization with local flows of renewable natural capital to the manipulation and control of large flows of non-renewable natural capital that originate outside of the farm. These substitutions have the following important implications for sustainability of world food production in the future.

ENERGY FLOW IN AGROECOSYSTEMS

Energy and the Evolution of Agricultural Society

A brief overview of the evolution of agricultural society illustrates the importance of the various substitutions between human and natural capital. Contrary to a common stereotype, the transition from hunting and gathering systems to sedentary agriculture did not occur because agricultural society guaranteed a more secure supply of food and an easier way of life, or because hunters and gatherers lacked the knowledge to make the transition. Agriculture is, in fact, more intensive than most hunting and gathering, and people therefore generally work longer hours in agricultural systems. Anthropologists and archaeologists show that some pre-agricultural societies had characteristics that are normally associated with more "advanced" agricultural societies—sedentism, relatively high population densities, long-term food storage systems, social stratification, and even incipient agricultural techniques. Pre-agricultural societies were endowed with substantial traditional knowledge about their surrounding ecosystems that provided their life support.

Because there was no clear and overriding advantage of the agricultural way of life, there had to be some factor or group of factors that "pushed" society in that direction. Some type of imbalance must have existed between the size of a population, its nutritional and lifestyle preferences, and the life support capacity of its food production system. The transition to an agricultural way of life was caused by a combination of energetic, demographic, climatic, social, and environmental factors (Minc and Vandermeer 1990). Favorable climatic changes in the post-Pleistocene period, the availability of domesticatable plant species, nutritional imperatives, and a variety of sociocultural factors were necessary conditions for the transition to an agricultural system. One of the principal driving forces was the steady onset of diminishing returns to hunting and gathering systems caused by the slow but continuous population growth throughout most of the inhabited regions of the world (Boserup 1965). Demographic stress slowly reduced the returns to, and eventually eliminated the possibility of, the strategy of expanding the area of resource exploitation that characterized many hunting and gathering societies' responses to population growth. The intensification of food production by artificially increasing the density of desirable plant species was one option available to societies confronted with that problem. It is likely that, over the long run, societies that chose to let their populations grow and intensify their agriculture out-competed other strategies for space and resources (Odum 1971; Cohen 1977).

While the transition to agriculture was caused by an array of factors, once adopted, further intensification was driven principally by energetic imperatives (Smil 1991). Agriculture has two preeminent advantages over pre-agricultural systems: (1) the ability to increase the amount of food produced per unit area, and (2) the ability to artificially manipulate yields. Intensification of agricultural production takes many forms, but they all have a common bio-physical basis—growing more food per hectare requires inputs of more energy per hectare. Humans obviously cannot increase the solar flux, which means that intensification requires the increased use of other renewable or non-renewable energy. Thus, the intensification of agriculture over the past 10,000 years, and especially in the last 100 years, was powered solely by increases in the quantity and/or quality of energy used per hectare. Various forms of renewable and non-renewable natural capital supplied that energy—a team of horses, a windmill, a diesel tractor, or a bag of fertilizer. Increasing the quantity or quality of energy used per hectare of land increases the quantity of the site's potential photosynthate that is converted to edible food, and thereby increases the density of population that can be supported. These principles are demonstrated by the analysis of the energetics of the evolution of cropping intensification from shifting cultivation to bush fallow, short fallow, regular annual cropping, and highly intensive multicropping (Pimentel and Pimentel 1979; Smil 1991).

The Green Revolution—Re-Allocating Work Between Human and Natural Capital

The large increases in the yields of the world's major grain crops in the last half-century are often attributed to the Green Revolution, a shining example of the triumph of science over nature. But it is important to understand the biophysical basis of the Green Revolution, which is based on (1) the large *subsidy* of solar energy with non-renewable natural capital (fossil fuels) and human capital (tractors, plant breeding), and (2) the *substitution* of certain forms of renewable natural capital in agriculture (allelopathy, nutrient cycles) with non-renewable natural capital and manufactured capital (mechanized weed control, fertilizers). In essence, what we have done is re-allocate the work required to produce edible plant biomass between various forms of human and natural capital, with a lion's share of the work now being done by the non-renewable and human capital components.

From this perspective it is important to state what the Green Revolution is *not*. Advances in human ingenuity have not significantly altered the energy conversion process that sustains nearly all life on the planet—the efficiency with which sunlight is converted to plant biomass. The gross photosynthetic efficiency of crops is less than natural systems they have replaced, systems that developed over millions of years (Odum 1971). Genetic engineering and plant breeding have had little impact on the efficiency of the photosynthetic process by which radiant energy in sunlight is converted into chemical energy in carbohydrate (Hall 1990). In North America, for example, the average rate of production of natural vegetation is about 2400 kg/ha, with about 0.1% of the solar energy converted to biomass. The most intensively fertilized, irrigated, and high-yield varieties of corn and wheat can achieve yields of 14,000 kg biomass/ha and 6,750 kg biomass/ha, respectively. This represents only 0.2%–0.5% of solar energy converted to plant biomass. Most crops are less productive than corn and wheat, so the overall efficiency of biomass in the form of crops probably averages close to that of natural vegetation (Pimentel et al. 1978).

What plant breeding and the increase in energy subsidy *has* improved is the harvest ratio—the increase in the percentage of total crop substance in organs that humans demand, such as seeds, pods, leaves, stems, tuber, roots (see Figure 11.1). Boosting the harvest ratio relies on re-allocating the work a planet needs to accomplish in order to survive and grow—the acquisition of water and nutrients from the soil, growth, the production of secondary compounds to defend against pathogens, and competition with weeds. The re-allocation of work between environmental energy and economic energy works in two important ways: (1) the removal or easing of the environmental factor(s) that most limit channeling primary production into harvestable seed, and (2) the replacement of plant or ecosystems functions that are necessary for survival, but which have been deliberately compromised, or inadvertently eliminated, by plant breeding and/or agronomic practices.

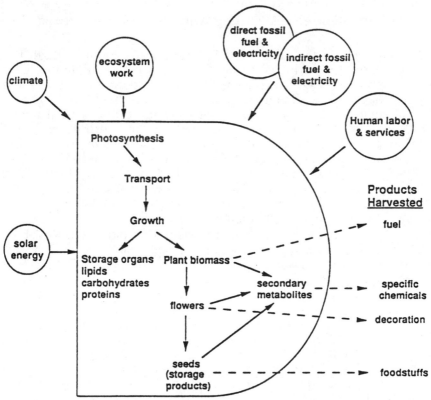

Figure 11.1. The allocation of energy flow in crops from photosynthesis to the formation of crop products (modified from Woolhouse 1984).

Limiting factors are overcome principally by the direct and indirect use of fossil fuels in industrial agriculture. Fossil fuels are used to pump groundwater or divert surface water for irrigation, erasing the water constraints on photosynthesis in regions where precipitation is the principal limiting factor. Natural gas is the principal feedstock and energizer in the production of nitrogen fertilizer that erases the constraints on production in regions where nitrogen availability is the principal limiting factor. Without the external fossil fuel subsidy, the amount of photosynthate that is diverted to seed is limited because the plant must divert significant energy/work to adapt to the stress of low water or nutrient availability. For example, desert plants must divert much of the solar energy they capture to support water-storing adaptations, deep roots, and mechanisms for opening stomata only at night when transpiration losses are minimized. But when some of the work of water supply is shifted to external sources of energy, more of the internal energy of the plant can be channeled into seed production (or whatever part of the plant has economic value).

The goal of plant breeding is to take full advantage of this re-allocation. For example, in the absence of external weed control, a tall crop with large, light-excluding leaves and an aggressive root system has a competitive advantage (Evans 1980). If the weeds are controlled by a herbicide or mechanical plowing, a shorter plant with smaller leaves and root system can be selected for through breeding, freeing assimilates for storage in seeds or other harvested organs. The same principal applies to other plant functions. The provision of irrigation water and fertilizers allow for the selection of plants with smaller root systems. Plants use some of their internal energy to produce secondary metabolites that provide resistance to disease and insect predation. The provision of external control reduces the need to channel photosynthate to the production of those compounds, creating the opportunity for more production of the desired product.

There are significant consequences of substituting the work done by plants, a form of renewable natural capital, with non-renewable natural capital and human capital. Plants that are bred to respond to fossil fuel inputs become highly dependent on them for their success and survival. Traditional varieties evolve around, and are better adapted to, the environmental conditions of their region. High yield varieties (HYV) are adapted to inputs of fertilizers, mechanization, and other fossil fuel-based inputs across many different regions. As a result, HYV of cultivars are frequently inferior to traditional varieties under low-input conditions (Evans 1980). In many developing nations without access to sufficient amounts or types of energy subsidies (fertilizers, irrigation), yields are far below the levels achieved in the industrial world.

Another consequence is that the total energy cost of plant production in industrial agroecosystems is much higher than natural systems. Plants have evolved over billions of years to capture and allocate their internal use of solar energy at a rate and efficiency that maximizes their competitive advantage. Despite the enormous increase in the stock of cultural capital, such as knowledge about plant biology, we still know surprisingly little about the genetic mechanisms that control basic plant processes such as photosynthesis, respiration, transport, flowering, and seed production (Woolhouse 1984), at least at the level that would enable us to boost photosynthetic efficiency. Therefore, human intervention in the internal energy flow of plants is less energy efficient than the natural allocation systems they replace or modify. Of course, this could change in the future—genetic engineering is still a young field, and future advances could lead to new ways of boosting yields without increased use of fossil fuels.

Weed control is a good example of the higher energy cost of human control (Jackson 1991). If some of the solar energy captured by a plant is channeled to the production of an herbicide in the roots to control weeds, then weed protection comes from within the plant. That use of solar energy has an opportunity cost in the form of a reduction in yield (i.e., solar energy stored in seed). Alternatively, protection could come in the form of mechanical weed control (plowing), but in traditional agriculture the energy penalty of that substitution is clear—a certain fraction of the energy harvested must be fed to the horses that supply the power for the mechanization. In this case the energy cost of external control is clearly higher, and in industrial agriculture it is certainly much higher.

Energy costs are higher in industrial agroecosystems also because of the strong forces of succession. The farther the installed agroecosystem is from the natural system, the more energy is required to suspend the forces of succession. The agroecosystems produced and maintained in industrial agriculture frequently bear little resemblance to the natural system in terms of structure, function, and nutrient cycling. Coffee plantations on deforested tropical land or rice production on a drained coastal wetland are two examples. Through the process of evolution, diverse natural systems have developed "staying powers" and self-developing mechanisms such as the stabilization of soils and geochemical cycles, and pest and weed control (Odum 1971). Highly simplified agroecosystems such as grain monocultures lack such staying powers and would be quickly out-competed if humans did not intervene with external subsidies of energy, usually in the form of non-renewable natural capital. In fact, humans have deliberately removed much of the natural ability of most cultivars to defend against pests and succession through hybridization and genetic manipulations that purposefully channel more photosynthate into seed rather than to root development, nutrient cycling, and various defense mechanisms. That substitution replaces the natural work done by solar energy with fossil fuels in the form of pesticides, fertilizers, insecticides, and fuel used to pump groundwater and power tractors.

THE IMPACT ON NATURAL CAPITAL

The large flow of energy, materials and environmental services through agroecosystems is mirrored by the depletion and degradation of many types and quantities of natural capital at a variety of spatial scales. The environmental impact of industrial agriculture can be categorized generally as impacts that deplete non-renewable natural capital, and impacts that deplete or degrade renewable natural capital.

The Depletion of Non-renewable Natural Capital

Foremost of these impacts is that food production for a large and increasing fraction of the world's population is now driven principally by stock resources instead of flow resources (i.e., non-renewable natural capital rather than renewable natural capital). In populous nations such as India, unsubsidized solar energy cannot generate a sufficient quality and quantity of calories to sustain that nation's 890 million people (and growing at 2.07% per year). Stocks of crude oil, natural gas, and phosphate rock are depleted to produce the fuel, fertilizers, and other inputs necessary to sustain the level of productivity that many nations now depend on. The dependence on stock resources insures that the energy cost of those inputs, and hence the food grown with them, will rise as the energy cost of extracting oil and phosphate increase with their depletion (Cleveland 1991). The dependence on oil and gas also ties the financial plight of many farmers to the world price of oil that is unstable and unpredictable. The energy price shocks had an enormous economic impact on farmers throughout the world. Combined with stable or falling grain prices, the energy price shocks precipitated the U.S. farm crisis in the 1980s, and had a devastating impact on farms that used large amounts of energy to pump groundwater in arid regions of the nation.

The Depletion of Renewable Natural Capital

The depletion of RNC occurs in two forms: the direct physical depletion of an ecosystem good or service, and the conversion of a renewable form of natural capital to a non-renewable form. Examples of direct depletion are the loss of natural allelopathy of a plant through plant breeding or agronomic practices, or the eradication of a species of insect that provides biological pest control. The ecosystem service of nutrient cycling is depleted when root systems are diminished deliberately by plant breeding that maximizes above ground production.

Renewable natural capital is converted to a non-renewable form when it is used or degraded at a rate faster than it is regenerated by natural processes. Examples of that process are soil erosion, soil degradation, and groundwater mining. The rate of soil erosion in most agroecosystems in the world is substantially greater than the rate of soil formation. Row cropping, the elimination of fallow periods, mechanization, the extension of cultivation to marginal lands, the harvest of crop residues, and other factors intensify soil erosion. Similarly, irrigation effectively mines underground water in regions where withdrawal exceeds the natural recharge. Water quality is significantly impaired in regions where soil erosion is high and where agricultural chemicals are washed into surface waters or leached into groundwater.

HOW THE DEPLETION OF NATURAL CAPITAL INCREASES THE ENERGY COST OF FOOD PRODUCTION

The human response to the depletion of agricultural natural capital often accelerates or reinforces the existing problem, both in agriculture and in other sectors of the economy. To maintain crop yields, the reduction of soil fertility caused by erosion requires an increase in fertilizer application to maintain production. In fact, about 50% of the fertilizer applied to U.S. farmland simply replaces nutrients lost by soil erosion (Pimentel et al. 1987). This establishes a positive feedback loop between soil erosion and the stocks of crude oil and phosphate rock (see Figure 11.2). The increase in fertilizer use requires the mining of more natural gas and phosphate rock to replace the nitrogen and phosphorous that wash from the soil. The extraction of natural gas and phosphate rock accelerates the depletion of the remaining stocks of those resources, which, in turn, increases the energy and resource cost of natural gas and phosphate. In fact, the energy cost to extract a cubic foot of natural gas and a ton of phosphate rock is increasing (Cleveland 1991), so their use to offset soil depletion simply accelerates their energy cost, and thereby the indirect energy cost of the crops grown on the depleted soil.

Another example of that positive feedback system is the depletion of groundwater. The exploitation of groundwater for irrigation and municipal purposes at rates greater than the rate of recharge has caused sharp declines in the water level in many U.S. aquifers and other regions. The energy cost to lift water is a direct function of depth to water level. Thus, the depletion of groundwater natural capital initiates a positive feedback loop in which more energy is extracted to lift the same quantity of water a greater distance, which depletes fossil fuel resources, which in turn enhances the increase in the energy cost of pumping, and so on.

This is occurring on a grand scale in the southwestern United States where the explosion of population and economic activity in an arid environment depletes water resources at an unprecedented rate. The water level in the aquifer that supports the Tucson, Arizona region is dropping steadily due to overdrafts. As a result, the energy cost to pump water increased from about 1.2 to 2.5 kwh per 1000 gallons from 1930 to 1990 (Azary 1992). In response to that depletion, an aqueduct was constructed to divert surface water from the Colorado River. The energy cost of that water, which powers a 1200-foot lift over a 330-mile journey, is about 10 kwh per 1000 gallons, about 5 times the cost of local groundwater. That exploitation strategy dramatically increases the overall scale of Tucson's water supply system, from a few hundred square miles (the recharge area of the Tucson aquifer) to many thousands of square miles (the drainage basin of the Colorado River). Thus,

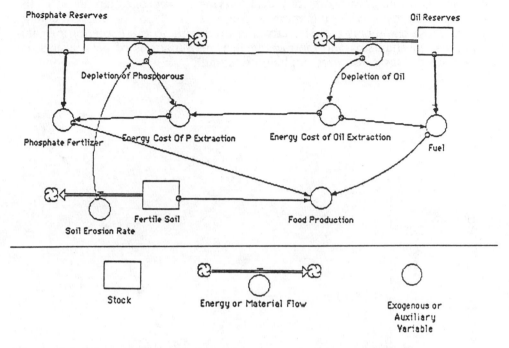

Figure 11.2. The positive feedback between soil erosion, the depletion of phosphate for fertilizer production, the depletion of oil reserves, and the energy cost of food production.

economic activity in the greater Tucson area, including agriculture, now owes its existence to the exploitation of a resource base that covers a significant fraction of the entire western United States.

A third example of the positive feedback between the depletion of natural capital, energy use, and human response is pest control. Control of pests and diseases with human-directed energy is essential in industrial agriculture because the natural self-defense mechanisms of most crops has been altered or erased through plant breeding and agronomic practices, and because the reduced diversity of monocultures automatically makes them easier targets for pests and diseases. But pesticides frequently eliminate a critical form of natural capital—the "free" pest control provided by the natural system, mostly through predacious insects that feed on the pests. Insect pests are generally less susceptible to pesticides than are their predators, and their populations tend to be large (Ehrlich and Ehrlich 1991). As a result, natural predators are eradicated more rapidly than the target pest species, and the latter develop resistance to the pesticides. As shown in Figure 11.3, the number of insects that are resistant to at least one pesticide is increasing rapidly (Bull 1982). Pest control can only be maintained by spraying larger amounts of pesticides, or by scientists in the agro-chemical industry trying to stay one step ahead of the evolving pests by developing more exotic and lethal chemicals. Of course, most of those chemicals are derivatives of crude oil, so the depletion of natural defense mechanisms and the race against the mutations of pest insects put us back on the same resource depletion treadmill.

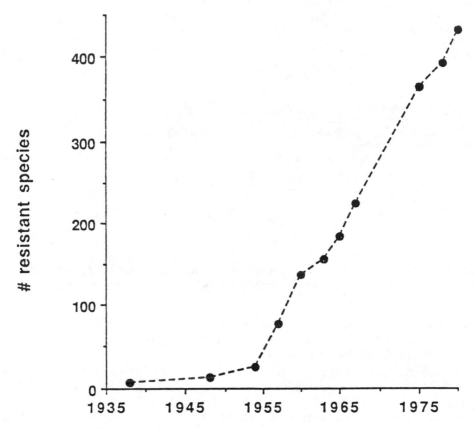

Figure 11.3. The number of insect species that have developed resistance to one or more pesticide (Data from Bull 1982).

The vicious cycle can be broken in some cases, or at least its effects can be mitigated to a degree. In the United States, for example, the rate of soil erosion from cropland decreased from about 18 to 16 t/ha/yr in the past two decades (David Pimentel, personal communication). Increased conservation tillage and the retirement of some marginal land were the principal reasons for the decline. A tripling of real oil prices in the 1970s and 1980s was an important catalyst for the adoption of conservation tillage techniques, such as disk and plant tillage and slot planting, because they require 60%–70% less direct fuel use. The improvement is encouraging, but if the soil formation rate in agriculture is only about 1 t/ha/yr, soil is still being lost about 16 times faster than it is being formed. More importantly, while there has been some improvement in soil conservation on some lands, certain agricultural practices are intensifying erosion on other land. No-till helps reduce erosion on some land, but the elimination of crop rotations and larger monocultures accelerates erosion on other lands.

The situation is bleaker in developing nations for a variety of reasons. Population pressure forces cultivation on to increasingly marginal, and hence highly erodible lands, while at the same time intensive, fossil fuel-based monocultures are encouraged by host governments and international lending agencies. The situation is exacerbated by the increasing tendency of the rural poor to collect and burn crop residues for cooking and heating. About one-third of the biomass fuel in the world is now crop residues (David Pimentel, personal communication). Removal of the residues accelerates erosion because they form an important protective cover for the soil surface (Pimentel et al. 1981). As a result, soil erosion in many developing nations is much worse than in industrial nations. China averages about 43 t/ha/yr, and Zimbabwe about 50 t/ha/yr (Pimentel et al. 1987).

CASE STUDIES: THE UNITED STATES AND INDIA

The shift from renewable natural capital to non-renewable natural capital and manufactured capital is illustrated with case studies of the agricultural sector in the United States and India. The experience in these nations is instructive because one is a highly industrialized nation that led the development of fossil fuel-based agriculture, while the other is a newly industrializing nation that turned to industrial agriculture a quarter century ago to eliminate the periodic famine that plagues a large and growing population. The experience in the United States and India also demonstrates the significant impact industrial agriculture has on its natural capital base.

The Shift to Manufactured and Non-Renewable Capital

Time series of some of the major renewable and non-renewable forms of natural capital used in the two countries are shown in Figures 11.4 and 11.5. The rapid shift towards manufactured capital in the early part of this century is clear in the United States, as evidenced by the 10,000-fold increase in the number of tractors used from 1910 to 1930. The shift towards non-renewable natural capital is exemplified by the huge increase in the use of fuels derived from crude oil, and by the increase in the use of inorganic fertilizers. Note that from 1910 to 1990 there was also a sharp drop in the use of human labor, while the quantity of land cultivated was essentially constant. The huge energy subsidy of labor and land that resulted from these substitutions is the principal cause of the prodigious increase in agricultural output and productivity characteristic of the United States.

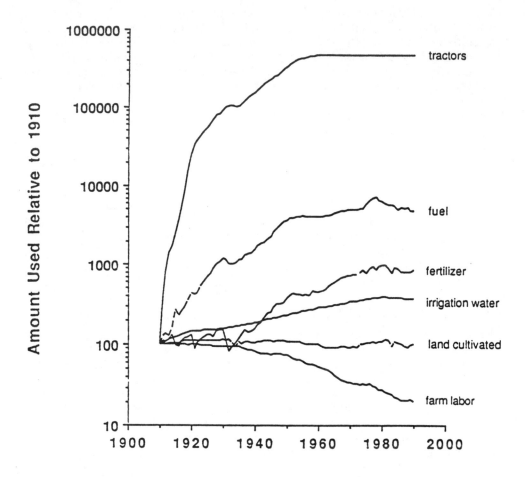

Figure 11.4. Inputs of various forms of natural and human capital to U.S. agriculture, 1910–1990.

The trend in Indian agriculture is generally the same, except that the substitutions did not begin in earnest until the 1960s. Manufactured capital and non-renewable capital are increasing at a rapid rate in India, while the quantity of labor and land use is relatively constant. The increase in agricultural output generated by these substitutions boosted per capita food production by more than 30% from 1970 to 1988. Note one important difference between the United States and India. The use of tractors, fuel, and fertilizer in the United States is clearly in, or at least is approaching, the saturation phase. The use of the those inputs in India is clearly still in a rapid growth phase.

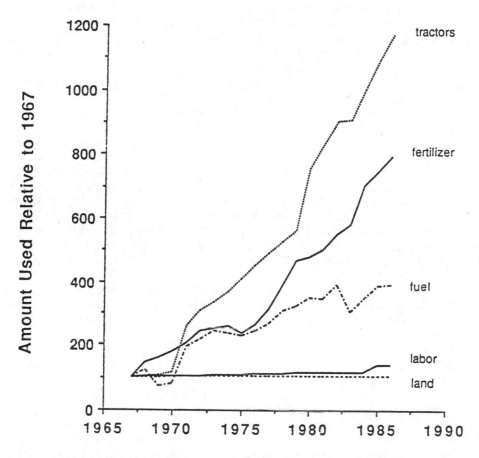

Figure 11.5. Inputs of various forms of natural and human capital to Indian agriculture, 1967–1988.

The Depletion of Natural Capital in Indian Agriculture

As India benefits from increases in food production powered by fossil fuels, so to does it pay the environmental price that characterizes nearly all industrial agroecosystems in the world. In the case of India, the environmental problems are magnified by a high population density and a growing population that produce intense competition for land and water resources from non-agricultural purposes, such as drinking water. The combination of these factors is depleting important sources of agricultural natural capital such as fertile soil, groundwater, and genetic plant diversity. That depletion raises the specter of a population that will be substantially larger at the turn of the century than when the Green Revolution started, but which will inherit a resource base that has been stripped of much of its life-support capacity.

About 6 billion tons of soil are eroded from India's land each year (Narayana and Sastry 1985). Estimates of the average rate of erosion vary from 7 to 75 t/ha/yr, with many estimates falling in the range of 30 to 40 t/ha/yr (see Table 11.1). Some of the highest rates of erosion are on the fertile soils in Northeastern India. These rates are extraordinarily fast compared with the slow rate of natural soil formation. Many interrelated factors contribute to soil erosion in India. Most of the 24 million ha of new agricultural land brought under cultivation from 1951 to the mid-1980s was marginal, highly erodible soil (Venkateswarlu 1985). Deforestation of watersheds increases runoff and thereby accelerates erosion and flooding on agricultural and other types of land. The catastrophic deforestation and erosion problems in the Himalayan watersheds are legendary. Deforestation also reduces the supply of fuelwood, causing the increased harvest of crop residues for fuel, which accelerates erosion. The combined effects of deforestation and erosion have many deleterious results. Fertilizer use increases to replace lost nutrients. Throughout all of India, the land area subject to annual flooding tripled between 1960 and 1984, and is now equivalent to more than 20% of the entire land area (Brown and Young 1990). Sedimentation rates in the catchment areas of some of India's major reservoirs are 3 to 4 times the predicted rates, and may shorten considerably the useful life of those reservoirs (Narayana and Sastry 1985).

Soil conservation programs are underway to combat the problem. Prime Minister Rajiv Ghandi established the National Wastelands Development Board in 1985 to address the general problem of land degradation, and to restore some highly eroded lands. By the end of the Sixth Five Year Plan, about 29.4 million ha were covered under some type of soil conservation program (Venkateswarlu 1985). There are, however, severe barriers to the adoption of such programs, not the least of which is the fact that a large majority of Indian farmers are extremely poor, lack sufficient capital to invest in conservation programs, and do not have ready access to the technological and communication infrastructure that disseminate the conservation programs. Moreover, the poor farmers hold much of the marginal, erosion-prone land, on which yields are already extremely low. The marginal decline in yields due to erosion are likely to be very small in these cases, increasing the reticence of the farmer towards soil conservation (Venkateswarlu 1985).

The pattern of water use has significantly impaired the land and water resource base in India, and as a result, poses a serious challenge to one of the most heavily irrigated agricultural systems in the world. Groundwater depletion is a serious problem in many regions where withdrawals now exceed recharge. Overdrafts accelerated in the 1970s when tube wells powered by diesel engines and electric motors rapidly replaced open dug wells. The adoption of tube wells was encouraged by financial assistance for mechanized water withdrawal systems. Individual case studies are representative of the national problem. The water table in large parts of Tamil Nadu fell 25–30 meters in the 1970s (Jayal 1985). In Maharashtra, 77 out of 1,481 watersheds have rates of withdrawal that exceed rates of recharge (Jayal 1985). In that region, competition for limited water supplies has developed between local subsistence farmers and larger farms that grow heavily watered commercial crops such as sugarcane.

Table 11.1. Estimates of Recent Soil Erosion Rates in India

Study	Region or Soil Type	Erosion Rate (t/ha/yr)	Comments
Lal et al. (1989)	national average	75	
Pimentel et al. (1987)	national average	25–30	assumes 6 billion tons lost per yr., 60-70% on cultivated land
	Deccan black soil region	40–100	
Narayana and Sastry (1985)	national average for red soils	4–10	sheet erosion on 72 million ha
	national average for black soils	11–43	sheet erosion on 88 million ha
	national average	>33	gully erosion on 4 million ha
	national average	>80	hill erosion on 13 million ha
Venkateswarlu (1985)	national average	7	
	Upper Gangetic alluvial plains	3	
	Lower Gangetic alluvial plains	14	
	Assam Valley	28	
	Western coastal region	39	

BACK TO THE FUTURE: SUSTAINABLE AGRICULTURE

The Green Revolution has been a double-edged sword, as the experience of India clearly shows. The poor ecosystem management strategies that evolved in industrial agriculture set in motion a series of complex modifications and feedbacks that have depleted the renewable and non-renewable natural capital basis of food production. The consequences of that depletion were less significant when fertile land and high quality fossil fuel deposits were abundant, and when population demands were small. The situation today demands new strategies of supplying food to a growing population. Cheap oil is a principal barrier to the implementation of new strategies because it can be used in the short run to mask the long-term effects of the depletion of other forms of natural capital. The good news is that many of the components of sustainable agriculture do not require massive technical innovation. Indeed, reducing the depletion of natural capital in agriculture will require a return to many forms of human capital that guide traditional agriculture. Practices such as inter-cropping, polyculture, biological pest control, traditional shifting cultivation, and agroforestry have centuries of experience of producing food in ways that maintain the natural capital basis of agriculture. Appropriate economic and institutional incentives that will guide agriculture towards these practices are required while insuring the production of sufficient food to meet the world's growing population.

REFERENCES

Altier, M. A. 1990. Why study traditional agriculture? In Agroecology, eds. C. R. Carroll, J. H. Vandermeer, and P. Rosset. New York: McGraw-Hill.

Azary, I. 1992. Groundwater depletion in Tucson, AZ. Paper presented at the second international meeting of the International Society of Ecological Economics, Stockholm, Sweden.

Berkes, F., and C. Folke. 1992. A systems perspective on the interrelations between natural, human-made, and cultural capital. *Ecological Economics* 5: 1–8.

Boserup, E. 1965. The Conditions of Agricultural Growth. New York: Aldine.

Brown, L. R., and J. E. Young. 1990. Feeding the world in the nineties. In State of the World, ed. L. R. Brown. New York: W. W. Norton.

Bull, D. 1982. A Growing Problem: Pesticides and the Third World. Oxford: Oxfam.

Buringh, P. 1989. Availability of agricultural land for crop and livestock production. In Food and Natural Resources, eds. D. Pimentel and C. W. Hall. New York: Academic Press.

Cleveland, C. J. 1991. Natural resource scarcity and economic growth revisited: economics and biophysical perspectives. In Ecological Economics: The Science and Management of Sustainability, ed. R. Costanza. New York: Columbia Univ. Press.

Costanza, R., and H. E. Daly. 1992. Natural capital and sustainable development. *Conservation Biology* 6: 37–46.

Ehrlich, P. R., and A. H. Ehrlich. 1991. Healing the Planet. Reading: Addison Wesley.

Evans, L. T. 1980. The natural history of crop yields. *American Scientist* 68: 388–97.

Gliessman, S. R., ed. 1990. Agroecology: Researching the Ecological Basis for Sustainable Agriculture. New York: Springer-Verlag.

Hall, A. E. 1990. Physiological ecology of crops in relation to light, water, and temperature. In Agroecology, eds. C. R. Carroll, J. H. Vandermeer and P. Rosset. New York: McGraw-Hill.

Jackson, W. 1991. Nature as the measure of sustainable agriculture. In Energy, Economics, Ethics: The Broken Circle, eds. F. H. Bormann and S. Kellert. New Haven: Yale Univ. Press.

Jayal, N. D. 1985. Emerging pattern of the crisis in water resource conservation. In India's Environment: Crisis and Responses, eds. J. Bondyopadhya, N. D. Jayal, U. Schoettli and C. Singh. Dehra Dun: Natraj Publishers.

Minc, L. D., and J. H. Vandermeer. 1990. The origin and spread of agriculture. In Agroecology, eds. C. R. Carroll, J. H. Vandermeer and P. Rosset. New York: McGraw-Hill.

Narayana, V. V. D., and G. Sastry. 1985. Soil conservation in India. In Soil Erosion and Conservation, eds. S. A. El-Swaify, W. C. Moldenhauer and A. Lo. Ankeny: Soil Conservation Society of America.

Odum, E. P. 1989. Ecology and Our Endangered Life Support Systems. Sunderland, MA: Sinauer.

Odum, H. T. 1971. Environment, Power and Society. New York: Wiley.

Pimentel, D., et al. 1987. World agriculture and soil erosion. *BioScience* 37: 277–83.

Pimentel, D., L. E. Armstrong, C. A. Flass, F. W. Hopf, R. B. Landy, and M. Pimentel. 1989. Interdependence of food and natural resources. In Food and Natural Resources, eds. D. Pimentel and C. W. Hall. New York: Academic Press.

Pimentel, D., M. A. Moran, S. Fast, G. Weber, R. Bukantis, L. Baillet, P. Bovering, S. Hindman, and M. Young. 1981. Biomass energy from crop and forestry residues. *Science* 212: 1110–15.

Pimentel, D., et al. 1978. Biological solar energy conversion and U. S. energy policy. *BioScience* 28: 376–82.

Pimentel, D., and M. Pimentel. 1979. Food, Energy, and Society. New York: Wiley.

Smil, V. 1991. General Energetics: Energy in the Biosphere and Civilization. New York: Wiley.

Venkateswarlu, J. 1985. Ecological crises in agroecosystems. In India's Environment: Crisis and Responses, eds. J. Bondyopadhya, N. D. Jayal, U. Schoettli and C. Singh. Dehra Dun: Natraj Publishers.

Woolhouse, H. W. 1984. Genetic engineering to modify energy flow in agriculture. In Energy and Agriculture, ed. G. Stanhill. Berlin: Springer-Verlag.

12 THE EMERGY OF NATURAL CAPITAL

Howard T. Odum
Center for Environmental Policy, Environmental Engineering
University of Florida
Gainesville, FL 32611

ABSTRACT

This chapter addresses several notions that have to do with the evaluation of natural capital, that is, the assets in the natural world that have required energy over a period of time to be built and organized. Soils, for example, require photosynthetic production by plants and the accumulation of the resulting plant matter over a period of time to build a useful soil layer for a human activity like agriculture. Calculating the natural energy over time required to build a given ecological asset is one means of trying to determine its replacement value, independent of the vagaries of using market values that depend on social desires at a given moment. The methods for determining the direct and indirect solar energy, or Emergy, required to make a service or product have been under development for some years and can be used in environmental evaluations. In conducting these evaluations, all energies are expressed in equivalent units, that is, conversion factors have been developed to express the ability of one form of energy to do work in comparison to other forms.

INTRODUCTION

An important aspect of emergy analysis is that it tries to evaluate the contributions of nature that are not identified by the economic system. In human economies, an exchange of goods and services from one sector to another is accompanied by countercurrent exchange of money. This is also true of some exchanges of the human economy with the natural world, (e.g., the taking of fish from the oceans is usually accompanied with some payment). However, there are many contributions from nature, that is, work that is done for the human economy that does not have any monetary exchange associated with it. These services, then, are not valued in the accounting schemes that humans use to assess economic activities. The emergy approach seeks to value both those transactions that have money flows associated with them, as well as these other contributions from nature that are not recognized in the usual exchanges involved in the economy. For transactions that have associated money, emergy evaluations include two calculations, one for the human services contributed and one for the environmental resources used up.

One important point to recognize is that the natural world has built up and accumulated impressive storages over very long periods of time—solar energy being the basis for much of this stored capital. Fossil fuels, phosphates, abundant forest land, tropical rain forests, large pelagic fish stocks, and soils are some examples. The human species in the 20th century has had the good fortune and technology to exploit these accumulated storages with increasing rapidity. The human species in undertaking this behavior may be obeying the "maximum power principle," which in its simplest form states that those systems that maximize power (rate of doing work) in the competition for energy resources are the ones that win out and succeed. Consequently, systems may not necessarily be seeking behavior to maximize efficiency during times of plentiful resources, rather they may seek to get as much as they can in the struggle for growth and survival. Recently, as the human community recognizes the finiteness of the environmental storages that are an important component of our prosperity, calls for management for sustainability have been heard. The usual notion is that we can somehow reach a steady state in which the size of our economy reaches a relatively constant size, one which can be maintained by a consumption of resources that is sustainable. But we may find, as with many natural ecosystems, that sustainability will involve oscillations, with human economies building during times when environmental storages are depleted, and declining during times when environmental storages recover and grow again.

This chapter explores these notions, presents emergy calculations for various components of natural capital, and presents some simple simulations for oscillating systems of man and nature.

STORAGE, EMERGY AND TRANSFORMITY

The literature on evaluation of nature is extensive, much of it reporting ways of estimating market values of the storehouses and flows in environmental systems. In recent approaches to environmental evaluation (Repetto 1992), monetary measures were sought for the storages of nature. Others have used the simple physical measures of stored resources, especially energy.

Shown in Figure 12.1a is a storage of environmentally generated resources. Energy sources from the left are indicated with the circular symbol. Energies from sources are used in energy transformation processes to produce the quantities stored in the tank. Following the second law, some of the energy is degraded in the process and is shown as "used energy" leaving through the heat sink, incapable of further work. Also due to the second law, the stored quantity tends to disperse, losing its concentration. It depreciates, with some of its energy passing down the depreciation pathway and out through the used energy heat sink.

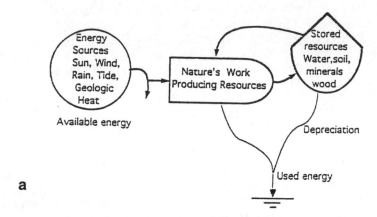

Energy Basis for Wealth

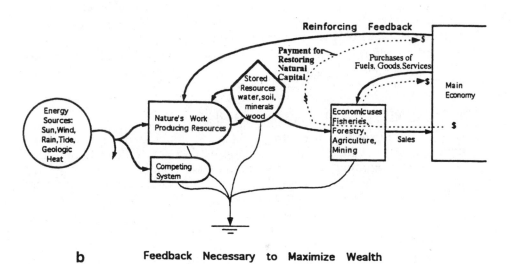

Figure 12.1. Systems diagrams of natural capital and its relationship to economic use.
(a) Resource production, storage, and depreciation; (b) Economic use of
stored resources.

To build and maintain the storage of available resources, work requiring energy use and transformation has to be done. Work is measured by the energy that is used up, but energy of one kind cannot be regarded as equivalentto energy of another kind. For example, one joule of solar energy has a smaller ability to do work then one joule of energy contained in coal, since the coal energy is more concentrated than the solar energy. A relationship between solar and coal energy could be calculated by determining the number of joules of solar energy required to produce one joule of coal energy. The different kinds of energy on earth are hierarchically organized with many joules of energy of one kind required to generate one joule of another type. To evaluate all flows and storages on a common basis, we use solar emergy (Odum 1986; Scienceman 1987) defined as follows:

> Solar emergy is the solar energy availability used up directly and indirectly to make a service or product. Its unit is the solar emjoule.

Although energy is conserved according to the first law, according to the second law, the ability of energy to do work is used up and cannot be reused. By definition, solar emergy is only conserved along a pathway of transformations until the ability to do work of the final energy remaining from its sources is used up (usually in interactive feedbacks).

Solar transformity is defined as follows:

> Solar transformity is the solar emergy required to make one joule of a service or product. Its unit is solar emjoules per joule.

For example, environmental contributions to storages of spruce forest wood in Sweden are estimated to be about 3800 solar emjoules per joule of wood produced (Doherty, Nilsson, and Odum 1992). Emergy calculations make estimates and comparisons of the magnitude of the work involved in the creation and maintenance of storages and flows of energy, matter, and information in the system of man and nature.

Solar transformity is calculated from data on energy flows in real networks by evaluating all the inputs observed contributing to an energy flow including those of nature and those from the economy.

The emergy calculations below are based on an analysis of the major energy flows of the biosphere (Odum 1988). For detailed calculations, see Odum (1984, 1987) and (Odum, Odum, and Blissett 1987).

The importance of any flux or storage is determined as the proportion that its emergy is of the total annual emergy flux of the economy within which it is found (for convenience it is usually a national economy). In order to help people visualize emergy contributions, we have sometimes expressed the proportion of total national emergy in dollars of GNP:

> EM$ (Macroeconomic dollar value) of a flux or storage is defined as its solar emergy value divided by the emergy/money ratio for an economy for that year.

Calculated in this way, the macroeconomic value of an environmental resource is usually much higher than its market value. Whereas market values are what is important to the small scale transactions of individuals and businesses, macroeconomic values are suggested as the proper evaluation for public welfare and maximizing overall wealth and prosperity. One value should not be substituted for the other.

DIFFERENTIAL EQUATIONS FOR EMERGY EVALUATION OF A STORAGE

Emergy of a storage is defined with the equations given in Figure 12.2. Emergy storage is not affected by the inherent depreciation required from a storage (energy flow to the heat sink). The energy dispersed in depreciation is a necessary part of the process of storing emergy. The transformity of the storage at any time is the emergy of the storage divided by the energy stored.

When dynamic models are simulated, there are equations for the state variables that may be calibrated in energy, mass, or monetary units as may be appropriate for each. After the model is running in the usual way, equations that calculate the emergy and transformity of each storage and flow can be added and values included in tabular or graphical output.

Quantitative Definitions

Energy Stored $= Q$; $dQ/dt = J - k1*Q - k2*Q$

EMERGY Stored $= E$:

   If $dQ/dt > 0$ then $dE/dt = Tr_j*J - Tr_Q*k_2*Q$

   If $dQ/dt = 0$ then $dE/dt = 0$

   If $dQ/dt < 0$ then $dE/dt = Tr_Q*dQ/dt$

   where transformity of $Q = Tr_Q = E/Q$

   and the transformity of $J = Tr_j$

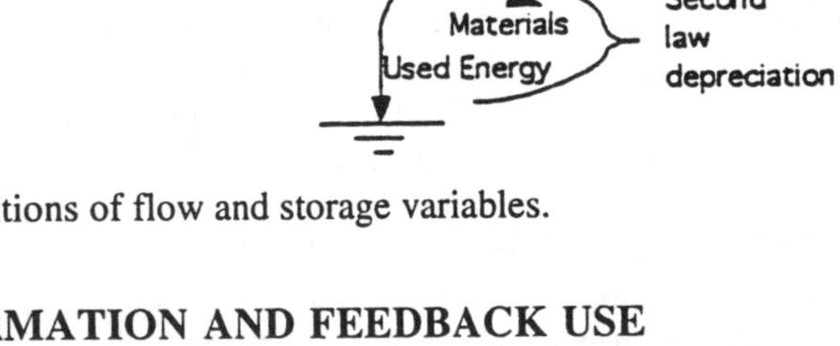

Figure 12.2. Definitions of flow and storage variables.

CAPITAL FORMATION AND FEEDBACK USE

In both ecological and economic systems developing stored assets as part of an accelerating growth pattern is observed. Accompanying growth is the feedback from storage to interact and stimulate the production process mul-

production in both ecology and economics. The simplest of general systems minimodels shows the kinetics and energetics of storage and process. The typical curve of growth with time is sigmoid because of the source limitation.

As systems develop storages, the turnover time increases, that is the storage per unit flux increases. Percent depreciation is less. Those who see maximizing storage (e.g., biomass in an ecosystem or assets in an economy) as an objective of succession and growth plot storage as a function of production and interpret the greater mass per unit production as increased efficiency and achievement. Their hypothesis is that systems organize to maximize storage (biomass, capital, money, stored wealth, etc.).

An alternative view is that there is an optimum storage that maximizes production. Only over a small range does building storage increase efficiency.

This shows how both viewpoints about storage, production and growth are alternate ways of looking at storage development (capital formation), even in the simplest of systems designs that emerge with self-organization for maximum intake and feedback of power.

ECONOMIC USE INTERFACES, EMERGY, AND MARKET VALUES

Figure 12.1b shows the interface between environment and economy, with work contributed by the environment and by human inputs that are purchased, such as fuels, goods, and services. Systems with products of no direct economic value are competing for available energy sources. There is emergy value added from purchased inflows. The further transformed products may be stored, available for use at yet higher levels in the economic part of the hierarchy.

When the resources from the environment are abundant, little work is required from the economy, costs are small and prices low. But this is when the contribution of real wealth is greatest. This is when everyone, assuming the human population is not large, has abundant resources from the natural environment.

When the resources are scarce, acquisition costs are higher, prices are high, and the market puts a high value on the product. But this is when there is little net contribution of the resource to the economy, real, natural wealth is scarce and standards of living low. In this sense market values are inverse to real wealth contributions from the environment and cannot be used to evaluate environmental contributions or environmental impacts.

With emergy evaluation all pathways are measured on the same basis, namely the solar emergy previously used up. Contributions from nature and those by humans are evaluated in common units.

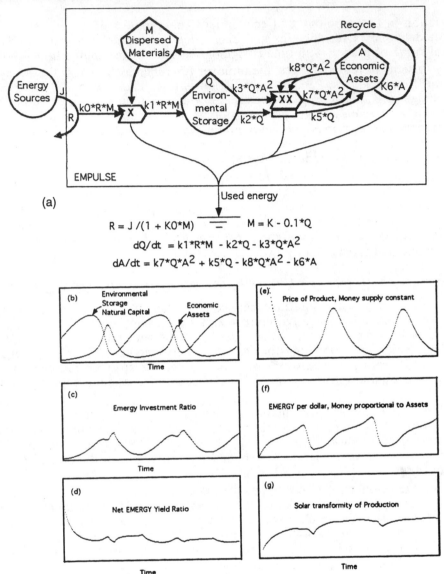

(a)

$$R = J /(1 + K0*M)$$
$$M = K - 0.1*Q$$
$$dQ/dt = k1*R*M - k2*Q - k3*Q*A^2$$
$$dA/dt = k7*Q*A^2 + k5*Q - k8*Q*A^2 - k6*A$$

Figure 12.3. Simulation model for examining the behavior of emergy analysis indices during oscillations. (a) systems diagram and equations; (b) oscillation of storages with alternating dominance of production and consumption; (c) emergy investment ratio; (d) net emergy yield ratio; (e) price of economic products when the money supply is held constant; (f) emergy/currency ratio when money supply is kept in proportion to economic assets; (g) solar transformity of economic products.

ECONOMIC REINFORCEMENT OF ENVIRONMENTAL PRODUCTION

With the environmental use pattern in Figure 12.1b, the desirable resources are used up by growth of demand. As stored resources become scarce, prices rise, encouraging further depletion. As the desired system is reduced, competing production systems prevail that are not in economic use. Note the competing system in Figure 12.1b. For example, in the Gulf of Maine off the U.S, coast, once-abundant bottom-feeding fishes have been over harvested. Their place has apparently been taken over by less-edible swarms of dogfish (*Squalus acanthias*).

The Atlantic White Cedar (*Chamaecyparies thyoides*) in the eastern United States was a preferred tree for harvesting smooth-grain timber in colonial times and is now scarce, its place taken by other wetland species. However, more sustainable forest systems have persisted in Sweden over several hundred years by human practices that stimulated the regeneration processes.

To maintain or restore an environmental storage, some reinforcement has to be applied to the desirable production subsystem feeding back from the economic users (Figure 12.1b). If money is paid to people so that they contribute, the reinforcement of the environmental producers, it is an investment in natural capital. I used a computer simulation model to demonstrate the way a system like that of Figure 12.1b, but without reinforcing feedback, eliminates the desirable stock, while the system in Figure 12.1b, with a feedback reinforcement, sustains the storage and continues the desired fishery production (Odum 1993).

UNDERESTIMATING NATURAL CAPITAL WITH MONETARY EVALUATION

Some authors use the monetary cost of replacing a resource storage as a measure of its value. This is incorrect, underestimating the wealth required for replacement, because the free environmental contributions are not included. For example, the figures by Repetto (1992) for natural capital are only for the human contributed part of those resources. Notice in Figure 12.1b the way the emergy for the economic feedback to help restore natural capital is only part of the emergy required. The rest is derived from the energy sources on the left.

Thus, evaluations of natural capital with monetary cost underestimate the contributions to the system's real wealth. If sustainable natural capital is the objective, underestimating their value will further contribute to their loss. Without feedback reinforcement from the human economy the environmental producers that are necessary for maximum economic production are pulled down by the exponential growth tendencies of the consumers, thus diminishing the wealth on which the buying power of money is based.

HIERARCHICAL DISTRIBUTION OF NATURAL CAPITAL

Because many joules of energy of one kind converge in any energy transformation, producing fewer joules of the next energy type, all the processes
are part of an energy hierarchy (Figure 12.4). In the hierarchy from the very
small to the very large ,there are storages at each level. Small storages affect
small areas and turnover rapidly, whereas the larger storages affect larger
areas and have long replacement times, lower depreciation rates, slower
turnover times, larger sizes, and larger territories of support and influence that
go with higher position in hierarchies.

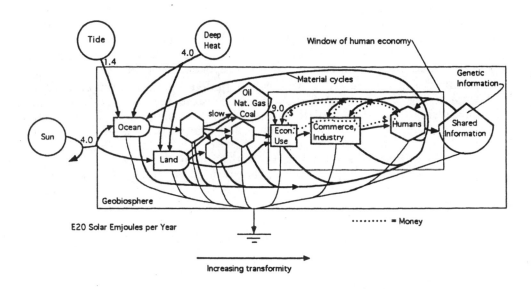

Figure 12.4. Highly aggregated diagram of the global hierarchy of energy transformation,
the main emergy sources, and the intermediate position of circulating
money. Numbers are annual flux of solar emergy inputs to the global
system.

Within the geobiosphere of the earth (Figure 12.4), input energies from the
sun, from the tides, and from the heat sources deep in the earth interact as a
single coupled system with a network of processes that converge with the
production and maintenance of storages of globally-shared information.
Solar transformities increase from left to right in Figure 12.4 along the series
of successive energy transformations. Solar transformity measures the
position in the energy hierarchy and indicates the appropriate range of utility.
Some emergy storages of the global system are given in Table 12.1.

Circulation of money provides more efficient processing on the scale of human beings and their businesses. Accumulating storages of monetary capital goes with the phases of accelerating growth, and facilitate the autocatalytic incorporation of the environmental resource storages into the consumer frenzy and its phases of rapid growth that characterize periods of development in human economies.

Table 12.1. Emergy of some Global Storages (Natural Capital)* Possible Orders of Magnitude

Item	Replacement time, years	Stored Emergy*, sej	Macroeconomic Value, 1992 Em$**
World infrastructure***	100	9.44 E26	6.3 E14
Freshwaters	200	1.89 E27	1.26 E15
Terrestrial ecosystems	500	4.7 E27	3.1 E15
Cultural & technol. information	1 E4	9.44 E28	6.3 E16
Atmosphere	1 E6	9.44 E30	6.3 E18
Ocean	2 E7	1.89 E32	1.25 E20
Continents	1 E9	9.44 E33	6.3 E21
Genetic information of species	3 E9	2.8 E34	1.86 E22

* Product of annual solar emergy flux, 9.44 E 24 sej/yr and order of magnitude replacement times in column 1.
**See notes in A.
***Highways, bridges, pipelines, etc.

PHYSICAL VERSUS MONETARY CAPITAL IN A MACROECONOMIC SYSTEM

The accumulation of storage provides the means for operating and controlling a system and for expanding a system, or starting new systems. Accumulating capital is a way of controlling a system, expanding a system or starting new systems. In an economy, however, the flows of money are a coupled counter-current to the flow of goods and services.

Behaviors of money and real wealth assets are on very different time scales. Money has a rapid flux, and storages of money have a rapid turnover time. Storages of real wealth may be large relative to flows and depreciation so that turnover times are much longer. The turnover time of the main features of human society, such as the infrastructure and the information shared in culture, may be 50–300 years. The turnover time of capital accumulations of money may be 1–5 years.

Whereas it is often customary to refer to real storages using the money measures of its formation, in a more complete systems approach to economics, money storages and fluxes are kept separate from those of real wealth, and the various coupling relationships carefully studied. Storage of wealth and the physical wealth are both represented by tank symbols (storage unit symbol). Dynamic models and computer simulations of ecological-economic systems thus avoid the confusion of using one variable to represent another, since the roles and processes are different. Simulation models of macroeconomics of this type were given previously (Odum 1983; 1987).

CHARACTERISTICS OF OSCILLATING REGIMES

The real world is observed to pulse and oscillate. There are oscillating steady states. In most systems, including those which people are part of, storages are observed to fill and discharge as part of oscillations. Some are chaotic. Maximizing power is a hypothesis that may account for the hierarchical storages and oscillations. Systems that transform energy into forms capable of autocatalytic feedback reinforcement increase both the intake of energy and the efficiencies of its feedback use. There is an optimum loading for maximum power transformation in many kinds of processes (Odum and Pinkerton 1955; Curzon and Ahlborn 1975). Loading is measured by the ratio of output force or concentration to the input force or concentration. One of the reasons oscillating steady states (repeating oscillation) displace steady ones may be that energy transformations have better loading ratios between inputs and outputs if there is an alternation in the growth of interconnected storages (i.e., producers and consumers; Q and A in Figure 12.3).

A number of simple models generate the essence of the observed alternation of production and pulsing consumption. Models that self organize for maximum power in their growth and oscillations should include recycling of materials, competition of consumer pathways, autocatalytic reinforcement, and feedback interactions from storages of transformed energy. For our purposes here, the model in Figure 12.3a (Alexander 1978) makes clear what kind of dynamic design relationships may be consistent with long-term power maximization.

Figure 12.3b shows the alternation of production and consumption of the model that Richardson and Odum (1981) and Richardson (1988) showed is self-organizing for similar power even when different algorithms of spatial organization of dispersed production and hierarchically centered consumption are included. In the complex real world the small oscillations are nested within the larger ones; these give large scale patterns to the smaller ones while filtering and absorbing the small oscillations.

Growth and succession in oscillating ecological and/or economic systems can be visualized as a short segment within the longer continuing oscillations (Figure 12.5). The growth and leveling portions of the curve somewhat represent the old concepts of succession and climax, but now it is necessary to consider the stage of coming down after which there is regrowth again. In the simplest of pulsing models (Figure 12.3a), there are three storages: (1) the dispersed materials, (2) the assembled resource products, and (3) the storage of temporary assets that are part of the consumer frenzy.

According to the pulsing paradigm, there is a period when net production of primary products is positive, with some growth of consumers. Then, when the product of production storage and consumer storage reaches a threshold that accelerates autocatalytic and higher order pathways, the system shifts to a temporary net consumption of products as the consumer assets make a temporary surge. In the case of the global economy, the accumulated products are the environmental resource, and the consumer storages are our economic, informational, and cultural assets.

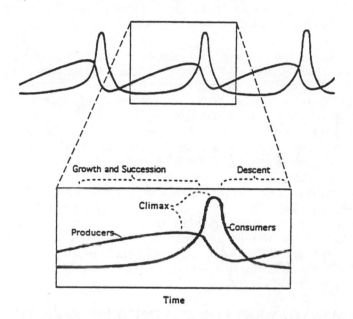

Figure 12.5. Succession, climax and descent seen as a window of time within a power-maximizing oscillatory steady state (Odum 1983, p. 445).

In the past we have sought power-maximizing, thermodynamic limits for efficiencies, transformities, investment ratios, etc. If power is maximized by oscillating systems with storages filling and discharging along the levels of hierarchy, then we have to recognize that energy transformation parameters have a varying baseline, with different values appropriate at different stages in the oscillatory cycle.

Figure 12.3 b–e shows the oscillation of some energy characteristics of the model as it alternately fills storages of natural and economic assets. The top right (Figure 12.3e) was obtained by holding the money supply constant. Prices of production are minimum when the rate of use of resources is largest, when wealth is greatest. The property of decreasing cost per unit resource has been misinterpreted by Simon and Kahn (1984) as evidence of the unimportance of resource shortages to economic vitality. The simulation shows the way resource determinism generates the curves of those authors. The declining cost part of the curve occurs when the resources are being transformed into assets that help process more. The declining cost is the action of resources feeding back to capture more resources. Some of this is resource emergy in the form of technology that increases efficiency (as said by Kahn and Simon) but these are resource-based and depreciate away when their emergy support lessens.

In Figure 12.3f, money was added in proportion to assets, so that there was no inflation. Towards the end of the rapid growth period the emergy/dollar ratio falls, as has been observed in highly developed countries (Huang and Odum 1991).

Even the efficiencies oscillate. The solar transformity varies (Figure 12.3e), becoming less efficient in times of strong feedback acceleration. Higher transformity (lower efficiency) is thermodynamically selected for during the transfer of storage to higher levels, because of the trade-off that favors a lower-than-maximum efficiency for maximizing power.

The net emergy yield ratio is defined as the ratio of the emergy of the yield of a production process divided by the feedback emergy from the economy in the process. It is a useful measure of a primary energy source's ability to support other processes. The net contribution is higher when natural resource storages are large and decreases with exploitation.

The emergy investment ratio is defined as the ratio of the purchased emergy feeding back from the economy to the free contributions of emergy from the environment. It measures the intensity of economic development relative to environmental use. With the development of assets in the period of frenzied growth, the loading on the environment is greater, and the intensity of feedback investment is larger.

STORAGE MANAGEMENT FOR MAXIMUM PROSPERITY

If this oscillating pattern is the normal one, then sustainability concerns managing, and adapting to the frequencies of oscillation of natural capital that perform best. Sustainability may not be the level "steady-state" of the classical sigmoid growth curve but the process of adapting to oscillation. The human economic society may be constrained by the thermodynamics that is appropriate for each stage of the global oscillation.

In the simplified models presented here, there is an appropriate time in the temporal cycle at which natural storages are built and a time when they are to be consumed. If the model is appropriate, there needs to be a changing public policy to maximize power and production for each stage in the cycle. There is an appropriate efficiency possible at each stage. The transformities also vary in a cycle. The model suggests the way our current realms may be conforming to what is needed for the stage in the thermodynamic cycle.

When this model is simulated with coefficients that accumulate products for a long time, the consumer pulse is very sharp, catastrophic in its severity and with high transformity. The larger emergy required over a longer time of storage achieves a commensurate effect by delivering a consumer pulse in a shorter time. What is regarded as catastrophe at smaller scale of time and space is the normal cycle in the window of the larger scale of time and space. Examples are fires, floods, hurricanes, earthquakes, and volcanoes.

Depending on the indices used, the global economy of 1992 appears to be at or near its climax crest that was reached by using up of natural storages lower in the hierarchy. If so, we go now to a period when the consumer storages come down, and there is a time of net production of environmental products after which another consumer frenzy can repeat.

SUMMARY

By evaluating the emergy of storages in environment and within the economy, decisions can be made for the short run as to what policies maximize the economy by utilizing appropriately the available capital of several kinds. By measuring emergy and its derived macroeconomic value, uses can be made commensurate with the work required for replacement, thus helping maximize economic performance.

For the longer run, the models of oscillating storage give insights on when to use and when to conserve. Conservation policy should work towards that economic use and feedback reinforcement that oscillates for maximum performance in the long run and the short run.

REFERENCES

Alexander, J. F., Jr. 1978. Energy Basis of Disasters and the Cycles of Order and Disorder. Ph.D. dissertation, Environmental Engineering Sciences. Gainesville: Univ. of Florida.

Coultas, C. L., and E. Gross. 1975. Distribution and properties of some tidal marsh soils of Apalachee Bay, Florida. *Soil Science Society of America Proceedings* 39 (5): 914–19.

Curzon, F. I., and B. Ahlborn. 1975. Efficiency of a Carnot engine at maximum power output. *American Journal of Physics* 43: 22–4.

Doherty, S. J., H. T. Odum, and P. O. Nilsson. 1992. Systems Analysis of the Solar Emergy Basis for Forest Alternatives in Sweden. Draft of final report to the Swedish State Power Board. Garpenberg, Sweden: College of Forestry.

Huang, S-L., and H. T. Odum. 1991. Ecology and economy: Emergy synthesis and public policy in Taiwan. *Journal of Environmental Management* 32: 313–33.

Odum, H. T. 1983. Systems Ecology. New York: Wiley.

———. 1986. Emergy in ecosystems. In Environmental Monographs and Symposia, ed. N. Polunin. New York: Wiley.

———. 1987. Models for national, international, and global systems policy. In Economic-Ecological Modeling, eds. L. C. Braat and W. F. J. Van Lierop. New York: Elsevier .

———. 1993 Systems of tropical forest and economic use. In press. In Tropical Forests, ed. A. E. Lugo. Centenniel Volume. Tropical Forestry Institute. Rio Piedras, PR: U.S. Forest Service.

Odum, H. T., E. C. Odum, and M. Blissett. 1987. Ecology and Economy: "Emergy" Analysis and Public Policy in Texas. Policy Research Project Report #78. Lyndon B. Johnson School of Public Affairs. Austin: The Univ. of Texas.

Odum, H. T., and R. C. Pinkerton. 1955. Time's speed regulator: the optimum efficiency for maximum power output in physical and biological systems. *American Scientist* 43 (2): 331–43.

Repetto, R. 1992. Accounting for environmental assets. *Scientific American* (June): 94–100.

Richardson, J. F. 1988. Spatial patterns and maximum power in ecosystems. Ph.D. dissertation, Environmental Engineering Sciences. Gainesville: Univ. of Florida.

Richardson, J. R., and H. T. Odum. 1981. Power and a pulsing production model. In Energy and Ecological Modeling, eds. W. J. Mitsch, R. W. Bosserman and J. M. Kloptatek. Amsterdam: Elsevier.

Scienceman, D. 1987. Energy and Emergy. In Environmental Economics, eds. G. Pillet and T. Murota. Geneva, Switzerland: Roland Eimgruber.

Simon, J. L., and H. Kahn. 1984. The Resourceful Earth, A Response to Global 2000. Oxford, U.K.: Blackwell.

13 MODELING OF ECOLOGICAL AND ECONOMIC SYSTEMS AT THE WATERSHED SCALE FOR SUSTAINABLE DEVELOPMENT

V. Krysanova and I. Kaganovich
Institute of Economics, Estonian Academy of Sciences
7 Estonia Avenue
EE 0101 Tallinn, Estonia

ABSTRACT

Interrelations between ecological and economic systems and the possibility of their sustainable development can be analyzed by computer models. Relatively large watersheds, with their natural boundaries, provide a convenient framework to investigate effects of human activities on ecosystems through simulation modeling. Such models may be used for estimating the productive capacity of renewable resources and various strategies of resource use to determine how to manage water and land resources for ecologically sustainable development.

Several approaches are proposed for watershed modeling in an ecological economic framework: (1) combining of economic and ecological models, ecologization of Leontiev-type input-output models, and simulation modeling of watershed loading with economic evaluation. Historical analysis is important for comparing past and current management patterns and for estimating the rates of important processes. Analysis in the simulation modeling framework reveals the close interdependence between ecological and economic components of the watershed system and the significant economic dependence of human societies on natural capital.

NEGATIVE CONSEQUENCES OF NATURAL RESOURCES USE

Due to the stability of food chains formed as a result of long-term evolution, natural ecosystems are relatively stable. They are able to resist, within reasonable limits, changes in external conditions and sharp variations in population density. Humans have converted the whole environment into their own "food chain" through industrialized use of natural resources and have become "the top consumers" of the world. The size of the human population does not seem dependent on the external factors and carrying capacity of the biosphere. So far the exhaustion of the richest supplies of material and energy resources has not hampered the industry development. On the contrary, it has even stimulated the search for new resources and ways of nature exploitation. Achievements of science and engineering in the 20th century have created the illusion of infinite resource-energy potential. As a result, an anthropocentric world view dominates in ordinary consciousness, science, technology, and economics.

Reality, however, gives us more reasons to examine such a world view. Ninety-six to ninety-eight percent of the extracted raw materials are dispersed into the environment as wastes (Yablokov et al. 1985). The quantity of wastes from anthropogenic sources is estimated as 10^{12} t, being equal half of the total weight of living organisms in the biosphere (while the biomass of humans constitutes only 0.01% of the total mass of biota). Radical changes in nutrient cycling due to agriculture intensification have increased nitrogen and phosphorus concentrations in lakes and coastal waters. Since the beginning of the 20th century the flow of phosphorus from land to water has increased 40 times (Koplan-Dix et al. 1980). This is the domain where the activity of Homo Sapiens is on the same level with the activity of biosphere. This shows that society's view of modern economy as being highly effective is based on a tremendous underestimation of ecological factors.

When free or cheap natural resources and environmental services are available, the technological development is determined almost exclusively by economic interest to achieve maximum productive output. Even limited natural reserves are considered as infinite or abundant. Technologies, which involve intensive exploitation of high quality natural resources, while ignoring their depletion and pollution of the environment, gain preference. Due to high inertness of the biosphere processes, consequences of anthropogeneous impacts can only be revealed after a long period, in unexpected sites, and often unconnected to the initial cause.

Our approach to analyzing interrelations between ecological and economic systems and to identifying possibilities of their sustainable development is through computer modeling. Systems analysis and simulation are potentially the best tools for understanding the functions and management of integrated ecological economic systems. R. Costanza (1991) proposed to establish a hierarchy of goals for ecological economic planning and management on global, regional, and local levels. In our opinion, ecological economic management at the middle level could also be analyzed for the territories of medium-sized drainage basins (800–2000 km²) through systems analysis and modeling. Drainage basins (or watersheds), including their valuable ecological and essential economic systems (see Figure 13.1), provide an appropriate framework for investigating environmental constraints of human activities.

The main purpose of watershed modeling usually is to investigate the direct and indirect effects of human activities on the ecosystem of a main waterbody. More generally, this modeling approach allows evaluation of the human impact on all of the watershed's ecological systems—the main waterbody, small lakes, and wetlands and their retention capacity, as well as the productive capacity of forests and other renewable resources. Various strategies of resources use and environmental protection may be analyzed. This chapter will develop a conceptual framework for a watershed's ecological and economic systems analysis through modeling that determines how to manage water and land resources for ecologically sustainable development.

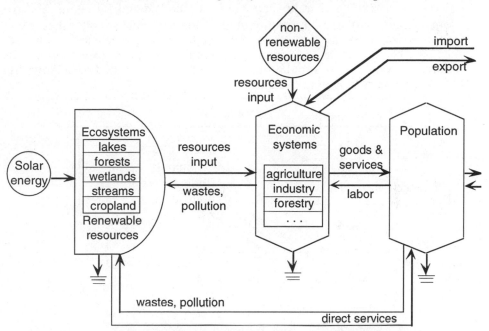

Figure 13.1. Interrelations of ecological and economic systems on a watershed.

THE DRAINAGE BASIN AS AN ECOLOGICAL ECONOMIC SYSTEM

Every drainage basin (or watershed) has definite geographical boundaries. Moreover, watersheds usually have a hierarchical structure—the drainage basin of a main river consists of drainage basins of tributaries of first order, which, in turn, includes drainage basins of tributaries of second order.

In most cases a drainage basin can be considered a semiclosed system relative to water and biogeochemical cycles, as it has definite boundaries and is amenable to measurement inputs-outputs (i.e., it possesses the necessary characteristics of the system). Figure 13.2 presents the main pathways of water and chemical compounds transport to a waterbody. Of course, a watershed's biogeochemical cycles are not fully isolated from the environment because precipitation and evapotranspiration, atmospheric deposition of nutrients and pollutants, export and import of materials generated by economy are *external to watershed flows*. On the other hand, nonpoint pollution from cultivated fields and forests, water discharge from municipal and industrial sewage treatment plants, and river discharge to lake or sea—being *internal to watershed flows* (Figure 13.2)—are more important, if problems of anthropogeneous nutrient loading or problems of water resources management are investigated (i.e., it depends on the problem under study). If the problem is created by global environmental factors, exogenous to the drainage basin (like long-range transport of SO_2), a watershed cannot be considered a system, and the problem cannot be solved by watershed modeling. But in all other cases (eutrophication of freshwater lakes and sea bays, water erosion, water resources use, heavy metals and toxic pollutants discharge), a watershed can be considered a system, and simulation modeling can be applied as a powerful tool to determine management strategy.

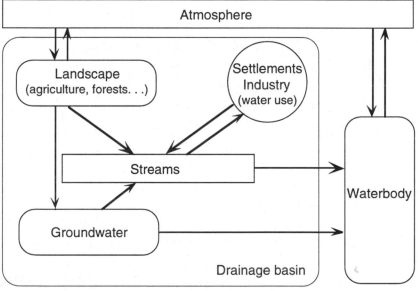

Figure 13.2. Water and chemical compounds transport pathways to a waterbody.

Quite often a watershed's natural boundaries do not coincide with artificial administrative or national borders. Sometimes national boundaries can be barriers to ecologically reasonable watershed management. AM. Jansson (1991a) specifically addressed the significance of open boundaries for sustainable development in the Baltic Sea region. Differences between natural and administrative borders sometimes can create problems with data processing, because statistical information is usually collected according to artificial administrative borders. This raises special concerns regarding spatial data like land use. In such cases the Geographical Information System (GIS) is a suitable tool for solving the problem. GIS is an organized system of computer hardware, software, and geographical and statistical data designed to efficiently capture, store, update, manipulate, analyze, and display geographically referenced information. Certain complex spatial operations are possible with a GIS (such as the processing of several layers of spatial data) that would be very difficult, time consuming, or otherwise impractical.

Although modeling of smaller watersheds is undoubtedly useful because a finer resolution can be applied and more detailed processes studied, analysis in an ecological economic framework is more appropriate for medium-sized drainage basins. As a rule, they incorporate a diversity of valuable ecological (lakes, wetlands, terrestrial ecosystems) and economic systems (agriculture, forestry, fishery, recreation), and that is why their integrated analysis is of great interest. Like islands (Jansson 1985), river basins provide convenient systems for testing ideas regarding the human impact on the carrying capacity of natural resources.

Figure 13.1 shows solar energy entering ecosystems in the watershed, including lakes, forests, wetlands, streams, and cropland. In terms of economy these elements are equivalent to renewable resources. Non-renewable (if any) and renewable resources, as well as labor, are exploited by the economic systems, including agriculture, industry, forestry. Consumer goods and services, produced by the economic system, are directed to the population. Direct environmental services are "transferred" from ecosystems to the population. The process of economic production is accompanied by generation of heat (leaving the system), wastes, and pollution feeding back to ecosystems. Wastes and pollution from towns and settlements are also directed to the environment. Connection with systems outside the watershed is realized through energy flows, import-export, migration of people and, to some extent—by pollution flows (such as pollution from the atmosphere). But these latter examples are not considered here. Biogeochemical cycles embrace the whole variety of ecosystems and economic systems at the higher-than-watershed level, so they are not shown in Figure 13.1.

Let us consider more thoroughly the role of some ecosystems—lakes, wetlands, and forests—as well as economic activity for the functioning of a watershed as a semiclosed system. The importance of freshwater lakes and shallow sea bays for fishing, water supply, recreation, and microclimate control can hardly be overestimated. Their most obvious ecological value consists in regulating the water regime, processing nutrients, and maintaining biological diversity. As the focal system of the watershed, the main waterbody receives water and accumulates various pollutants from the whole drainage basin. Eutrophication and toxification of shallow waterbodies are significant large-scale problems around the world. Urbanization, industrialization, intensification of agriculture have all increased the nutrient content in waterbodies. Negative consequences of eutrophication include excessive algae blooms, oxygen depletion, and decreased fish stocks and catches. From an economic point of view, the most obvious consequences of shallow waters pollution relate to the recreational value and fisheries. Methods of modeling of lake ecosystems are quite advanced (Jørgensen 1976; Jørgensen 1982; Jørgensen et al. 1986). However, modeling of a lake ecosystem, together with its catchment area, is an important task because it allows estimation of not only the critical loading on a lake, but also the influence on total load of different technical measures and land use changes.

Significant attention is now directed to the role of wetlands as nutrient sinks in the landscape. Besides, wetlands are very important habitats for flora and fauna. Some species are adapted to the specific conditions of wetlands and depend on them for their survival. Other species use wetlands as refuges, because their original habitat was lost due to human activity. At the same time, wetlands are among the most threatened ecosystems in the world. Intensification of agriculture and forestry in 20th century has led to extensive drainage of wetlands. For example, in Denmark and Sweden about 90% of the wetland area has been drained (Rydlov et al. 1991). Because of drastic changes in nitrogen and phosphorus cycling the drainage of wetlands decreased the nutrient retention capacity within river basins and increased nutrient loading to waterbodies. From an economic point of view, wetlands may be among the most cost-effective means of reducing nutrient loading to lakes and sea bays (Andréasson-Gren 1991). That is why the protection of relatively untouched wetlands and the restoration or construction of artificial wetlands has become an important management issue. Though many efforts have been made recently to study the retention capacity of wetlands of different types, such knowledge still needs systematization and verification in a watershed modeling framework.

Forests are important for the control of clean air, land runoff, and soil erosion. They are also important as habitats for a diverse community of plants and animals. Negative human impacts on forests are (1) air pollution, (2) overexploitation, and (3) clear-cutting and conversion to agriculture land. On the other hand, forest replantation and air pollution control are examples of positive anthropogeneous impacts on forests.

Economic systems can be analyzed according to the economic subsectors of agriculture, industry, forestry, fishery, tourism, and recreation. If the resource aspect of the system is of special concern, the consumption of non-renewable and renewable resources should be analyzed, as well as the generation of pollution by each subsector of the economy. There is great interdependence between the economic and ecological systems in a watershed (Figure 13.1): clean streams and lakes are necessary for fisheries and recreation; fertile soils and satisfactory water regime are required for agriculture; clean air is vital to forest growth. Meanwhile, industrial wastes and pollutants influence the state of ecosystems.

ANALYSIS IN HISTORICAL RETROSPECT

A historical analysis is important for comparing management patterns of the past and today and for estimating the rates of important processes. Over the centuries, immediate human demands have influenced management goals, including water use for agricultural, domestic, and industrial consumption. The ecological significance of rivers and waterbodies has been neglected. Rivers have been considered simply as conduits and receptacles of wastes. Rivers have been straightened, and deepened, and the vegetation has been removed from riversides. In some European countries (Denmark and southern part of Norway) most small rivers have disappeared during the last century. Wetlands have been considered worthless areas, and many of them have been drained and transformed into agricultural land and forests. The negative consequences of these activities have only recently started to be fully understood. In a historical analysis of landscape transformations for the Swedish island of Gotland (Jansson 1991b), it was concluded that changes in land use (five-fold expansion of agricultural areas compared with pre-industrial times) and negative impacts on water resources will constrain the future development of the island's economic system.

Analysis of historical watershed land-use in retrospect is required for successful watershed modeling. GIS greatly enhances analysis of land use. Retrospective analysis may, for example, be useful for establishing reasonable distribution of land-use patterns and recommending reconstruction of wetlands and riparian zones along river courses. Also, in the case of macrophyte-type eutrophication, comparing maps from different time periods would allow one to estimate the rate of plant invasion.

MODELING OF A WATERSHED AS AN ECOLOGICAL ECONOMIC SYSTEM

Computer simulation in an ecological economic framework can be used not only to reveal the impact of human activities on the functioning of natural ecosystems (one of the aims of ecological models), or to estimate the demand of renewable and non-renewable resources and waste production (that can be done by means of economic models), but to reinforce integration of methods and modeling techniques.

A large number of watershed models has been presented in the scientific literature using different approaches for descriptive, management, and planning purposes (David et al. 1979; Heidke et al. 1986; review in: Krysanova, Luik 1989). Models of drainage basins can be classified according to (1) watershed size, (2) sources of pollution, (3) level of complexity and (4) types of submodels used. Methods vary from simple empirical models based on regression analysis technique to discrete and continuous simulation models. Planning and management models are most often used to estimate interrelations between environmental protection and agricultural production purposes. The modeling approach is especially useful in evaluating nonpoint source pollution (Haith 1982). A good review on methodological aspects of watershed modeling aimed in the control of eutrophication is given by S. O. Ryding and W. Rast (1990).

Simple watershed models focused mainly on ecological issues provide estimates of pollutant or nutrient loads reaching a lake or a waterbody by determining their contribution from different subwatersheds. Such models can also be considered management models, if the estimated load is used to select the best management practice. The next level is to combine the watershed load submodel, river submodel, and waterbody ecosystem submodels (Haith 1982; Krysanova et al. 1989b). Calculation of model scenarios and estimation of their ecological-economic efficiencies can be considered as the management submodel. Management models can also include a planning watershed model, a waterbody model, and cost estimates of various nutrient control measures, which are used in optimization frameworks (Somlyódy 1982). An interesting method of landscape computer modeling aimed at predicting changes in land cover patterns has been developed by R. Costanza et al. (1990).

But the most promising approach is developing combined ecological-economic models, where interdependencies of both ecological and economic issues can be studied. Due to the complexity of the problem there are still not many such case studies in literature. One excellent example is the ecological economic study of the island of Gotland, which includes models of agriculture, water quality, and energy (Zucchetto, Jansson 1985), though it was not specifically designed as a watershed model.

Braat and Steetskamp (1991) proposed an aggregated ecological economic model for analysis of regional development as a semiclosed system. They confirmed that the long-term welfare level in the stabilized phase depended strongly on maintaining a productive capacity of the renewable resources (Forrester 1971; Odum 1983; Braat, Steetskamp 1991). Because river basins can be considered semiclosed systems, these results should hold for them as well. Consequently, when constructing ecological economic models of a watershed, one should consider the productive capacity of renewable resources. So, water and nutrient cycles in a watershed, as well as the state of ecosystems, are of primary importance.

In the opening chapter of the book *Ecological Economics: The Science and Management of Sustainability,* R. Costanza, H. E. Daly, and J. A. Bartholomew (1991) raised a number of major research questions concerning ecological economic modeling. We shall try to address some of these questions, including a most crucial one regarding the integration of modeling mechanisms and tools used in economic and ecological models. In our opinion, there are several possible ways to model watersheds in an ecological economic framework:

- couple economic and ecological submodels at the level input-output,

- include ecological constraints and objectives in Leontiev-type economic input-output models, or

- include economic evaluation in simulation watershed models, focused mainly on watershed loading.

Let us consider more thoroughly these three approaches.

COUPLING OF ECONOMIC AND ECOLOGICAL MODELS

In a watershed scale, combining economic models for describing the exploitation of valuable resources with ecological simulation models for estimating the effects of human activities and constraints for economic development is one possible approach. For example, in the aforementioned research project on Gotland (Zucchetto, Jansson 1985), though not specifically designed as a watershed model, submodels of agriculture, water quality, and energy were combined. A Leontiev-type input-output model estimated the total resource use and generation of wastes and ecological economics:modelingpollution from the economic activities. Waste loadings were used as inputs for ecological simulation models, and the results of water quality models were later used in optimization models to determine environmental constraints on resource use or pollution outlets. Results from these optimization models were used to recommend changes in the regional economy to counteract environmental degradation.

ECOLOGIZATION OF LEONTIEV-TYPE INPUT-OUTPUT MODELS

Reality gives many reasons to change *anthropocentric* views and objectives to *biocentric* ones. Natural resources and the environment can no longer be considered simply as constraints, limiting the level and rate of reaching traditional economic goals. Instead, sustainable natural resources and environment are becoming basic objectives, which force us to alter and adapt the techniques of input-output models.

Economic activity on watershed can be analyzed for economic sectors as consumers of non-renewable and renewable resources and producers of consumer goods and wastes. If some kinds of resources (water, timber, mineral resources) are of special interest, they can be analyzed by input-output models as well. Ecologization means that ecological constraints and objectives are widely used in these input-output models.

An example of such an approach is the series of models (Kaganovich 1977; Kaganovich et al. 1982; Barabaner et al. 1990) for the investigation of resources use in the industrial North-East region of Estonia as a region of ecological disaster. The dynamic input-output model of Kantorovich-type (1965) was used to analyze the alternative strategies of resources use. The analysis was focused on mixed strategy, minimizing the total sum of labor, capital and resource expenditures. This work was oriented to the solution of debatable practical problems. It was necessary to assess:

- the final for the whole system economic and ecological characteristics of proposed variants of development;

- the values of power capacity of power-producing units for different variants of their location in the region, providing that concentrations of pollutants in towns and settlements will be below the permitted level; and

- possible consequences of realization of the industrial programs for environment and agriculture.

Different technologies were analyzed from the points of view of intersectoral problems and connections in the industrial complex, regional (infrastructure, labor resources), ecological and environmental factors (resources use, pollution, cost of purification, loss of agricultural land). The regional ecological aspect was represented by characteristics of natural resources and other elements of environment involved in the production activity. Multi-variant optimization calculations were carried out in a dialogue routine with interval-setting of uncertain parameters. These operations used a database which included characteristics of ecological situation and background level of pollution in towns. The last version of the model put more attention to the resource-use incentives that would meet the requirements of sustainable development (Kaganovich 1992). For that purpose the integral income from production activity and natural capital expenditures, taking into account the discount rate, have been compared.

SIMULATION MODELING OF WATERSHED LOADING WITH ECONOMIC EVALUATION

As biocentric views come to the forefront, modeling of ecological processes becomes more important. In our case, the attention has focused on the nutrients and attempts to restore natural cycles, and human impacts and the ways to minimize them. Modeling of watershed loading is very important, as the loading is a concentrated expression of effects of human activities in the watershed scale. Different types of human activities are commonly responsible for pollution—agriculture, industry, and domestic wastes. Different pollutants can be analyzed by similar methods. But most often nutrient loadings (nitrogen and phosphorus) are of concern, being the main causes of the accelerating eutrophication in last decades.

Our experience is based on the modeling of two medium-sized drainage basins in Estonia, the agricultural watershed of the Matsalu Bay (3500 km²) and the Estonian part of the Lake Peipus watershed (16,300 km²) (Krysanova et al. 1989b; Krysanova, Luik 1989a; Krysanova 1991). Simulation modeling of nutrient flows in watersheds reveals the spatio-temporal dynamics of nutrient contribution from different sources and subwatersheds to a waterbody, and helps develop an eutrophication control strategy based on load reduction. Let us consider some important methodologies for modeling watershed loading and general practical recommendations resulting from such a model.

Simulation modeling of nutrient flows on a watershed can occur in several essential stages:

- *spatial resolution* of the watershed,
- creation of *a data base,*
- identification of *major nutrient sources* in the watershed,

- *estimation of point and nonpoint* sources,

- assessment of the *retention capacity* in watershed scale, and

- *analysis of nutrient control measures and cost-benefit analysis.*

Identification of spatio-source distribution of nutrient load based on land use data and hydrochemical samples is a main problem in modeling medium-sized and large watersheds. It requires solving the problem of spatial disaggregation of a watershed. Its solution depends to a large extent on the availability and representativity of territorial information. Division into subwatersheds could be determined by sampling sites for water quality measurements. More thorough disaggregation is possible, if more frequent measurements are available at least for one smaller representative subwatershed.

After that, collection, quality assurance and completion of data are needed. Main blocks of data cover water quality, land use, and economic activities. Calculation of the annual nutrient load for subwatersheds based on water quality measurements is necessary for source apportionment. Different methods could be proposed for that, including linear interpolation of water quality data. Data on economic activities in a watershed include (1) the number of habitants for towns and settlements, (2) the number of animals in livestock farms, (3) the capacity and state of sewage treatment plants and industrial water treatment facilities, and (4) the state of animal manure depots.

All the data should be collected in accordance with the chosen spatial structure. Obtaining this data is difficult because statistical information is usually collected according to "artificial" administrative borders. However, the application of GIS can facilitate spatial data processing. A GIS for a watershed consists of a digitized topographical map and several "layers" of spatial data, including coastlines (if any), river networks, borders of the watershed and subwatersheds, lakes, land use data, and a soil map.

The major nutrient sources in a watershed can be identified from a preliminary analysis of the data base. Potential nutrient point sources include:

- municipalities (STP—sewage treatment plants),

- industries, and

- livestock farms.

Potential nonpoint sources include:

- dry and wet deposition from the atmosphere,

- urban nonpoint runoff (e.g., storm water),

- drainage from agricultural land, and

- drainage from forests, grassland and wetlands.

The relative importance of different sources varies with different watersheds.

Traditionally, estimation of pollution loading was emission-related and focused on point sources because reliable assessment of nonpoint sources of pollution and source division of riverine load is difficult. For example, the estimates of total load of BOD$_7$, total nitrogen and total phosphorus to the Baltic Sea made by the Helsinki Commission of the Management of the Baltic Sea (HELCOM) do not make a source division; they simply divide the total load into three components: direct discharges from municipal sewage, direct discharges from industrial sewage, and total river transport into the Baltic Sea. The latter includes upstream agricultural, municipal, and industrial sources, plus deposition from the atmosphere.

Estimation of nutrient inputs from point sources is relatively easy, because most sewage treatment plants and factories routinely measure their effluent discharges. In recent decades the increased awareness of the importance of nonpoint sources of nutrient pollution (Karlsson et al. 1988) has captured researchers' attention. The intensified use of commercial fertilizers during the last decades and increased nutrient leaching from cultivated soils has made them a major nutrient source in many watersheds.

An appropriate method for estimating nonpoint source loads is *indirect measurements*. The concept of *unit area load* (UAL) (Ryding, Rast 1989) or the similar concept of load from *elementary area of pollution* (EAP) (Krysanova et al. 1989b) can be applied. Unit loads represent the quantity of nutrients generated per unit area, per unit time for each *type* of nonpoint nutrient source in the drainage basin (such as arable land with a special type of soil, or forest). UAL could be measured directly in specially chosen field plots with specific land use activity or could be obtained from simulation based on river sampling in a small representative sub-basin. The corresponding formulas, used in the Matsalu Model for submodels, simulating water and nutrient cycling in the EAP (Krysanova et al. 1989b; Krysanova, Luik 1989a), can be used in other case studies as well. It is possible to use UALs estimated for another basin, if landscape and climatic conditions, as well as management patterns in agriculture, are similar in both cases. To estimate the load from a certain nonpoint source of pollution, the UAL is multiplied by the total area of the corresponding land use activity in the subwatershed.

Retention of nutrients in a watershed is an important factor influencing the magnitude of the total nutrient load on a waterbody. Wetlands can act as nutrient filters between agricultural land and surface waters. That is why the total load at the rivermouth is not simply the algebraic sum of estimated inputs from a number of sources. Retention in soils, wetlands, river courses, and lakes can be differentiated.

A more comprehensive understanding of nutrient retention for the watershed as a whole is needed. Presumably, the simulation modeling framework can provide such insight. Usually there are several uncertain parameters in the total nutrient load estimation, including retention coefficients for each subbasin. In such a situation, interval presentation is recommended for the uncertain parameters. Later, the levels of confidence should be determined for the different components (i.e., the estimates of loads from subwatersheds and the estimates of loads from different sources of pollution). After that, the verification procedure could be a step-by-step iteration process, in which the magnitude of uncertain parameters will be determined to reach a satisfactory comparison of simulation results and to accomplish the spatio-source distribution of load. For example, the total dissolved losses of phosphorus (LPD) are obtained as

$$LPD = \sum_i \sum_j PD_{ij} \; R_i + \sum_i \sum_k S_{ik} \; PDF_{ik} \; RF_i$$

where i =   number of subwatershed,

j =   number of point source of pollution,

k =   number of soil type,

PD_{ij} =   dissolved phosphorus losses from point source (kg),

R_i =   retention coefficient for subwatershed i for point sources,

S_{ik} =   area of soil type k in subwatershed i, (ha),

PDF_{ik} =   dissolved phosphorus losses from fields with soil type k, (kg/ha), and

RF_i =   retention coefficient for subwatershed i for nonpoint sources.

After that, analysis of possible nutrient control measures in the drainage basin could reveal potential cost-effective measures to reduce the load from the most important sources. Once more the difference between point and nonpoint source pollution becomes critical, because it is more difficult to estimate the cost of correcting the damage created by drainage from croplands. Different measures in the agricultural sector, as well as in the construction, improvement and increase in the capacity of sewage treatment plants, should be analyzed. For example, the restoration of wetlands can be a low-cost ecologically effective alternative for reducing nitrogen load (Andréasson-Gren 1991). A synthesis of the load calculations and the analysis of control measures provides estimates of ecological efficiencies (reduction in the total load to waterbody) and costs for different measures or so-called ecological economic efficiencies. They can be used to predict changes of the total load in relation to various management patterns and to develop the cost-effective strategy of eutrophication control. In general, such cost-benefit analysis based on the results of modeling can be important in formulating sustainable management strategies for watersheds.

What practical recommendations have been derived from the results of modeling obtained in our case studies? Simulation modeling of the agricultural watershed of the Matsalu Bay (Krysanova et al. 1989) confirmed that within currently accepted technological scheme in agriculture, it is possible to decrease nutrient load by 25% without reducing crop production. The necessary measures include maintenance of recommended agrotechnical terms in application of fertilizers, in-soil application of mineral fertilizers, partial replacement of mineral fertilizers by animal manure (for light soils), and vegetation strips along river courses. The general recommendations are

- possibly lower fertilizer application rates, adjusted to crop requirements,

- vegetation cover of fields during as long period as possible (e.g., successive cropping), and

- establishment and reconstruction of vegetated buffering systems between cultivated land and surface waters.

The second watershed of the Lake Peipus is larger than the Matsalu watershed; it has a variety of sources of nutrient pollution: intensive agriculture and insufficiently treated or untreated sewage waters in towns, and low capacity of animal manure depots. The main purpose of modeling was to establish spatio-source distribution of the total phosphorus load from the Estonian part of the watershed to the Lake Peipus, and to estimate ecological economic efficiencies of different phosphorus control measures. The estimates for nonpoint sources of pollution were derived from the Matsalu model, as climatic conditions and agriculture practice are similar in both areas. The results of modeling (Krysanova 1991) suggested a set of measures, the application of which will reduce the load below the critical level. In general, the improvement of the state of the Lake is possible if similar efforts are undertaken simultaneously in the Russian part of the whole drainage basin.

WATER RESOURCES MANAGEMENT

This section presents some examples of actual water resources management —what kinds of policies are applied in different countries, what institutions are established, and what attempts are undertaken to manage at a watershed scale. This section concludes with possible ways to apply the results of modeling to management at a watershed scale. While the Council of Europe declared River Basin Management as a guiding principle of the European Water Charter in 1968, this principle has been realized in only a few countries. For example, France and Sweden have adopted opposing strategies in their management of water resources (Gustafsson 1989).

The French strategy is based on a combination of economic incentives and governmental regulatory policies, with the emphasis on incentives. Municipalities, industries, and farmers pay a water use charge and effluent charge—the latter is based on the volume discharged as well as on the level of pollution (weighted taking into account selected components of pollution). By this, the Polluter Pays Principle (PPP) is introduced. Moreover, in France six River Basin Agencies were established as principal water institutions with the authority to assess water charges for their regional budgets. Seventy percent of the total sum obtained through charges is used to decrease discharge of pollutants to waterbodies. As a whole, river basin management in France is treated as a complex socio-political process, involving water interest groups, citizens, administrators, and politicians. In some other countries (Great Britain, Poland) there are attempts to reform their water management institutions to meet the requirements of the natural river basins.

The Swedish model of water resources management is of pure "command and control" type. The general policy is to fix tough discharge standards for firms applying for an operation permit, so that these firms are forced to construct efficient equipment for sewage treatment. So, pollution is legally allowed up to a standard discharge level. Due to lack of the complex river basin approach, nonpoint sources of pollution stay practically uncontrolled. The Swedish Environment Protection Board fixes and supervises governmental regulation to be followed at the lower levels, and regional planning institutions are practically nonexistent. Decision-making occurs on an administrative level. As a result, the perspectives of long-term water planning are unclear (Gustafsson 1989).

Estonia represents the approach that relys on both command and control (permits, fines) and market-based instruments (user charges and pollution fees) to enforce environmental regulations. Fees for water pollution (BOD, suspended particulates, petroleum products, phenols, phosphorus, nitrogen, sulphates, fats) are collected from agricultural, municipal, and industrial water users. Half of the revenues from fees and fines are retained in the local environmental protection funds. The Ministry of Environment has overall responsibility for the management of natural resources, including water resources. However, market-based policies work better in a market economy. In Estonia, the institutional framework of a market economy will eventually develop, but it will take time before market rules become really effective. So, the use and effectiveness of administrative policies should not be neglected in water resources management. Nevertheless, the development of river basin management strategies is widely recognized as a high priority concern for the future.

The results of modeling can be important in attempting to manage at a watershed scale. It seems that France, with its experience in water resources management, could easily adopt this idea. An integrated modeling approach incorporating the elements of GIS would be especially useful. Land use and water quality maps, combined with load estimates for different sources of pollution, and evaluation of spatio-source distribution of total nutrient load to waterbody, can be important instruments for planning and management. Calculation of dynamic scenarios would be useful to check the effect of the control measures on the total loading. The effect of water protection policies could also be assessed in the modeling framework. In our opinion, the technique of integrated watershed modeling would help formulate a sustainable river basin management strategy for the future.

FROM NATURE PROTECTION TO SUSTAINABLE DEVELOPMENT

This section outlines some perspectives of natural resources use. Analysis in the simulation modeling framework reveals the close interdependence between ecological and economic components of the watershed system and significant economic dependence of human societies on natural capital. It reveals the value of natural resources even if they are not used.

Under present conditions, when the negative consequences of resources used have become obvious, the anthropocentric world view should change to a biocentric one. Due to growing public recognition of the threat of global ecological catastrophe, trends indicate changes in the paradigm of natural resources use in developed countries. Namely, parameters of the environment can no longer be considered simply as constraints, limiting the level and rate of reaching of traditional goals. Instead, they are becoming basic objectives. The problem of mankind's survival is currently of great social interest. Nature's services can not be taken for granted as before. General recognition of the ideology of survival, its implementation in human consciousness, could reconstruct individual and social functions of preference, disseminate them to ecological goods and services, and *turn natural and artificial utilities into comparable values in minds and at the market.* Ecological services should be considered as primary consumer goods, competitive in relation to the novelties of domestic technique and luxury goods. This will mean the reorientation of social consciousness to the ideology of ecocentrism (biocentrism). The chance to survive will improve, if the protection and reconstruction of environment becomes more profitable than its exploitation.

Confirmation of ideology of ecocentrism supposes that the concept of economic damage from the environmental pollution should be changed as well. Generally accepted concepts correspond to the prevailing anthropocentric approach—the damage is equivalent to expenditures on prevention and compensation of the effects of ecological disturbances on people and property. The estimates made according to this scheme are usually underestimated: the value of losing ecological services is deliberately higher than direct economic losses. Ecocentrism could transfer the problem of economic damage from the pollution of environment to another plane: the disturbance of the vitality of biocoenosises should become the indication of damage. In the long run, prevention of biocoenosises from depression most fully corresponds to fundamental and long-term interests of people. First of all, the state of flora and fauna is the indicator of changes of chemical state of geospheres, which are potentially dangerous for people. Secondly, given the interconnections of all elements of biota, disturbances in one section will adversely affect, earlier or later, the whole system, including people.

From the position of ecocentrism, the technologies that change the status quo of the environment should be considered dangerous and detrimental. Full ecological expenditures are of infinite value, as infinite is the usefulness of diminishing natural reserves. Therefore, high prices for their losses will follow as the ecological consciousness of society and its ability to give up momentary needs for future ones are developed.

As the biocentric conception becomes more firmly established, the demand for ecological goods and services and ecologically safe production will rise. The production of goods, involving emission of pollutants, will become more expensive and will be rejected from the market through economic sanctions. As a result, the technosphere should become more adapted to natural cycles of materials and energy. Ideally, nature does not need protection. It has enough possibilities and time for effective self-defense. The question is whether mankind, in its efforts to survive, is able to switch its spiritual and material energy from exploitation to the protection of the stability of biosphere.

REFERENCES

Andréasson-Gren, I. M. 1991. Costs for nitrogen source reduction in a eutrophicated bay in Sweden. In Linking the Natural Environment and the Economy: Essays from the Eco-Eco Group, eds. C. Folke and T. Käberger. Dordrecht: Kluwer.

Barabaner, N., I. Kaganovich, and A. Laur. 1990. Modeling and Analysis of Resources Use in the Mining-Industry Complex (at the example of oil-shale complex). (In Russian). Tallinn: Institute of Economics, Estonian Academy of Science.

Braat, L. C., and I. Steetskamp. 1991. Ecological economic analysis for regional sustainable development. In Ecological Economics: the Science and Management of Sustainability, ed. R. Costanza. New York: Columbia Univ. Press.

Costanza, R. 1991. Assuring sustainability of ecological economic systems. In Ecological Economics: the Science and Management of Sustainability, ed. R. Costanza. New York: Columbia Univ. Press.

Costanza, R., H. E. Daly, and J. A. Bartholomew. 1991. Goals, agenda and policy recommendations for ecological economics. In Ecological Economics: the Science and Management of Sustainability, ed. R. Costanza. New York: Columbia Univ. Press.

Costanza, R., F. H. Sklar, and M. L. White. 1990. Modeling coastal landscape dynamics. *BioScience* 40 (2): 91–107.

David, L., L. Telegdi, and G. van Straten. 1979. A Watershed Development Approach to the Eutrophication Problem of Lake Balaton (A multiregional and multicriteria model). CP-79-16. Laxenburg: International Institute for Applied System Analysis.

Forrester, J. W. 1971. World Dynamics. Cambridge, MA: Wright-Allen.

Gustaffson, J. E. 1989. A new perspective in river basin management. In River Basin Management, V. Proceedings of an IAWPRC Conference. Oxford: Pergamon Press.

Haith, D. A. 1982. Models for Analyzing Agricultural Nonpoint-Source Pollution. RR-82-17. Laxenburg: International Institute for Applied System Analysis.

Heidke, T. M., M. T. Auer, and R. P. Canale. 1986. Microcomputers and water quality models: access for decision makers. *Journal of Water Pollution Control Fed.* 58 (10): 960–6.

Jansson, AM. 1985. Natural productivity and regional carrying capacity for human activities on the island of Gotland, Sweden. In Economics of Ecosystems Management, eds. D. O. Hall, N. Myers and N. S. Margaris. Dordrecht: W. Junk Publishers.

———. 1991a. On the significance of open boundaries for an ecologically sustainable development of human societies. In Ecological Economics: the Science and Management of Sustainability, ed. R. Costanza. New York: Columbia Univ. Press.

————. 1991b. Ecological consequences of long-term landscape transformations in relation to energy use and economic development. In Linking the Natural Environment and the Economy: Essays from the Eco-Eco Group, eds. C. Folke and T. Käberger. Dordrecht: Kluwer.

Jørgensen, S. E. 1976. A eutrophication model for a lake. *Ecological Modelling* 2: 147–65.

————. 1982. Modeling the eutrophication of shallow lakes. In Ecosystem Dynamics in Freshwater Wetlands and Shallow Water Bodies, 2. Proceedings of the International Science Workshop. Moscow: CIP GKNT.

Jørgensen, S. E., L. Kamp-Nielsen, and L. A. Jørgensen. 1986. Examination of the generality of eutrophication models. *Ecological Modelling* 13: 251–66.

Kaganovich, I. 1977. On the complex analysis of territorial-industrial problems with ecological factors. (In Russian). *Ekonomika i matematicheskije metody* 13 (5): 998–1007.

————. 1992. Negative consequences of resource's use: economic aspect. (In Russian). Proceedings of the Estonian Academy of Science, Humanitarian and Social Sciences. 41 (2): 95–110.

Kaganovich, I., A. Laur, A. Maamagi, and K. Tenno. 1982. Dialogue Analysis of Regional Production Programs. *Angewandte Systemanalyse* 3 (4): 161–6.

Karlsson, G., A. Grimvall, and M. Lowgren. 1988. River basin perspective on long-term changes in the transport of nitrogen and phosphorus. *Water Resources* 22: 139–49.

Kantorovich, L. V., and V. L. Makarov. 1965. Optimal models of the perspective planning. (In Russian). In Application of Mathematics in Economic Research, 3: 7–87. Moscow: Mysl.

Koplan-Dix, I. S., and A. Stravinskaja. 1980. Anthropogeneous Impact on Small Lakes. (In Russian). Leningrad: Nauka.

Krysanova, V. 1991. Interactive simulation modeling for environmental planning on watershed. In Environment, Energy and Natural Resource Management in the Baltic Region, eds. J. Fenger, K. Halsnaes, H. Larsen, H. Schroll and V. Vidal. Third International Conference on Systems Analysis. Copenhagen: Nordic Council of Ministers, Nord 1991:48.

Krysanova, V., and H. Luik, eds. 1989a. Simulation modeling of a system "watershed-river-sea bay." (In Russian). Tallinn: Valgus.

Krysanova, V., A. Meiner, J. Roosaare, and A. Vasilyev. 1989b. Simulation modeling of the coastal waters pollution from agricultural watershed. *Ecological Modelling* 49: 7–29.

Odum, H. T. 1983. Systems Ecology: An Introduction. New York: Wiley.

Ryding, S-O., and W. Rast, eds. 1990. The control of eutrophication of lakes and reservoirs. Vol. 1, Man and the Biosphere Series. Paris: UNESCO and the Parthenon Publishing Group.

Rydlöv, M., H. Hasslöf, K. Sundblad, K. Robertson, and H. B. Wittgren. 1991. Wetlands—Vital Ecosystems for Nature and Societies in the Baltic Sea Region. WWF Report to the HELCOM Ad Hoc High-Level Task Force. Washington, DC: WWF.

Somlyódy, L. 1982. A Systems Approach to Eutrophication Management with Application to Lake Balaton. Paper presented at the Third International Conference on State-of-the-Art in Ecological Modeling, Colorado State Univ., May 24–28.

Yablokov, A. V., and S. A. Ostroumov. 1985. Levels of animate nature protection. (In Russian). Moscow: Nauka.

Zucchetto, J., and AM. Jansson. 1985. Resources and Society: A Systems Ecology Study of the Island of Gotland, Sweden. New York: Springer-Verlag.

14 MULTIPLE USE OF ENVIRONMENTAL RESOURCES: A HOUSEHOLD PRODUCTION FUNCTION APPROACH TO VALUING NATURAL CAPITAL

Karl-Göran Mäler, Ing-Marie Gren, & Carl Folke
Beijer International Institute of Ecological Economics
Royal Swedish Academy of Sciences
Box 50005
S-10405 Stockholm, Sweden

ABSTRACT

Environmental resources such as ecosystems generate crucial ecological services to households. The value of these services is seldom reflected in market prices. In this chapter we develop a methodology, based on the household production function, which makes it possible to consider the economic value to individuals of such "unpriced" services. For this purpose we distinguish between values revealed by markets and values not revealed by markets, and focus on the values revealed by markets. The value revealed by markets is defined as the change in expenditures on marketed goods and services due to a change in the supply of an environmental resource. This approach is applied to the economic valuation of mangrove ecosystems.

INTRODUCTION

Natural capital is in many respects becoming a scarce resource for economic development. Large parts of the natural capital are unpriced, but may nevertheless be of fundamental value for human well-being. Hence, there is a need to include these values in economic analysis. By valuation we mean the assignment of a monetary value to economic goods and services and in particular, to environmental resources. For most conventional goods and services, this valuation is done by the market. For others, the valuation is done by pricing boards (e.g., agricultural pricing), by trade unions, by central banks, etc. However, there exist a substantial number of goods, services, and factors of production that are not assessed an economic value at all. The purpose of economic valuation techniques is to identify the "correct" prices for such goods and services in order to decrease the differences between private and social values.

Not all environmental resources can be economically valued at present (although in theory we can always define a value). It is well known that environmental degradation can lead to domestic and international political instability and even to military conflicts. At present, there simply does not exist any technique that can be used to value such consequences. Another example is obvious in the effect on community cooperation from environmental programs. The value from cooperation and participation as such cannot be measured and should not be valued because of the arbitrary character of the results. In spite of the very fast development of methods for valuing environmental resources during the last fifteen years, there are still areas that can not be handled by these techniques (Johansson 1987). There still are intangibles, although these have been reduced substantially.

The reason why valuation is important is simply that unless a resource is assessed a value, it tends to be mismanaged, and in particular, overexploited. One could, of course argue that political decisions on the use of the resource, without any explicit valuation, can lead to efficient management of the resource. This is true, but it overlooks that the political decision reflects, at least implicitly, a value assessment of the resource. The starting point for modern valuation techniques based on economic theory is that it should be preferable in a democracy that these value assessments are made explicit and be based on the consequences for those concerned.

In the literature on economic valuation of environmental resources a distinction between non-use and use values is often made which can be traced back to Krutilla (1967). The theoretical basis for this distinction will be briefly discussed in this chapter. It will then be shown that for various reasons this classification of values may not always be appropriate, especially when attempting to put a money value on multifunctional ecosystems, which include several of the ecological services. The main purpose of this chapter is therefore to suggest another classification, which is based on the possibilities of *observing revealed behavior* (i.e., values revealed by markets and values not revealed by markets). The analysis is then focused on values revealed by markets, and it is shown how the household production function can be used to measure part of the value of a multifunctional resource through the values of the ecological services that the household actually uses. This approach is then analytically applied to the mangrove ecosystems which are known for their rich production of ecological services.

The chapter is organized as follows: the first section gives a formal description of the measurement of individual and social welfare. Then we consider the welfare measurement of only one individual, and in particular, look at the distinction between values revealed by markets and values not revealed by markets. Next, multiple use of a resource is considered, and the notion of household production functions is introduced. In the final section the idea of integrating household production functions and ecological modeling is illustrated.

WELFARE MEASUREMENTS

Quite often one can find references to potential welfare and various compensation criteria in environmental economics literature. A change is said to increase the potential welfare if it is such that the winners from the change can compensate the losers and still be better off, regardless of whether the compensations are made (if the compensations are made, then we have the usual Pareto criterion). Quite often it is almost automatically assumed that an increase in potential welfare is desirable. This is nonsense, and that it is nonsense has been known for more than forty years. First of all, the compensation criteria are logically inconsistent and can easily lead to non-transitive ranking of alternatives (Graaf 1957). Transitivity implies that if a good A is preferred to B which, in turn, is preferred to C, then good A is preferred to C. Non-transitivity could then mean that good C, is preferred to A. But more important is the implicit assumption that an increase in potential welfare is equal to an increase in actual welfare. That means that it does not matter who looses and how much they loose from the change. Even if all benefits go to the most wealthy individuals and if the already poverty-stricken part of the population is being made worse off, it is still regarded as an increase in welfare if the rich can compensate the poor even if they do not pay any compensations.

The only way to make a judgment whether or not welfare has increased is to explicitly introduce ethical considerations on the distribution of welfare. This can be done through a social welfare function. We know from Arrow's impossibility theorem that it is impossible to derive the social welfare function from individual preferences in a way that is consistent with some reasonable conditions concerning efficiency, non-dictatorship, choice of alternatives and completeness of preferences (Arrow 1970). However, A. Sen has argued that there quite often exists information that would make aggregation of individual preferences possible. One can therefore interpret the welfare function either as representing an aggregate of individual preferences (given Sen's assumptions on information) or representing the preferences of some individual or organization.

Consider a change in society from A to B. This change implies that the individual utility levels will change from u_h^A to u_h^B, and that the social welfare will change with $\Delta W = W(U_g^B, ..., U_h^B) - W(U_g^A, ..., U_h^A)$. If $\Delta W > 0$, then social welfare will increase in society according to the applied welfare function. The objective of valuation is to find out whether or not this net result is positive. This raises at least three questions:

- How do individual utilities change?
- How can the social welfare function be identified and estimated?

- How is the set of individuals defined?

The objective of this chapter is to analyze the theory behind the estimation of individual utility changes, and we will therefore not discuss the last two questions. However, with respect to the third question, there is no reason why the set of individuals should be restricted to the present generation. If present people have preferences over the welfare of future generations, the future generations should of course be included in the set of individuals.

INDIVIDUAL VALUES OF ENVIRONMENTAL RESOURCES REVEALED BY MARKETS AND NOT REVEALED BY MARKETS

The first step in building a theory of individual values is to find ways by which the preferences of individuals can be described. The standard approach in economic theory is to postulate a utility function for a typical household. Although a household generally consists of several individuals, it is assumed that the preferences of these individuals can be described by one utility function. Let then the utility function for an arbitrary household be (and as we are considering only one household, the subscript h will not be needed)

$$U = U(x_1, x_2, ..., x_n, S)$$

where x_i is the amount of the private (marketed) good i consumed by the individual and S is the supply of an environmental resource (soil, clean air, clean water, forest etc.).

The household's behavior can be described as maximizing the utility subject to the budget constraint

$$px = I$$

where p is price vector, x the vector of net consumption of marketed goods, and I the wealth of the household. Maximization generates the Marshallian demand functions

$$x = x(p, S, I).$$

The demand for a market good is thus a function of the prices of the good in question and on other market goods, the supply of the environmental resource and the income. Depending on the relation between the market good and the environmental resource two different cases can be identified: a change in the supply of the environmental resource does or does not affect the demand for the market good. Let us start by considering the case when a change in the supply of the resource will not affect the market behavior of the household. A necessary and sufficient condition for this is that the utility function is additively separable in x and S^1.

---

1. Sufficiency is immediate. Necessity can be proved be calculating the partial derivative of x_i with respect to S for each i at set these expressions equal to zero. This will yield a system of

$$U = \phi(x) + \psi(S)$$

In this case, we shall say that there is no use value associated with the environmental resource. The idea behind this interpretation is that the use of a resource should be reflected in the demand for marketed goods and services. One example is the collection of fuelwood which requires equipment such as axes. The use of a lake for recreational purposes, say fishing, requires rods.

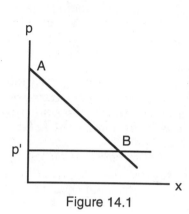

Figure 14.1

Furthermore, whatever the use of the resource may be, the use will require time and the time allocated to the use of the resource will obviously affect the time spent on other marketed goods and services and therefore, on the demand for these. The relation between the use of the environmental resource and the marketed good or service may be either positive or negative as is discussed later. However, when a change in the resource does not affect market behavior, it seems reasonable to talk of pure non-use value of the resource. It should, however, be noted that a current non-use value may turn into a use value in the future when there is more knowledge available about the ecosystem production of services. This definition of non-use value differs from the one that is being discussed in the literature (Mäler 1978; Smith 1987).

In the literature use value is usually defined by reference to a market good which represents the use of the resource. The use value of the resource is then defined as the consumer surplus for that good or the area p'AB in Figure 14.1. Thus, the use value is defined as if the market good and the environmental resource are weak complements (to be defined in Appendix C). The non-use value is then defined as the residual between the total value and the use value.

However, this would exclude many situations in which one could speak of a use value. For example, cleaner air may mean that I do not have to clean my windows as often. Should the value of this reduction in cleaning expenses be regarded as a use value or non-use value? It is hard to believe that cleaning expenses and cleaner air are weak complements (in fact much more likely to be substitutes) and that cleaning expenses would therefore be non-use values seems to be odd.

---

partial differential equations in the unknown function U, the solution of which is a function additive separable in x and S.

A much more natural approach would be to associate a use value to the value of market goods that are affected by a change in the resource. However, there are obvious cases when this definition would also seem counter-intuitive. For example, for most economists, use value is associated with using up resources. Forestry, agriculture, and hunting all provide examples on use values associated with using up resources, and the use values usually associated with these examples are covered within the conventional definition of use values.

The definition proposed here is much wider and would include values that would not correspond to using up resources. However, it is theoretically quite difficult in defining the use of a resource (unless one would accept that use of a resource simply means that an individual has a positive desire for the resource). Perhaps the best way out is to do away with the concept of use and non-use values and instead talk of values revealed by markets and values not revealed by markets. *Values revealed by markets would then correspond to the value derived from observed market behavior, while values not revealed by markets could never be revealed from observing individual behavior on markets.* This is the terminology that we use here. The advantage of these definitions is that they are *completely related to observable variables.*

We can try to make this precise in the following way. Let us consider the second case when a change in the supply of the environmental resource affects the market behavior, i.e., when $\partial x_i/\partial S \neq 0$ for at least one i. Then the utility function will be additively separable in a function of x and S and a function only of S:

$$u = \psi(x, S) + \phi(S).$$

Obviously, this representation is not uniquely defined. However, by adding the condition that ψ cannot be made additively separable in a function of x and S and a function of only S by any monotonically increasing function, the representation is obviously unique.

The value associated with the occurrence of S in ψ will be interpreted as the value revealed by markets and the value associated with the occurrence S as the value not revealed by markets. As shown in Appendix A, this can be made more precise by using the expenditure function which defines the minimum expenses for an individual for obtaining a certain level of utility. The value from a change in the supply of an environmental resource is defined as the corresponding change in all expenditures which would make the utility constant. *The value revealed by markets is then the change in expenditures on market goods and services needed to keep $\psi(x, S)$ constant at a certain change in the supply of the environmental resource.* See Appendix A for a formal definition.

We will subsequently deal only with the estimation of the value revealed by markets as defined above. A few comments on the estimation of the value not-revealed by markets may, however, be in order. First, the only reason why values not revealed by markets are interesting is due to the fact that their existence shows that values revealed by markets will be biased estimates of the total value. We cannot take for granted that an estimated value revealed by markets is an appropriate estimation of the total value. First, the size (and the sign) of a possible value not revealed by markets must be analyzed.

In the approaches to be discussed there is no need to make a distinction between the value revealed by markets and the value not revealed by markets. The value not revealed by markets can never be estimated from observed market behavior since it is defined in such a way that it does not affect such behavior. Therefore, the estimation of it must use other techniques— techniques that can be used to estimate the total value directly. One way is to use contingent valuation to estimate the total value. Another approach would be to carry out *real* valuation studies (i.e., non-hypothetical) to obtain estimates of the total value. Both these approaches are discussed in the literature, and there is no need to dwell on them here.

However, there is one more approach which has been very rarely (or perhaps never) used. Usually, there are possibilities for households or other agents to make voluntary contributions to protect a resource. When Mount Carmel National Park outside Haifa burned down, voluntary contributions were forthcoming. Similarly, voluntary contributions from Swedes for the planting of trees in Kenya are quite common. It should be possible to use information on such contributions in order to get a better picture of the economic value of the resource. Appendix B shows that given information on individual contributions and incomes, bid-function can be estimated econometrically. This bid-function can then be used to calculate the social value of a change in the supply of the resource. To the best of our knowledge, this approach has only been used by Mäler (1989) in an analysis of acid rains in Europe. However, it seems possible and even fruitful to extend the method to more general problems.

MULTIPLE USE OF RESOURCES

Since we only study values revealed by markets, it is assumed that the utility function can be written

$$U = \psi (x, S)$$

and that ψ cannot be decomposed linearly in a function of only S and a function of x and S. However, it may still be possible to decompose ψ in other ways. Assume that it is possible to write $\psi (x, S)$ as $U = \psi z, q_1, ..., q_k)$, where

$$q_j = f_j(y^j, S) \quad \text{and} \quad x = z + \sum y^j.$$

We will interpret f_j as a household production function, using the vector y^j of marketed goods and services and S as inputs and producing q_j. Z is the consumption of marketed goods and services that is not used as inputs in conjunction with the environmental resources. One example is the use of trans-portation services in order to reach a certain recreation site. We will in the next section give many examples on different production functions and even extend the analysis and the formalism to dynamic situations.

There are several arguments in favor of using a household production function approach in this context. One is that one could regard the produc-tion function more stable than the utility function. Even if preferences for different goods change over time because of various exogenous factors, it is less likely that the production function will change, except for changes in the technical knowledge. Furthermore, one could reasonably assume that the pro-duction function is constant across different individuals, while preferences are not. Therefore, the problem of aggregation over individuals are much smaller within a household production function approach than if one would be working completely within a utility function framework. Next, information on the production function may be obtained through observing technologies which in general will be simpler than by observing preferences. Finally, all established approaches to value resources through observations on market be-havior—artificial markets, travel cost methods, hedonic methods, etc.—can be characterized through à priori assumptions on the production function.

Using this idea of household production function, it is now possible to ex-press the value revealed by markets of a change in the environmental resource in terms of production functions and the values of the outputs $q_1, ..., q_k$. This is done by recognizing that for each level of production of q_j the household faces costs for buying the market goods which are used as inputs. The total expenditures for obtaining a certain utility level for the household then con-sist of two parts; expenses for buying market goods which are used for con-sumption, z, and costs for producing the environmental output q_j.

When there is a change in the supply of environmental resources, a house-hold responds by adjusting the use of inputs for producing the household output q_j. This adjustment may also change the quantity of q_j which, in turn, affects the consumption of other marketed goods and services. If the change in environmental resources implies that the production of the output goes up, we would expect that the expenses for market goods used in consumption go down. One example is a creation of a recreation place at a lake which in-creases the use of transports to this site which probably implies that the de-mand for other marketed goods and services are decreased. This change in expenses on market goods used in consumption is here interpreted as the gross value of the change in the supply of the environmental resource. However, when there is a change in the supply of the environmental resource, not only the gross value is changed but also the costs for producing the envi-ronmental output. *The value of the change in the supply of the environmental resource is then the net effect or "profit" for the household, which is the change in gross value minus the change in costs for producing the output.* Thus, in order to estimate the value revealed by markets of environmental re-sources, we need to estimate the household profit function. That includes the estimation of two functions—the valuation function and the cost functions. In Appendix C the occurrence of non-separability in the production of the different environmental outputs is analyzed.

AN APPLICATION TO VALUATION OF THE MANGROVE SWAMP

Coasts in the tropics which receive suspended sediment (mud) from river mouths, and where water depths are shallow, typically have mangrove swamp forest, consisting of stilted trees. The mangrove prop-roots serve to trap sediment from ebb and flood tidal currents, so that the land is gradually extended seaward. Mangroves commonly consist of several shoreward belts of the red, the black, and the white mangrove (Strahler 1973).

The mangrove provides for a number of functions or ecological "services" to local human communities (Hamilton et al. 1989; Lal 1990; Ruitenbeck 1991). These include the protection of the shore from erosion from the waves, the supply of fuelwood, and the entrapment of nutrients in tidal pools which then can be used by fish for breeding. On the other hand, mangrove swamps can be transformed into intensive shrimp farms or be used for feeding the pulp industry with raw materials (Primavera 1993). In both these cases, the other services of the mangrove swamps will irreversibly be lost. The basic economic question is to make the appropriate trade-off between preservation or transformations of the mangroves. In order to make such a trade-off, the value of preservation must be estimated, and it seems that the use of the household production functions is an appropriate approach.

Let us illustrate the use of the household production function approach for a stylized mangrove system. What follows should only be seen as a very rough sketch of one way of modeling a complex ecological system and the services it is providing the local community. We will then identify three different services: provision of fuelwood, supply of fish from the sea and protection of the shore from erosion. These services are denoted S_2, S_3 and S_4 in that order. S_2 is interpreted as the total stock of the mangrove system, measured as wood biomass. S_3 is the fish population, and S_4 is the land area allocated to cultivation of the cereal. The only input besides the services from the natural capital is labor. The households then use these inputs for collecting fuelwood, harvesting fish, and cultivating cereals. These outputs are, in turn, used as inputs for producing the consumption good food. The following household production functions can then be identified:

$$q_1 = f_1(y_{11}, y_2, y_3, y_4) \qquad \text{(production of food)}$$

$$q_2 = f_2(y_{12}, S_2) \qquad \text{(collection of fuelwood)}$$

$$q_3 = f_3(y_{12}, S_3) \qquad \text{(harvesting of fish)}$$

$$q_4 = f_4(y_{14}, S_4) \qquad \text{(cultivation of cereal)}$$

where y_{1i} denotes the allocation of labor time to production of q_i and y_i is the input of q_i in the production of food. If the local community is completely closed, the total quantities of collected fuelwood and harvested fish would be used as inputs in producing food, i.e., $y_i=q_i$ for i=1,2. We would expect most of the cereal harvested to be used as inputs in food production but some could also be used as seeds. If the community can trade fuelwood and fish at fixed prices p_2 and p_3 expressed in units of food, we could export the difference between the production and the use of these inputs, $v_i=q_i-y_i$, which would give incomes from exports corresponding to $p_2v_2+p_3v_3$. These incomes can then be used to enhance the consumption of food. Let us assume that this is the case.

The welfare implication for the representative household is then a function of food and other market goods used for consumption, x. The utility function is then written as

$$u = u(x, q_1+p_2v_2+p_3v_3)$$

Assuming the above production functions and the assumption of efficiency, we can easily derive the cost functions

$$C_i(p_2, p_3, S_2, S_3, S_4, q_i) \text{ where } i = 1, 2, 3, 4$$

Since the services provided by the mangroves are, in this model, ultimately used for consumption of food, it is the cost for producing food, C_1, that is the interesting cost function. It gives the cost of producing the quantity q_1 of food per unit of time, given that the prices for fuelwood and fish are p_2 and p_3, that the ecological services provided by the mangroves are given by S_i at each point of time, and that the area for cereal production is S_4.

Remember from the foregoing section that the gross value of a resource can be expressed as the total expenditures on affected market goods used in consumption. Because of our assumption that all prices are expressed in unit of food, the valuation function is extremely simple:

$$v(q_1, q_2, q_3, q_4) = q_1,$$

and the total net value of the services from the mangroves is simply the value minus the costs which is written as

$$\pi = q_1 - C_1(p, S, q_1).$$

However, the interesting observation is now that the two different functions S_2, S_3, and the land area allocated to cereal production S_4 are not independent of each other. They are all tied together through the ecosystem, and it is necessary to have a good understanding of the functioning of the ecosystem in order to value the services. Let us therefore posit a very simple model of the mangrove swamp (we now need the area S_5 covered by the mangroves)

$$dS_2/dt = g_2(S_2, S_5) - q_2, \qquad \text{(growth of the mangroves)}$$

$dS_3/dt = g_3(S_2, S_3, S_5) - q_3,$ (growth of the fish population)

$dS_4/dt = g_4(dS_2/dt),$ (protection of the shoreline)

$dS_5/dt = -h$ (mangrove areas converted to shrimp farms)

The first equation states that the mangroves have a natural growth that is determined by the stock of mangroves and the land area they can expand on. The second equation states that the growth of the fish population is determined by the natural growth which, in turn, depends on the size of the population, the size of the mangroves and the size of the land area (including pools, etc.). The g_4-function in the third equation is such that a reduction in the size of the mangroves, i.e., in S_2 would imply a reduction in S_4 while a constant size of the mangroves would keep the land unchanged. Finally, the last equation represents the conversion of mangrove lands to shrimp farms.

We now have a set of household production functions, describing the use of the services provided by the mangroves and a model of how these services are interrelated to each other and to their uses. What is further needed before this machinery can be applied is some ideas on how the system has been and will be managed.

Usually such ecosystems as mangroves have been communally managed for very long periods. The conventional wisdom is that when there are no well-defined private property rights, the resource will be inefficiently managed. However, there is now ample evidence, both theoretical and empirical, that local common property resources can be efficiently managed (Berkes 1989; Dasgupta and Mäler 1991). It is therefore natural to assume that the system has and will be managed efficiently (unless there are very big changes in the system). Such efficient management can be represented as the solution to the intertemporal problem where the discounted sum of current and future streams of utility are maximized which, in our simple example, is written as

$$\text{Max} \int_0^\infty u(q_1 + p_2 v_2 + p_3 v_3) e^{-rt} dt$$

Here is r the social rate of discount and should be determined with respect to the growth rates of all natural capital assets utilized by the community as well as by the productivity of other kinds of capital.

The model is now completely specified, and we can work out the details of the optimizing problems (which obviously is necessary if we want to apply this machinery to the estimation of the welfare effects from an increase in the area allocated to shrimp farms (i.e., an increase in h). It should, however, be noted that even for such a simple problem as this, the optimization problem is quite complicated with four state variables. It is, therefore, natural to focus on

steady states which really are what we are looking for if we are looking sufficiently far into the future.

SUMMARY AND CONCLUSIONS

The main objective of this study has been to suggest a method to put monetary value on multifunctional ecosystems. For this purpose an alternative classification of values was given where a distinction was made between values revealed by markets and values not revealed by markets. The main advantage
with this classification as compared to the commonly used division between
use and non-use values is that the estimation of value is based on observed
behavior. The household production function was then used to show that the
value revealed by markets of a multifunctional ecosystem can be estimated if
we have a good knowledge of (1) the households' combination of different
market goods and the environmental resource in order to obtain certain ecological services, and (2) the ecosystem's production of various ecological
services used as inputs in the production of the household outputs.

Obviously, this approach to the estimation of the welfare effects from different schemes of managing resources is based on the possibility of getting
experts with different scientific background to work together to construct the
essential building stones.

- Ecologists are needed to delineate the resource system under study and to model
  it in a convenient way.

- Ecologists, economists and anthropologists are needed to identify and estimate
  the household production functions that have the services from the resource system as inputs.

- Economists are needed to identify the economic environment within which the
  valuation is going to be relevant.

It should be noted however that even if we are successful in all the above-
mentioned respects, we will not be able to put a monetary value on the entire
ecosystem in question for at least two reasons. First, only the production of
ecological services are included in the valuation and not the build up and
continuous functioning of the system required to produce the services.
Second, we do not currently know (scientifically) about all ecological services
that are actually produced and even less about how they are sustained over
time. Thus only part of the ecosystem is valued. But, as compared to the
commonly used valuation methods based on hypothetical markets, it is made
explicit which environmental services are subject to valuation.

APPENDIX A: VALUES REVEALED BY MARKETS AND VALUES NOT REVEALED BY MARKETS

Let us interpret S as a vector of services provided by an ecosystem which is
assumed to be exogenous given. The expenditure function is defined from

m = min px

s t $\psi(x, S) + \phi(S) = U$

which yields

m = $\xi(p, S, U-\phi(S))$.

Consider now a change in the supply of the resource from S' to S". The total value to the household from this change is defined as the corresponding change in expenditures which would keep the utility constant, i. e.,

$$TV(S' \rightarrow S") = \xi(p, S', u-\phi(S')) - \xi(p, S", u-\phi(S")) =$$

$$= \xi(p, S', u-\phi(S')) - \xi p, S', u-\phi(S")) + \quad (= NMRV)$$

$$+ \xi(p, S', u-\phi(S")) - \xi(p, S", u-\phi(S")) \quad (= MRV).$$

The terms in the first row define the values not revealed by markets (NMRV) while the terms on the second row define the values revealed by markets MRV). Note that the separation of terms above is not unique. Instead of adding and subtracting $\xi(p, S', u-\phi(S"))$, we could equally well have added and subtracted $\xi(p, S", u-\phi(S'))$, with the result that the value revealed by markets and the values not revealed by markets would have been different. However, this ambiguity should not surprise. It is equivalent to using the compensating variation or the equivalent variation and should not be of serious concern.

APPENDIX B: BID-FUNCTIONS

Assume the utility functions are identical among all the M households, but that their incomes differ. Let the utility function of a typical household be

$u(y_m, S)$

where y_m is consumption in general for household m and S is the resource. Suppose that it has been proposed to increase S, but that the increase will be determined by the size of voluntary contributions. Assume household i offers to pay R_i. The total contribution will then be $\Sigma_i R_i$. With that total contribution, the total resource will be $S = \mu(\Sigma_i R_i)$, where $\mu(.)$ is an increasing function. In this situation, the households would be free riding and the total bid is surely biased downward. Assume, however, that the resulting bids correspond to a Nash non-cooperative equilibrium. Then R_m would be determined from

max $U(y_m - R_m, \mu(\Sigma_i R_i))$.

Setting the partial derivative with respect to R_m equal to zero gives

$-U'_1(y_m - R_m, \mu(\Sigma_i R_i)) + U'_2(\mu(\Sigma_i R_i)\mu'(\Sigma_i R_i)) = 0.$

From this we see that R_m will be a function of the parameters of the utility function and the income y_m. Such a bid-function may be estimated econo-metrically, and from the parameters in the bid-function, the parameters of the utility function may be identified. For example, when the utility function is a Cobb-Douglas function, the bid function becomes

$$R_m = \frac{\langle 1-\alpha \rangle \mu' y_m - \alpha \sum i \neq_m Ri}{\langle 1-\alpha \rangle \mu' - \alpha}$$

If this function can be estimated, the parameter a of the utility function can be estimated and the true social value of the change in the resource can be calculated.

APPENDIX C: THE HOUSEHOLD PROFIT FUNCTION

First, corresponding to each production function f_j corresponds a cost function C_j, defined by

$$C_j(p, S, q_j) = \min p y_j$$

s. t. $f_j(y^j, S) = q_j$

This cost function gives the cost to the household to produce q_j units of output j when the household can buy the input commodities y^j at prices given by the price vector p.

Second, define the compensated demand functions $z(p, q_1, ..., q_k, u)$ from

$$\min pz$$

s.t. $u(z, q_1, ..., q_m) = u.$

Finally, let us introduce the notation

$$v(p, q) = -pz.$$

$v(p, q)$ will be interpreted as the valuation function of the outputs $q_1, ..., q_k$. In fact, pz tells us the necessary expenditures on consumption goods in order to sustain utility level u whenever the output of household production is the vector q. If one of the q goes up, we would expect pz to go down (the neces-sary consumption to achieve utility u will go down if the household has more of a household product). By introducing a minus sign in the definition above, we can be assured that an increase in a household output will correspond to an increase in the value of v. However, the minus sign makes the valuation function negative, but that is not a real problem. What will matter in the future are changes in the household production, and they will be valued correctly by the v function. Anyone who is worried about the sign of the valuation function may obviously add a constant (for example pz(p, 0, ... 0) on the right hand side of the definition above in order to obtain a positive valuation function.

Let us now carry out the following calculations, in which we start with the definition of the expenditure function and relate that function to the valuation and cost functions. We have

$$m = \xi(p, S, U) =$$

$$= \min_q \{pz(q) + \Sigma C_j(p, S, q_j)$$

and therefore

$$MRV = \xi(p, S', U) - \xi p, S'', U) =$$

$$\min_q \{pz(q) + \Sigma C_j(p, S', q_j) -$$

$$\min_q \{pz(q) + \Sigma C_j(p, S'' U) = \text{(let q' be the minimizing argument in the first brackets}$$
    and q'' the same in the second bracket) =

$$\{pz(q') + \Sigma C_j(p, S', q_j')\} -$$

$$\{pz(q'') + \Sigma C_j(p, S'', q_j'')\} =$$

$$\{v(p, q'') - \Sigma C_j(p, S'', q_j'')\} -$$

$$\{v(p, q') - \Sigma C_j(p, S'. q_j')\} =$$

$$\pi(p, S'', U) - \pi(p, S', U).$$

Thus the value revealed by markets of a change in S from S' to S'' is the resulting difference in the "household" profit $\pi(p, S, U)$. It can be shown that all known estimation methods of values revealed by markets that are based on market behavior are based on ad hoc assumptions on the household production functions (or which is the same thing the household cost functions C_j).

From a practical point of view it is interesting to know whether the valuation function is additive separable in the q's. Unfortunately, this is in general not the case. Remember that $v(p, q)$ was defined from

$$v = - pz(p, q)$$

where $z(p, q)$ was defined from

$$\min pz$$

$$\text{s.t. } u(z, q) = u$$

The first order conditions for this are
$$p_i = \mu \Sigma \, \partial u / \partial z_j$$

It now follows that $\partial v/\partial q_i = \mu \; \partial u/\partial q_i$ and therefore

$$\partial v^2/\partial q_i \partial q_j = \mu \; \partial u^2/\partial q_i \partial q_j + \partial \mu/\partial q_j.$$

The first term is zero only if the utility function is additive separable in q_i and q_j. μ can be interpreted as the marginal monetary value of utility (i.e., the inverse of the marginal utility of income. The second term is zero only if the marginal utility of income is constant and independent of a change in household output, something that does not seem likely. Thus, in general the v function will not be separable in the various household outputs. However, there is a natural case in which both these conditions will hold, namely when the outputs can be sold on a market at given prices. In that case, with only two outputs q_1 and q_2, $v(p, q) = p_1 q_1 + p_2 q_2$. In fact, we will later see that this is behind some of the more important techniques for estimating the value of a resource.

In spite of the rather negative result on the separability of v in general, on the margin, the valuation of the different outputs can be made independently of each other. Because of the envelope theorem, the two first terms cancel, and we have that the marginal change in the household profit function is equal to the marginal value minus the marginal cost of the resource.

Finally, we should note that there is the possibility that a change in S will not change the output vector q. Let us look at that possibility in more detail. In particular, let us consider only one use of the resource so that q is a scalar. It is easily seen that q is independent of S if and only if

$$\partial C/\partial q \partial S = 0$$

and it can be shown that this is the case if and only if there is a market good, say x_1, so that

$$q = f(x_1 + a(S), x_2, ..., x_n).$$

This is known as weak substitutability between x_1 and S (Mäler 1985). The idea is that the marginal rate of substitution of the resource for the market good is only a function of the resource availability.

REFERENCES

Arrow, K. J. 1970. Social Choice and Individual Values. Edinburgh: Oliver and Boyd.

Berkes, F., ed. 1989. Common Property Resources. London: Belhaven.

Dasgupta, P., and K-G. Mäler. 1991. The environment and emerging development issues. In Proceedings of the World Bank Annual Conference on Development Economics. Washington, DC: The World Bank.

Graaf, J. de V. 1957. Theoretical Welfare Economics. London: Cambridge Univ. Press.

Hamilton, L. S., J. Dixon, and G. O. Miller. 1989. Mangrove forests: an undervalued resource of the land and of the sea. In Ocean Yearbook 8, eds. E. M. Borgese, N. Ginsburg and J. R. Morgan. Chicago: The Univ. of Chicago Press.

Johansson, P-O. 1987. The Economic Theory and Measurement of Environmental Benefits. London: Cambridge Univ. Press.

Lal, P. N. 1990. Conservation or Conversion of Mangroves in Fiji. Occasional Papers of the East-West Environment and Policy Institute. Paper No. 11, U.S. East-West Center.

Mäler, K-G. 1978. Some Thoughts on the Distinction Between User and Non-user Values of an Environmental Resource. Stockholm: Stockholm School of Economics.

———. 1985. Welfare economics and the environment. In Handbook of Natural Resource and Energy Economics, eds. A. V. Kneese and J. L. Sweeney. Amsterdam: Elsevier.

———. 1989. The acid rain game. In Valuation Methods and Policy Making in Environmental Economics, eds. H. Folmer and E. van Ierland. Amsterdam: Elsevier.

Primavera, H. J. 1993. In press. A critical review of shrimp pond culture in the Philippines. *Review of Fisheries Science.*

Ruitenbeck, J. H. 1992. Mangrove Management: An Economic Analysis of Management Options with a Focus on Bintuni Bay, Irian Jaya. Environmental Management Development in Indonesia Project (EMDI), Jakarta, Indonesia.

Smith, V. K. 1987. Nonuse values in benefit cost analysis. *Southern Economic Journal* 54: 19–26.

Strahler, A. 1973. Introduction to Physical Geography. New York: Wiley.

15 STRATEGIES FOR ENVIRONMENTALLY SOUND ECONOMIC DEVELOPMENT[1]

Faye Duchin and Glenn-Marie Lange
Institute for Economic Analysis
New York University
269 Mercer Street, 2nd Floor
New York, NY 10003

ABSTRACT

This chapter presents a detailed evaluation of a well-known approach to sustainable development, one implicit in the Brundtland Report. A string of plausible assumptions about what can be done to promote sustainable development over the next several decades is transformed into a well-defined scenario about the possible actions that might be taken in all parts of the world economy. The scenario is optimistic in its assumptions about savings in energy and materials based on extensive recycling of metals and paper, fuel-efficient vehicles, and a variety of other measures.

The economic results of the scenario and the effects on energy use and related emissions of carbon dioxide and oxides of sulfur and nitrogen are calculated for sixteen geographic regions and reported for each of four groups of regions. The scenario is found to be feasible from an economic point of view, and is especially attractive for industrialized regions, but results in a significant expansion of pollution. In other words, we believe that the recommendations of the Brundtland Report cannot achieve the stated economic and environmental objectives simultaneously.

The conceptual framework developed for this analysis consists of a World Model and a World Database. The model describes each economy in terms of physical stocks and flows as well as related costs and prices. The database has been extended for this study using the information collected in ten case studies that cover the range of energy-intensive and material-intensive activities. This modeling framework can now be used to analyze alternative

---

1. This chapter is based on a forthcoming book, *Ecological Economics, Technological Change, and the Future of the Environment* (Oxford University Press, 1994). Financial support was provided by the Norwegian government and the United Nations. We wish to acknowledge the major contributions of Knut Thonstad of the Norwegian Central Bureau of Statistics and Annemarth Idenburg of the University of Twente (the Netherlands). The computer work was carried out with the support of the U.S. National Science Foundation at the Cornell Center for Supercomputing.

scenarios that are far bolder in their potential to reduce emissions than the one that has been investigated here.

INTRODUCTION

This chapter attempts to tell several different stories. In many ways the most important consists of our practical conclusions about what needs to be done to forestall increasingly serious environmental problems. This story leads us to examine technological considerations governing how a wide range of economic activities are carried out in different parts of the world now and how they might be done differently in the future. We then go on to assess the economic and environmental consequences of following a particular path over the next several decades. To do this we utilize a framework that makes it possible to "weigh ends and means together in order to set [objectives] that are reasonable in relation to the efforts necessary to achieve them" (Sagoff 1988). We believe that too much attention is generally given to targets and too little to the possible means of achieving them. Here the emphasis is on the means and their relation to the ends.

The possible significance of this work, however, lies not only in its specific conclusions but also in their relationship to the other story lines. The conclusions we reach are based on asking a new set of questions, developing the concepts, and building the analytic framework needed to address them. We believe that the new field of Ecological Economics, with its focus on issue-oriented rather than single discipline-based research, can provide fertile soil for this kind of analysis. We try to make our case convincing enough that more analysts, as well as informed citizens, will want to join us in asking these kinds of questions. We hope that the conclusions of our practical story will reinforce our more abstract arguments about methodology.

The large-scale use of the technologies developed since the beginning of the Industrial Revolution, and especially since World War II, has placed considerable stress upon the environment mainly in the industrialized economies. This is due in part to steeply increasing rates of extraction and processing of natural materials, accompanied by the generation of a wide range of waste products. In addition, the new technologies have involved the dissipation of naturally occurring materials and the synthesis of materials which are not found in nature. Widely dispersed materials, whether natural or synthetic, are not easy to collect. Even if collected, many of the new materials cannot be reused or recycled at reasonable cost after their initial useful life is over.

Poverty, especially coupled with the rates of population growth typical of many developing countries, also generates characteristic patterns of pollution. Mainly associated with improperly treated human sewage and the degradation of land, these problems tend to be localized although they affect large numbers of people. Today's concerns about the global environment are a response to the massive scale on which modern technologies are now utilized and the certainty that they will spread to the developing countries in the decades ahead.

At the same time, modern technology is responsible for the elevated material standard of living that has been achieved by a significant fraction of the earth's population. As we approach the 21st century, most parts of the developing world will be attempting to improve their material standards of living both in absolute terms and relative to the industrialized countries. This effort will involve the development of new technologies, particularly those which have a biological basis, and the widespread diffusion of both old and new technologies.

People in rich and poor countries alike value increased levels of comfort which tend to be associated with increased material throughput. Two good examples are the personal automobile and air-conditioning. There are significant differences in their use in different countries at the present time. In the United States both are used even more intensively than in other affluent societies. The differences are attributable to cultural preferences and historical circumstances as well as population density and climate. The automobile is already very popular in developing countries, and air-conditioning will be increasingly sought after, especially in cities with a tropical climate.

While increased affluence based on the further spread of modern technologies can be expected to place even greater pressures on the global environment, it is also true that new technologies can be vastly more efficient and cleaner in the use of energy and materials and the generation of waste, than the ones they displace. The potential for these benefits may still be largely untapped because until now, the costs of raw material inputs and the penalties for generating pollution and degrading the environment have been low relative to likely future costs and penalties.

Another factor that could reduce future pressures on the environment is the prospect for selectively curtailing some types of activities and expanding others that are more environmentally benign. For example, industrialized countries appear to have reached a number of plateaus—population growth and the use of materials tend to level off as affluence increases. Many activities run into limits, as in the case of traffic congestion or the exhaustion of landfills, because they elicit social responses before irreversible physical barriers are breached. Over the next several decades, reliance on benign technologies and prudent practices is likely to be increasingly prevalent.

This chapter explores some important aspects of the relationships among increasing affluence, pollution, and technological choices. The analysis considers the situation in the world economy over the next several decades.

THE SIGNIFICANCE OF THE BRUNDTLAND REPORT

The Brundtland Report, *Our Common Future* (World Conference on Environment and Development 1987), popularized the definition of "sustainable development" as humanity's ability "to ensure that it meets the needs of the present without compromising the ability of future generations to meet their own needs" (p. 8). To achieve this outcome, the Report continues, "the international economy must speed up world growth while respecting the environmental constraints" (p. 89) by the appropriate management of technology and social organization. The Brundtland Report is comprehensive in identifying the areas of economic and environmental problems, and it describes different types of technological and organizational measures that might be taken in each individual area to contribute toward sustainable development.

The Brundtland Report is an important point of departure for the study described in this chapter because it reflects a progressive and moderate position about how both economic prosperity and preservation of the environment could be achieved. It implicitly argues that they are mutually supportive in that we cannot have one without the other. According to this view, people will not be willing or able to afford the preservation of the environment unless they have a high material standard of living, nor will they be able to maintain that standard of living (let alone continually increase it) unless they assure the continued provision of environmental services. Of course, it is also true that the two objectives can compete with each other, because meeting environmental goals generally requires resources that could otherwise be allocated for growth (which could be achieved at least over the short-term), and vice versa. The Brundtland Report takes the position that both economic and environmental objectives can be achieved if reasonable choices are made regarding technology and social organization. This chapter, and the book on which it is based, undertake to evaluate this proposition.

The Brundtland Report has had to reconcile a relatively wide range of views—it sometimes sounds like and is the work of a committee, but this feature is also its great strength. The members, many of whom are influential individuals in different parts of the world, were willing and able to reach a consensus. Thus the conclusion that we will present—that the position taken in the Brundtland Report is not realistic—underscores the need for significant rethinking about how to achieve environmentally sound economic development in both rich and poor countries.

To make the challenges of sustainable development concrete, we focused in this prototype study on global environmental problems, in particular on emissions of carbon dioxide. These are directly related to the use of energy which plays a crucial role in economic development. The emphasis on fuel use made it easy to examine the emissions of oxides of sulfur and nitrogen as well.

FRAMEWORK FOR THIS STUDY

In order to analyze the position of the Brundtland Report, the numerous recommendations directly or indirectly related to energy and air pollution that are sprinkled throughout it needed to be assembled, and further developed with more detail and concreteness, into an integrated scenario that is free of at least surface contradictions: we have called this the "Our Common Future" (OCF) Scenario. For comparison, a Reference Scenario was constructed in which no technological change takes place after 1990. (Subsequently additional scenarios, comprised of alternative technological and organizational assumptions, were also developed.) An analytic framework was needed to evaluate the economic and environmental implications and the feasibility of the scenario. The model and database we have used—called the World Model and the World Database—are well-suited to this task because, in the assumptions that are inevitably built into them, they represent a moderate view of how an economy does and can work. Technological choices for each sector of the economy are described in the database in physical terms such as an engineer might use. This characteristic makes it natural to relate them directly to material inputs and to the generation of pollutants and other wastes. They are also represented by the model in value terms, allowing for a direct relationship with changes in income and wealth. The first version of this modeling system was constructed by Wassily Leontief and his colleagues (Leontief, Carter, and Petri 1977).

THE ECONOMIC MODEL

Building the scenario required a systematic, quantitative description of the technologies currently in use in the different parts of the world economy and identifying the kind of options intended by the Brundtland Report but not generally explicitly identified. This information was then incorporated into the World Database, which is a repository of information about technological options, a generalized input/output database that can now be used for various purposes. The bulk of the World Database has been built over a number of years; while it is crude in many ways, its prior existence made this study possible.

To analyze the scenario we made use of an input/output model of the world economy, the World Model. The basic mathematical structure of the current version of the model is given in Figure 15.1. The mathematical equations for the price and income models were formulated in the course of this project.[2] It divides the world into sixteen geographic regions, each described in terms of about fifty interacting sectors. Regions are linked within each time period by the trade of commodities and flows of capital and economic aid; they are linked over the period from 1980 to 2020 by the accumulation of capital and of international debt or credit. Use of energy and materials are directly represented, and flows of pollutants have been incorporated. Public and private consumption and sector-level investment are also represented, both in terms of detailed goods and services and in the aggregate. The output of agri-

2. Duchin, F., G. Lange, and T. Johnson. 1990. *Strategies for Environmentally Sound Development: An Input-Output Analysis*, Third Progress Report to the United Nations, Contract #PRS/CON/66/89, March.

cultural products and of minerals and emissions of pollutants are measured in physical units; most other quantities are measured in constant U.S. prices. The database describes the inputs and outputs associated with alternative technologies specified by the OCF Scenario.

A standard economics textbook might describe the objective of such an exercise as choosing those alternatives, and in the process, determining the corresponding "right" prices, that assure optimal social welfare for a global economy in equilibrium. This theoretical position, and the actual models that are based on it (general equilibrium models), are appealing for their comprehensiveness and simplicity of objective. However, any theory or model of an economy will have its shortcomings, and some of the major ones of this perspective are that a real economy is never in equilibrium, that social welfare is not the same as economic welfare, that even economic welfare is not sufficiently well-defined to lend itself to a single measure, and that most environmental problems cannot reasonably be associated with a money price that would make it possible to simply add them in to this single measure, because any such price would be arbitrary. The challenge we have tried to face is to go beyond criticism of mainstream economic models and beyond qualitative description alone, to build a formal framework that is capable of representing the activities of a real economy and arrive at quantitative results while avoiding these shortcomings.

The World Model and Database provide a starting point for this undertaking. While the theory and the model are open to criticism, mainly for all that is still left out, we believe they are well-suited to this kind of investigation for a number of important reasons.

The World Model does not attempt to determine a unique and optimal path to sustainable development but simply to evaluate the implications of a set of technical and organizational choices that are made outside the model (by the Brundtland Report, in this case). Economic well-being is gauged by a number of variables, not by GDP alone, and many environmental variables are measured in physical units (e.g., tons of carbon emissions) with no attempt or need to put a price on them. There is not a single bottom-line; weighing the relative importance of the economic and environmental outcomes is considered to be a social and political responsibility. We believe this is a realistic position about what an economist can hope to contribute to these questions.

The theoretical framework of the World Model and Database is well-suited to this type of inquiry. The empirical content of the database—that is, the description of the technological choices—takes on an importance (in terms of the allocation of research effort) that it is not accorded in frameworks in which all the major relations are represented by mathematical equations. In addition, the openness to fundamental multidisciplinary collaboration is greater because it is recognized as a major avenue for the further development of the theory.

In a conventional economics textbook, the prominent policy questions address only monetary and fiscal options—the most familiar example being the design of a carbon tax, in fact an "optimal" carbon tax. These kinds of instruments are clearly important and necessary. But the design of a carbon tax and the assessment of its effects are most fruitful *following* a different type of analysis such as that described below. The first stage of analysis needs to focus on the implications of the technological options for achieving specific targets. If the objectives still appear desirable, the economic incentives needed to make the shift voluntary can be explored in later stages.

CASE STUDIES AND DATABASE

The content of the *Our Common Future* (OCF) Scenario was developed in the course of carrying out ten case studies that provide the bulk of the data for this project. In order to examine likely future changes in emissions of carbon dioxide and oxides of sulfur and nitrogen, we focused our attention on the following sectors, most of which were selected because they are energy-intensive.

- Household Use of Energy
- Transportation
- Electric Power Generation
- Industrial Energy Conservation
- Construction
- Metal Fabrication
- Paper
- Chemicals
- Electronics

The technologies of greatest interest to us would promote development by their efficient use of energy and materials, their potential for low pollution, and the likelihood that they could be commercially implemented on a significant scale over the next three decades.

Data collection for each case study covered one or several related sectors (e.g., iron and steel) and required identifying the technological choices faced both in producing the output (in the iron and steel sector itself) and in using it (in the metal-fabricating sectors). The description of each technological alternative, (e.g., electric arc technology), involves quantifying the major inputs per unit of output. This technology is of particular interest because it makes extensive use of scrap iron and steel, significantly reducing energy requirements per ton of finished product. A much larger share of the energy for this process, however, is provided by electricity rather than by direct use of fossil fuels.

The case studies cover likely future changes in the use of energy by households and for transportation, in electricity generation, and for industrial production; they also examine pollution control options. We have made assumptions about the changing use of materials in the processing and fabricating industries, as well as for construction. The changes in current-account use of energy, materials, and other inputs are achieved either costlessly, through current-account substitutions of some inputs for others, through investment in new types of capital goods, or through some combination of these. In the case of investment, we assume that the new capital goods are phased in through the expansion of production that relies on new technology or through the normally planned replacement of capacity in place. In general, we have not assumed accelerated replacement of operational capital, although the consequences of this avenue for modernization can be easily explored in other scenarios. We have also made assumptions about population growth and urbanization, trade shares in world markets, and the rate of introduction of electronic equipment with attendant improvements in overall productivity.

In the large, material-intensive construction industry, the share of maintenance and repair construction is projected to grow by about 50% in all regions over the next several decades with compensating reductions in the share of new construction. By 2020, the share of maintenance in total construction will reach 50% in high income Europe and North America, for example, while in the developing countries of Africa, at the other extreme, it attains 15%. This assumption is significant because replacement and maintenance construction activities are considerably less material-intensive than new construction, although they require increased inputs of chemical products, mainly paints.

We expect an increased level of fabrication for inputs to construction, especially in the developing countries. Examples of this are the substitution of ready-mixed or pre-fabricated concrete for concrete mixed on site, an increase in the degree of fabrication of metal inputs, and a substitution of processed wood for wood in the rough state.

With income growth and urbanization in developing countries, there will be a significant expansion in the construction of buildings, which use more cement, concrete, and wood, relative to infrastructure and engineering construction, which are metal-intensive. The anticipated increase in fossil fuel prices will reduce the competitiveness of steel for bearing structures relative to cement and wood. Copper is expected to face increased competition from fiber optics, plastics, and aluminum. Aluminum will increasingly be produced in developing regions with cheap hydroelectric power and will substitute for many other materials.

The combined effect of these assumptions about construction is a reduction in the physical quantity of raw materials required per average structure. In the construction industry as a whole, we anticipate that inputs of steel per unit of output will decline by 25% in most developed regions and by about 50% in most developing regions by 2020. Copper inputs should decline by about 1/3 in most regions. Aluminum inputs are expected to increase by be-

tween 15% and 33%, depending on the region. For most regions cement inputs per unit constructed are assumed to decline by 15%–25%, while inputs of
wood (in terms of raw wood content) are assumed to remain unchanged. All
these changes contribute toward a substantial indirect savings of energy per
unit of output of the construction sector.

For the details about the construction case study, and for all other case
studies, the reader is referred to chapters 6–13 of Duchin and Lange (1994).
In addition to the technological assumptions, we have assumed that levels of
economic activity (as measured by GDP) will increase by about 2.8% a year
for the entire world (see Table 15.1); this is consistent with the view underlying the Brundtland Report. Future population is based on the medium projections of the United Nations. Additional assumptions were made about future
energy prices, about each region's shares of world exports for all goods and
services, and about the quantity of foreign aid.

Table 15.1. Projected Annual Rates of GDP Growth for World Model Regions 1990 to
2020

	1990 to 2000	2000 to 2010	2010 to 2020
High Income North America	2.4%	2.2%	1.9%
Newly Industrializing Latin America	4.3	4.1	3.9
Low Income Latin America	3.7	3.8	4.0
High Income Western Europe	2.2	2.1	1.9
Medium Income Western Europe	3.3	2.9	2.2
Eastern Europe	2.4	2.5	2.9
Soviet Union	2.1	2.2	2.2
Centrally Planned Asia	5.0	4.6	4.2
Japan	3.5	3.1	2.5
Newly Industrializing Asia	6.2	4.5	3.5
Low Income Asia	4.9	4.9	5.0
Major Oil Producers	3.3	3.3	3.4
North Africa and Other Middle East	3.2	3.2	3.2
Sub-Saharan Africa	3.1	3.4	3.9
Southern Africa	1.2	1.4	2.0
Oceania	2.2	2.2	2.3
World	2.9	2.8	2.7

Source: Prepared for this study by UN/DIESA and Central Bureau of Statistics of
Norway.

EVALUATION OF THE *OUR COMMON FUTURE* SCENARIO

Under the assumptions of the *Our Common Future* Scenario, clean and efficient modern technologies associated with the use of energy and materials are
adopted in all parts of the world economy over the next several decades. The
timing and the specific technologies are different for the different regions, but

overall the assumptions appear rather optimistic. The results are analyzed first from an economic, and then from an environmental point of view.

From an economic point of view the scenario is attractive in that it is cost-saving: global consumption (at the overall rates of economic growth that are approximately those of the Brundtland Report) is higher than under the Reference Scenario (see Table 15.2). The advantages are very unevenly distributed, however, and it is a matter of judgment whether or not most of the developing countries are better off. For these countries the new technologies represent a capital-intensive path; the capital is paid for by holding down consumption and by increased debt and aid relative to the Reference Scenario. This outcome is feasible only if these loans and aid would actually be forthcoming.

In North America, Japan, Eastern Europe, and medium income Western Europe, the OCF Scenario makes it possible to maintain or increase consumption because investment requirements are reduced. These reductions result from significant savings in energy and materials and associated capital requirements. These regions also increase their volumes of exports relative to imports (i.e., the balance of trade is higher) and improve their international credit (or reduce their debt) under the OCF Scenario.

A second set of regions is also able to increase consumption relative to the Reference Scenario, but these regions cannot finance this solely through savings in investment. Southern Africa, Oceania, and low income Asia increase their imports and their debt.

The OCF Scenario assumes modern, generally more capital-intensive technology, so it is not surprising that in some regions investment is higher than under the Reference Scenario. (Consumption still grows significantly over the time horizon of the scenarios under the OCF Scenario). This is true for the former Soviet Union, which experiences virtually no change in consumption relative to the Reference Scenario, and in centrally planned Asia, low income Latin America, and northern Africa, where consumption is lower than under the Reference Scenario. In most of these regions, more investment requires increased net imports and increased debt. In Eastern Europe, however, it is possible to increase net exports and reduce the debt.

The OCF Scenario makes it possible to increase consumption while also increasing investment relative to the Reference Scenario in several resource-rich regions—newly industrializing Latin America, the oil-rich Middle East, low-income Asia, and sub-Saharan Africa. But these regions have to increase net imports, reducing their credit positions.

From an environmental point of view, the *Our Common Future* Scenario reduces emissions of the three pollutants that are tracked significantly below what they would be under the Reference Scenario. But the problem is that these emissions still increase substantially between 1990 and 2020; in particular,

Table 15.2.   Consumption, Investment, and International Exchange under the OCF
Scenario Relative to the Reference Scenario for World Model Regions in
2020

	Consumption (constant prices)	Investment (constant prices)	Balance of Trade (constant prices)	International Credit (current prices)
High Income North America	+	-	+	+
Newly Industrializing Latin America	+	+	-	-
Low Income Latin America	-	+	-	-
High Income Western Europe	-	-	+	+
Medium Income Western Europe	=	-	+	+
Eastern Europe	+	-	+	+
Former Soviet Union	=	+	-	-
Centrally Planned Asia	-	+	+	+
Japan	+	-	+	+
Newly Industrializing Asia	=	-	+	+
Low Income Asia	+	+	-	-
Major Oil Producers	+	+	-	-
North Africa and Other Middle East	-	+	=	-
Sub-Saharan Africa	+	+	-	-
Southern Africa	+	=	-	-
Oceania	+	-	-	-
World	+	-	=	+

Notes: 1. Consumption, investment, and the balance of trade are measured in constant 1970
US prices. International credit is measured in current US relative prices. Improved
credit includes the case of reduced debt.
2. The entry + means that the value is higher for the OCF Scenario than for the
Reference Scenario; – indicates a lower value; = indicates no change. If the last two
columns have different signs, the terms of trade changed for that region relative to
other regions. Note that GDP is the same under both scenarios.

Source: Institute for Economic Analysis.

carbon emissions nearly double (see Table 15.3). In addition, the locus of
pollution shifts decisively from the rich countries, where historically most of it
has originated, to the developing countries where most of the world's people
live and where most future increase in population will take place. Material use
and emissions are still much lower than in the rich countries on a per capita
basis. However, increases are likely in the future since people in developing
regions can be expected to aspire to the material standards of those in the
developed regions. The feedback on the economy of potential environmental
damage due to these elevated levels of emissions has not been taken into
account in this study in part because the relationships are not yet well
understood.

Table 15.3.   Emissions of Carbon Dioxide, Sulfur Oxides, and Nitrogen Oxides under the *Our Common Future* Scenario in 1980 through 2020

Carbon Dioxide (10^6 metric tons of carbon)					
Levels of Emissions	1980	1990	2000	2010	2020
Rich, Developed Economies	2,592	2,790	2,877	2,859	2,799
Newly Industrialized Economies	249	339	573	738	997
Other Developing Economies	678	1,180	1,906	2,605	3,515
Eastern Europe & Former USSR	1,211	1,323	1,435	1,548	1,733
World Total	4,730	5,632	6,791	7,740	9,044
Regional Distribution					
Rich, Developed Economies	0.55	0.50	0.42	0.37	0.31
Newly Industrialized Economies	0.05	0.06	0.08	0.10	0.11
Other Developing Economies	0.14	0.21	0.28	0.34	0.39
Eastern Europe & Former USSR	0.26	0.23	0.21	0.20	0.19
World Total	1.00	1.00	1.00	1.00	1.00
Per Capita Emissions (kg/person)					
Rich, Developed Economies	3,067	3,070	2,978	2,671	3,073
Newly Industrialized Economies	492	552	795	899	1,091
Other Developing Economies	250	352	463	532	624
Eastern Europe and Former USSR	3,210	3,269	3,351	3,433	3,681
World Total	1,066	1,067	1,089	1,078	1,122
Sulfur Oxides (10^6 metric tons of SO_2 equivalent)					
Levels of Emissions	1980	1990	2000	2010	2020
Rich, Developed Economies	51.3	45.4	42.0	37.6	34.6
Newly Industrialized Economies	5.4	8.0	13.1	15.4	19.9
Other Developing Economies	18.9	28.8	40.7	49.8	59.2
Eastern Europe & Former USSR	49.5	45.0	41.3	37.4	33.6
World Total	125.0	127.2	137.1	140.2	147.3
Regional Distribution					
Rich, Developed Economies	0.41	0.36	0.31	0.27	0.24
Newly Industrialized Economies	0.04	0.06	0.10	0.11	0.14
Other Developing Economies	0.15	0.23	0.30	0.36	0.40
Eastern Europe & Former USSR	0.40	0.35	0.30	0.27	0.23
World Total	1.00	1.00	1.00	1.00	1.00
Per Capita Emissions (kg/person)					
Rich, Developed Economies	61	50	43	37	33
Newly Industrialized Economies	11	13	18	19	22
Other Developing Economies	7	9	10	10	11
Eastern Europe & Former USSR	131	111	96	83	71
World Total	28	24	22	20	18

Table 15.3. Continued

Nitrogen Oxides (10^6 metric tons of NO_2 equivalent)					
Levels of Emissions	1980	1990	2000	2010	2020
Rich, Developed Economies	35.2	34.8	35.4	34.4	31.8
Newly Industrialized Economies	4.9	6.3	10.8	14.5	20.2
Other Developing Economies	9.6	16.5	25.6	35.1	48.9
Eastern Europe & Former USSR	19.3	20.6	22.3	24.2	26.5
World Total	69.0	78.2	94.1	108.2	127.4
Regional Distribution					
Rich, Developed Economies	0.51	0.45	0.38	0.32	0.25
Newly Industrialized Economies	0.07	0.08	0.12	0.13	0.16
Other Developing Economies	0.14	0.21	0.27	0.32	0.38
Eastern Europe & Former USSR	0.28	0.26	0.24	0.22	0.21
World Total	1.00	1.00	1.00	1.00	1.00
Per Capita Emissions (kg/person)					
Rich, Developed Economies	42	38	37	34	30
Newly Industrialized Economies	10	10	15	18	22
Other Developing Economies	4	5	6	7	9
Eastern Europe & Former USSR	51	51	52	54	56
World Total	16	15	15	15	16

Source: Institute for Economic Analysis

Certain conclusions about long-term, global strategies and policy implications follow directly from these results. It appears that the economic and environmental objectives of the Brundtland Report cannot simultaneously be achieved. To the extent that our results about the physical reality are convincing, no amount of social organization and political will can make the *Our Common Future* Scenario work. This leaves two paths for action, and we feel that it is important to pursue both of them.

TWO DIRECTIONS FOR THE FUTURE

A scenario about much bolder technological and social changes might bring the basic objective of the Brundtland Report—sustainable development—into reach. Going well beyond the aim of more fuel-efficient vehicles, it may be necessary to replace a large portion of private automobiles in developed countries by practical, convenient systems of public transport. New designs for communities would reduce the need for motorized transport. Developing countries could adopt similar approaches instead of pursuing the route of expansive highway networks, suburbanization, and unrestrained use of individual cars.

Another opportunity for technological innovation is the large-scale, region-specific reliance on renewable sources for energy and materials. This is a far-reaching proposition because it would involve not only the phasing in of materials with significantly different properties than those they would displace, but also significant changes in the use of land and potential competition with agriculture. Of course, developing countries already make significant use of renewable materials (biomass) but generally in relatively unprocessed forms. Research and development in this area is likely to be promoted by the emerging field of engineering, Industrial Ecology.

Industrial Ecology is a set of ideas that has taken root in the engineering community in the United States. It involves life-cycle planning of products for durability, reuse, and recyclability. At the corporate level, major objectives are better customer acceptance and greater efficiency in satisfying present and potential future environmental regulation. These concepts need to be generalized by other institutions, with a global mission on a longer-term perspective, to identify and promote promising technological options that might not otherwise be explored.

The areas that have just been described are also in the spirit of the Brundtland Report; we call them bold because they differ in two major ways. First, they require a far greater break with present practices which means that they would be harder to achieve both socially and politically, in part because particular interests would be threatened at least in the short-term. Second, they require breakthrough achievements—not so much scientific ones, like realizing superconductivity or fusion at room temperature—but achievements in technical design, like innovative transportation plans, which are expensive to develop and for which there is not evident institutional and financial support. It is not clear who will take the initiative for a second Brundtland Report along these lines, but we hope that the results of our analysis can help by establishing the need for such an undertaking.

Modernization and economic growth lead to the transformation of traditional social relations. This process is rarely smooth and is implicated at least indirectly in the rise of phenomena like religious fundamentalism and urban misery. These issues have not been addressed in the present chapter; hopefully, they will be situated as part of the challenge of sustainable development as the U.N. system gears up for the Social Development Summit that will be held in 1995.

APPENDIX A: WORLD MODEL

(1) $\quad [(I - A + \delta - T \, (\hat{B} \, (\hat{G} + \hat{R}))] \, x = c + go + e$

(2) $\quad GDP = p' \, (c + go + e + T \, [\hat{B} \, (\hat{G} + \hat{R})] \, x - \delta x$

(3) $\quad c = c\alpha + POP \cdot \beta$

(4) $\quad go = \gamma \cdot GDP \cdot h$

(5) $\quad e = \hat{S} \, w$

(6) $f = F \tilde{x}$

(7) $BOP = \bar{p}' (e - x)$

(8) $SAV = GDP - p' [(T (\hat{B} (\hat{G} + \hat{R}) + \delta c) x - e_c]$

Definition of variables and parameters:

A	n x n matrix of current-account inputs per unit of output
δ	n x n diagonal matrix of import coefficients[3]
T	n x n matrix of the structure of gross investment per dollar of investment
$\hat{B}$	n x n diagonal matrix of capital requirements per unit of output
$\hat{G}$	n x n diagonal matrix of anticipated sectoral growth rates
$\hat{R}$	n x n diagonal matrix of rates of replacement of capital
x	n x 1 vector of output
c	n x 1 vector of consumption
go	n x 1 vector of government spending
e	n x 1 vector of exports
GDP	scalar
p	n x 1 vector of constant U.S. prices
POP	scalar population (exogenous)
α, β	n x 1 vectors governing consumption by item as a function of total consumption and population, respectively
γ, θ	scalars (percent)
h	n x 1 vector, composition of government spending
$\hat{S}$	n x n diagonal matrix of export shares
w	n x 1 vector, size of world trade pools
f	(n+m) x 1 vector of emissions
F	(n+m) x n matrix of emission coefficients
BOP	scalar, balance of payments
$\bar{p}$	n x 1 vector of current U.S. prices
SAV	scalar, volume of savings

---

3. All imports have been treated in the World Model as competitive imports.

$\tilde{x}$ x extended by additional entries to represent use of fuel by households and government

δ_c, e_c components of δ and of e corresponding to flows of capital and aid

Figure 15.1.   World Model Representation of One Region for One Time Period.

REFERENCES

Duchin, F., and G. Lange. 1994. In press. Ecological Economics, Technological Change, and the Future of the Environment. Oxford, U.K.; New York: Oxford Univ. Press.

Leontief, W., A. Carter, and P. Petri. 1977. Future of the World Economy. Oxford, U.K.; New York: Oxford Univ. Press.

Sagoff, M. 1988. The Economy of the Earth. Cambridge, U.K.; New York: Cambridge Univ. Press.

World Commission of Environment and Development. 1987. Our Common Future (published as the Brundtland Report). Oxford; New York: Oxford Univ. Press.

16 SEA-LEVEL RISE & COASTAL WETLANDS IN THE U.K.: MITIGATION STRATEGIES FOR SUSTAINABLE MANAGEMENT

R. K. Turner,[1] P. Doktor[2] and N. Adger[1]

1. *Centre for Social and Economic Research on the Global Environment*
   University of East Anglia, Norwich (UEA)
   University College, London (UCL)
2. *Environmental Appraisal Group*
   School of Environmental Sciences
   University of East Anglia, Norwich

ABSTRACT

This chapter surveys a range of sustainability perspectives and applies them in the context of potential global warming, its consequences and policy responses in coastal zones. A strong sustainability (SS) approach (underpinned by the precautionary principle) is contrasted with a weak sustainability (WS) approach (buttressed by the cost-benefit principle). It is possible to distinguish these two approaches and their preferred policy response options at both the macro and micro-policy level. At the latter level, WS requires only a "business as usual" sea and coastal defense strategy until the level of uncertainty is reduced. SS requires that a full range of sea-level ruse mitigation options (from retreat to full protection) be appraised and that the most cost-effective option is identified.

A case study analysis, based on the East Anglian Coastal Zone (England), is outlined and the cost-benefit results are reported. The results indicate that although not all the assets at risk in the zone have been valued in economic terms, a precautionary-protection plus manage retreat-strategy is the most cost-effective option.

INTRODUCTION

Speculations on Weak and Strong Sustainability

According to one recent critique of "sustainability," it remains unclear what concept of "sustainable development" (SD) can be both morally acceptable and operationally meaningful (Beckerman 1992). The most accepted definition of SD is that offered by the Brundtland Commission. SD is a multifaceted concept that has generated an extensive, on-going, cross-disciplinary debate. A number of salient elements of the SD debate can be identified, in order to ascertain the range of contrasting views that currently exist between economists, scientists, and philosophers.

Intragenerational and Intergenerational Equity

SD is future-oriented in that it seeks to ensure that the welfare of future generations at least equals that of current generations. It is therefore, in economic terms, a matter of intergenerational equity, and not just efficiency. The distribution of rights and assets across generations determines whether the efficient allocation of resources sustains welfare across human generations (Howarth and Norgaard 1992). The ethical argument is that future generations have the right to expect an inheritance sufficient for generating a level of welfare no less than that enjoyed by the current generation (i.e., some sort of intergenerational social contract).

SD also has a poverty focus, which in one sense, is an extension of the intergenerational concern. Daly and Cobb (1989) have argued that families endure over intergenerational time. To the extent that any given individuals are concerned about the welfare of their descendants, they should also be concerned about the welfare of the present generation from whom the descendants will inherit. Accordingly, a concern for future generations should reinforce and not weaken the concern for current fairness. Ethical consistency demands (despite the trade-offs involved) that if future generations are to be left the means to secure equal or rising per capita welfare, the means to maintain and improve the well-being of today's poor must also be provided. Collective rather than individual action is required in order to effect these socially desirable intra- and intergenerational transfers.

The degree of concern for the welfare of future generations that is ethically required of the current generation is another controversial matter. Six positions seem possible: (1) moral obligations to the future exist, but the welfare of the future is less important than present welfare, (2) moral obligations to the future exist and the future's welfare is almost as important as present welfare, (3) a discounting procedure is acceptable only after the imposition of pre-emptive constraints on some forms of economic development, (4) obligations to the future exist, and the future is assigned more weight than the present, (5) rights and interests of future people are exactly the same as contemporary people, and (6) there is no obligation all on the present to care about the future.

Specification of the Sustainability Inheritance Asset Portfolio

Economists have contributed to the SD debate the idea that the depletion of environmental resources (source and sink resources) in pursuit of economic growth is akin to living off capital rather than income (Victor 1991). SD is then defined as the maximum development that can be achieved without running down the capital assets of the nation, which are its resource base. The base includes man-made capital, natural capital, human capital, and moral (ethical) and cultural capital. Victor identifies four schools of thought on the "environment as capital" issue: (1) the mainstream neoclassical school, (2) the London school (after Pearce, Barbier, Markandya, and Turner), (3) the post-Keynsian school, and (4) the thermodynamic school (after Boulding, Georgescu-Roegen, Daly, Perrings, and Common). This spectrum of views ranges from a position of *very weak sustainability* through *very strong sustainability* (Klassen and Opschoor 1990). In practice these various sustainability paradigms are less clearly defined and overlapping.[1]

Very Weak Sustainability (Solow-Sustainability)

This sustainability rule merely requires that the overall stock of capital assets should remain constant over time. The rule is, however, consistent with any one asset being reduced as long as another capital asset is increased to compensate. This approach to sustainability is based on (1) a Hicksian definition of income, (2) the principle of constant consumption (buttressed by a Rawlsian maximum justice rule operating intergenerationally), (3) production functions with complete substitution properties, and (4) the Hartwick rule governing the reinvestment of resource rents (Common and Perrings 1992).

Thus, following Hicks, income is the maximum real consumption expenditure that leaves society as well endowed at the end of a period as at the start. The definition therefore presupposes the deduction of expenditures to compensate for the depreciation or degradation of the total capital asset base that is the source of the income generations (i.e., conservation of the value of the asset base). Assuming a homogeneous capital stock (perfect substitution possibilities), the Hartwick rule states that consumption may be held constant in the face of exhaustible resources only if rents deriving from the efficient use of those resources are reinvested in reproducible capital.

It is possible to derive an intuitive weak sustainability measure or indicator (in value terms) for determining whether a country is on or off a sustainable development path (Pearce and Atkinson 1992). Thus a nation cannot be deemed sustainable if it fails to save enough to offset the depreciation of its capital assets. That is,

$$WSI > 0 \qquad \text{if } S > \partial K \qquad\qquad (1)$$

where WSI is a sustainability index, S is savings and ∂K is depreciation on capital. Dividing through by income (Y) we have,

$$WSI > 0 \text{ if } (S/Y) > (\partial K/Y) \qquad\qquad (2)$$

---

1. One of our referees has argued that in fact there is no difference between weak and strong sustainability. We believe that there is an important distinction, but that it involves psychological and motivational factors.

or   WSI > 0 if $(S/Y) > [\partial_m/Y + (\partial_n/Y)]$ (3)

where ∂_m is depreciation on man-made capital, ∂_n is depreciation on natural capital, and natural capital and man-made capital are substitutable.

Weak Sustainability (Modified Solow-Sustainability)

Perrings (1991) and Common and Perrings (1992) argue that the technological assumptions (substitution possibilities) of the weak sustainability approach violate scientific understanding of the evolution of thermodynamic systems and ecological thinking about the complementarity of resources in system structure and the importance of diversity in system resilience.

The London school has also modified the very weak sustainability approach by introducing an upper bound on the assimilate capacity assumption, as well as a lower bound on the level of natural capital stocks that can support SD assumption, into the analysis (Barbier and Markandya 1990; Pearce and Turner 1990; Klassen and Opschoor 1990). The concept of critical natural capital (e.g., keystone species and keystone processes) has also been introduced to account for the non-substitutability of certain types of natural capital (e.g., environmental support services, and man-made capital). Thus the requirement for conservation of the value of the capital stock has been buttressed by constraints aimed at the preservation of some proportion and/or components of natural capital stock in physical terms.

This modified Solow-sustainability thinking implies a sustainability constraint which will restrict, to some degree, resource-using economic activities. The constraint will be required to maintain populations/resource stocks within bounds thought to be consistent with ecosystem stability and resilience. To maintain the instrumental value (benefits) humans obtain from healthy ecosystems, the concern is not preservation of specific attributes of the ecological community but rather the management of the system to meet human needs, support species and genetic diversity, and enable the system to adapt to changing conditions.

A set of physical indicators will be required in order to monitor and measure biodiversity and ecosystem resilience. As yet there is no scientific consensus on how biodiversity should be measured. Measuring genetic diversity presents the least difficulty, but measuring species diversity is more problematic. Measures of species richness, taxonomic richness, and richness of genera or families have all been investigated, and none are without difficulties. Problems associated with measuring biodiversity at the community level are even greater, but such measures would be very useful to policy-makers if the aim is conservation at the ecosystem level. Community classification schemes can be developed globally from biogeographic provinces down to regions within a country. Ecoregion classification is based primarily on the physical environment. A minimum set of twenty-two indicators (including species richness, species risk index, and community diversity) has been proposed for wild and domesticated species (Reid et al. 1992).

For some commentators sustainability constraints of this type represent expressions of the precautionary principle (O'Riordan 1992) and are related to safe minimum standards (SMS) (Bishop 1978). Toman (1992), quoting the work of Norton, has suggested that the SMS concept is a way of giving shape to the intergenerational social contract. Given irreversibility and uncertainty about the impact of economic activities on ecosystem performance, SMS posits a socially determined dividing line between moral sustainability imperatives and the freeplay of resource trade-offs. To satisfy the intergenerational social contract, the current generation might rule out in advance (depending on the social opportunity costs involved) actions that could result in damage impacts beyond a certain threshold of cost and irreversibility. Social and not individual preference values will be part of the SMS setting process (Turner 1988a).

Strong Sustainability (Ecological Economics Approach)

The weaker versions of sustainability are consistent with a declining level of environmental quality and natural resource availability as long as other forms of capital are substituted for natural capital. Similarly, if the imposition of SMS is judged to impose too high a social opportunity cost given the inevitable uncertainties involved in conservation benefits forecasting, environmental quality will decline over time.

A variety of analysts note limitations in the economic calculus underlying the weak sustainability rules (Sagoff 1988; Brennan 1992; Ehrlich and Ehrlich 1992). Many ecosystem functions and services can be adequately valued in economic terms, but others may not be amenable to meaningful monetary valuation. Critics of conventional economics have argued that the full contribution of component species and processes to the aggregate life-support service provided by ecosystems has not been captured in economic values (Ehrlich and Ehrlich 1992). Nor has the prior value of the aggregate ecosystem structure (life-support capacity) been taken into account in economic calculations—indeed it is probably not fully measurable in value terms at all. There is the risk that as environmental degradation occurs, some life-support processes and functions will be systematically eroded, increasing the vulnerability (reduced stability and resilience) of the ecosystem to further shocks and stress.

On this strong sustainability view, it is not sufficient to just protect the overall level of capital, rather natural capital must also be protected, because at least some natural capital is non-substitutable. The lack of a comprehensive set of meaningful monetary value estimates for environmental resources also supports the strong sustainability viewpoint buttressed by the precautionary principle. Thus the strong sustainability rule requires that natural capital be constant, and the rule would be monitored and measured via physical indicators. The case for this "strong" view is based on a number of factors: (1) presence of uncertainty about ecosystem functioning and its total service value, (2) presence of irreversibility in the context of some environmental resource degradation and/or loss, (3) the loss aversion felt by many individuals when environmental degradation processes are at work, and (4) the criticality (non-substitutability) of some components of natural capital.

While akin to SMS, the strong sustainability (SS) rule is not the same since what is stressed in the latter approach is the combination of factors (e.g., irreversibility and uncertainty), not their presence in isolation. Further, SMS says conserve unless the benefits foregone (social opportunity costs) are very large. SS says, whatever the benefits foregone, some "critical," natural capital losses are unacceptable (i.e., constant "aggregate" natural capital, not constant natural capital for each asset).

SS need not imply a steady-state, stationary economy, but rather changing economic resource allocations over time which are not sufficient to affect the overall ecosystem parameters significantly (i.e., beyond the point where the resilience of the system or key components of that system are threatened). A certain degree of "decoupling" of the economy from the environment should therefore be possible via technical change, substitution of intermediate natural capital for natural capital, and environmental restoration investment in a "moderated" growth scenario.

In practice, the difference between WS and SS will probably be less clear-cut, becoming more a difference of motivation and psychology. Given the social, economic, and scientific uncertainties, SS advocates would wish to invoke the precautionary principle in order to conserve the maximum stock of "critical" and "other" natural capital. A safety-margin mentality would be dominant.

Very Strong Sustainability (Stationary State Sustainability)

The very strong sustainability perspective reduces calls for a steady-state economic system based on thermodynamic limits and the constraints they impose on the overall scale of the macroeconomy. The rate of matter and energy throughput in the economy should be minimized. The second law of thermodynamics implies that complete recycling is impossible (even if it were socially desirable), and the limited influx of solar energy poses an additional constraint on the sustainable level of production in an economy (the solar influx potential is a matter of some dispute). Daly (1991) has defined the "scale effect" as the scale of human impact relative to global carrying capacity. For him, the greenhouse effect, ozone layer depletion, and acid rain all constitute evidence that we have already gone beyond a prudent "plimsoll line" for the scale of the macroeconomy. Zero economic growth and zero population growth are required for a zero increase in the "scale" of the macroeconomy. Supporters of the steady-state paradigm would, however, emphasize that "development" is not precluded and that social preferences,

community-regarding values and generalized obligations to future generations can all find full expression in the steady-state economy as it evolves
(Hirsch 1976; Daly and Cobb 1989).

A Systems and Co-evolutionary Perspective for Sustainability

The adoption of a systems perspective emphasizes that economic systems are
underpinned by ecological systems and not vice versa. There is a dynamic
interdependency between economy and ecosystem. The properties of biophysical systems are part of the constraints set which bound economic activity. The constraints set has its own internal dynamics that react to economic
activity exploiting environmental assets (extraction, harvesting, waste disposal,
and non-consumptive uses). Feedbacks then occur which influence economic
and social relationships. The evolution of the economy and the evolution of
the constraints set are interdependent—"co-evolution" is thus a crucial concept (Norgaard 1984; Common and Perrings 1992).

Norton and Ulanowicz (1992) advocate a hierarchical approach to natural
systems (which assumes that smaller subsystems change according to a faster
dynamic than do larger encompassing systems) as a way of conceptualizing
problems of scale in determining biodiversity policy. For them, the goal of
sustaining biological diversity over multiple human generations can only be
achieved if biodiversity policy is set at the landscape level of the ecosystem.
Ecosystem modeling must take place on a scale appropriate to the crucial dynamic that supports the sustainability goal. This dynamic, which they term the
"autopoietic" (self-making) feature of ecosystems, supports and sustains
species across generations. The protection of the health of these landscape-
level processes should therefore be the central goal of biodiversity policy.
Individual species are valued for their contribution to a larger dynamic, and
significant financial expenditure may not always be justified to save ecologically marginal species. The goal for policy should be to protect as many
species as possible, but not all.

Ecosystem health (stability and resilience or creativity), is intuitively useful
to help focus attention on the larger systems in nature and away from the
special interest groups (Norton and Ulanowicz 1992). The full range of public and private instrumental and intrinsic values all depend on protection of
the health of larger-scale ecological systems. Thus when a wetland is
disturbed or degraded, the impacts of the disturbance on the larger level of
the landscape must be determined. A successful policy will encourage a
patchy landscape.

Because the component parts of a system are dependent on the existence
and functioning of the whole, putting an aggregate value on wetlands and
other ecosystems is more complicated than has previously been supposed in
the economics literature. The total wetland is the source of *primary value*
(PV). The existence of the wetland structure (all its components, their interrelationships and the interrelationship with the abiotic environment) is prior to

the range of functions and services that the system then provides. So in this sense there is a primary value source (aggregate life support service value, or "glue" value)[2] over and above the combined value of the function/service values. These latter values we term *total secondary values* (TSV), and they are conditional on the continued "health" of the ecosystem.

More formally:

an ecosystem in aggregate possess primary value = PV = e ("glue" value);

a "healthy" ecosystem provides a range of valuable functions and services (secondary values) = TSV;

therefore total ecosystem value (TV) > TSV;

	TSV =	TEV (total economic value);
	TEV =	UV + NUV;
and	UV =	DUV + IUV + OV
	NUV =	EV + BV;
therefore,	TEV =	(DUV + IUV + OV) + (EV + BV)
So,	TV =	(TEV, e)
and	TV =	(TEV, 0) = 0
and	TV =	(0, e) $\geq$ 0
where	e =	"glue" value of the ecosystems
	UV =	use value
	DUV =	direct use value
	IUV =	indirect use value
	OV =	option value
	NUV =	non-use value
	EV =	existence value
	BV =	bequest value
	TV =	total ecosystem value

The concept of *total economic value* (TEV) has two limitations. TEV may fail to fully encapsulate the total secondary value (TSV) provided by an ecosystem, because in practice, some of the functions and processes are diffi-

---

2. Robert Costanza has suggested that our "e" (glue) value might also include the redundancy or "insurance" capacity of ecosystems (i.e., the retention of the capacity for the emergence of new keystone species and processes as systems respond to unexpected shocks and change over time). Thus "e" could encompass a safety-margin concept (and therefore further support for the precautionary principle and SS position) which would cover the uncertain redundancy capacity, as well as our uncertainty over the "true" potential of ecosystem functions and services.

cult to analyze (scientifically and economically) and value in monetary terms. But in addition, TEV also fails to capture the TV of ecosystems, because it ignores the "glue" value of the system itself. The very existence of the system makes available the range of individual functions and services from which we derive instrumental and non-instrumental value (Turner 1992).

Sustainability, Cultural Capital Depletion and Sustainable Livelihoods

Co-evolution is a local process, so local human subsystems serve as a basis for the discussion of evolution in ecological economics (Berkes and Folke 1992). Traditional ecological knowledge may be important and therefore, cultural diversity and biological diversity may be equally fundamental to long-term societal survival. Diverse cultures encompass not just diverse environmental adaptation methods and processes, but also a diversity of worldviews (technocentrism and ecocentrism) that support these adaptations. The conservation of this rapidly diminishing pool of social experience in adaptability (cultural capital) may be as pressing a problem as that of conserving biological diversity. Cultural adaptations to the problem of maintaining an environmental balance constitute as important a reservoir of information as the genetic information contained in species currently threatened with extinction (Perrings, Folke, and Mäler 1992).

The most convincing body of evidence for human self-organizational ability may be found in the literature of common property-resources (Ostrom 1990). Institutions, co-evolution history, and traditional ecological knowledge are all components of the dilemma. A large number of self-regulating regimes governing access to common property resources exist, and their scope in limiting the level of economic stress on particular ecological systems is clearly very wide (Bromley 1992).

Any sustainability strategy for the future will have to confront the question of how a vastly greater number of people can gain at least a basic livelihood in a manner which can be sustained. Many of the livelihoods will have to endure in environments which are fragile, marginal, and vulnerable (Chambers and Conway 1992).

From the viewpoint of policy development, the aim must be to promote sustainable livelihood security via vulnerability reduction. Both public and private action is required for vulnerability reduction—public action to reduce external stress and shocks through, for example, flood protection, and prevention and insurance measures; and private action by households, which add to their portfolio of assets and repertoire of responses so that they can respond more effectively to loss limitation. Chambers and Conway (1992) offer a range of monetary value and physical indicators for sustainable livelihood monitoring and assessment of trends (e.g., migration and off-season opportunities, rights and access to resources, and net asset position of households).

Sustainability and Ethics

Many commentators acknowledge that traditional ethical reasoning faces a number of challenges in the sustainable development debate (Regan 1981; Norton 1987). Ecological economists would argue that the systems perspective demands an approach that places the requirements of the system above those of the individual. This will involve ethical judgments about the role and rights of present individuals versus the system's survival and therefore the welfare of future generations. We again argue that the poverty focus of sustainability highlights the issue of intragenerational fairness and equity.

So "concern for others" is an important issue in the debate. Given that individuals are, to a greater or lesser extent, self-interested and greedy, sustainability analysts explore the extent to which such behavior could be modified and how to achieve the modification (Turner 1988b; Pearce 1992). Some argue that a stewardship ethic—weak anthropocentrism (Norton 1987)—is sufficient for sustainability (i.e., people should be less greedy because other people, including the world's poor and future generations, matter), and greed imposes costs on these other people. Bioethicists would argue that people should also be less greedy because other living things matter and greed imposes costs on these other non-human species and things. This would be stewardship on behalf of the planet itself (Gaianism) in various forms up to "deep ecology" (Naess 1973; Turner 1988a).

The degree of intervention in the functioning of the economic system deemed necessary and sufficient for sustainable development also varies across the spectrum of viewpoints. Supporters of the steady-state economy (extensive intervention) would argue that at the core of the market system is the problem of "corrosive self-interest." Self-interest is seen as corroding the very moral context of community that is presupposed by the market. The market depends on a community that shares such values as honesty, freedom, initiative, thrift, and other virtues whose authority is diminished by the individualistic philosophy of value (consumer sovereignty) of conventional economics. If all value derives only from the satisfaction of individual wants, then nothing remains to restrain self-interested, individualistic want satisfaction (Daly and Cobb 1989).

Depletion of *moral capital* may be more costly than the depletion of other components of the total capital stock (Hirsh 1976). The market does not accumulate moral capital—it depletes it. Consequently, the market depends on the wide system (community) to regenerate moral capital just as much as it depends on the ecosystem for natural capital.

Individual wants (preferences) have to be distinguished from needs. For humanistic and institutional economists, individuals do not face choices over a flat plane of substitutable wants, but a hierarchy of needs. This hierarchy of needs reflects a hierarchy of values which cannot be completely reduced to a single dimension (Swaney 1987). Sustainability imperatives therefore represent high order needs and values.

We now apply the principles of the weak and strong sustainability paradigms to aspects of the problem of coastal zone management in the context of climate change.

CLIMATE CHANGE INDUCED SEA-LEVEL RISE (SLR): COST-BENEFIT VERSUS THE PRECAUTIONARY APPROACH

It is at the global (macro-policy) level that the distinction between weak and strong sustainability can be most clearly drawn, at least in the context of potential global warming and its consequences and responses. Two contrasting views previously outlined illustrate the different types of policy responses—the *precautionary approach* (strong sustainability paradigm) and the *cost-benefit approach* (weak sustainability paradigm). From the *precautionary principle* viewpoint, there is a pressing need to introduce measures that will reduce overall energy consumption, reduce greenhouse gases emissions per unit of energy consumption or GNP, shift to low-CO_2 fuels, and reduce atmospheric emissions of GHGs. Advocates of the *cost-benefit approach* point out that these drastic measures should not be implemented without any serious attempt to weigh the economic costs and benefits of climate change or alternative control strategies (Nordhaus 1991).

From this cost-benefit viewpoint, estimation of the *greenhouse damage function* (i.e., the social costs of climate change such as ecosystem change, changing agricultural patterns, flooding, and erosion of coastal zones) and the *abatement cost function* (i.e., the social costs of pollution emission reduction, energy conservation investments, fuel switching, and sea defense and coastal protection measures and structures) is a high priority and must precede radical policy changes.

Quantification of the greenhouse damage and abatement cost functions would help determine an economically efficient climate change response strategy that maximizes net economic welfare. A fundamental policy question in terms of a social trade off is at issue—how much current sacrifice is acceptable in order to constrain the rate of consumption damage from climate change in the future?

Estimates of both emission control costs and damages are highly uncertain and incomplete, but one recent cost-benefit study has concluded that climate change is likely to produce a combination of gains and losses with no strong presumption of substantial net economic damages (Nordhaus 1991). A 3°C rise in global mean surface temperatures and consequent climate change is tentatively estimated to produce economic damages equal to between a 0.25% of the 1981 U.S. GNP and 2% of GNP. Control cost estimates range from $2 billion per year for a 10% reduction in total emissions to $19 billion per year for a 50% reduction. It turns out that the efficient degree of control of GHGs would be insignificant in the case of high control costs, low damage estimates and high discounting. At the opposite end of the spectrum, high damage estimates and no discounting would require a degree of GHG control close to 33% of all emissions (Nordhaus 1991).

Global-mean sea-level projections have all been falling as modeling capabilities and scientific knowledge increase. One very recent study has incorporated new GHG emission scenarios, the effects of CO_2 fertilization feedback from stratospheric ozone depletion, and the radiative effects of sulphate aerosols into models yielding new projections for radiative forcing of climate and for changes in global-mean temperature and sea-level (Wigley and Raper 1992). Changes in temperature and sea-level are significantly reduced (20%–30%) below the levels published by the Intergovernmental Panel on Climate Change (IPCC) in 1990. Thus the best guess SLR project is reduced from 66cm by 2100 to 48cm by 2100 (19cm by 2050). Nevertheless, even this sea-level rise (SLR) is still at a rate roughly four times that estimated for the past century.

The Nordhaus (1991) cost-benefit study can be criticized because it extrapolates damage and abatement functions modeled on U.S. economy conditions across the rest of the world. In his study, SLR is likely to cost some $50 billion in capital costs (protection costs and losses of low-lying land) to the U.S. economy.

Ayres and Walter (1991) have argued that the Nordhaus analysis significantly underestimates future damage. They produced their own estimate of the costs of sea-level rise (this item accounts for 92% of the total costs identified by Nordhaus). Using their assumptions on land values, land losses, refugee resettlement costs, and coastal defense costs, total costs come to $18.5–$21 trillion. Annualized, this represents 2.1%–2.4% of gross world income, nearly ten times higher than Nordhaus's "central" estimate for total losses and slightly outside his range of error. Ayres and Walter conclude that a cost of $30–$35 per tonne of CO_2 (equivalent) is more realistic than Nordhaus's $3.30 ("central" estimate), just to account for the effects of SLR on "vulnerable" economies like Egypt and Bangladesh.

However, both studies add the estimated costs of coastal protection to those of the loss of low-lying land, and, in the case of Ayres and Walter, to the cost of resettling refugees. Table 16.1 summarizes the reported SLR damage estimates from both studies. Although it is logical to include SLR in the damage function, it does not seem obvious why the costs of preventing the major impacts of SLR should be added to the costs when the rise occurs.

Supporters of the precautionary approach would argue that the consequences of sea-level rise (direct and indirect) are both extensive and significant. Some 60% of the world's population live in coastal areas—3 billion people—and more than 40 of the world's largest cities are located within the coastal zone (Turner, Kelly, and Kay 1990). Further, there could already be a commitment to significant sea-level rise as a result of historic changes in greenhouse gas concentrations due to delays in the responses of the climate, oceans, and ice masses.

Table 16.1. Estimates of Aggregate World Costs of Climate Induced Sea-Level Rise

Category of sea level rise impact	Nordhaus (1991)	Ayres and Walter (1991)
Coastal protection	20,000 km coast	0.5–1m km (world coast) *$5 m/km
Loss of property and land	4,000 sq miles	500m ha *$30,000/ha
Resettlement of refugees	—	100m refugees *$1000
Total capital cost (world economy)	$405 billion	$18,500–21,000 billion

Because of the uncertainties in both the science and the economics of global warming the precautionary viewpoint would advocate a "safety-margin" approach to policy. Thus, for example, policymakers should plan for SLRs that have only a 10%–15% chance of being exceeded (i.e., a target rate of 70cm SLR by 2050). Another precautionary argument in favor of prompt action in the sea-level rise context stems from the analysis of the concept of *vulnerability* (Hewitt 1984; Kasperson, Dow, Golding, and Kasperson 1990; and Turner, O'Riordan, and Kemp 1991). Vulnerability concerns the degree to which a system may react adversely to the occurrence of an acute or chronic hazard, such as SLR. Conversely, resilience in a system is a measure of the system's capacity to absorb, adapt, and recover from the occurrence of a hazardous event. Vulnerability has both a biophysical and a socio-economic dimension, and it is clear that many coastal regions, including major cities, are already facing problems owing to SLR. These problems have been made worse by the existence of three types of inter-related "failures": information failure and lack of systems thinking; market failure; and intervention failure concerning projects and policies in coastal zones (Turner, Kelly, and Kay 1990). Typical of these high risk zones would be the Mississippi delta, the Nile delta, the Po delta, and Southern Bangladesh.

Decreasing the susceptibility of coastal regions to SLR by addressing existing coastal zone management problems more effectively than has occurred in the past should be an immediate priority. Both the cost-benefit and precautionary principle approaches would favor such action. However, the PP would require much more extensive moves and would regard such mitigation measures as the barest minimum that should be done as a precautionary response to global warming. An accelerated rate of degradation of coastal marine environments may be anticipated in the next few decades as a consequence of policy failures, population growth, and current trends in socio-economic patterns of development. The potential impacts of SLR will not only accentuate current coastal problems but will also add extra constraints to sustainable utilization and development of coastal zones (Pernetta and Elder 1992).

SLR COASTAL WETLANDS AND COASTAL ZONE MITIGATION STRATEGIES

SLR Mitigation Options

A distinction has to be drawn between cost-benefit analysis of the greenhouse effects where SLR is part of the damage function and abatement costs are in terms of reduction of the causes (GHG emissions abatement). (Nordhaus 1991; Ayres and Walter 1991), and SLR within a *partial analysis*. In this latter analytical mode, the damage function includes the losses associated with SLR, and the abatement function takes SLR as exogenous and includes a range of mitigation strategies and their costs (Adger and Fankhauser 1993).

The difference between WS and SS at this level of analysis and policy is as follows: from the WS viewpoint, only a "business as usual" sea and coastal defense strategy would be required, until global uncertainties are reduced. Existing defense systems would be maintained and improved only as actual SLR changed over time; from the SS viewpoint, a full range of mitigation options from retreat to full protection of coastal zones should be appraised, allowing for the most cost-effective option to be identified and implemented in advance of actual SLR changes (induced by global warming).

While it is probably the case that, from a coastal biodiversity conservation perspective, curbing the rate and magnitude of global warming would be optimum, there is still a need to deploy partial analysis to identify SLR mitigation options that would help coastal wetlands and species survive (Reid and Trexler 1991). This requirement is reinforced by the on-going losses suffered by coastal wetlands (due to pollution, population, and development pressures) and by the probable SLR that the world is already committed to because of the past GHG emissions.

One of the most important aspects of SLR from the standpoint of biodiversity is the rate rather than the magnitude of rise. IPCC (1990) predicted rates of rise three to six times higher than these experienced last century and even the latest less pessimistic study puts its best guess SLR rate at four times the historical rate (Wigley and Raper 1992). Rates of SLR expected under conditions of global warming are probably at the very limit of wetlands' capacity to keep pace (migration and formation).

Nevertheless, natural capital assets such as wetlands are only one component of the total capital assets stocks in coastal zones that needs to be managed in an integrated and sustainable way. The problem is that different SLR mitigation strategies produce non-uniform effects across the spectrum of capital assets. A study by Park et al. (1989) in the United States, for example, has forecasted that, given a 50cm SLR by 2100, more than 26% of coastal wetlands in the East and the Gulf states of the United States would be lost if all dry areas are protected. But only 14% of wetlands would be lost if just the currently diked areas were protected from inundation. Some 17% of wetlands would be lost if all residential and commercial areas were to be protected.

There would also be a baseline wetlands loss of 11% at the current rate of SLR.

United States sources quoted in Reid and Trexler (1991) tend to argue that a *pure retreat option* (i.e., not a "do-nothing" process but one based on zoning and development planning controls) throughout coastal zones would, in principle, best protect coastal ecosystems. This is because those ecosystems would be free to migrate landward without human impediment. In practice, it is of course recognized that the financial and political implications of this option are such as to make it untenable. But from a U.K. perspective, it is not even clear that pure retreat would be optimal in ecosystem conservation terms. Many of the U.K.'s most prized coastal wetlands (at least on the eastern seaboard) are semi-natural freshwater systems which require continued protection from saline intrusion. It may be that in the U.K., in some areas, a trade-off will have to be made between saltmarshes trying to retreat landward and freshwater wetlands requiring hard or soft engineering protection from saline intrusion.

A *complete protection strategy* (engineering solution) would be expensive in capital construction and maintenance cost terms. It would have a significant direct negative effect on wetlands and other coastal ecosystems and processes, and also an indirect negative effect on ecosystems as new development activities, encouraged by the new level of protection, entered the zone and competed for space.

Reid and Trexler (1991) advocate an intermediate strategy, a *managed retreat, adaptive process* which would include actions to make adaptation to future SLR easier, thereby ensuring the continued conservation of coastal ecosystems. The central objective of this strategy would be to institutionalize the assurance that current and future development will make way as necessary for migrating ecosystems. As they put it, this option allows the continuing economic use of coastal drylands, while simultaneously keeping them available for coastal ecosystems when the time comes (Reid and Trexler 1991).

Weak and Strong Sustainability Paradigms and SLR Mitigation Options

A reactive strategy largely based on engineering solutions (hard and soft defense structures) constructed in response to actual SLR would, we argue, be in line with the weak sustainability viewpoint. This strategy could also include a "no net loss" wetlands principle, buttressed by the assumption that future wetland loss due to SLR could be offset by wetland substitution—habitat creation, relocation, and restoration (Buckley 1989). Or a safe minimum standards (SMS) approach might be adopted to protect wetlands unless the social opportunity costs of doing so were deemed unacceptable.

The limitation of this reactive strategy is that the wetlands substitution process may in practice be quite restricted (i.e., artificially created wetlands will not be able to provide all the services and functions that complex natural systems do). Secondly, without anticipatory policies, development pressure (in all its manifestations, such as pollution and recreation pressure) may well increase in the coastal zone, exacerbating wetland conversion/degradation problems. Thirdly, more engineering structures in the "wrong" places will impede the migration of saltmarshes (although freshwater systems may be protected).

Overall, it seems that natural capital cannot be adequately conserved by means of the relatively quick technological fixes that offer protection to cities and other coastal developments. The Norfolk Broads multi-functional wetland on the eastern coast of England illustrate the complexity of the wetland management and protection task (Turner and Jones 1991). The Broads is largely a freshwater system under continual threat from North Sea flood risk. Engineers have long held the view that the best flood alleviation plan would focus on a tidal barrier at the mouth of the only river estuary leading to the main Broads wetland—the Yare River Barrier plan. As originally conceived the barrier would only be closed when the wetland and surrounding areas were threatened by North Sea surge tides and consequent flooding and saline intrusion. However, recent drought, over abstraction, and water pollution problems in the region have all served to produce low river flows and increased salinity in the wetland. It is now recognized that the barrier will have to be used (if it is constructed) as only one component in a complete water management regime. The barrier will have to be supplemented by storage/washland areas in order to combat saline intrusion.

In line with the *strong sustainability paradigm* would be a proactive strategy based largely on a managed retreat policy. The policy would operate through a range of enabling measures as follows:

- permitting and controlling of coastal property development, through specifications on building construction types (easily moveable) and bans on bulkheads;

- purchasing of wetlands or "flowage easements" by government or conservation groups;

- issuing assurance bonds (Costanza and Perrings 1990), and

- creating an insurance fund policy, to be paid out to landowners when property has to be abandoned.

Given the range of capital assets present in coastal zones and the fact that different SLR mitigation strategies will provide a non-uniform protection effect across all assets, a package of policies will be required. Engineering solutions will be deployed in densely populated and developed coastal regions to protect human capital and financial investments. They may also protect freshwater coastal wetlands that often provide a range of high-value functions and services.

Where coastal uplands appear particularly important to the migration or formation of wetlands in the face of SLR, retreat may be allowed. The purchase of land or the purchase of "flow easements" by NGOs or government would aid this strategy. Alternatively, changes in the planning laws might be considered.

East Anglian Coastal Zone: Case Study

The basic approach of the study was to combine information about the physical hazard with data on assets at risk to produce a physical/numerical and then economic estimation of the impact of SLR (Turner 1991). The study area comprised the East Anglian coast between Hunstanton and Felixstowe. The main risks posed by SLR in this area (hazard zone) result from increased flooding and coastal erosion. The output of the latest global climate models and their predictions of SLR were used as baselines for the estimation of the future potential physical hazard.

The estimated mean SLR values were adjusted to account for local geological subsidence in order to produce regional rises in sea-level by the year 2050. Four different SLR values were used in the analysis (20, 40, 60, and 80cm), which reflects the uncertainty inherent in the climate models.[3] The historic return periods of water levels and rates of coastal retreat were then adjusted upwards to account for the predicted sea-level changes.

The next task was to establish what assets were actually at risk. The flood hazard zone was defined as all coastal areas below 5m AOD. Areas at risk from coastal erosion were assumed to be limited to land adjacent to an eroding coastline. These zones were found to contain many different types of assets, both natural and man-made. A detailed survey was therefore carried out to establish the numbers and location of properties, total population, area and type of agricultural land, transport infrastructure, number of waste sites and nature conservation sites, together with historical sites, landscape resources, and recreation resources.

The impact of SLR upon the hazard zone depends greatly on how the coastline is managed. Three different stylized management approaches representing a range of possible responses to SLR were therefore produce (1) abandon defenses, (2) maintain defenses at their current physical height, and (3) improve defenses. These responses were defined according to the extent and effectiveness of defenses that they represented. Efforts were then made to model actual feasible changes in flood return periods and coastal erosion rates by 2050.

Having established the change in risk faced by coastal assets as a result of estimated future SLR, the final analytical step was to express the consequences of such changes in economic terms. The first approach adopted was to consider the value of the assets at risk. This was done by considering the proportion of the national gross domestic product (GDP) represented by the assets

---

3. The 80 cm value was added as a "worst case" figure not based on the climate model predictions.

within the hazard zone—a measure of what is at risk rather than a measure of the economic damage cost or lost social value due to SLR.

The second approach to economic evaluation, however, tried to estimate direct damage and loss via a partial social cost benefit (SCB) model. The model was restricted to a consideration of the two "active" management response options (i.e., Maintain and Improve). The social cost-benefit analysis (SCBA) was "partial" because it only considered the economic benefit (damage cost avoided) of protection properties and agriculture.[4] The study also examined various methods designed to value a wide range of ecological, amenity, and recreational assets.

The total cost equations for the maintain defense (status quo) option and the improve defense option are given below:

$$TCd_{sq}(t) = 1 \cdot Cd_m(t) + 1 \cdot Cd_r(t) + Cd_{rb} + C_{BN}(t)$$

$$TCd_{new}(t) = CW_r(t) \cdot L \cdot h' + C_{sd}(t) \cdot L \cdot h' + 1 \cdot Cd_m(t) + Cd_{er}^{(t-n)} + Cd_{rb} + C_{BN}(t)$$

where

TCd_{sq}	= total costs of maintaining current defenses
TCd_{new}	= total costs of improving defenses in line with SLR
L	= length of defense
Cd_m	= maintenance costs
Cd_r	= capital replacement costs
Cd_{rb}	= repair beach costs
C_{BN}	= beach nourishment costs
CW_r	= wall raising costs
Cd_{er}	= early defense replacement costs
h'	= height of defense raising
C_{sd}	= costs of soft defenses
C_{hd}	= costs of hard defenses.

In the actual case study, defense system cost calculations under the three response option scenarios were simplified.

On the damage costs side, the following elements were analyzed:

In the case of fixed (non-moveable) assets like agricultural land and property, the value of the potential *erosion/inundation losses* was expressed as (adjusted for different flood probabilities and sea-level rises P (Δh(t)):

---

4. These two asset categories were the most amenable to monetary evaluation.

$$y_{e(t)}^{agr} = \int_{t=Tn}^{t=T} y^{agr} \bullet L_a^{agr}(t) \bullet e^{-r(t)} dt$$

where, y^{agr} =   average annual loss of agricultural land value, expressed as rental value, market value of the land or net margin output value

L_a^{agr} =   abandoned agricultural land, ha

T_n =   threshold point in time when flooding frequency is deemed to be equivalent to inundation; or erosion occurs

T =   project time horizon end date, 2050

r =   discount rate

and,

$$y_{e(t)}^{p} = \int_{t=Tn}^{t=T} y^{p} \bullet P_a(t) \bullet e^{-r(t)} dt$$

where, y^p =   annual income loss from property, based on capital values

P_a =   number of abandoned properties

Flooding damage costs suffered by fixed assets such as agricultural enterprises and property were expressed as (adjusted for different flood probabilities and sea- level rises P (Δh (t))):

$$y_f^{agr} = \int_{t-1}^{t=Tn,nT} [y^{agr} \bullet L_{f(t)}^{agr} + yP \bullet P_{f(t)}] \bullet e^{-r(t)} dt$$

where y_f^{agr} =   average annual output value loss from agricultural land (net margins); in the case of saline flooding output loss is experienced for up to 3 years after each event

L_f^{agr} =   flooded agricultural land, ha

y^p =   depth/damage function data; for property flooding damage value

P_f =   number and type of properties flooded

T_n =   point in time when frequency of flooding reaches inundation /conditions, if this occurs before 2050.

Even if the assets at risk in the hazard zone are moveable, there is still an economic cost involved for reconstruction of the asset elsewhere and the relocation process:

$$TC_{rL}(t) = K_{rc} + \sum_{i=1}^{n} C_{rL(t)}^{i} + C_{rL}^{pop}(t) \cdot e^{-r(t)} dt$$

where TC_{rL} = total costs of relocation process, for moveable assets with positive residual lifetimes at T (2050)

K_{rc} = capital costs of asset reconstruction

C_{rL}^{i} = relocation costs for industrial/commercial activity type i

C_{rL}^{pop} = relocation/disruption costs for residential population (households).

Thus the total economic costs of the do nothing, abandon defenses strategy would be :

$$\int_{t=1}^{t=T} y_{e(t)}^{agr}(t) + y_{c}^{P}(t) + y_{f(t)}^{agr,P} + TC_{rL}(t)$$

The total benefits of adopting either the maintain defenses or the improve defenses strategies would be determined by the magnitude of the damage costs and losses avoided (i.e., the difference between the damage costs and losses incurred in the abandon defenses situation and those incurred in either of the two "defend" responses). The net benefits of each response strategy can be computed by subtracting the capital and maintenance defense system costs from the value of total benefits (expressed as damage costs avoided).

The economic cost-benefit decision is then to accept that response strategy that maximizes the present value of net benefits (PVNB):

$$\max t \, (B_t - C_t) \cdot e^{-r(t)} > 0$$

where B_t = total benefits (damage costs and losses avoided)

C_t = capital and operating costs of defense system

r = discount rate.

This basic economic analysis will be more complicated in hazard zones which are already "defended" from erosion/flooding by various sea defense and coastal protection structures and systems. These "defenses" will not collapse immediately but will be subject to progressive failure over an uncertain period of time. Other complications also include the current "status" of the asset inventory in the zone (i.e., the quality, age and condition of property, vintages of plants and equipment in industrial enterprises, and the quality and carrying capacities of natural ecosystems).

The economic value of property, and plants and equipment, for example, will vary with their replacement date and consequent remaining residual economic lifetimes. For assets which will be fully depreciated by 2050, we need to estimate what, if any, accelerated depreciation cost (due to an earlier-than-expected replacement date) is applicable, given SLR. It is also important to try and distinguish between marginal and non-marginal asset changes. Thus the threat (let alone the actual loss) of extensive loss of land/property and/or loss of unique irreplaceable structures and natural systems may stimulate macroeconomic effects within the national economy. There may also be an indirect cost incurred by people who suffer flooding or inundation in the form of psychological damage (i.e., increased stress, anxiety, and related illnesses). Environmental goods and services pose special valuation problems.

The SCBA was based upon a mathematical model which simulated the timing and magnitude of benefits and costs associated with the four SLR predictions between the years 1990 and 2050. In the analysis the benefit of an active response strategy is taken to be avoided flood damage and erosion loss. This is evaluated by comparing the flood and erosion damage in the Maintain and Improve responses to that likely to be experienced in an Abandon response strategy (i.e., the baseline case). The costs of the active responses depend on actual flood and erosion prevention measures (i.e., hard and soft engineering structures).

Overall the SCBA showed that in the flood hazard context, active response options were economically efficient. For the erosion hazard, however, the picture is less clear, and on a region-wide basis, erosion prevention may not be cost-effective. Overall both the GDP approach and the SCBA approach served to highlight the fact that the end-results of the appraisal are just as sensitive to changes in socioeconomic parameters, (economic growth rates and discount rates) as they are to changes in physical parameters (SLR).

One factor prevalent throughout the entire study is uncertainty. This results from the inadequacy of, and interrelationships between, both data and scientific knowledge. Uncertainty characterizes climate modeling, magnitude of physical hazard, characteristics and value of assets, and effect, characteristics, and costs of defense measures. In many cases where uncertainty exists, a simplifying assumption based on nominal values was sometimes used (e.g., the change in effectiveness of erosion defenses with SLR). In other cases, a range of values has been used assuming that the true value, although unknown, will probably fall within the range. An example of this approach at the macro level is the use of four SLR scenarios. At the micro level within the designated flood planning units in the zone, the range of possible future flood characteristics was assumed to be bounded by the conditions related to two possible flooding scenarios, A and B (derived from local historical flood event data). This approach is similar to a form of sensitivity testing.

Even with complete knowledge of all parameters the value of a SCBA is not just the production of a single NPV-or NCB-ratio economic efficiency result, which then determines policy. SCBA plays a wider heuristic role which is important in helping to understand how system interrelationships work—the system in this case being the impacts and responses to SLR, their costs and benefits. Sensitivity testing and the heuristic role played by SCBA is most important given the inevitable uncertainties involved in the SLR context. It enables the decision maker to grasp which are the important and influential parameters operating with the system and to understand the implications of making changes within the system, and so aids rational decision making.

The basic model was therefore recalibrated in order to consider how the original hazard/response model (Bateman et al. 1991) could be further improved and in addition, to carry out more sensitivity testing. The results are summarized in Table 16.2.

Table 16.2. Cost-Benefit Analysis of Coastal Defense Strategies in East Anglia (U.K.)

Response Strategies	SLR (cm by 2050)			
	20	40	60	80
		(Net benefits £ million)		
Maintain current defenses (flood hazard)	1002–1008	960–970	912–924	848–867
Improve defenses (flood hazard)	1067–1070	1055–1058	1048–1050	899–900
Maintain current defenses (flood + erosion hazard)	963–969	928–938	910–922	862–881
Improve defenses (flood + erosion hazard)	1021–1024	1008–1001	1021–1023	873–874

Source: Adapted from Doktor, Turner, and Kay (1991).

CONCLUSIONS

The concept of strong sustainability is complex and includes the interaction of critical natural capital, with cultural and human capital, all of which infer a strong emphasis on equity (both inter- and intra-generational). It also goes beyond a safe minimum standards approach in that any losses to "critical" natural capital and significant losses to "other" natural capital are unacceptable, both because of uncertainty over which elements of natural capital are critical and because of the moral imperative of intergenerational equity. The concept of ecosystem primary value has been introduced, which, in the context of wetlands, is explained by the different components of the wetlands, their interrelationships, and the interrelationship with the abiotic environment. The partial nature of the total economic valuation of ecosystems is a further argument in support of the SS and precautionary approach.

The case study of damage to coastal and wetland ecosystems in the U.K. due to potential sea-level rise caused by climate change and natural processes has been used to illustrate the differences between strong sustainability with precautionary prescriptions and weak sustainability prescriptions. The results illustrate that, although all the elements are not valued in economic terms, in the East Anglia case study, a set of precautionary actions and managed retreat is desirable to maintain the natural capital stock of the hazard zone, rather than an abandon strategy which tends to be advocated elsewhere.

ACKNOWLEDGMENTS

CSERGE is funded by the U.K. Economic and Social Research Council. We are grateful to Charles Perrings, Dennis King, Robert Costanza, Jason Shogrun, and David Pearce for helpful comments on an earlier version of this chapter. All remaining errors are of course our responsibility.

REFERENCES

Adger, W. N., and S. Fankhauser. 1993. Economic analysis of the greenhouse effect: optimal abatement level of strategies for mitigation. *International Journal of Environment and Pollution* 3: 104–19.

Ayres, R. U., and J. Walter. 1991. The Greenhouse Effect: damages, costs and abatement. *Environmental and Resource Economics* 1: 237–70.

Barbier, E. B., and A. Markandya. 1990. The conditions for achieving environmentally sensitive development. *European Economic Review* 34: 659–69.

Bateman, I. et al. 1991. Economic Appraisal of the Consequences of Climate-Induced Sea Level Rise: A Case Study of East Anglia. Report to Ministry of Agriculture Fisheries and Food. Environmental Appraisal Group, School of Environmental Sciences. Norwich: Univ. of East Anglia.

Berkes, F., and C. Folke. 1992. A systems perspective on the interrelations between natural, human-made and cultural capital. *Ecological Economics* 5: 1–8.

Beckerman, W. 1992. Economic growth and the environment: whose growth? Whose environment? *World Development* 20: 481–96.

Bishop, R. C. 1978. Economics of endangered species. *American Journal of Agricultural Economics* 60: 10–18.

Brennan, A. 1992. Moral pluralism and the environment. *Environmental Values* 1: 15–33.

Bromley, D. W. 1992. The commons, common property and environmental policy. *Environmental and Resource Economics* 2: 1–18.

Buckley, G. P., ed. 1989. Biological Habitat Reconstruction. London: Belhaven.

Chambers, R., and G. Conway. 1992. Sustainable Rural Livelihoods: Practical Concepts for the 21st Century. Discussion Paper 296. Sussex: Institute of Development Studies.

Common, M., and C. Perrings. 1992. Towards an ecological economics of sustainability. *Ecological Economics* 6: 7–34.

Costanza, R., and C. A. Perrings. 1990. A flexible assurance bonding system for improved environmental management. *Ecological Economics* 2: 57–76.

Daly, H. E. 1991. Towards an environmental macroeconomics. *Land Economics* 67: 255–9.

Daly, H. E., and J. B. Cobb. 1989. For the Common Good: Redirecting the Economy Towards Community, the Environment and a Sustainable Future. Boston: Beacon Press.

Doktor, P., R. K. Turner, and R. C. Kay. 1991. Economic Appraisal of the Consequences of Climate-Induced Sea Level Rise: Extension Study. Report to National Rivers Authority (Anglian Region). Environmental Appraisal Group, School of Environmental Sciences. Norwich: Univ. of East Anglia.

Ehrlich, P. R., and A. H. Ehrlich. 1992. The value of biodiversity. *Ambio* 21: 219–26.

Hewitt, K., ed. 1984. Interpretation of Calamity. London: Unwin Hyman.

Hirsh, F. 1976. Social Limits to Growth. London: Routledge.

Howarth, R. B., and R. B. Norgaard. 1992. Environmental valuation under sustainable development. *American Economic Review Papers and Proceedings* 82: 473–7.

Intergovernmental Panel on Climate Change (IPCC). 1990. Climate Change: The IPCC Scientific Assessment. Cambridge: Cambridge Univ. Press.

Kasperson, R. E., K. Dow, D. Golding, and J. X. Kasperson, eds. 1990. Understanding Global Environmental Change: The Contributions of Risk Analysis and Management. Center for Technology, Environment and Development. Worcester, MA: Clark Univ.

Klassen, G. K., and J. B. Opschoor. 1990. Economics of sustainability or the sustainability of economics: different paradigms. *Ecological Economics* 4: 93–116.

Naess, A. 1973. The shallow and the deep, long range ecology movement: a summary. *Inquiry* 16: 95–100.

Nordhaus, W. D. 1991. To slow or not to slow: the economics of the greenhouse effect. *Economic Journal* 101: 920–37.

Norgaard, R. B. 1984. Coevolutionary development potential. *Land Economics* 60: 160–72.

Norton, B. G. 1987. Why Preserve Natural Variety? Princeton: Princeton Univ. Press.

Norton, B. G., and R. E. Ulanowicz. 1992. Scale and biodiversity policy: a hierarchical approach. *Ambio* 21: 244–9.

O'Riordan, T. 1992. The Precaution Principle in Environmental Management. CSERGE GEC Working Paper 92-03, CSERGE. Norwich; London: Univ. of East Anglia; Univ. College London.

Ostrom, E. 1990. Governing the Commons: The Evolution of Institutions for Collective Action. Cambridge: Cambridge Univ. Press.

Park, R. A. 1989. The Effects of SLR on U.S. Coastal Wetlands and Lowlands. Report No. 164. Indianapolis: Holcomb Research Institute.

Pearce, D. W. 1992. Green economics. *Environmental Values* 1: 3–13.

Pearce, D. W., and G. D. Atkinson. 1992. Are National Economies Sustainable? Measuring Sustainable Development. CSERGE GEC Working Paper 92-11, CSERGE. Norwich; London: Univ. of East Anglia; Univ. College of London.

Pearce, D. W., and R. K. Turner. 1990. Economics of Natural Resources and the Environment. Hemel Hempstead; London: Harvester Wheatsheaf.

Pernetta, J. C., and D. L. Elder. 1992. Climate, sea level rise and the coastal zone: management and planning for global change. *Ocean and Coastal Management* 18: 113–60.

Perrings, C. 1991. The Preservation of Natural Capital and Environmental Control. Paper presented at the Annual Conference of EARE, Stockholm.

Perrings, C., C. Folke, and K-G. Mäler. 1992. The ecology and economics of biodiversity loss: the research agenda. *Ambio* 21: 201–11.

Reid, W. V. J. A. McNeely, D. B. Tunstall, D. A. Bryant, and M. Winegrad. 1992. Developing Indicators of Biodiversity Conservation. Draft, unpublished document. Washington, DC: World Resources Institute.

Reid, W. V., and M. C. Trexler. 1991. Drowning the National Heritage: Climate Change and U.S. Coastal Biodiversity. Washington DC: World Resources Institute.

Regan, T. 1981. The nature and possibility of an environmental ethic. *Environmental Ethics* 3: 19–34.

Sagoff, M. 1988. The Economy of the Earth: Philosophy, Law and the Environment. Cambridge: Cambridge Univ. Press.

Swaney, J. 1987. Elements of a neo-institutional environmental economics. *Journal of Environmental Issues* 21: 1739–79.

Toman, M. A. 1992. The difficulty in defining sustainability. *Resources* 3–6.

Turner, R. K. 1988a. Wetland conservation: economics and ethics. In Economics, Growth and Sustainable Environments, eds. D. Collard, D. W. Pearce and D. Ulph. London: Macmillan.

———., ed. 1988b. Sustainable Environmental Management: Principles and Practice. London: Belhaven.

————. 1991. Economic Appraisal of the Consequences of Climate Induced Sea Level Rise: Executive Summary. Report to Ministry of Agriculture Fisheries and Food. Environmental Appraisal Group, School of Environmental Sciences. Norwich: Univ. of East Anglia.

————. 1992. Speculations on Weak and Strong Sustainability. CSERGE GEC Working Paper 92-26, CSERGE. Norwich; London: Univ. of East Anglia; Univ. College of London.

Turner, R. K., M. Kelly, and R. Kay. 1990. Cities at Risk. London: BNA International.

Turner, R. K., and T. Jones, eds. Wetlands, Market and Intervention Failures. London: Earthscan.

Turner, R. K., T. O'Riordan, and R. Kemp. 1991. Climate change and risk management. In Climate Change: Science, Impacts and Policy, eds. J. Jäger and H. L. Ferguson. Cambridge: Cambridge Univ. Press.

Victor, P. A. 1991. Indications of sustainable development: some lessons from capital theory. *Ecological Economics* 4: 191–213.

Wigley, T. M. L., and S. C. B. Raper. 1992. Implications for climate and sea level of revised IPCC emissions scenarios. *Nature* 357: 293–300.

17 NATURAL CAPITAL AND THE ECONOMICS OF ENVIRONMENT AND DEVELOPMENT[1]

Edward B. Barbier
Department of Environmental Economics and Environmental Management
University of York
Heslington, York YO1 5DD U.K.

ABSTRACT

In recent years economic analysis of the natural resource management and environmental problems of developing countries has increased considerably. This analysis treats the environment as an "asset" or form of "natural capital" that must be managed to sustain economic development. The crucial decision is how much to "draw down" stocks of natural capital and to reinvest in other economic assets (i.e., reproducible, man-made capital, foreign assets, and human resources), in order to meet both current and future economic demands. Thus the value of natural resources as economic assets in the development process depends on the present value of their economic welfare potential compared to those of other assets available to the developing economy. It follows that an "optimal" strategy for a developing economy with sufficient natural resources would be to draw down its stock of natural capital to finance economic development by reinvesting the proceeds in other assets that are expected to yield a higher economic return. However, there are obvious problems in applying this strategy to developing countries. Major difficulties include (1) ecological limits on the substitutability for the environmental functions of natural capital with man-made capital, (2) some environmental values may not be known or reflected in the "prices" of resources, (3) markets may be distorted, (4) resource extraction may be inefficient, and finally, (5) rents may not be reinvested in other assets but dissipated and dispersed. The following chapter discusses these difficulties in more detail, drawing on empirical evidence from developing economies. The chapter concludes by suggesting that efficient and sustainable management of natural resources requires improved policies and accounting procedures to reflect the actual value of natural capital in the development process.

---

1. I am grateful to comments supplied by Charles Perrings, Stephen Viederman, and an anonymous referee.

INTRODUCTION

As part of the *Agenda 21* preparations for the U.N. Conference on Environment and Development in June 1992, the UNCED Secretariat made the following statement:

> In the last two decades, there has been some progress through conventional economic policy applied in parallel with environmental policy. It is now clear that this is not enough, and that environment and development must be taken into account at each step of decision making and action in an integrated manner (UNCED 1992b).

A recent review of progress in economic research on natural resource degradation in developing countries in the 1980s has also shown important research initiatives in "resource accounting" methods: (1) the impact of economic policies on the environment, particularly pricing policies for natural resource products and resource-based activities; (2) the effects of "user incentives" such as common property rights, land tenure and distribution, intrahousehold division of labor, perceptions of risk, and degree of participation in deciding how individuals should manage the resources available to them; and (3) improvements in the methodology of economic valuation of environmental impacts through the appraisal of the benefits of environmental preservation and the costs of degradation (Barbier 1991b).

In short, there is a rapidly emerging "new" sub-discipline of *economics of environment and development*, and there is an even more rapidly expanding demand for policy analysis that applies environmental economics to development problems. In the near future, this analysis will be concerned with the following issues:

- proper economic valuation of natural resources,

- integration of environmental considerations into economic planning and policy-making,

- economic incentives for sustainable resource management,

- the role of trade and international economic policies in resource management, and

- the relationship between poverty, population growth, and the environment.

An overriding feature of recent economic analysis in environment and development has been to treat the environment as an "asset" or form of "natural capital" that must be managed to sustain economic development. The crucial decision is how much to "draw down" stocks of natural capital to reinvest in other economic assets (i.e., reproducible, man-made capital, foreign assets, and human resources), in order to meet both current and future economic demands. This decision is of paramount importance to many low-income, developing countries, which are usually characterized as having an abundance of natural capital and too little human and man-made capital. Similarly, poor people within developing countries are considered to be overly dependent on natural resources and primary production, because of the lack of economic opportunities in other sectors of the economy and insufficient (man-made) capital formation.

The following chapter addresses the role of natural capital in the economics of environment and development. The first issue is, given the development priorities of the South, whether loss of natural capital is an economic problem for developing countries. The next two sections examine this issue in more detail by exploring the economic significance of natural capital depletion for development and the implications for resource-dependent, low-income economies in particular. A second issue is the relationship between natural capital and poverty—the existence of a poverty-environment "trap." The available evidence suggests that there are many misconceptions over this issue, which will be briefly discussed and, hopefully, clarified. Finally, the fundamental issue in the end to be addressed by policymakers is the design of appropriate economic incentives for efficient and sustainable natural resource management. Unfortunately, as presented in the conclusion of this discussion, these are often distorted in development by policy and market failures.

IS NATURAL CAPITAL DEPLETION AN ECONOMIC PROBLEM IN DEVELOPING COUNTRIES?

To answer this question requires addressing two additional issues:

- are natural resources an efficient form of retaining "wealth" for developing countries?

- are the opportunity costs of natural resource degradation for developing countries greater than the benefits gained?

Although apparently different, the questions are inherently related and represent two aspects of the same fundamental economic problem concerning natural capital depletion in developing countries.

Like other "assets" in developing economies, environmental resources can be seen as a form of "natural" capital. That is, they have the potential to contribute to the long-run economic productivity and welfare of developing countries. Thus the value of a natural resource as an economic asset depends on the present value of its income, or welfare, potential. However, in any growing economy there will be other assets or forms of wealth that yield income. Any decision to "hang on" to natural capital, therefore, implies an opportunity cost by foregoing the chance to invest in alternative income-yielding assets, such as man-made "reproducible" capital. If natural resources are to be an "efficient" form of holding on to wealth, then they must yield a rate of return comparable to or greater than that of other forms of wealth. In other words, an "optimal" strategy for a developing country would be to "drawn down" its stock of natural capital to finance economic development by reinvesting the proceeds in other assets that are expected to yield a higher economic return. Under such circumstances, natural resource depletion is not an economic problem but is economically justified; it should proceed up to the point where the comparative returns to "holding on" to the remaining natural capital stock equals the returns to alternative investments in the economy. If the latter always exceeds the former, then in the long run even complete depletion of natural capital is economically "optimal." The only economic problem remaining is that this process of depletion of natural capital and its transformation into other economic assets is done in the most efficient manner possible.

The idea that economic well-being, or welfare, may not be affected and may even be enhanced if the rents derived from depleting natural capital such as tropical forests, wetlands, biodiversity, energy, minerals, and even soils, are reinvested in reproducible capital has been around for some time in economics literature. For example, what is now known as the "Hartwick-Solow rule" states that reinvestment of the rents derived from the intertemporally efficient use of exhaustible natural resources are reinvested in reproducible, and hence non-exhaustible, capital will secure a constant stream of consumption over time (Solow 1974 and 1986; Hartwick 1977). Similarly, basic renewable resource economic theory suggests that, for slow-growing resources such as tropical forests, it may, under certain conditions, be more economically optimal to harvest the resource as quickly and efficiently as possible and reinvest the rents in other assets that increase much faster in value. Similarly, if the harvesting costs are low, or the value of a harvested unit is high, then the resource may also not be worth retaining today (Clark 1976; Smith 1977).

However, there are obvious limits to the applications of the above rules to the environment and development problems in the South. For example, the Hartwick-Solow rule assumes that there is sufficient substitutability between reproducible, man-made and natural capital over time, such that they effectively comprise a single "homogeneous" stock. Moreover, the above rules assume that (1) all economic values are known and reflected in the "prices" of resources, (2) markets are undistorted, (3) resource extraction is efficient, and (4) rents are reinvested in other assets in the economy. As a consequence, more recent theories now stress the limits to "substitution" between many forms of natural and man-made capital, even for developing countries interested in "drawing down" their natural capital stock in favor of investing in other forms of capital (Barbier 1989a; Barbier and Markandya 1990; Bojö, Mäler and Unema 1990; Pearce, Barbier and Markandya 1990). From an ecological standpoint, a major shortcoming is the failure to consider the economic consequences of the loss in "resilience"—the ability of ecosystems to cope with random shocks and prolonged stress—that results from natural capital depletion (Common and Perrings 1992; Conway 1985; Conway and Barbier 1990; Holling 1973).

For example, certain functions of a tropical forest, such as its role in maintaining micro-climates, protecting watersheds, providing unique habitats, and supporting economic livelihoods of indigenous peoples, may be irretrievably lost when the forest is degraded or converted. Often the economic values of these and other functions of tropical forest are not properly accounted for in decisions concerning forest use. There is uncertainty over many of these values, such as the forest's role in maintaining biodiversity and global climate. Finally, there is little evidence to suggest that tropical forest resources are currently being exploited efficiently, nor that the rents earned from activities that degrade or convert the forests are being reinvested in more "profitable" activities. Rather, market and policy failures are often rife, and economic rents tend to be dissipated and misused.[2]

In addition, the degradation of resources in many low-income countries is often occurring in "fragile" environments—semi-arid rangelands, "marginal" cropland, coastal ecosystems, converted forest soils. If resource degradation affects the "resilience" of these environments to respond to further stresses (e.g., soil erosion, devegetation) and random shocks (e.g., drought, pest attacks), then the economic livelihoods of peoples dependent on these environments can be affected. If carrying capacities and ecological tolerance limits are transgressed, then collapse of these economic-environmental systems is a possibility (see Barbier 1989a and 1990a; Conway and Barbier 1990; Perrings 1990; Scoones 1993).

A further problem facing the conventional efficiency conditions driving the Hartwick-Solow rule is that it does not take into account the potential threat posed by population growth to the sustainability of economic development. The conventional rule suggests that for future generations to have the same aggregate level of real consumption expenditure as the present generation, the aggregate value of the capital stock must be maintained. However, in developing countries where populations are currently expanding rapidly, the concern should be with *per capita* rather than with aggregate real consumption expenditure. Thus maintaining the per capita value of natural assets should also be the relevant sustainability issue. Moreover, as per capita natural capital is "depreciated" both by use and population growth in developing economies, more resources need to be devoted to the preservation of the asset base where population is growing than where it is not.[3]

In summary, although it is theoretically possible that the current rates of resource depletion in developing regions are economically "optimal," it is unlikely. Too often, decisions concerning natural capital depletion and conversion are made without considering the opportunity cost of these decisions.[4]

---

2. For further discussion of these issues in relation to tropical deforestation, see Barbier (1992b).
3. I am grateful to Charles Perrings for drawing my attention to this point.
4. However, the methodology and its applications for incorporating environmental costs and benefits in investment and planning decisions in developing countries does exist, whether at the level of national accounts (e.g., see Ahmad, El Serafy, and Lutz 1989; Repetto et al. 1989; TSC/WRI 1991), investment projects (Dixon et al. 1988), or individual resource systems, such as tropical wetlands (Barbier, Adams, and Kimmage 1991; Barbier 1993; Ruitenbeek 1991), tropical forests (Barbier 1991a and 1992b; Peters, Gentry, and Mendelsohn 1989; Ruitenbeek 1989), drylands (Bojö 1991; Dixon, James, and Sherman 1989 and 1990), wildlands (Dixon and Sherman 1990; Swanson and Barbier 1992),

In short, we do not know whether it is worth "holding on" to the natural capital as an economic asset because we do not know or bother to take into account the potential economic benefits that it yields. As a result, decisions will always be biased towards environmental degradation because the underlying assumption is that the foregone benefits provided by natural capital in developing countries are necessarily negligible. If this is not the case, as much evidence suggests, then current levels of resource depletion in developing countries may not be optimal but "excessive." Hence, depletion of natural capital may indeed be an economic problem for developing countries—a problem that stems not just from the inefficient transformation of natural capital into other forms of capital but also from its "unsustainable" management for economic welfare.

RESOURCE-DEPENDENT DEVELOPING ECONOMIES

So far it has been argued that if current development strategies are to be considered *sustainable*, then net depreciation of the natural asset base should be compensated by investment in renewable (human or physical) capital, and where this is not possible for some natural assets and ecological functions, then environmental degradation should be avoided. Moreover, if natural resource management for development is to be *efficient*, then the benefits of depleting natural capital and its transformation into other economic assets must exceed the costs.

Unfortunately, there is little evidence to suggest that current development strategies are either efficient or sustainable. One of the major reasons for this is the failure to appreciate the critical role of natural capital in economic development. Natural resource management is usually accorded a low priority on the development agenda, and even basic facts about the natural asset base are ignored. If governments of developing economies are to be convinced that sustainable and efficient management of their resource base is essential to economic development, then they must perceive the role of natural resources in supporting the economy as a whole.

Tables 17.1–17.8 indicate the high degree of *resource dependency* of many low- and lower-middle-income developing countries. Many of these economies are directly dependent on natural resource products for the overwhelming majority of their exports (Tables 17.1 and 17.2). In most cases, export earnings are dominated by one or two commodities. Resource dependency has been a feature of these economies over the past 25 years, and for most low-income countries, has remained a persistent feature since the mid-1960s. More lower-middle income economies have reduced their resource dependency with time, but this will clearly be a long-term process. Careful man-

---

agroforestry (Anderson 1987; Barbier 1992a), and watersheds (Easter, Dixon, and Hufschmidt 1986; Gregersen et al. 1987).

Table 17.1. Low Income Economies with High Export Concentration in Primary Commodities a/

Contribution of Primary Commodities to Total Exports b/	Export Share in 1980/81	Export Share in 1965	Main Export Commodities b/ 1		2	
over 90%				%		%
Uganda ($280)	100	100	100	Coffee	95.6	Tea	c/0.3
Eq. Guinea ($410)	100	91	NA	Cocoa	34.4	Coffee	3.1
SAO Tome & pr. ($490)	99	100	NA	Cocoa	95.5	Copra	1.8
Ethiopia ($120)	99	99	99	Coffee	53.8	Hides	15.3
Rwanda ($320)	99	99	100	Coffee	75.5	Tea	c/10.8
Yemen PDR ($430)	99	NA	94	NA		NA	
Zambia ($290)	98	99	100	Copper	83.0	Cobalt	5.4
Burkina Faso ($210)	98	85	95	Cotton	32.6	Livestock	e/26.8
Nigeria ($290)	98	99	97	Petroleum	87.3	Cocoa	4.7
Liberia ($450) c/	98	98	97	Iron Ore	63.4	Rubber	e/16.1
Ghana ($400)	97	98	98	Cocoa	51.1	Gold	20.3
Mauritania ($480)	97	99	99	Fish	65.8	Iron Ore	33.3
Niger ($300)	96	98	95	NA		NA	
Somalia ($170)	95	99	86	Meat	39.7	Banana	34.5
Zaire ($170)	93	94	92	Copper	35.8	Coffee	e/11.2
Sudan ($480)	93	99	99	Cotton	30.3	Livestock	24.4
Togo ($370)	92	85	97	Phosphate	36.2	Cotton	12.6
Comoros ($440)	c/92	e/86	NA	Cloves	41.7	Vanilla	c/33.3
Lao PDR ($180)	90	e/100	NA	Timber	51.7	Electricity	19.0
over 80%							
Chad ($160)	c/89	e/96	97	Cotton	69.4	Hides/Skins	e/3.8
Myanmar ($210) c/	89	d/81	NA	Rice	32.7	Teak	e/32.2
Guinea-Bissau ($190)	c/87	d/71	NA	Cashewnut	73.3	Groundnut	c/6.7
Guyana ($420)	c/87	NA	NA	NA		NA	
Madagascar ($190)	84	92	94	Coffee	26.6	Cloves	5.7
Malawi ($170)	83	93	99	Tobacco	62.8	Tea	10.3
Burundi ($240)	83	96	95	Coffee	82.6	Tea	5.0
Kenya ($370)	83	88	94	Coffee	26.2	Tea	c/21.9
Tanzania ($160)	81	86	87	Coffee	31.4	Cotton	12.7
over 70%							
Maldives ($410)	c/77	e/70	NA	Fish	c/57.1		
Benin ($390)	74	96	95	Cotton	13.4	Fuel	9.4
Indonesia ($440)	71	96	96	Petroleum	40.0	Rubber	5.0
Mali ($230)	70	83	97	Cotton	36.9	Livestock	29.0
over 60%							
C.A.R. ($380)	60	74	46	Diamonds	40.9	Coffee	18.9
over 50%							
Sri Lanka ($420)	57	79	99	Tea	25.9	Rubber	c/7.0

Notes: a/ Low-income economies are those with per capita incomes of $545 or less in 1988. U.S. dollar figure at country listed indicates GNP per capita in 1988. b/ Contributions to the value of total merchandise exported 1988, unless indicated. c/ 1987 value. d/ 1981–83 average value. e/ 1984 value.
Source: Based on various editions of the following World Bank documents: *World Development Report; Trends in Developing Countries; Commodity Trade and Price Trends; African Economic and Financial Data.*

Table 17.2. Lower Middle-Income Economies with High Export Concentration in Primary Commodities a/

Contribution of Primary Commodities to Total Exports b/		Export Share in 1980/81	Export Share in 1965	Main Export Commodities b/			
				1		2	
	over 90%				%		%
Bolivia ($570)	97	100	95	Gas	40.1	Tin	13.9
Papua N.G. ($810)	95	100	90	Gold	37.7	Copper	28.9
Ecuador ($1120)	93	93	98	Petroleum	44.8	Fish/Shrimp	19.0
	over 80%						
Yemen A.R. ($640)	89	49	100	Oil	93.7		
Honduras ($860)	89	89	96	Bananas	39.0	Coffee	21.0
Congo ($910)	89	94	37	Oil	71.6	Timber	15.6
Cote d'Ivoire ($770)	88	90	95	Cocoa	25.7	Coffee	13.1
Cameroon ($1010)	88	97	94	Petroleum	48.9	Coffee	12.2
Paraguay ($1180)	88	NA	92	Cotton	10.3	Timber	2.5
Chile ($1510)	85	90	96	Copper	48.4	Agriculture	13.2
	over 70%						
Panama ($2120)	79	91	98	Petroleum	31.7	Banana	1d/7.9
Peru ($1300)	78	83	99	Copper	12.9	Zinc	8.8
Senegal ($650)	75	81	97	Fish	26.9	Groundnut	14.8
Columbia ($1180)	75	72	93	Coffee	30.2	Oil	17.0
Syria ($1680)	75	NA	90				
Egypt ($660)	74	92	80	Oil	64.4	Cotton	6.5
Dominican Rep. ($720)	74	81	98	Nickel	31.2	Sugar	20.5
El Salvador ($940)	71	63	83	Coffee	60.6	Fish	c/3.5
	over 60%						
Guatemala ($900)	62	71	86	Coffee	34.8	Banana	7.6
Zimbabwe ($650)	60	63	85	Tobacco	21.5	Gold	13.1
Costa Rica ($1690)	60	68	84	Coffee	30.4	Banana	18.8
	over 50%						
Malaysia ($1940)	55	80	94	Rubber	9.8	Palm Oil	8.4
Jordan ($1500)	53	57	81	Minerals	38.6	Food	10.2
Brazil (42160)	52	59	92	Soya	9.4	Coffee	6.8
Morocco ($830)	50	72	95	Ph. Acid	16.8	Phosphate	13.3

Notes: a/ Low-income economies are those with per capita incomes of $2160 or less in 1988. U.S. dollar figure at country listed indicates GNP per capita in 1988. b/ Contributions to the value of total merchandise exported 1988, unless indicated. c/ 1987 value. d/ 1984 value.

Source: Based on various editions of the following World Bank documents: *World Development Report; Trends in Developing Countries; Commodity Trade and Price Trends; African Economic and Financial Data.*

agement of the natural resource base may be necessary to maintain the "capital" required for this transition and to achieve long-term development goals. Running down the natural resource base today through inefficient and unsustainable exploitation could jeopardize development efforts by reducing future export earning potential, as well as needlessly wasting current earnings.

The failure to manage resources efficiently and sustainably also increases vulnerability to the economic stresses imposed by external debt. External debt as a percentage of the gross national product (GNP) and debt servicing as a proportion of GNP and exports have risen substantially in virtually all low- and lower-middle income resource-dependent economies (see Tables 17.3 and 17.4). For these economies the ability to meet debt repayments and to induce further economic development will depend on the continued successful exploitation of their natural resource bases. Without sustainable management, the debt burden may severely constrain development efforts.

The resource base in poor resource-dependent economies is far from static. Although comparison of land classifications across countries is fraught with difficulties, the most notable change over the last fifteen years in most economies is the decline in forest area and the increase in cropland (see Tables 17.5 and 17.6). Much of the forest land has presumably been lost to agricultural conversion, with fuelwood and fodder gathering a factor in some areas and depletion for timber operations important in major producing countries. Some empirical explorations of the causes of deforestation, particularly from frontier agriculture expansion, are discussed further.

Expansion of cropland clearly appears to follow the classic agricultural extensification pattern (see Tables 17.7 and 17.8). Particularly in low-income resource-dependent economies, low levels of fertilizer use and yield changes suggest very little agricultural intensification, with the notable exception of Indonesia, Sri Lanka, and to some extent, Kenya, Benin, Ghana, Nigeria, and Zambia. A worrying trend in some poor sub-Saharan economies is the fall in agricultural yields, particularly for cereals. Many other economies show substantial yield increases only because yields were abysmally low in the mid-1970s. Some economies that have increased agricultural intensification—for example Indonesia, Sri Lanka, and Kenya—also face rapid population growth and severe constraints on cropland availability, as indicated by extremely low levels of cropland per capita. In general, most of the low-income resource-dependent economies have long since extended cropland beyond their stock of favorable land with good chemical and physical properties for agriculture. Those countries that do still have "excess" favorable land face difficulties in cultivating it, as much of this land occurs in areas with climates and growing seasons (e.g., arid/semi-arid zones) that prevent or severely constrain rainfed cultivation.

Cropland use and productivity in resource-dependent lower middle income economies follow similar patterns as in the low-income economies (see Table 17.8). In general, fertilizer use and yields and yield changes appear higher in the lower-middle-income groups compared to the low-income economies, suggesting a slightly higher level of agricultural intensification. However, like the low-income countries, many lower-middle-income economies also display low cropland availability per capita and a low ratio of favorable to total cropland. Climatic conditions again limit cultivation of much of the "excess"

Table 17.3. Debt and Debt Service Ratios in Resource-Dependent Low-Income
Economies a/

	External Total Debt as Percentage of GNP b/		Debt Service as Percentage of GNP			
			GNP		Exports	
	1970	1988	1970	1988	1970	1988
over 90%						
Uganda (100)	7.3	34.3	0.5	1.0	2.9	14.0
Eq. Guinea (100)	c/140.2	120.1	NA	NA	c/6.0	28.5
Sao Tome & Pr. (99)	d/54.3	146.0	NA	NA	d/4.7	251.0
Ethiopia (99)	9.5	50.6	1.2	4.3	11.4	37.4
Rwanda (99)	0.9	25.5	0.2	0.7	1.5	9.6
Yemen PDR (99)	d/85.3	199.4	NA	10.8	d/5.1	46.5
Zambia (98)	37.5	116.7	4.6	4.9	8.0	14.2
Burkina Faso (98)	6.6	43.4	0.7	2.0	7.1	11.9
Nigeria (98)	4.3	102.5	0.7	7.0	7.1	25.7
Ghana (97)	22.9	44.6	1.2	4.0	5.5	20.6
Mauritania (97)	13.9	196.2	1.8	11.9	3.4	21.6
Niger (96)	5.0	66.0	0.4	5.6	4.0	32.6
Somalia (95)	24.4	185.2	0.3	0.4	2.1	4.9
Zaire (93)	9.1	118.0	1.1	2.8	4.4	6.9
Sudan (93)	14.8	74.6	1.7	0.6	10.6	9.5
Togo (92)	16.0	81.6	1.0	7.0	3.1	18.3
Comoros (92)	c/94.3	e/94.4	NA	NA	d/1.9	e/3.2
Lao PDR (90)	c/50.2	153.5	NA	1.8	c/12.6	143.5
over 80%						
Chad (89)	9.9	33.2	0.9	0.7	4.2	2.7
Myanmar (89) c/	d/25.3	e/45.7	NA	NA	d/20.3	e/59.2
Guinea-Bissau (87)	d/118.4	e/271.0	NA	NA	d/0.0	e/92.0
Madagascar (84)	10.4	192.7	0.8	9.3	3.7	39.0
Malawi (83)	43.2	85.9	2.3	4.6	7.8	19.0
Burundi (83)	3.1	69.8	0.3	3.3	2.3	25.1
Kenya (83)	26.3	58.5	3.0	5.7	9.1	25.3
Tanzania (81)	20.7	140.1	1.6	3.0	6.3	17.8
over 70%						
Maldives (77)	c/71.6	44.2	NA	NA	c/16.9	e/5.9
Benin (74)	15.1	49.3	0.7	1.0	2.5	5.4
Indonesia (71)	30.0	61.7	1.7	11.5	13.9	39.6
Mali (70)	71.4	100.8	0.2	2.5	1.4	14.2
over 60%						
C.A.R. (60)	13.5	53.3	1.7	1.1	5.1	5.9
over 50%						
Sri Lanka (57)	16.1	61.6	2.1	4.8	11.0	17.6

Notes: a/ Percentage figure after each country listed indicates contribution of primary commodities to total exports. Low income economies are those with GNP per capita of $545 of less in 1988. b/ Total debt includes public, publicity guaranteed and private nonguaranteed debt. c/ 1980 value. d/ 1984 value. e/ 1987 value.
Source: Based on World Bank, *World Development Report 1990* and World Bank, *Trends in Developing Economies 1989.*

Table 17.4. Debt and Debt Service Ratios in Resource-Dependent Lower Middle- Income Economies a/

	External Total Debt as Percentage of GNP b/		Debt Service as Percentage of GNP			
			GNP		Exports	
	1970	1988	1970	1988	1970	1988
over 90%						
Bolivia (97)	49.3	114.9	2.5	5.6	12.6	32.9
Papua N.G. (95)	33.4	64.2	4.8	15.6	24.5	30.9
Ecuador (93)	14.8	94.2	2.2	5.7	14.0	21.4
over 80%						
Yemen A.R. (89)	c/35.1	41.7	NA	3.4	c/1.3	16.0
Honduras (89)	15.6	68.3	1.4	7.2	4.9	28.6
Congo (89)	46.5	205.0	3.4	13.1	11.5	28.7
Cote d'Ivoire (88)	19.5	135.1	3.1	12.4	7.5	31.9
Cameroon (88)	12.6	27.0	1.0	4.6	4.0	27.0
Paraguay (88)	19.2	36.4	1.8	5.0	11.8	24.6
Chile (85)	32.1	79.3	3.9	7.9	24.5	19.1
over 70%						
Panama (79)	19.5	81.2	3.1	0.2	7.7	0.2
Peru (78)	37.3	56.1	7.0	1.3	40.0	8.7
Senegal (75)	15.5	63.6	1.1	5.2	4.0	19.3
Columbia (75)	22.5	42.1	2.8	8.0	19.3	42.3
Syria (75)	10.8	25.0	1.7	2.6	11.3	21.1
Egypt (74)	22.5	126.7	4.8	4.4	38.0	16.6
Dominican Rep. (74)	23.9	77.3	2.7	5.8	15.3	14.4
El Salvador (71)	17.3	31.5	3.1	3.3	12.0	18.8
over 60%						
Guatemala (62)	6.5	28.3	1.6	4.5	8.2	27.2
Zimbabwe (60)	15.5	37.3	0.6	8.2	2.3	27.9
Costa Rica (60)	25.3	89.2	5.7	7.7	19.9	19.9
over 500%						
Malaysia (55)	10.8	56.3	2.0	16.5	4.5	22.3
Jordan (53)	22.9	94.0	0.9	19.6	3.6	31.9
Brazil (52)	12.2	29.6	1.6	4.5	21.8	42.0
Morocco (50)	18.6	89.8	1.7	6.5	9.2	25.1

Notes: a/ Percentage figure after each country listed indicates contribution of primary commodities to total exports. Lower middle-income economies are those with GNP per capita of $2160 or less in 1988.b/ Total debt includes public, publicity guaranteed and private nonguaranteed debt. c/ 1980 value.

Source: Based on World Bank, *World Development report 1990* and World Bank, *Trends in Developing Economies 1989.*

Table 17.5. Land Use Changes in Resource Dependent Low-Income Economies, 1975–87 a/

	Total Land area 1987 ("000 km^2)	Land use ('000 km^2) and % change since 1975–77							
		Cropland		Pasture		Forest		Other b/	
		85–87	%	85–87	%	85–87	%	85–87	%
Uganda (100)	200	67	21.4	50	0.0	58	-7.9	25	-21.3
Eq. Guinea (100)	28	2	0.0	1	0.0	13	0.0	12	0.0
Ethiopia (99)	1,101	139	1.4	451	-1.1	275	-3.5	236	5.9
Rwanda (99)	25	11	18.6	4	-27.3	5	-5.5	5	1.8
Yemen PDR (99)	333	1	9.2	91	0.0	15	-6.1	226	0.4
Zambia (98)	741	52	3.8	350	0.0	293	-3.0	46	18.3
Burkina Faso (98)	274	31	19.4	100	0.0	68	-8.1	75	1.3
Nigeria (98)	911	313	3.8	210	0.8	146	-17.0	243	7.5
Ghana (97)	230	29	5.5	34	-2.8	84	-7.7	84	8.4
Mauritania (97)	1,025	2	0.0	393	0.0	150	-0.9	481	0.3
Niger (96)	1,267	35	32.0	93	-8.4	25	-19.1	1,113	0.5
Somalia (95)	627	9	3.0	289	0.0	89	-5.3	241	2.0
Zaire (93)	2,268	66	8.2	92	0.0	1,756	-1.9	353	8.7
Sudan (93)	2,376	125	1.8	560	0.0	471	-6.2	1,220	2.4
Togo (92)	54	14	1.0	2	0.0	14	-26.3	24	25.3
Lao PDR (90)	231	9	5.9	0	0.0	1	-11.1	6	6.7
Guinea-Bissau (87)	28	3	15.8	11	0.0	11	0.0	3	-11.9
Chad (89)	1,259	32	4.0	450	0.0	131	-5.7	647	1.0
Madagascar (84)	582	31	8.5	340	0.0	149	-9.7	62	28.1
Malawi (83)	94	24	4.3	18	0.0	44	-13.2	8	272.2
Burundi (83)	26	13	5.3	9	6.3	1	9.6	3	-32.8
Kenya (83)	567	24	5.6	37	-1.0	37	-7.5	469	0.4
Tanzania (81)	886	52	2.9	350	0.0	425	-2.7	58	21.6
Benin (74)	111	18	3.3	4	0.0	37	-12.0	51	9.5
Indonesia (71)	1,812	211	8.1	440	0.0	180	0.1	868	1.3
Mali (70)	1,220	21	4.6	300	0.0	86	-4.5	814	0.4
C.A.R. (60)	623	20	5.1	30	0.0	358	-0.3	215	0.0
Sri Lanka (57)	65	19	-1.0	4	0.0	17	-2.2	24	2.5

Notes: a/ Percentage figure after each country listed indicates contribution of primary commodities to total exports. Low-income economies are those with GNP per capita of $545 or less in 1988. b/ Other land includes uncultivated land, grassland not used for pasture, built-on areas, wetlands and roads.

Source: World Resources Institutes/UNEP/UNDP, *World Resources 1990–91.*

Table 17.6. Land Use Changes in Resource Dependent Lower Middle-Income Economies, 1975–87 a/

	Total Land area 1987 ("000 km^2)	Land use ('000 km^2) and % change since 1975–77							
		Cropland		Pasture		Forest		Other b/	
		85–87	%	85–87	%	85–87	%	85–87	%
Bolivia (97)	1,084	34	3.0	268	-1.2	558	-1.3	224	4.5
Papua N.G. (95)	453	385	8.9	1	-18.5	383	-0.5	65	3.0
Ecuador (93)	277	26	1.4	49	61.5	121	-19.6	81	14.9
Yemen A.R. (89)	195	14	0.3	70	0.0	16	0.0	95	0.0
Honduras (89)	112	18	5.9	25	8.3	36	-18.5	33	18.5
Congo (89)	342	7	2.7	100	0.0	212	-0.9	22	9.1
Cote d'Ivoire (88)	318	36	22.4	30	0.0	69	-42.1	183	31.2
Cameroon (88)	465	70	7.2	83	0.0	250	-4.2	63	11.2
Paraguay (88)	397	22	71.2	192	26.0	166	-20.4	18	-25.2
Chile (85)	749	56	4.0	119	1.7	87	0.0	487	-0.8
Panama (79)	76	6	4.5	13	8.3	40	-7.0	17	11.3
Peru (78)	1,280	37	12.8	271	0.0	694	-3.5	278	8.1
Senegal (75)	193	52	0.6	57	0.0	59	-4.2	24	3.7
Columbia (75)	1,039	53	3.2	398	7.3	515	-5.5	73	1.5
Syria (75)	184	56	1.3	83	-3.1	5	16.2	40	25.3
Egypt (74)	995	25	-7.5	0	0.0	0	0.0	970	0.2
Dominican Rep. (74)	48	15	13.2	21	0.0	6	-3.1	7	-18.9
El Salvador (71)	21	7	8.9	6	0.0	1	-35.4	6	-1.0
Guatemala (62)	108	18	10.2	14	7.9	41	-16.4	36	17.4
Zimbabwe (60)	387	28	9.1	49	0.0	199	0.0	111	-2.0
Costa Rica (60)	51	5	6.1	23	34.1	16	-22.9	6	-16.4
Malaysia (55)	329	44	3.1	NA	NA	152	-0.3	167	57.9
Jordan (53)	89	4	5.5	6	0.0	134	0.0	6	0.0
Brazil (52)	8,457	767	22.7	1,670	6.4	5,604	-4.2	415	0.3
Morocco (50)	446	84	9.0	209	7.7	52	0.2	101	-17.9

Notes: a/ Percentage figure after each country listed indicates contribution of primary commodities to total exports. Low-income economies are those with GNP per capita of $545 or less in 1988. b/ Other land includes uncultivated land, grassland not used for pasture, built-on areas, wetlands and roads.

Source: World Resources Institutes/UNEP/UNDP, *World Resources 1990–91*.

Table 17.7. Cropland Use and Productivity in Resource-Dependent Low-Income
Economies a/

	Cropland ha ('000 ha) 1987	Cropland ha per capita 1989	Fertilizer use		% Yield changes 76–78 to 86–88		Favorable land index c/
			(kg/ha) 85–87	% change b/	Cereals	Roots	
Uganda (100)	6.705	0.38	0	0	6	29	8.8
Eq. Guinea (100)	230	0.53	0	0	NA	-12	9.1
Ethiopia (99)	13,930	0.30	4	100	22	-9	d/105.9
Rwanda (99)	1,120	0.16	1	100	0	-12	8.1
Yemen PDR (99)	119	0.05	13	86	12	12	d/2,753.8
Zambia (98)	5,208	0.64	16	33	19	3	11.4
Burkina Faso (98)	3,140	0.36	5	67	34	18	d/206.9
Nigeria (98)	31,335	0.29	10	400	43	6	28.5
Ghana (97)	2,870	0.20	10	150	12	6	9.7
Mauritania (97)	199	0.10	7	0	47	45	d/14,723.1
Niger (96)	3,540	0.51	1	0	-2	-10	d/1,362.5
Somalia (95)	933	0.13	3	-25	76	-5	d/366.2
Zaire (93)	6,690	0.19	1	-50	15	8	18.2
Sudan (93)	12,478	0.51	4	-33	-24	-24	d/599.3
Togo (92)	1,431	0.43	7	250	-1	-24	19.3
Lao PDR (90)	901	0.23	2	200	66	3	4.1
Guinea-Bissau (87)	335	0.35	1	0	42	14	0.0
Chad (89)	3,205	0.58	2	0	20	25	d/966.5
Madagascar (84)	3,067	0.26	4	33	2	5	71.7
Malawi (83)	2,377	0.29	17	70	-9	-28	46.2
Burundi (83)	1,332	0.25	2	100	5	-4	5.0
Kenya (83)	2,420	0.10	46	109	4	20	d/252.4
Tanzania (81)	5,230	0.20	9	50	34	-18	57.0
Benin (74)	1,840	0.40	37	95	10	9	12.3
Indonesia (71)	21,220	0.12	100	270	45	26	35.5
Mali (70)	2,076	0.23	15	200	26	-3	d/1,674.0
C.A.R. (60)	2,005	0.71	1	0	48	58	39.9
Sri Lanka (57)	1,887	0.11	106	116	49	50	25.9

Notes: a/ Percentage figure after each country listed indicates contribution of primary commodities to total exports. Low-income economies are those with GNP per capita of $545 or less in 1988. b/ Percentage change in annual average fertilizer use since 1975/77. c/ Index denotes (favorable land/total cropland)*100, where favorable land denotes land with good chemical and physical properties relevant for crop fertility and management, as defined by the fertility capability classification (FCC) system. d/ Much of the favorable land in these countries occur in areas with climates and growing seasons (e.g., arid/semi-arid zones)

Source : World Resources Institutes/UNEP/UNDP, *World Resources 1990–91.*

Table 17.8.   Cropland Use and Productivity in Resource-Dependent Lower Middle-Income Economies a/

	Cropland ha ('000 ha) 1987	Cropland ha per capita 1989	Fertilizer use		% Yield changes '76–78 to '86–88		Favorable land index c/
			(kg/ha)85–87	% change b/	Cereals	Roots	
Bolivia (97)	3,399	0.48	2	100	19	-12	254.3
Papua N.G. (95)	386	0.10	30	50	5	0	NA
Ecuador (93)	2,646	0.25	34	8	7	-35	75.6
Yemen A.R. (89)	1,360	0.18	13	86	2	46	d/184.7
Honduras (89)	1,785	0.36	19	36	54	121	35.2
Congo (89)	679	0.35	5	25	15	18	0.0
Cote d'Ivoire (88)	3,640	0.30	9	-36	14	29	3.8
Cameroon (88)	6,995	0.64	7	133	13	7	27.6
Paraguay (88)	2,176	0.52	5	400	20	10	493.1
Chile (85)	5,580	0.43	46	130	90	44	79.6
Panama (79)	575	0.24	58	35	30	10	28.0
Peru (78)	3,725	0.17	43	13	30	15	362.1
Senegal (75)	5,225	0.73	4	-56	15	8	59.7
Columbia (75)	5,318	0.17	81	65	11	9	95.0
Syria (75)	5,630	0.47	42	180	41	28	15.2
Egypt (74)	2,560	0.05	347	85	18	31	d/817.9
Dominican Rep. (74)	1,474	0.21	47	-8	33	-1	30.8
El Salvador (71)	733	0.14	111	-26	11	39	34.5
Guatemala (62)	1,865	0.21	62	24	20	43	25.1
Zimbabwe (60)	2,769	0.29	56	8	-6	29	32.2
Costa Rica (60)	526	0.18	166	31	13	-11	16.0
Malaysia (55)	4,380	0.26	154	126	2	6	6.5
Jordan (53)	414	0.10	34	113	126	39	d/716.9
Brazil (52)	77,500	0.53	49	20	32	5	29.2
Morocco (50)	8,462	0.35	36	57	39	22	d/107.1

Notes: a/ Percentage figure after each country listed indicates contribution of primary commodities to total exports. Low-income economies are those with GNP per capita of $545 or less in 1988. b/ Percentage change in annual average fertilizer use since 1975/77. c/ Index denotes (favorable land/total cropland)*100, where favorable land denotes land with good chemical and physical properties relevant for crop fertility and management, as defined by the fertility capability classification (FCC) system. d/ Much of the favorable land in these countries occur in areas with climates and growing seasons (e.g., arid/semi-arid zones) that prevent or severely constrain rainfed cultivation.

Source: World Resources Institutes/UNEP/UNDP, *World Resources 1990–91.*

favorable land in the North African and Middle Eastern countries, whereas social, economic and climatic constraints on frontier agricultural expansion also limit exploitation of the additional favorable land available in the South American countries. Clearly, much cultivation of marginal—or ecologically fragile—land occurs in resource-dependent lower-middle-income economies.

Thus the continuing dependence of most of the world's poorest economies on their resource base should make environmental management a high priority as a development concern. This is particularly true given that past economic policies and investments have led to rapid changes—frequently with adverse economic consequences—in resource stocks and patterns of use. Demographic trends have often worsened the relationship between population and resource carrying capacity in many regions. Continuing agricultural extensification into marginal lands have increased the susceptibility of economic systems and livelihoods to environmental degradation.

As mentioned elsewhere, a re-appraisal of demographic and agricultural policies to take into account (1) the new realities of resource-carrying capacity constraints and (2) the need to improve the potential of marginal lands while sustaining the productivity of high potential lands, is required (Conway and Barbier 1990; Barbier 1989b; FAO 1990; Leonard et al. 1989; Pearce 1991; Pretty et al. 1992; Repetto 1987; Repetto and Holmes 1983). With regard to population-environment linkages, one useful approach is to view the role of population growth in terms of increasing the "scale" of human demands on limited natural systems (Foy and Daly 1989). Another approach is to extend economic theories of household resource allocation to determine the extent to which increased family size is a cause and/or effect of environmental degradation (Dasgupta 1992; Repetto and Holmes 1983).

To summarize, this section has argued through the use of basic indicators that the natural asset base of the poorest, resource-dependent economies is being rapidly run down. Yet these economies remain in a fundamental state of "underdevelopment." In short, development is essentially "unsustainable" because net depreciation of the natural asset base (and any increase in population) is not being compensated by investment in renewable (human and physical) capital. Clearly, a higher priority should be placed on efficient and sustainable management of the natural resource to maintain the "capital" required for this transition and to achieve long-term development goals. As will be discussed in the following section, rethinking from an economic perspective the relationship between poverty and environmental degradation in developing countries is also a necessary component of this new development thinking.

POVERTY, ENVIRONMENT, AND DEGRADATION

If developing countries are not using their resources sustainably, then either resource users within these countries are not being confronted by the social opportunity cost of resource use, or they are constrained by limited asset holdings and/or income, or both. In the next section, the factors influencing inefficient resource use (i.e., divergence between private and social opportunity cost) are explored. This section discusses how resource endowments—especially conditions of poverty—influence resource utilization.

A common assumption is that the effectiveness of public policies and economic incentives in controlling environmental degradation in developing economies is limited by the existence of a poverty-environment "trap." Given the presence of both high levels of poverty and environmental degradation in developing countries, it is tempting to conclude that poverty *causes* environmental degradation. However, poverty and environmental degradation may be positively correlated, but correlation does not imply causation. If anything, recent evidence suggests that poverty-environmental linkages cannot be reduced to simple unidimensional cause-effect relationships (Barbier 1989b; Conway and Barbier 1990; Dasgupta 1992; Hardoy, Mitlin, and Satterthwaite 1992; Jagannathan 1989; Pearce and Warford 1993; Pretty et al. 1992).

There are numerous complex factors that influence poor peoples' perceptions of the environment and their behavior towards natural resource management. These range from the economic distortions arising from policy and market failures, to underlying labor and capital endowments and constraints (including pressures to increase family size), to access to alternative employment and income-earning opportunities, to institutional and legal factors such as tenure or access security, property rights, and delivery systems. Often what is perceived to be a *direct* link between poverty and environmental degradation proves to be an *indirect* link under careful analysis. Public policies and other factors often affect the incentive structures and redirect capital and labor flows between sectors and regions, with adverse consequences for the poor and their ability or willingness to manage resources sustainably (Jagannathan 1989).

As poor people have little or no access to capital and must rely on family or low-skilled labor for earning income, it would make no economic sense for them to degrade any "natural" capital at their disposal. Many studies have revealed that poor people and communities are often acutely aware of the essential role of natural resources in sustaining their livelihoods, and equally, of the costs and impacts of environmental degradation.[5] This would suggest that, *ceteris paribus*, there exist tremendous *incentives* for the poor to manage and sustain the stock of natural capital at their disposal in order to maintain or enhance both their immediate and future livelihood options. Where they choose to degrade their environment —and there may be rational grounds for doing so under certain circumstances—it is because changing economic and social

---

5. For further discussion of the role of natural resources in the livelihood security of the poor, see Chambers (1988), Pretty et al. (1992), Richards (1985), and Wilson (1988).

conditions have altered the incentive structures of the poor, including perhaps their control over or access to essential resources.

Thus from an economic perspective, simply observing that poor people are "driven" to degrade the environment—even when this appears to be the case—is not helpful. Designing appropriate policy responses to alleviate problems of poverty and environmental degradation, therefore, requires careful analysis of the *determinants* of individual behavior. Such an analysis would clarify the factors leading them to degrade their environment, their responses to environmental degradation, and the incentives required to induce conservation.

Where further analysis reveals that poverty is not the direct "cause" of environmental degradation, design of appropriate policy responses will nevertheless be affected by poverty's indirect role. The response of poor people and communities to incentives encouraging sustainable resource management may be affected by special factors influencing their behavior, such as high rates of time preference induced by greater risk and uncertainty over livelihood security, labor, and capital constraints, insecure tenure over and access to resources, imperfect information and access to marketed inputs, and a variety of other conditions and constraints (Barbier 1989b and 1990b; Binswanger 1980; Pender and Walker 1990).

Moreover, the poor are not a homogenous group. The work of Lipton (1983 and 1988) highlights how the "ultra" or "core" poor, marginally poor, and the non-poor in developing countries all differ in terms of demographic, nutritional, labor market, and asset-holding characteristics.[6] A recent study in Malawi highlights how the marginally poor and the "core" poor face different incentives and constraints in combating declining soil fertility and erosion, which is a serious problem afflicting smallholder agriculture (Barbier and Burgess 1992b).

In Malawi, female-headed households constitute a large percentage (42%) of the "core-poor" households. They typically cultivate very small plots of land (< 0.5 ha) and are often marginalized onto the less fertile soils and steeper slopes (> 12%). They are often unable to finance agricultural inputs such as fertilizer, to rotate annual crops, to use "green manure" crops or to undertake soil conservation. As a result, poorer female-headed households generally face declining soil fertility and lower crop yields, further exacerbating their poverty and increasing their dependence upon the land. The special constraints and needs of poor female-headed households must be carefully considered when designing economic policies to alleviate poverty and control

---

6. The exact numbers and composition of poor and ultra-poor will obviously vary by country and region; however, as an approximate indication Lipton (1988) suggests that the ultra-poor can be defined as those at significant risk of income-induced caloric undernutrition and the poor as those with sufficiently low income to be at risk of hunger but not undernutrition, with the former usually falling in the bottom 10%–20% income category (e.g., the "poorest quintile").

land degradation. Otherwise, an important subset of the rural population will not respond fully to policy measures and incentives to improve environmental management, and the problem of land degradation may continue unmitigated.

A worrying trend in developing economies is the concentration of the poorest groups in "ecologically fragile" zones—areas where environmental degradation or severe environmental hazards constrain and even threaten economic welfare. As indicated by Leonard (1989) around 470 million, or 60% of the developing world's poorest people, live in rural or urban areas that can be classified as "ecologically fragile."[7] Around 370 million of the developing world's poorest people live in "marginal" agricultural areas. These less favorable agricultural lands, with lower productivity potential, poorer soils and physical characteristics, and more variable and often inadequate rainfall, are easily prone to land degradation due to overcropping, poor farming practices, and inadequate conservation measures.

The result is that the economic livelihoods and welfare of the poorest income groups in low potential areas are at greater risk from increasing environmental degradation. It is this risk, combined with the impact of public policies, institutions, and investments on the economic incentives that the poorest face that may have the most profound—and often perverse—effects on the willingness and ability of the poorest groups to counteract degradation.

Another major "poverty reserve" in developing countries is peripheral urban areas, or "squatter" settlements. Recent evidence from West Java and Nigeria confirms that the informal employment sector and settlements around urban and semi-urban settlements are often the preferred "open access" resource for the poor (Jagannathan 1989). Although precise estimates are scarce of how many of the 1.3 billion urban dwellers in developing countries live in "squatter" settlements, it is common for between 30%–60% of the population in large cities to live either in illegal settlements or in tenements and cheap boarding houses. In smaller urban centers of fewer than 100,000 inhabitants—which contain about three quarters of the developing world's urban population—the proportion of people living in illegal settlements may be smaller than in the large cities. However, the proportion living in areas with inadequate infrastructure or services may be as high or even higher in smaller compared to larger urban centers (Cairncross, Hardoy, and Satterthwaite 1990). As a consequence, the economic welfare of a substantial and growing number of the poorest urban dwellers is threatened by the environmental haz-

---

7. The "poorest people" are defined by Leonard (1989) as the poorest 20% of the population in developing countries. In commenting on the data presented by Leonard, Kates (1990) argues that it is too simplistic to equate all land of low agricultural potential and squatter settlements with "areas of high ecologically vulnerability." Thus "while there is good reason to expect an increasing geographic segregation of the poor onto the threatened environments, both the purported distribution of the hungry and the actual state of environmental degradation needs to be examined much more carefully."

ards and health risks posed by pollution, and inadequate housing, sanitation, water, and other basic infrastructure services.[8]

The concentration of the poorest groups in developing countries in "ecologically fragile" areas suggests that it is the welfare of the poorest in developing economies that is at the greatest risk from continued environmental degradation. As argued by Kates (1990), throughout the developing world the poor often suffer from three major processes of *environmental entitlement loss*:

1. the poor are *displaced* from their traditional entitlement to common resources by development activities or by the appropriation of their resources by richer claimants;

2. the remaining entitlements are *divided* and reduced by their need to share their resources with their children, or to sell off bits and pieces of their resources to cope with extreme losses (crop failure, illness, death), social obligations (marriages, celebrations) or subsistence; and

3. the resources of the poor are *degraded* through excessive use and by failure to restore or to improve their productivity and regeneration—a process made worse by the concentration of the poor into environments unable to sustain requisite levels of resource use.

As a result of these processes, the economic livelihoods of the poor become even more vulnerable to the risks posed by environmental degradation, and their ability and willingness to manage resources sustainably may become even more constrained.

Moreover, statistics indicate that the poorest groups in the relatively more affluent developing regions of Latin America and Asia are even more concentrated in ecologically fragile areas than the poorest groups in Africa (Leonard 1989). This would suggest that the problem does not easily disappear with economic growth and rising national income.

In short, whether poverty is directly or only indirectly the "cause" of environmental degradation seems less relevant a concern when compared to the implications of pervasive environmental degradation for the livelihoods of the poor. This is the real "poverty-environment" link that should be the focus of development efforts. If poverty alleviation is an ultimate aim of economic development, then efficient and sustainable environmental management is a necessary means for achieving this goal. Yet past economic policies and investments in developing countries have led to rapid changes in resource stocks and patterns of use—frequently with adverse economic consequences, particularly for the poor. It is to this issue that the next section now turns.

---

8. Further examples of the impact of urban environmental problems and sub-standard living conditions on the welfare of the urban poor can be found in Hardoy and Satterthwaite (1989), Hardoy, Cairncross, and Satterthwaite (1990), and Hardoy, Mitlin, and Satterthwaite (1992).

PUBLIC POLICIES AND NATURAL RESOURCE MANAGEMENT

Unsustainable and inefficient resource use in developing countries can often be attributed to individual resource users not facing the social opportunity costs of their actions. That is, the private costs of actions leading to environmental degradation do not reflect the full social costs of degradation, in terms of the foregone environmental values. This section summarizes the underlying causes, focusing particularly on the problem of inadequate government policies.

Much excessive degradation of the environment and natural resources in developing countries is thought to result from individuals in the marketplace and by governments not fully recognizing and integrating environmental values into decision-making processes. If markets fail to adequately reflect environmental values, *market failure* is said to exist. Typical examples include poorly defined or no property rights, the presence of externalities (e.g., pollution or "public goods"), imperfect market competition, and divergence between public and private rates of time preference. *Institutional failure*, such as lack of formal markets, poor information flows and inadequate regulatory bodies and government services, may also exacerbate these problems. In addition, where government decisions or policies do not fully reflect environmental values, there is *policy* or *government failure*. Throughout the developing world, the existence of poorly formulated input and output pricing policies, insecure land titling and registration, tax thresholds and rebates, cheap and restricted credit facilities, overvalued exchange rates, and other policy distortions, has exacerbated problems of natural resource management. By failing to make markets and private decision makers accountable for foregone environmental values, these policies may contribute to market failure. At worst, the direct private costs of resource-using activities are subsidized and/or distorted, thus encouraging unnecessary environmental degradation.

The result of market, policy, and institutional failures is a distortion in economic incentives. There are several reasons for this outcome.

First, the market mechanisms determining the "prices" for natural resources and products derived from conversion of natural resource systems *do not automatically take into account wider environmental costs*, such as disruptions to ecological functions, assimilative capacity, amenity values, and other environmental impacts or foregone *option* and *existence values*—i.e., the value of preserving certain natural environments, species, and resources today as an "option" for future use or simply because their "existence" is valued. Nor do market mechanisms account for any *user cost*—the cost of foregoing future direct or indirect use benefits from resource depletion or degradation today.

In addition, even the direct costs of harvesting resources or converting natural resource systems *are often subsidized and/or distorted by public policies*. As a result, individuals do not face even the full *private costs* of their own actions that degrade the environment. Unnecessary and excessive degradation ensues.

If public policies are to be re-directed to achieve efficient and sustainable management of natural resources in developing countries, major changes are required. Economic valuation of the environmental impacts arising from market and policy failures is essential for determining the appropriate policy responses. Often, however, insufficient data and information exist to allow precise estimation of the economic costs arising from market and policy failures. In most cases, cost estimates as orders of magnitude and indicators of the direction of change are sufficient for policy analysis.[8]

However, with many natural resource problems in developing countries we are not even at this state of *optimal ignorance* to begin designing appropriate policy responses. In the face of such uncertainty we should be humble in our public policy prescriptions. Even the standard economic tool of "improved pricing policy" should be invoked with caution. In most developing countries there is little empirical understanding of the linkages from price changes to short- and long-term supply and demand responses to natural resource impacts. The situation is complicated by the presence of underemployment, informal and incomplete markets, labor and capital constraints, and above all, the problem of widespread poverty. Thus we are often ignorant of the impact of public policies on the economic incentives faced by individual producers and households for managing natural resources, particularly in the case of the poorest groups who are on the "margin" of the formal economic and social systems.

This problem occurs frequently in the economic analysis of *dryland degradation*. The term "drylands" is usually applied to all arid and semi-arid zones, plus areas in the tropical sub-humid zone subject to the same degradation processes that occur on arid lands. Accounting for about one-third of global land and supporting a population of 850 million, the world's drylands are rapidly being degraded through population growth, over-grazing, cropping on marginal lands, inappropriate irrigation and devegetation. Yet these areas are being asked to support increasing numbers of the world's poorest people. The process of dryland degradation is often referred to as "desertification," where the productive potential of the land is reduced to such an extent that it can neither be readily reversed by removing the cause nor easily reclaimed without substantial investment.

However, there are few economic studies of the costs of dryland degradation.[9] Even further behind—and more controversial—is the analysis of the effects of economic and resource management policies on dryland degradation in developing countries. This is often attributed to the superficial identifi-

---

9. See, for example, the case studies in Pearce, Barbier, and Markandya (1990) and the discussion of tax and regulation policies in developing countries in Anderson (1990).

10. For recent examples of such studies, see Bishop and Allen (1989), Bishop (1990), and Lallement (1990).

cation of the causes of desertification and to the frequently poor identification of the reasons behind the failures of dryland projects (Nelson 1988). Although the majority of "causes" are attributable to population growth and natural events, dryland degradation is also symptomatic of an agricultural development bias that distorts agricultural pricing, investment flows, research and development (R&D) and infrastructure towards more "favored" agricultural land and systems (Barbier 1989b).

Where drylands "development" is encouraged, it is usually through the introduction of large-scale commercial agricultural schemes that can conflict with more traditional farming and pastoral systems. Frequently, appraisal of the environmental impacts of these large-scale schemes has shown that the investments should be modified, and in some cases, should not proceed at all. For example, in the arid zone of northern Nigeria, upstream irrigation development investments on the Hadejia and Jama'are rivers and their tributaries threaten traditional floodplain agriculture and pastoralism downstream. The assumption behind these investments is that the benefits of the upstream irrigation developments far exceed the benefits of the floodplain system, which are routinely ignored in project analysis of these developments. However, a recent analysis of the net benefits of the Hadejia-Jama'are floodplain shows that benefits are considerable, particularly in comparison with one major irrigation project, the Kano River Project (Barbier, Adams, and Kimmage 1991). The results suggest that the opportunity cost of diverting water to upstream developments in order to preserve floodplain benefits could be high, and these costs must be considered carefully in any analysis of these developments.

The complexity of social, economic, and environmental relationships confronting dryland management is often formidable. A common misconception is that the extension of private property rights, commercial agriculture, and markets will "automatically" solve dryland management problems in the long run. At the same time, not all dryland farmers and pastoralists, even in the most distant and resource-poor regions, are totally isolated from agricultural markets. Virtually all subsistence households require some regular market income for cash purchases of some agricultural inputs and basic necessities; many farmers and pastoralists provide important cash and export crops. As a result, alterations in market conditions—whether from changes in policies, climatic conditions, R&D innovations, or other factors—do have a significant impact on the livelihoods of rural groups in dryland areas. Understanding their responses to these changing market conditions is a crucial aspect of the dryland management problem.

In many instances, market responses are governed by the complex economic-environmental relationships found in dryland management systems, which can lead to unintended results. For example, the argument that higher slaughter prices in a dryland pastoral economy will, *ceteris paribus*, encourage higher rates of offtake, and so lower levels of grazing, has been questioned (Perrings 1990). Higher slaughter prices can reduce both the real product cost of herd maintenance and the real product cost of carrying capacity, and these have opposite effects on grazing pressure. In situations where the cost of carrying capacity (a grazing fee) is zero, higher slaughter prices will not reduce rangeland degradation. Similarly, a study of Arabic gum production in the Sudan indicates that fluctuations in the real price of gum and its price relative to those of other agricultural crops have had important impacts on farmers' cropping patterns, diversification strategies and decisions to re-plant gum— with important consequences for the Sudan's gum belt (IIED/IES 1990; Barbier 1992a). Even though it is economically profitable *and* environmentally beneficial to grow gum, it is only when these economic incentives are properly dealt with by the government will rehabilitation of the important gum belt of the Sudan take place.

Soil erosion and land degradation are not confined just to drylands and other marginal lands; the problem is pervasive throughout all agricultural systems, degraded forest lands, public and privately owned lands, and large and small holdings in the Third World. Designing appropriate policy responses to control soil erosion and land degradation for all types of cropland in developing countries is again hampered by the data limitations and the lack of microeconomic analyses of farmers' responses to erosion and incentives to adopt conservation measures.

The limited evidence that does exist suggests that relationships—such as the effects of agricultural input and output pricing on farm-level erosion—are complex and difficult to substantiate.[10] Nevertheless, there are some indications that subsidies for non-labor inputs, notably inorganic fertilizers, can artificially reduce the costs of soil erosion to farmers and, on more resource-poor lands, substitute for manure, mulches, and nitrogen-fixing crops that might be more appropriate. On the other hand, the *inaccessibility* of inorganic fertilizers—e.g., shortages caused by rationing cheap fertilizer imports—can actually lead to sub-optimal application and encourage farming practices that actually increase land degradation. Similarly, the relationship between erodibility and profitability of different cropping systems needs to be carefully analyzed, particularly in relation to changing *relative prices* of different crops and changes in *real producer prices and incomes* over time. More complex incentive effects arise from the relationships between erosion and the availability of labor, off-farm employment, population pressure, tenure and access to frontier land, the development of post-harvesting capacity and other complementary infrastructure, and the availability of credit at affordable interest rates.[11]

A tentative conclusion is that there are often strong economic incentives determining farmers' decisions to invest in soil conservation. Farmers will

---

11. See Barbier and Burgess (1992a) for a recent review.
12. For relevant case studies see Barbier (1989a, ch. 7); Barbier (1990b); Barbier and Burgess (1992b); Bojö (1991); Mortimore (1989); Phantumvanit and Panayotou (1990); and Southgate, Sierra, and Brown (1989).

generally not modify their land management practices and farming systems unless it is in their direct economic interest to do so. Such modifications are expensive and may involve risk. The risk may be further exacerbated by the price fluctuations for the various crops found in farming systems (Barbier 1991c). Unless soil erosion is perceived to be a threat to farm profitability, or alternatively, unless changes in land management lead to at least some immediate economic gains, farmers will be less willing to bear these substantial costs. In addition, the more productive or profitable the land use, the more farmers will be able to maintain and invest in better land management and erosion control practices. Higher productivity and returns will also mean that farmers can afford to maintain terraces and other conservation structures and to continue with labor-intensive erosion control measures. On the other hand, poorer farmers dependent on low-return cropping systems, such as maize or cassava, may be aware that soil erosion is reducing productivity but may not be able to afford necessary conservation measures. At the other extreme, farmers with very profitable crops that are extremely erosive, such as temperate vegetables on steep upper volcanic slopes with deep topsoils, may not consider soil conservation measures if their returns do not appear to be affected by soil erosion losses. Unfortunately, we still do not sufficiently understand the economic and social factors determining these incentives for soil conservation in most developing regions.[12]

An equally important challenge for future economic analysis is to examine the causes of large-scale land use changes and resource degradation. Thus clearing forest land for agriculture is thought to be the major cause of tropical deforestation. A number of economic studies have been launched, particularly in Latin America, to analyze the main factors inducing people to settle in and to clear "frontier" forest lands for agriculture.

For example, Binswanger (1989) and Mahar (1989) make the case for the role of subsidies and tax breaks, particularly for cattle ranching, in encouraging land clearing in the Brazilian Amazon. However, more recent analyses by Schneider et al. (1990) and Reis and Margulis (1990) emphasize the role of agricultural rents, population pressures and road building in encouraging small-scale frontier settlement. In the northern Brazilian Amazon, the total road network (paved and unpaved) increased from 6,357 to 28,431 km during 1975–88. A simple correlation between road density and the rate of deforestation shows that as road density increases, the rate of deforestation increases in larger proportions (Reis and Margulis 1990). Schneider et al. (1990) argue that these factors encouraging frontier agriculture ("nutrient mining") far outweigh the more publicized impacts of fiscal incentives for cattle ranching. A statistical analysis by Southgate, Sierra, and Brown (1989) of the causes of tropical deforestation in Ecuador indicates that colonists clear forest land not only in response to demographic pressure but also to

---

13. For recent reviews, see Barbier and Bishop (1993), Barbier and Burgess (1992a), and Southgate (1988).

"capture" agricultural rents and to safeguard their tenuous legal hold on the land.

Although there are an increasing number of case studies examining the factors behind tropical deforestation and agricultural frontier expansion, there have been few attempts to explore these linkages through statistical analysis. One such analysis by Palo, Mery, and Salmi (1987) for 72 tropical forest countries identified a strong link between tropical deforestation and population density, population growth, and increased food production. A study by Capistrano (1990) and Capistrano and Kiker (1990) examined the influence of international and domestic macroeconomic factors on tropical deforestation.[13] The econometric analysis indicates the role of high agricultural export prices in inducing agricultural expansion and forest clearing, as well as the influence of domestic structural adjustment policies, such as exchange rate devaluation and increased debt-servicing ratios. Comparative analysis of 24 Latin American countries also highlights the strong but indirect relationship between population pressure and frontier expansion—increasing numbers of urban consumers raise the demand for domestic production and hence for agricultural land—and the countervailing role of increased agricultural productivity and yield growth in slowing agricultural expansion (Southgate 1991). A statistical analysis by Burgess (1992) supports the hypothesis that population pressure and industrial roundwood production are positively associated with forest clearance in the tropics during the period 1980–85. Economic development, as represented by rising per capita GNP, and improvements in agricultural yields reduce forest clearing. However, the analysis also indicates that countries with relatively small forests to total area are depleting these forests at a high rate.

A recent study in Thailand highlights the complex linkages between agricultural crop prices, the relative returns from different crops, and the demand for land (Phantumvanit and Panayotou 1990). In Thailand, approximately 40% of the increase in cultivated land in recent years can be attributed to converted forest land. The most important factors affecting the demand for cropland, and thus forest conversion, appear to be population growth followed by non-agricultural returns, although agricultural pricing also has a significant influence. Higher aggregate real prices may have a slightly positive influence on the demand for cropland, and thus increased forest clearing; however, this direct effect may be counteracted by the indirect impact of higher agricultural prices on raising the productivity of existing land and increasing the cultivation of previously idle land, thus reducing the demand for new land from forest clearing. Changes in relative prices also influence the demand for

---

14. Capistrano (1990) and Capistrano and Kiker (1990) use changes in timber production forest area as a proxy for total deforestation. Thus their analytical results are more relevant to the deforestation of tropical timber production forests than to overall tropical deforestation. For an in-depth review of the impact of logging relative to other factors influencing tropical deforestation, see Amelung and Diehl (1992), Barbier et al. (1993), and Burgess (1992).

new cropland by affecting the relative profitability of land-saving as opposed to land-extensive cropping systems.[14]

The analysis of land clearing in Thailand also illustrates some of the potential linkages with agricultural input prices. For example, the relative returns to land-saving cropping systems can also be affected by the relative costs of inputs (e.g., fertilizer, seed, credit, irrigation, and agrochemicals) where these differ between land-saving and land-extensive systems. More importantly, investments in land productivity will also be affected by the costs of these inputs. Thus the Thailand case study suggests that lower input prices generally would reduce the demand for new cropland and forest conversion both directly and indirectly by making previously idle land more attractive to cultivate. Changes in relative input prices could also make a difference on agricultural extensification by affecting the relative returns to land-saving cropping systems.

Clearly, both microeconomic analysis of individuals' behavior and responses to environmental degradation, and macroeconomic analysis of the broader economic-environmental linkages affecting degradation will be required if coherent and effective public policies for natural resource management in developing countries are to be designed and implemented.

CONCLUSION

This chapter has sought to provide an overview of the role of natural capital in economic development. This role is the fundamental concern of a new discipline that has recently emerged and is growing rapidly—the economics of environment and development. As this discussion has highlighted, progress to date in this discipline has already produced one firm conclusion—efficient and sustainable management of natural resources requires improved policies and accounting procedures to reflect the actual value of natural capital in the development process. Further progress will require advances in the five areas described in the introduction.

However, in demonstrating the potential contribution of an economics of environment and development, this chapter has subsequently provided evidence that current rates of natural resource depletion are not promoting sustainable development, and the transformation of natural into human and/or physical capital does not appear to be efficient. Progress in this area will have to address two fundamental underlying causes: how current policy distortions and market failures allow resource users to avoid the social opportunity cost of environmental degradation and how environment-poverty linkages influence patterns of resource utilization.

---

15. The authors also examined the effects of relative price changes on land productivity and the cultivation of previously idle land, but found this relationship more difficult to estimate in such an aggregate analysis.

Tackling these issues will be the key challenge facing the economics of environment and development for many years to come. What we perceive to-day to be the "improved policies and accounting procedures" necessary for efficient and sustainable management of natural resources for development may not be what is required for tomorrow. This is particularly true given that economics must increasingly rely on improved ecological understanding of the basic environmental functions performed by natural and even human-modified systems before we can appreciate the true contribution of these functions to development. An economics of environment and development must rest on sound ecological underpinnings to be meaningful in its policy prescriptions.

Although interest in these topics is finally growing, especially in light of the U.N. Conference on Environment and Development, what is needed most is "political will." Thus, in formulating a strategy for integrating environment and development at the policy, planning, and management levels, the UNCED concluded: "The responsibility for bringing about changes lies with Governments in partnership with the private sector and local authorities, and in collaboration with national, regional, and international organizations, including in particular UNEP, UNDP, and the World Bank" (UNCED 1992a). Ultimately, however, the clients of development are not policymakers or managers, but all people comprising present and future generations.

REFERENCES

Ahmad, Y., S. El Serafy, and E. Lutz. 1989. Environmental Accounting for Sustainable Development. Paper presented at a UNEP-World Bank Symposium. Washington, DC: The World Bank.

Amelung, T., and M. Diehl. 1992. Deforestation of Tropical Rain Forests: Economic Causes and Impact on Development. Kiel, Germany: Institute of World Economics.

Anderson, D. 1987. The Economics of Afforestation. Baltimore: Johns Hopkins Univ. Press.

———. 1990. Environmental Policy and the Public Revenue in Developing Countries. Environment Working Paper No. 36. Washington, DC: The World Bank.

Barbier, E. B. 1989a. Economics, Natural-Resource Scarcity and Development: Conventional and Alternative Views. London: Earthscan.

———. 1989b. Sustainable agriculture on marginal land: a policy framework. *Environment* 31 (9): 12–17.

———. 1990a. Alternative approaches to economic-environmental interactions. *Ecological Economics* 2: 7–26.

———. 1990b. The farm-level economics of soil erosion: the uplands of Java. *Land Economics* 66 (9): 199–211.

———. 1991a. The Economic Value of Ecosystems, 2: Tropical Forests. LEEC Gatekeeper GK 91-01. London: London Environmental Economics Centre.

———. 1991b. Natural resource degradation: policy, economics and management. In Development Research: The Environmental Challenge, ed. J. T. Winpenny. London: Overseas Development Institute.

———. 1991c. The Role of Smallholder Producer Prices in Land Degradation: The Case of Malawi. LEEC Discussion Paper DP 91-05. London: London Environmental Economics Centre.

————. 1992a. Rehabilitating gum arabic systems in Sudan: economic and environmental implications. *Environmental and Resource Economics* 2: 19–36.

————. 1992b. Tropical deforestation. In Economics for the Wilds: Wildlife, Wildlands, Diversity and Development, eds. T. M. Swanson and E. B. Barbier. London: Earthscan.

————. 1993. Valuing tropical wetland benefits: economic methodologies and applications. *Geographic Journal* Part 1 (59): 22–32.

Barbier, E. B., W. M. Adams, and K. Kimmage. 1991. Economic Valuation of Wetland Benefits: The Hadejia-Jama'are Floodplain, Nigeria. LEEC Discussion Paper 91-02. London: London Environmental Economics Centre.

Barbier, E. B., and J. T. Bishop. 1993. In press. Economic and social values affecting soil and water conservation in developing countries. *Journal of Soil and Water Conservation*.

Barbier, E. B., and J. C. Burgess. 1992a. Agricultural Pricing and Environmental Degradation. Policy Research Papers. World Development Report. Washington, DC: The World Bank.

————. 1992b. Malawi—Land Degradation in Agriculture. Divisional Working Paper No. 1992-37. Washington, DC: The World Bank.

Barbier, E. B., J. C. Burgess, J. T. Bishop, B. A. Aylward, and C. Bann. 1993. The Economic Linkages Between the International Trade in Topical Timber Products and the Sustainable Management of Tropical Forests. Final Report: ITTO Activity PCM(XI)/4. Yokohama: The International Tropical Timber Organization.

Barbier, E. B., and A. Markandya. 1990. The conditions for achieving environmentally sustainable development. *European Economic Review* 34: 659–69.

Binswanger, H. P. 1980. Attitudes toward risk: experimental measurement in rural India. *American Journal of Agricultural Economics* 90: 395–407.

————. 1989. Brazilian Policies that Encourage Deforestation in the Amazon. Environment Department Working Paper No. 16. Washington, DC: The World Bank.

Bishop, J. T. 1990. The Costs of Soil Erosion in Malawi. Malawi Country Operations Division. Washington, DC: The World Bank.

Bishop, J. T., and J. Allen. 1989. The On-Site Costs of Soil Erosion in Mali. Environment Department Working Paper No. 21. Washington, DC: The World Bank.

Bojö, J. 1991. The Economics of Land Degradation: Theory and Applications to Lesotho. The Economic Research Institute. Stockholm: Stockholm School of Economics.

Bojö, J., K-G. Mäler, and L. Unemo. 1990. Environment and Development: An Economic Approach. Dordrecht: Kluwer .

Burgess, J. C. 1992. Economic Analysis of the Causes of Tropical Deforestation. LEEC Discussion Paper DP92-03. London: The London Environmental Economics Centre.

Cairncross, S., J. E. Hardoy, and D. Satterthwaite. 1990. The urban context. In The Poor Die Young: Housing and Health in Third World Cities, eds. J. E. Hardoy, S. Cairncross and D. Satterthwaite. London: Earthscan.

Capistrano, A. D. 1990. Macroeconomic Influences on Tropical Forest Depletion: A Cross-Country Analysis. Ph.D. dissertation. Food and Resource Economics Dept. Miami: Univ. of Florida.

Capistrano, A. D., and C. F. Kiker. 1990. Global Economic Influences on Tropical Closed Broadleaved Forest Depletion, 1967–85. Mimeo. Food and Resource Economics Dept. Miami: Univ. of Florida.

Chambers, R. 1987. Sustainable Livelihoods, Environment and Development: Putting Poor Rural People First. Discussion Paper 240. Institute of Development Studies. Brighton: Univ. of Sussex.

Clark, C. 1976. Mathematical Bioeconomics: The Optimal Management of Renewable Resources. New York: Wiley.

Common, M., and C. Perrings. 1992. Towards an ecological economics of sustainability. *Ecological Economics* 6: 7–34.

Conway, G. R. 1985. Agroecosystem analysis. *Agricultural Administration* 20: 3–55.

Conway, G. R., and E. B. Barbier. 1990. After the Green Revolution: Sustainable Agriculture for Development. London: Earthscan.

Dasgupta, P. S. 1992. Population, resources and poverty. *Ambio* 21 (1): 95–101.

Dixon, J. A., R. A. Carpenter, L. A. Fallon, P. B. Sherman, and S. Manipomoke. 1988. Economic Analysis of the Environmental Impacts of Development Projects. London: Earthscan.

Dixon, J. A., D. James, and P. B. Sherman. 1989. The Economics of Dryland Management. London: Earthscan.

————., eds. 1989. Dryland Management: Economic Case Studies. London: Earthscan.

Dixon, J. A., and P. B. Sherman. 1990. Economics of Protected Areas: Benefits and Costs. London: Island Press; Earthscan.

Easter, K. W., J. A. Dixon, and M. M. Hufschmidt. 1986. Watershed Resources Management: An Integrated Framework with Studies from Asia and the Pacific. Boulder, CO: Westview Press.

FAO. 1990. Sustainable development and natural resource management. In The State of Food and Agriculture, 1989, Part 3. Rome: FAO.

Foy, G., and H. E. Daly. 1989. Allocation, Distribution and Scale as Determinants of Environmental Degradation: Case Studies of Haiti, El Salvador and Costa Rica. Environment Department Working Paper No. 19. Washington, DC: The World Bank.

Gregersen, H. M., K. N. Brooks, J. A. Dixon, and L. S. Hamilton. 1987. Guidelines for Economic Appraisal of Watershed Management Projects. Rome: FAO.

Hardoy, J. E., S. Cairncross, and D. Satterthwaite, eds. 1990. The Poor Die Young: Housing and Health in Third World Cities. London: Earthscan.

Hardoy, J. E., D. Mitlin, and D. Satterthwaite. 1992. The future city. In Policies for a Small Planet, ed. J. Holmberg. London: Earthscan.

Hardoy, J. E., and D. Satterthwaite. 1989. Squatter Citizen: Life in the Urban Third World. London: Earthscan.

Hartwick, J. M. 1977. Intergenerational equity and the investing of rents from exhaustible resources. *American Economic Review* 66: 972–4.

Holling, C. S. 1973. Resilience and the stability of ecological systems. *Annual Review of Ecological Systems* 4: 1–24.

IIED/IES. 1990. Gum Arabic Rehabilitation Project in the Republic of Sudan: Stage I Report. London: IIED.

Jagannathan, N. V. 1989. Poverty, Public Policies and the Environment. Environment Working Paper No. 24. Washington, DC: The World Bank.

Kates, R. W. 1990. Hunger, Poverty and the Environment. Paper presented at the Distinguished Speaker Series, Center for Advanced Study of International Development, Michigan State Univ., Lansing, May 6.

Lallement, D. 1990. Burkina Faso: Economic Issues in Renewable Natural Resource Management. Agricultural Operations, Sahelian Department, Africa Region. Washington, DC: The World Bank.

Leonard, H. J., M. Yudelman, J. D. Stryker, J. O. Browder, A. J. De Boer, T. Campbell, and A. Jolly. 1989. Environment and the Poor: Development Strategies for a Common Agenda. New Brunswick: Transaction Books.

Lipton, M. 1983. Labour and Poverty. World Bank Staff Working Papers No. 616. Washington, DC: The World Bank.

————. 1988. The Poor and the Poorest: Some Interim Findings. World Bank Discussion Paper No. 25. Washington, DC: The World Bank.

Magrath, W. B., and P. Arens. 1987. The Costs of Soil Erosion on Java—A Natural Resource Accounting Approach. Washington, DC: World Resources Institute.

Mahar, D. 1989. Government Policies and Deforestation in Brazil's Amazon Region. Washington, DC: The World Bank.

Mortimore, M. 1989. The Causes, Nature and Rate of Soil Degradation in the Northernmost States of Nigeria and an Assessment of the Role of Fertilizer in Counteracting the Processes of Degradation. Environment Department Working Paper No. 17. Washington, DC: The World Bank.

Nelson, R. 1988. Dryland Management: The Desertification Problem. Environment Department Working Paper No. 8. Washington, DC: The World Bank.

Palo, M., G. Mery, and J. Salmi. 1987. Deforestation in the tropics: pilot scenarios based on quantitative analysis. In Deforestation or Development in the Third World, eds. M. Palo and J. Salmi. Division of Social Economics of Forestry. Helsinki: Finnish Forestry Research.

Paris, R., and I. Ruzicka. 1991. Barking Up the Wrong Tree: The Role of Rent Appropriation in Tropical Forest Management. Environment Office Discussion Paper. Manila: Asian Development Bank.

Pearce, D. W. 1991. Population growth. In Blueprint 2: The Greening of the World Economy, ed. D. W. Pearce. London: Earthscan.

Pearce, D. W., E. B. Barbier, and A. Markandya. 1990. Sustainable Development: Economics and Environment in the Third World. London: Edward Elgar.

Pearce, D. W., and J. J. Warford. 1992. World Without End: Economics, Environment and Sustainable Development. Oxford: Oxford Univ. Press for The World Bank.

Pender, J. L., and T. S. Walker. 1990. Experimental Measures of Time Preference in Rural India. Draft paper, Food Research Institute. Stanford: Stanford Univ. and ICRISAT.

Perrings, C. 1990. Stress, Shock and the Sustainability of Optimal Resource Utilization in a Stochastic Environment. LEEC Discussion Paper DP90-04. London: London Environmental Economics Centre.

Peters, C., A. Gentry, and R. Mendelsohn. 1989. Valuation of an Amazonian rainforest. *Nature* 339: 655–56.

Phantumvanit, D., and T. Panayotou. 1990. Natural Resources for a Sustainable Future: Spreading the Benefits. Prepared for the 1990 TDRI End-Year Conference on Industrializing Thailand and its Impact on the Environment, Chon Buri, Thailand, December 8–9.

Pretty, J., I. Guijt, I. Scoones, and J. Thompson. 1992. Regenerating agriculture: the agroecology of low-external input and community-based development. In Policies for a Small Planet, ed. J. Holmberg. London: Earthscan.

Reis, E., and S. Margulis. 1990. Options for Slowing Amazon Jungle-Clearing. Paper presented at the Conference on Economic Policy Responses to Global Warming, Rome, September 5–7

Repetto, R. 1988. Economic Policy Reform for Natural Resource Conservation. Environment Department Working Paper No. 4. Washington, DC: The World Bank.

————. 1987. Population, resources, environment: an uncertain future. *Population Bulletin* 42 (2): special issue.

Repetto, R., and M. Gillis, eds. 1988. Public Policies and the Misuse of Forest Resources. Cambridge: Cambridge Univ. Press.

Repetto, R., and T. Holmes. 1983. The role of population in resource depletion in developing countries. *Population and Development Review* 9 (4): 609–32.

Repetto, R., W. B. Magrath, M. Wells, C. Beer, and F. Rossini. 1989. Wasting Assets: Natural Resources in the National Income Accounts. Washington, DC: World Resources Institute.

Richards, P. 1985. Indigenous Agricultural Revolution. London: Hutchinson.

Ruitenbeek, H. J. 1989. Social Cost-Benefit Analysis of the Korup Project, Cameroon. Prepared for the World Wide Fund for Nature and the Republic of Cameroon. London: World Wide Fund.

———. 1991. Mangrove Management: An Economic Analysis of Management Options with a Focus on Bintuni Bay, Irian Jaya. Prepared for EMDI/KLH, Jakarta, Indonesia.

Schneider, R., J. McKenna, C. Dejou, J. Butler, and R. Barrows. 1990. Brazil: An Economic Analysis of Environmental Problems in the Amazon. Washington, DC: The World Bank.

Scoones, I. 1993. Economic and Ecological Carrying Capacity: Applications to Pastoral Systems in Zimbabwe. In Economics and Ecology: Contributions to Sustainable Development, ed. E. B. Barbier. London: Chapman and Hall.

Smith, V. L. 1977. Control theory applied to natural and environmental resources: an exposition. *Journal of Environmental Economics and Management* 4: 1–24.

Solow, R. M. 1974. Intergenerational equity and exhaustible resources. *Review of Economic Studies. Symposium on the Economics of Exhaustible Resources*, 29–46.

———. 1986. On the Intertemporal Allocation of Natural Resources. *Scandinavian Journal of Economics* 88 (1): 141–9.

Southgate, D. 1988. The Economics of Land Degradation in the Third World. Environment Department Working Paper No. 2. Washington, DC: The World Bank.

———. 1991. Tropical Deforestation and Agriculture Development in Latin America. LEEC Discussion Paper 91-01. London: London Environmental Economics Centre.

Southgate, D., R. Sierra, and L. Brown. 1989. The Causes of Tropical Deforestation in Ecuador: A Statistical Analysis. LEEC Discussion Paper DP89-09. London: London Environmental Economics Centre.

Swanson, T. M., and E. B. Barbier. 1992. Economics for the Wilds: Wildlife, Wildlands, Diversity and Development. London: Earthscan.

TSC/WRI. 1991. Accounts Overdue: Natural Resource Depreciation in Costa Rica. Washington, DC: World Resources Institute.

UNCED Secretariat. 1992a. Integrating environment and development in decision-making. Ch. 8 in Agenda 21, Part 1. New York: UNCED.

———. 1992b. Integration of Environment and Development in Decision Making. Report to the Preparatory Committee for UNCED. New York: UNCED.

Wilson, K. B. 1988. Indigenous Conservation in Zimbabwe: Soil Erosion, Land-use Planning and Rural Life. Paper submitted to the Panel Session Conservation and Rural People, African Studies Association of U.K. Conference, September 14.

18 CAN WE JUSTIFY SUSTAINABILITY? NEW CHALLENGES FACING ECOLOGICAL ECONOMICS[1]

Dennis M. King
Associate Director
Maryland International Institute for Ecological Economics
University of Maryland
Center for Environmental and Estuarine Studies
Solomons, Maryland 20688 USA

ABSTRACT

The important challenges facing ecological economics have moved from the realm of the ideological to the world of the practical. World leaders are beginning to understand that investing in natural capital is a prerequisite for sustainability. What decision-makers need now is more information about ecological economic linkages so they can reconcile the long-term goal of sustainability with near-term economic needs.

In some circumstances the conventional method of evaluating production and investment decisions—benefit-cost analysis (BCA)—can be expanded to include indirect measures of costs and benefits that result from ecological economic linkages; the results may lead to more sustainable near-term resource management decisions. In other cases the use of benefit cost analysis is impractical, and sustainable near-term production and investment decisions need to be justified on the basis of other criteria, such as safe minimum standards.

In this chapter an expanded benefit-cost framework is developed using six discrete pathways of potential project impacts. Two originate with *withdrawals* from and *emissions* to nature, and generate measures of costs and benefits that are based on an expanded view of *facts* about ecological economic linkages. Two other pathways develop market-based and non-market estimates of *costs* and *benefits* associated with ecological economic linkages, and are responsible for assigning *values*. The last two pathways deal with *distributional impacts* and *other socio-economic considerations* that reflect collective decisions about *equity* and *balance*.

1. Paper presented at the Second Conference of the International Society for Ecological Economics (ISEE), "Investing in Natural Capital, Stockholm," Sweden, August 3–6, 1992.

Where there is too much uncertainty about *facts,* or too much disagreement about *values, equity,* or *balance,* decision-making on the basis of benefit-cost analysis may be impractical. In such cases the level of protection granted to a particular natural resource and the amount of public intervention to control market forces that could otherwise result in further degradation of that resource can be based on safe minimum standards. A second evaluation framework is developed here that uses measures of ecological *importance* and *reversibility* to assign safe minimum standards.

INTRODUCTION

The Role of Economists

The fundamental question in all applied economic analysis is whether one state of the world is better than another. Such a question is difficult to answer even when the evaluation criteria are based on clearly defined community goals and values. When the analysis is expected to consider alternative states of the world with respect to the shared goals of mankind, species survival, intergenerational fairness, and religious, philosophical, moral, or humanistic notions of value, it is clear that economic analysis by itself is not enough. It is not surprising, therefore, that very few economists are ever asked for their professional opinion about how the world economy or the foundations of western civilization should be altered to deal with emerging ecological problems. Like most scientists, engineers, and other professionals, economists toil at less important tasks—measuring how changing states of the world might affect the clearly defined economic interests of specific industrial sectors, traders, businesses, nations, and so on. In fact, economic logic is hardly ever applied at a global scale and when it is, it has very little influence on anything, including the standards that most practicing economists are required by their jobs to adhere to—the rules for measuring cost and benefits. With few exceptions, the analytical focus and the indicators of performance used by economists are based on the decision criteria that are considered important by the employers of economists. Investors look at financial rates of return and payback periods, businessmen look at short-term sales and profits, governments look at GDP growth and employment, and so on. Practicing economists either adopt measures of costs and benefits that reflect these performance criteria, or they look for some other form of employment.

Expanding the Role of Economists

If economists are to incorporate the value of natural capital and the implications of losing natural capital into their analyses, they cannot do it alone. They need information that describes how changing stocks of natural capital can be expected to affect the streams of costs and benefits that are important to decision-makers. This requires facts, or at least well-articulated impressions, about economically meaningful bio-physical constraints and how ecological linkages can affect economic performance. At the present time conventional economic analysis does ignore many ecological costs. It is wrong, however, to infer that economists fail to comprehend that the economic system is a subset of the ecological system or that they adhere to an antiquated "neo-classical paradigm." The problem in most cases is that they cannot determine how ecological economic interactions should affect the measures of economic costs and benefits that are required of them in their work. An important long-term goal of ecological economics should be to convince industry and government decision-makers to invest in natural capital and expand the economic criteria and time horizons they use to make production and investment decisions. But an enormous amount of near-term good can be done by economists and scientists if they work together to illustrate how ecological economic linkages affect the measures of costs and benefits that industry and government leaders use to make decisions *now*.

This chapter addresses the direct application of ecological economic principles in project and policy analysis. The attention given to "sustainability" will probably have very little near-term impact on industry and government decisions. Production and investment alternatives will continue to be assessed on the basis of conventional (anthropomorphic) measures of benefits and costs and will continue to be selected on the basis of efficiency criteria and distributional effects (who gains and who loses), with, hopefully, passing consideration of "non-economic" criteria (everything else that is worthwhile). The chapter is intended to show scientists how their work can influence the range of benefits and costs that are used in economic analysis and to illustrate to economists how the results of scientific research can expand the range of conventional costs and benefits.

THE FOCUS OF APPLIED ECOLOGICAL ECONOMICS

Scientists deal with issues of fact (what is), and philosophers deal with issues of value (what should be). Economists, whether they choose to or not, must work with both. A sign of high quality economic analysis has always been that *facts* and *values* are separated as much as possible in empirical analysis and reporting. Figure 18.1 shows a simple decision matrix that helps explain why this separation becomes so difficult when economic analysis is focused on complex ecological economic problems.

	Values:	
	Agreement	Disagreement
Facts: Agreement	Consensus	Negotiation
Facts: Disagreement	Research	Political Leadership or Endless Research

Figure 18.1. Focus of applied research.

The established routine for initiating applied economic analysis is to identify stakeholders, find out where they disagree, separate factual questions from questions of value, and focus research on those factual or value-based questions that seem most important. In the simplest of the four cases shown in Figure 18.1, there is general agreement about the *facts* (e.g., the high risk of peregrine falcon extinction) and *values* (e.g., the peregrine falcon should be preserved at any reasonable cost). There is *consensus*, so the role of the economic analysis is simple—find the most cost-effective conservation strategy. Questions may arise about the distributional impacts of implementing the least-cost strategy—who should pay the cost—but in general, the focus of the economic analysis can be practical and non-controversial.

In more typical cases there may be either agreement about *facts* (e.g., declining water quality is threatening fish), but not *values* (fishery-related benefits are worth the cost of improving water quality). Or there may be agreement about *values* (fish resources are definitely worth protecting), but not *facts* (do fish need protection from over-fishing, from habitat degradation, or from poor water quality?). Where there is agreement about *facts*, but not *values*, the best chance at a solution is probably *negotiation* among "stakeholders." Economic analysis can be developed to support such efforts without becoming involved in underlying scientific issues. Where there is agreement about *values*, but not *facts*, a lasting solution will probably require more *research*, and the economic analysis may need to await the results of that work to be credible. Consider the issue of global warming, for example—until there is agreement about whether it is occurring and how fast, it will be difficult to determine the cost of responding to it.

Many modern environmental problems are so complicated and involve so many stakeholders that they fall in the more contentious category where there is disagreement about *facts* **and** *values*. As scientists begin to understand and describe the intricate web of ecological economic relationships that link natural and human systems, they reveal how few agreed-upon *facts* there really are. At the same time they demonstrate the all-pervasive nature of environmental "externalities" (e.g., consider the effects of acid rain, global warming, wetland loss) and how many different *values* are at stake in resource management decisions. In a growing number of cases, in other words, there are so many unresolved questions about *facts* and so many different *values* involved that it is difficult to focus applied economic analysis in ways that make sense.

EXPANDED BENEFIT COST ANALYSIS

Most economists agree *in principle* that BCA should be all inclusive—that it should include intangible non-marketed benefits and costs as well as market-based measures and should account for benefits and costs regardless of which social or economic group is affected.[2] *In practice,* however, time and budget constraints always limit the range of costs and benefits that can be considered in BCA. It is common for BCA to emphasize directly measurable benefits and costs that are important to decision-makers at the expense of benefits and costs that are more diffuse and difficult to measure or that may be important primarily to people other than the decision-maker. These decisions about which benefits and costs to include in the BCA significantly affect the results.

Reports on the results of a BCA usually begin or end with caveats that identify classes of costs and benefits that were excluded or undercounted and refer the reader to supplemental documents that describe "non-economic" project impacts. The importance of the BCA compared to these supplemental project impact documents provides a strong incentive to express as many important environmental and social impacts as possible in economic terms that can be used directly in the BCA. Recent advances in non-market valuation techniques, for example, permit changes in recreational opportunities, health, and ecological risks, aesthetics, and other classes of non-marketed benefits and costs to be included in BCA (Kopp and Smith 1993). To apply these techniques, however, the types of non-marketed products and services that are associated with various resources must be known—analysts need *facts* before they can assign *values.*

Several excellent references cover the fundamentals of BCA and describe how BCA should be applied to questions involving the environment and economic development (Bojo et al. 1992). Recent publications include sections that describe non-market valuation techniques, explain how to account for internal and external costs and benefits and how to select an appropriate discount rate. The underlying premise in all of this work, however, is that the products and services gained and lost over time as a result of a project or policy are known. This literature provides guidance for assigning and adding up values for known environmental products and services, not for conducting research into indirect streams of benefits and costs that might result from ecological economic linkages. This is a serious limitation because some of the most catastrophic environmental and economic impacts from past projects and policies were due to ecological economic linkages that could have been foreseen, but were never addressed in BCA that were too narrowly focused.

Figure 18.2 presents an expanded view of the research that is required for a comprehensive BCA focusing on fact finding about complex ecological economic linkages as well as assigning values. In this framework the consequences of any proposed project, program, or policy are shown to follow six discrete pathways. Pathways 1 and 2 trace *withdrawals* from and *emissions* to nature and deal mostly with *facts.* Pathways 3 and 4 include direct measures of *market-based costs and benefits* and indirect measures of *non-marketed*

---

2. The guidelines for conducting BCA in the United States, for example, are provided in Executive Order 12291, which describes a full array of use and non-use values that should be considered.

costs and benefits and deal mostly with *values*. Pathways 5 and 6 describe *distributional impacts* (who gains and loses) and effects on *socio-economic goals* other than efficiency, which are important supplements to conventional BCA.

Each frame in Figure 18.2 represents an area of applied research where specialists already have analytical models and understand basic cause and effect relationships. The difficulty in expanding BCA to include this work is that in applied science and in applied economics, there are few incentives for researchers working on one problem to provide results in formats that researchers working on other problems might require to contribute to a comprehensive accounting of costs and benefits. An agronomist working in Frame 1, for example, may establish that an irrigation project will result in the erosion of topsoil (a withdrawal), which a hydrologist working in Frame 4 would interpret as sedimentation reaching an adjacent water body (an emission). Without an established procedure for tracing the subsequent effects of sedimentation on fish populations and fishermen (Frames 2 and 3), on aquatic ecosystems (Frame 8), on measures of risk and value (Frames 12 and 13) and, ultimately, on estimates of economic damages (Frame 11), the full costs associated with sedimentation cannot be included in the BCA of the irrigation project.

These pathways of indirect and induced costs are excluded from most applied BCA because of time and budget constraints, but they often become all too apparent after the project is in place. A dedicated interdisciplinary effort to facilitate the flow of information from one research module to another could dramatically reduce the cost of tracing such impacts and result in a more comprehensive accounting of costs and benefits before a project is undertaken. Initially this effort would involve: 1) developing measurement and reporting standards, 2) establishing protocols for formatting data and communicating data quality, 3) estimating baseline cause-effect relationships, 4) determining guidelines for regional and site-specific adjustment, and in some cases, 5) establishing general purpose sets of "shadow prices" that can be assigned to specific types of emissions and withdrawals.

A New Focus

It is not possible to discuss here the specific ecological economic linkages that need to be addressed for a full accounting of benefits and costs, but two features of Figure 18.2 deserve special attention. First is the central role of the ecosystem models (Frame 8) in converting estimates of withdrawals and emissions into estimates of health and ecological risks and economic damages. Second is the enormous empirical burden that the expanded framework imposes on non-market valuation methods that are expected to convert estimates of health and ecological risks and environmental damage into money-based measures of costs and benefits. From the perspective of project analysis, the relative importance placed on these two research areas may be the most important difference between the emerging field of ecological economics and the established field of environmental economics.

In Figure 18.2 ecological economic research emphasizes flows into and out of the frame representing ecosystem models (Frame 8). Environmental economics, on the other hand, has tended to focus on environmental problems as sources of specific "externalities" that can be traced and assigned dollar values. Environmental economic research has emphasized population or species level impacts (Frames 2 and 3) and economic values associated with them (Frame 6), but has not focused on ecosystem level impacts. Two illustrations, one at the species/population level and one at the ecosystem level, will demonstrate why this difference in focus has such a significant effect on the outcome of BCA and why more integrated ecological economic analysis is needed.

Illustration #1: Chesapeake Bay Oysters

Using *conventional economics,* the value of oysters in the Chesapeake Bay at any given time is based on their harvested value to fisherman, processors, or consumers as measured by the income, employment, and sales revenues associated with the oyster harvest and associated multiplier effects. For example, ex-vessel value of the Chesapeake Bay oyster harvest in 1992 was $2.5 million, which after processing, generated business sales of about $7.5 million household income of about $13,125,000, and federal and state taxes of $4,593,750 and accounted for approximately 500 U.S. jobs.[3]

Using *environmental economics,* the treatment of oyster values would be expanded to include recreational as well as commercial fisheries, and might take account of "non-use" values attributed to oysters by people who do not currently fish or eat oysters but may wish to do so sometime in the future (option value), and by people who like to know that there are oysters in the bay (existence value) or that their descendants will be able to enjoy oysters (bequest value). Extending the notion of value in this way requires a significant amount of economic research involving household surveys of use and non-use values, but the focus would remain on the *direct* contribution of oysters to the satisfaction of revealed or expressed human needs.

An *ecological economics* analysis would take a much broader focus and consider the role of the oyster population in the overall ecosystem of the bay and what that is worth. The oyster population of the bay, before it was targeted by commercial fishermen, filtered all the water in the bay every three or four days, removing excess nutrients and maintaining extremely high water quality and unmatched biological productivity and diversity. The current oyster population of the bay is 1% of this historical level, takes 300–400 days

3. Preliminary figures based on personal communication with Maryland State Department of Natural Resources.

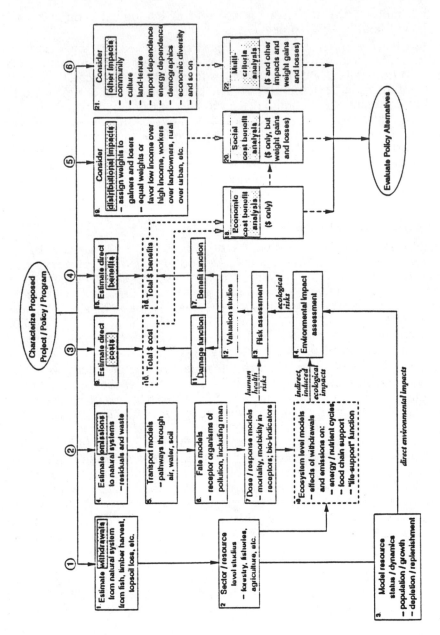

Figure 18.2. An expanded view of benefit cost analysis.

to filter the same amount of water, and cannot keep up with the increased nutrient load from expanding agricultural and residential development in the watershed. An ecological economic analysis of oyster values might focus on the resulting build up of nutrients in the bay that lead to frequent algae blooms and localized fish kills, or the multi-million dollar federal and state programs attempting to clean nutrients from the bay, or the cost of restrictions on local agriculture and commercial and residential development to control nutrient flows to the bay. This broader focus on oyster value might also include an evaluation of how excess nutrients in the bay have reduced light penetration to submerged aquatic vegetation resulting in a decline in critical habitat for finfish. The resulting decline and recent closure of important rockfish and bluefish fisheries in the bay, and the loss of associated jobs, incomes, sales, and tax revenues in those fisheries might also be included.

Ecological economic analysis might show that the real economic value of oysters—their highest and best economic use—is in their natural role as "the kidneys" of the Chesapeake Bay and not as a temporary source of direct income or recreational enjoyment for fishermen. The indirect and induced economic losses experienced by real estate owners and developers, agricultural interests, recreationalists, fishermen, and others as a result of the decline in Chesapeake Bay oysters are real. But, because there are so few tools of applied ecological economics, simple population and sector level impacts associated specifically with oysters and oyster fisheries have provided the only basis for analyzing the economics of managing this resource.

Illustration #2: Chesapeake Bay Wetlands

Using *conventional economic analysis,* the market value of a hectare of tidal wetland in the Chesapeake Bay area, with no restrictions on development, is approximately $10,000–$20,000. Understandably, the question posed most often by business and civic leaders when they are asked to support local wetland protection initiatives is whether the value of a hectare of undeveloped tidal wetlands is greater than or less than $10,000–$20,000. Environmental economists have answered parts of this question by measuring the value of undeveloped wetlands with respect to specific functions such as fish or waterfowl habitat, storm surge and wave protection, and aesthetics.[4] None of this work, however, has been successful at assigning dollar values to the full range of wetland functions listed in Table 18.1. To understand why, consider the flow of analysis described in Figure 18.3, which outlines only part of the work required to assign values to one relatively obscure wetland function—sediment trapping.

As Figure 18.3 illustrates, the principle source of economic values associated with the sediment trapping function of wetlands is reduced sedimentation

---

4. For a recent review of wetland valuation studies, see Scodari (1993).

Table 18.1. Wetland Functions and Values

Wetland Function	Types of Value
Groundwater Recharge/Discharge	Maintain healthy drinking water
Floodwater Storage Conveyance/Desynchronization	Reduces soil erosion, property damage
Shoreline Anchoring	Protects beaches, property, ecosystem
Storm Wave/surge Protection	Reduces erosion, property damage
Sediment trapping	Maintains aquatic ecosystems; reduces dredging requirements; maintains hydropower
Pollution Assimilation	Reduces treatment costs, improves public health
Nutrient Retention/Cycling	Maintains nitrogen balance
Fishery Habitat	Better commercial/recreational fishing; lower seafood prices; improved international balance of trade
Waterfowl Habitat	Better hunting, birdwatching, etc.
Habitat for Fur-bearers	Improved commercial and recreational opportunities
Food-chain Support	Off-site benefits to freshwater anadromous, marine fish, etc.
Agriculture - low intensity grazing	High productivity, low farm costs
Energy (peat, etc.)	Major subsistence energy source with some commercial value
Natural Products-timber, hey, etc.	Supply wholesale/retail markets
Microclimate Regulation	General life support; ill-defined economic linkages
Global Climate Regulation	General life support; ill-defined economic linkages
Carbon Cycling	General life support; significant but unknown economic linkages
Storehouse of Biodiversity	Direct, indirect and serendipity value of scientific, medical discoveries, genetic pools, seed banks
Active Recreation	Boating, swimming, etc.
Passive Recreation	Sight-seeing, birdwatching, etc.
Natural Laboratory Classroom	In-field research/teaching - kindergarten through adult
General Aesthetics	Open space, natural beauty, spiritual enrichment

Source (King 1991).

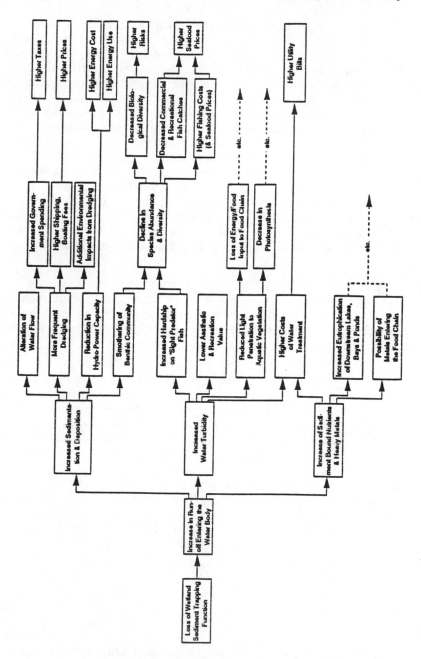

Figure 18.3. One pathways of wetland benefits: Sediment trapping.

of adjacent water bodies which generates two different kinds of beneficial impacts: (1) *reduced turbidity* in the receiving water, which increases overall water quality, biological productivity, and fishery-related benefits, and (2) *reduced sedimentation* of the water body, which means less routine dredging for navigation, higher productivity at hydropower facilities, fewer erratic shifts in stream flows, and more consistent patterns of storm water runoff that protects property, other wetlands, and coastal ecosystems. These cascading ecological economic impacts generate benefits to consumers, businesses, taxpayers, landowners, and recreationalists. Similar streams of indirect and induced benefits can be associated with other wetland functions listed in Table 18.1. In the absence of any generally accepted framework for conducting ecological economic research—for tracing wetland functions and assigning values to them—the time and money required for any researcher to incorporate the full range of wetland-related benefits in any single study has been prohibitive. The general approach to assigning *in situ* values to oyster and wetlands is the same, the difference is in the complexity of the ecological economic linkages.

Some Significant Obstacles

It is important to push for a full accounting of costs and benefits in BCA and to develop frameworks for tracing and measuring as many ecological economic impacts as possible. However, there are three potentially insurmountable problems that may make expanded BCA unworkable in any given situation.

- *First* is the cost in time and resources of conducting enough research to identify, trace, and measure the impacts of a proposed project on costs and benefits associated with the functions of a complex ecosystem. The transport and fate of a pollutant through air or water, the likely exposure of a receptor organism, dose-response relationships, changing ecosystem dependencies, and patterns of ecological succession and ascendancy, subsequent levels of resource abundance and availability, and associated changes in economic values are all important topics that require expensive and time-consuming research to document effectively. Associating any observed historical change in these variables with particular policies, programs, or projects and using them to forecast the future requires even more detailed research. Reliable general frameworks for integrating scientific research and assigning values can only go so far in controlling these costs.

- *Second* is the questionable validity and limited usefulness of non-market valuation methods; most of these methods are still in the development stage. The most popular one, and the only one capable of monetizing "non-use" values, is *Contingent Valuation* or CV analysis, and it is still more-or-less experimental.[5] CV relies on the results of household surveys of willingness-to-pay (WTP) to protect or willingness-to-accept payment (WTA) to give up benefits. It is the subject of ongoing controversy even when it is applied to familiar topics such as

---

5. For a thorough discussion of contingent valuation, see Carson and Mitchell (1991).

recreational fishing or hunting. The potential of CV to ascribe dollar values to the complex functions of resources or to the roles of "keystone" species is even more questionable. In parts of the world such as countries in Africa, Asia, and Latin America, where people are not accustomed to markets and are living in poverty or have a cultural aversion to inquiries about their values, CV may be completely useless.

- *Third,* and perhaps most troublesome, is the uncertainty involved in forecasting costs and benefits that rely on long chains of arguments that are based on complex ecological economic linkages. Significant uncertainty is bound to accumulate as we undertake this kind of research because physical, biological, chemical, and economic phenomena are difficult to measure, do not always take place in predictable ways, and have unstable lag structures. Some critical cause-effect relationships are also outside the experience of researchers and have not been studied long enough for them to have any real sense of their importance. To assess the value of ecological economic losses from the possible release of a single hazardous substance at a proposed project site, for example, the analysis would need to deal with the probability that a release will occur, the possible quantity that might be released, how concentrations might change as it disperses in air, water, or soil, how many and what types of organisms might be exposed, each organism's dose/response function, how each organism's response might affect the functioning of the ecosystem and eco-succession, how changing ecosystem conditions might affect measures of economic performance, and, finally, how people might react or adapt in ways to mitigate costs or replace benefits. The propagation of uncertainty as information is passed from one module to another in such an analysis could leave the analyst and decision-makers without any real confidence in resulting measures of costs, risks, or benefits.

In some instances, therefore, it may seem reasonable to forego attempts to justify investments in natural capital on the basis of expanded BCA and look for some other justification. The following section explores an approach that allows decisions about natural capital to be based on social trade-offs without any formal comparison of costs and benefits. It still requires decision-makers to evaluate costs and benefits, but it does not require that all costs and benefits be monitized.

ESTABLISHING SAFE MINIMUM STANDARDS [6]

Figure 18.4 summarizes a framework within which natural resources are aggregated into three groups on the basis of their *importance*, in terms of ecological and economic function, and the *reversibility* of resource use decisions that affect them. These two criteria determine the appropriate level of protection—the level of collective control exerted over normal economic forces that would otherwise result in the resource being depleted or lost. For sake of dis-

---

6. The concept of safe minimum standards was first introduced to the economics literature by Ciriacy-Wantrup (1952).

cussion, a specific type of public intervention into individual decision-making—prohibitions, command and control, market-based incentives—is suggested for each resource group.

At one extreme are natural assets that provide extremely important ecological functions and are virtually irreplaceable (e.g., coral reefs). These Group 1 resources appear near the top-left corner of Figure 18.4 and would be given the most protection such as prohibitions or sanctuaries. At the other extreme are resources that are ecologically less important and are relatively easy to restore or replace (e.g., deciduous forests). These Group 3 resources would appear near the bottom right corner of Figure 18.4 where the level of collective intervention is limited to promoting economic efficiency by ensuring "full cost" pricing of all inputs. Between these two extremes are resources that do not demand complete protection, but require active management and some level of collective control over individual decision making (i.e., fisheries and

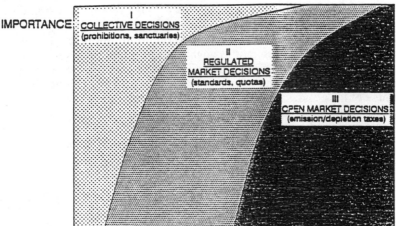

Importance—	Potential harm/risk to ecological system if the resource is lost.
Reversibility—	Potential for natural or engineered restoration if the resource is lost.
Examples:	High Importance/Low Reversibility—e.g., ozone layer, some wetlands, rain forest
	Low Importance/High Reversibility—e.g., field crops, forests, some fisheries
Collective Decisions—	override market forces and individual choice
Regulated Market Decisions—	command and control with individual choice constrained by engineering or performance standards or quotas
Open Market Decisions—	open markets and individual choice but with a full accounting of costs. Externalities are internalized. (Laissez Fair where no significant externalities occur).

Figure 18.4. Public intervention to protect natural capital.

forests). Normal market forces are allowed to guide individual decisions with regard to these resources, but within well-defined guidelines imposed in the form of quotas, minimum size limits, engineering and performance standards, and open and closed seasons and areas, and so on.

The criteria for deciding whether a given stock of natural capital should be considered completely off limits to individual interests (Group 1), or open to exploitation under strict "command and control" regulations (Group 2), or managed using only "market-based incentives" (Group 3) would need to be the subject of scientific, economic and political debate rather than technical arguments about specific ecological-economic linkages. The extinction of a fur seal population, for example, may not threaten the earth's life support system, but would be irreversible and so would rank near zero on both the ecological importance and reversibility scale. With the lines drawn as they are in Figure 18.4 if fur seals were in danger of extinction they would be in Group 1 and would be given full protection. This may be unacceptable to some people who, although willing to provide full protection to resources that provide essential life support functions, might believe that the benefits of protecting individual freedom and open markets exceed those of providing full protection to individual species. The following sections describe ongoing research that may provide opportunities for measuring *importance* and *reversibility* and could make the framework a practical tool for evaluating resource management alternatives.

Ecological Importance

Successful attempts to integrate ecological and economic research require that ecological systems be viewed as sets of processes rather than collections of resources. A new research area is emerging in what is known as " ecological health," or with a slightly different connotation, "ecosystem integrity" (Costanza, Norton, and Haskell 1992) which attempts to find indicators of the overall well-being of ecological systems. The fields of landscape ecology and systems ecology are also expanding rapidly and introducing many new ideas about interdependencies within and between ecosystems and how to measure them.[7] These fields offer insights about which specific attributes of an ecosystem and which aggregate measures of energy or carbon or nutrient flows should be used as standards to evaluate and compare ecosystems. For present purposes the main point is that the ecological *importance* of a resource, as used in Figure 18.4, should be based in some way on its contribution to ecological health. A few examples of relevant indices that are under development will help illustrate potential applications.

At the most practical level are methods of ranking ecosystems that rely on simple field observations and are already widely used. One such technique is known as Habitat Evaluation Procedure (HEP, U.S. Fish and Wildlife 1980),

---

7. Two new journals have appeared in the 1990s: *Restoration Ecology* and *Ecological Engineering*.

which employs weighted sets of Habitat Suitability Indices associated with particular species groups to rank the relative importance of ecosystems. Another is the Wetland Evaluation Technique, known as WET (Adamus et al. 1987), which focuses less on direct habitat values and more on physical functions such as groundwater recharge, floodwater retention and desynchronization, and storm wave and surge protection. It ranks wetlands on their capacity to protect property and nourish natural resources in surrounding areas. More elegant possibilities include an Index of Biological Integrity developed by James Karr, which focuses on aquatic ecosystems and includes twelve critical characteristics of fish communities (Karr 1992). This index, or the relative weights assigned to organisms included in it, might be useful for determining the ecological importance of specific bodies of water. Another similar measure is an Index of Ascendancy developed by Robert Ulanowicz which incorporates four characteristics of ecosystem performance: species richness, niche specialization, developed cycling and feedback mechanisms, and overall activity. His index can be applied to any type of terrestrial or aquatic ecosystem (Ulanowicz 1992).

A more holistic approach to evaluating the health of complex ecosystems was introduced recently by William Waide, who proposes that measures of ecosystem health or integrity with regard to wetlands in particular, should rely only on fundamental ecosystem theory until we have "a better appreciation of the dominant structure and functional attributes of wetlands viewed as an integrated ecosystem as well as the major biological, physical, and chemical determinants of ...macroscopic holistic properties." (Waide 1981). Waide favors a model based on two "stability vectors": *resistance*, which relates to the amount of elemental biotic and abiotic storage in the system and to long turnover times and extensive element recycling, and *resilience*, which requires the rapid turnover and recycling of elements. Figure 18.5 shows the relative ranking of different wetland types using these two attributes.

Reversibility of Ecological Damage

Man's ability to create or restore complex ecosystem functions is a controversial subject, but some ecosystems are definitely easier to replace or restore than others. At a basic level, ecosystem restoration can be evaluated on the basis of three factors: the *level* of ecological function that it is possible to restore, the *speed* at which ecological functions can be expected to recover, and the *cost* of the restoration effort. Not surprisingly, there is a positive relationship between the success of ecological restoration projects as measured by the *level* and *speed* of functional recovery and the amount of effort and resources committed to restoration activities as reflected in the *cost*. This relationship is the basis of the analytical framework displayed in Figure 18.6 (King 1992).

Using the framework displayed in Figure 18.6, an ecosystem is shown to be providing 100% of its potential ecological function until a proposed project

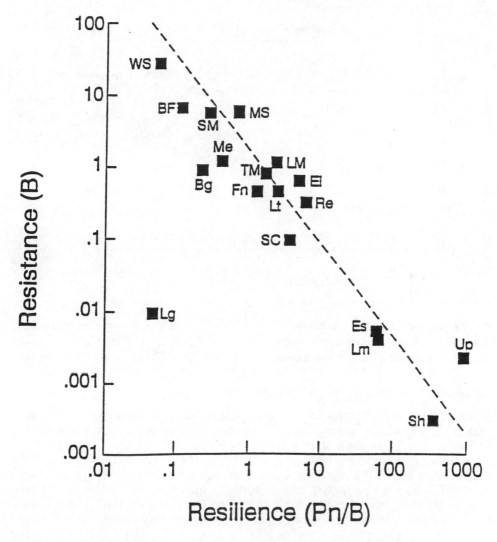

Figure 18.5. Relationship between wetland *resistance* and *resilience*. The relationship between wetlands resistance (B = total biomass (kg/C/m^2)) and resilience (turnover rate, P_n = net primary productivity): Ws, wooded swamp; BF, bog forest swamp; SM, salt marsh; MS, mangrove swamp; Me, meadows marsh; LM, lotic marsh; Bg, bog; TM, tidal marsh; Fn, fens; El, estuarine littoral; Lt, littoral; Re, marine reef; SC, streams, creeks; Lg, estuarine lagoons; Es, open estuary; Lm, limnetic; Up, coastal upwelling; Sh, continental shelf. Source: Waide (1981).

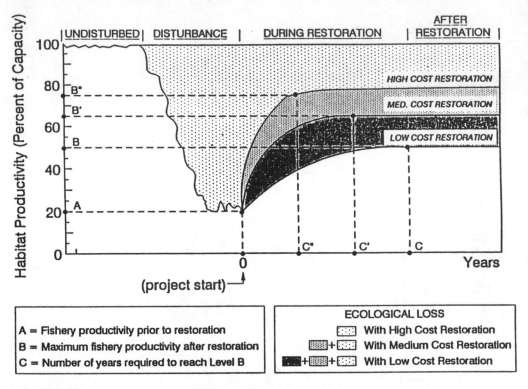

Figure 18.6. Evaluating the reversibility of resource losses.

or some other event results in an impact that reduces its ecological function in year 0 to A% of the natural level. Investments in ecological restoration that occur in year 1 are then shown to increase the level of ecological function from A% of the natural level to B% of the natural level after C years. The shaded area above the recovery curve represents the loss of ecological function and indicates that higher cost restoration improves the likelihood of success as reflected in higher attainable values of B (more function), lower attainable values of C (faster recovery), or both. Using this framework, the *technical feasibility* of restoring an ecosystem after it is degraded is reflected by the highest attainable value of B and the lowest attainable value of C. The *economic feasibility* is determined by the cost of providing this level of restoration. Some combination of *technical feasibility* and *economic feasibility* might provide the necessary information to rank the *reversibility* of an unwise resource allocation decision as depicted in Figure 18.4.

SUMMARY AND CONCLUSIONS

Sustainability is a long-term goal, but industry and government leaders need to make day-to-day production and investment decisions that are responsive to near-term economic needs. Applied ecological economics can help decision makers reconcile near-term and long-term strategies for achieving economic goals by illustrating how decisions that alter the quantity and quality of natural capital can change near-term economic performance.

Most production and investment decisions are justified using some type of Benefit Cost Analysis (BCA). However, time and budget constraints usually preclude a full accounting of costs and benefits associated with ecological economic linkages. Applied ecological economic research that provides a framework for evaluating ecological economic linkages within the context of BCA and facilitates the inclusion of indirect measures of ecological economic costs and benefits in BCA could have a significant influence on industry and government decision making. The overall goal of such research should be to reduce the time and money required to trace, estimate, and document indirect costs and benefits associated with decisions to invest or draw down natural capital. A useful first step is to establish protocols and procedures and measurement and reporting standards to help integrate biological, physical, and social science research in ways that generate information that is suitable for BCA. The framework for expanded BCA presented in Figure 18.2 is a general outline of the research modules that need to be integrated. A program aimed at expanding BCA along these lines might begin with the following tasks:

1. Characterize the analytical models available within each research module in terms of reliability, timeliness, data requirements, costs, and output format.

2. Develop standard measurement and reporting protocols and output formats for each research module based on the data and documentation requirements of subsequent research modules.

3. Determine which analytical models, assumptions, data, and empirical results are transferable from region to region or site to site, and what adjustment procedures are necessary.

4. Establish generally applicable procedures for communicating information about data quality, the expected range of error around parameter estimates, and the likely propagation of uncertainty as empirical results are passed from one module to another.

There will be situations where our inability to describe complex ecological economic linkages or assign values to ecological functions will make it impossible to justify protecting or investing in natural capital on the basis of any type of empirical BCA. In such situations, applied ecological economic research can contribute to the decision-making process in other ways. The analytical framework summarized in Figure 18.4, for example, can facilitate analysis and policy debate regarding the level of protection that should be given to different classes of natural capital without resorting to direct comparisons of costs and benefits. Using this framework safe minimum standards and the level of public intervention in individual decision making are based on the ecological *importance* of the resource in question and the *reversibility* of private production and investment decisions. Aggregate indices of "ecological health" or "ecosystem integrity" can be used to measure *importance* and the *technical feasibility* and *economic feasibility* of restoring lost ecological functions can be used to measure *reversibility*.

Investing in natural capital may mean foregoing the near-term economic payoffs from depleting or polluting resources that are difficult to restore or it may mean actually investing dollars to restore resources that can be restored. Assigning a resource to a particular spot on Figure 18.4 will help determine which it is and where we should put our priorities.

REFERENCES:

Adamus, P. R., E. J. Clarain, Jr., R. D. Smith, And R. E. Young. 1987. Wetland Evaluation Technique (WET), Vol. II: Methodology. U.S. Army Corps of Engineers Waterways Experiment Station, Vicksburg, MS. Operational Draft Technical Report Y87.

Bojo, J., K-G Mäler, and L. Unemo. 1992. Environment and Development: An Economic Approach. Boston: Kluwer.

Ciriacy-Wantrup, S. V. 1952. Resource Conservation. Berkeley: Univ. of California Press.

Costanza, R., B. Norton, and B. Haskell, eds. 1992. Ecosystem Health: New Goals for Environmental Management. Washington, DC: Island Press.

Karr, J. R. 1992. Ecological integrity: protecting Earth's life support systems. In Ecosystem Health: New Goals for Environmental Management, eds. R. Costanza, B. Norton and B. Haskell. Washington, DC: Island Press.

King, D. M. 1992. The economics of ecological restoration. In Natural Resource Damages: Law and Economics, eds. K. W. Ward and J. W. Duffield. New York: Wiley.

————. 1991. Wetland Creation and Restoration: An Integrated Framework for Evaluating Costs, Expected Results and Compensation Ratios. Report prepared for the Office of Policy Planning and Evaluation, U.S. Environmental Protection Agency.

Kopp, R., and V. Kerry Smith. 1993. Valuing Natural Assets: The Economics of Natural Resource Damage Assets. Washington, DC: Resources For the Future.

Mitchell, R. C., and R. F. Carson. 1991. Using Surveys to Value Public Goods: The Contingent Valuation Method. Washington, DC: Resources For the Future.

Scodari, P. 1992. Wetland Management Benefits. Report prepared for the Office of Policy Planning and Evaluation, U.S. Environmental Protection Agency, October.

U.S. Fish and Wildlife Service. 1980. Habitat Evaluation Procedure (HEP) Manual. ESM 102. Division of Ecological Services. Washington, DC: U.S. Department of the Interior, Fish and Wildlife Service.

Ulanowicz, R. E. 1992. Ecosystem health and trophic flow networks. In Ecosystem Health: New Goals for Environmental Management, eds. R. Costanza, B. Norton and B. Haskell. Washington, DC: Island Press.

Waide, W. 1981. Analysis of Selected Wetland Functions and Values. Report 81D-01, U.S. Army Corps of Engineers, Institute for Water Resources, Arlington, VA.

19 RENEWABLE RESOURCES AS NATURAL CAPITAL: THE FISHERY

Colin W. Clark[1] and Gordon R. Munro[2]
1. *Institute of Applied Mathematics*
2. *Department of Economics*
   The University of British Columbia
   Vancouver, BC V6T 1Z2

ABSTRACT

We discuss fishery resources as a form of natural capital. Using the analogy between the classical model of capital accumulation and a basic dynamic model of fishery exploitation, we discuss problems of both disinvestment and reinvestment in natural capital. In the case of a fishery, the initial disinvestment phase often involves unique problems, stemming from the common-property nature of the resource and from the high degree of uncertainty pertaining to a poorly observable, fluctuating biological population. As a result, excessive disinvestment in the resource often occurs, leading to the necessity of a subsequent reinvestment program.

Historically, many fishery management programs have been accompanied by extreme overcapitalization in terms of fishing capacity. Misallocation of both natural and conventional capital are then intertwined. One approach to overcoming these problems, using individual transferable catch quotas (ITQs), is discussed. The owner of an ITQ in essence owns a tradable share of the natural capital asset represented by the fish stock.

We suggest that the successful management of a renewable resource stock, such as a fish population, demands that the resource be treated as a valuable asset of natural capital.

INTRODUCTION

Nobel Prize winner and economist George Stigler has defined capital as anything other than human beings "which yields valuable services over an appreciable time." This definition of capital clearly encompasses all natural resources, as well as what might be termed "conventional" capital, (i.e., capital goods produced by human beings). In this chapter we focus on fisheries as an example of "natural" capital. We do so in part because of the familiarity that the authors have with the example. We do so as well, however, because the application of the economist's theory of capital and investment to fisheries has been developed to a greater degree than is true of most other natural resources. Nonetheless, we shall argue that the principles that apply to fishery resources as natural capital have a broad application to other natural resources.

While it shall be demonstrated that the economist's theory of capital and investment can be applied with some ease to the fishery, and can carry us some distance in the analysis of fisheries management problems, we shall argue as well that there are fundamental differences between the application of the theory to conventional capital and the application to fisheries as natural capital.

One important difference arises from the fact that in the analysis of investment in conventional capital, the absolute level of existing capital is almost never seen to be excessive. Indeed, under normal circumstances, the existing level of capital, in absolute terms, is usually below the optimum. The problem is one of optimal investment, not one of disinvestment.

In the case of fishery resources as natural capital, it is quite possible to have a situation in which the amount of natural capital is, from an economic point of view, excessive. The optimal depletion of capital in the form of a fishery resource thus becomes very much an issue. Unfortunately, achieving this optimal depletion often incurs extreme practical difficulties, partly resulting from the common-property nature of most fishery resources, and also because of the very great uncertainties invariably associated with fish population dynamics. In our view these difficulties only strengthen the need to consider fishery resources as natural capital, albeit with special, unique properties.

ELEMENTARY THEORY OF CAPITAL AND INVESTMENT AND ITS APPLICATION TO FISHERIES

We offer a model of conventional capital and investment that can be traced back at least to 1928 (Burmeister 1980). It is assumed that society is capable of producing but one good that can be used either as capital or for consumption purposes. Besides capital, the only other productive input is labor service provided by a stationary labor force. The problem before us is to determine the optimal level of capital and the optimal rate of investment, given that the objective is to maximize the collective utility from consumption of the members of the labor force over time. Consumption goods are assumed to be distributed equally among the members of the labor force, who have, in turn, identical utility functions. Finally, it is assumed that production technology is frozen.

We thus have:

$$\text{maximize} \int_0^\infty e^{-\delta t} \overline{L} U(c) dt \tag{1}$$

$$\text{subject to} \frac{dK}{dt} = F(K, \overline{L}) - \overline{L} c \tag{2}$$

and

$$K(0) = K_0, \qquad K \geq 0, \qquad c \geq 0 \qquad (3)$$

Here $K = K(t)$ represents the stock of capital, $F(\ )$ denotes the rate of productivity of capital (net of depreciation), $L = \overline{L}$ the labor force, $c = c(t)$ consumption of the individual worker, $U(c)$ the individual worker's utility function, and δ the social rate of discount.

In applying optimal control theory (or the classical calculus of variations), one can easily show that the above dynamic optimization problem possesses an optimal equilibrium solution $K = K^*$, determined by the famous Golden Rule of Capital Accumulation:

$$F'(K^*) = \delta \qquad (4)$$

where we write simply $F(K) = F(K, \overline{L})$, suppressing the constant labor input $\overline{L}$.

In the limiting case that $\delta = 0$, the Golden Rule states that investment in capital should proceed up to the point that sustainable consumption is maximized. This will occur at a finite level of capital, because of the stationary nature of the labor force and the unchanging nature of technology. The law of diminishing returns comes into play. We have throughout $F''(K) < 0$. Consequently, it is conceivable (if unlikely in practice) that investment in capital could proceed too far, with the result that we would enter a region in which $F'(K) < 0$. The marginal product of capital would be negative and the capital stock would indeed be excessive.

A comment on the optimal rate of investment is in order. If, as is normally assumed, $K_0 < K^*$, the optimal transition phase involves a varying positive rate of consumption, $c*(t)$:

$$0 < c*(t) < F(K(t)) \qquad (5)$$

The ultimate equilibrium level of capital K^* and consumption rate $c* = F(K^*)$ are reached asymptotically (the "catenary turnpike" theorem, Samuelson 1965). This is really a reflection of the fact that the individual worker is, in these models, assumed to be subject to diminishing marginal utility, $U''(c) < 0$, which leads to the consequence that rapid investment in the capital stock is costly.

Even the most casual observation of real world economies reveals that very few have stationary labor forces. Typically, the labor forces of these economies are growing steadily. Consequently, modern models of capital and investment, now referred to as growth models, have replaced the assumption of a stationary labor force with one in which the labor force is assumed to be increasing exponentially:

$$L = L_0 e^{rt}$$

If the production function $F(K,L)$ is homogeneous of degree 1, it is easy to see that equation 2, the rate of growth of K, now applies to the *per capita* capital stock K. The Golden Rule Equation (equation 4), similarly, applies to the optimal *per capita* level of capital.

It is possible for there to be overinvestment in the sense that the amount of capital on a *per capita* basis becomes excessively large, due to the law of diminishing returns. The possibility of the level of capital on an absolute basis being excessive is not to be given serious consideration. Indeed, optimality now implies that the level of total capital should grow exponentially at the rate r.

Casual observation also reveals that in most economies technology is not frozen, but rather is changing through time with a positive impact on output. Growth theorists commonly have dealt with technology by treating technological change as being labor augmenting in nature. Without going into the details, the net effect is similar to increasing the rate of growth of the labor force, thus reducing even further the possibility of the amount of capital in absolute terms ever being excessive.

Finally, we should take note of the fact that even more modern theories of growth question whether diminishing returns applies to capital, even on a per capita basis. If one accepts these newer theories, then the existence of an excessive amount of capital in absolute terms is all but inconceivable, even in the unlikely event that the labor force proved to be constant.[1]

Now consider the application of the theory of capital and investment to the fishery. The objective in managing the fishery is to maximize the net economic benefits from the fishery through time, where the net economic benefits are often expressed in terms of profits or resource rent.

Our basic (deterministic) dynamic fishery model takes the form (Clark and Munro 1975):

$$\text{maximize} \int_0^\infty e^{-\delta t} R(x,h)dt \tag{6}$$

$$\text{subject to} \frac{dK}{dt} = F(x, \overline{A}) - h \tag{7}$$

and

$$x(0) = x_0, \qquad x \geq 0, \qquad h \geq 0 \tag{8}$$

---

1. Of course, all this growth theory is predicated on the assumption that there are no exogenous "limits to growth" induced by the finite capacity of the earth's resource stocks to support indefinite growth of industrialization. If such limits do in fact exist, the possibility of over-capitalization, perhaps inadvertent, does arise. But this is beyond the scope of the present chapter. See also Munro and Scott (1985); Munro (1992).

where $x = x(t)$ denotes the fish biomass, $F(\ )$ denotes the natural growth rate of the fish population, $\overline{A}$ is a parameter representing the aquatic environment, h denotes the rate of harvest, $R(x,h)$ is the rate of net profit or resource rent resulting from harvesting, and δ is the social rate of discount.

The similarity of this model to the capital theory model with a stationary labor force (equations 1–3) is apparent. The fish biomass $x = x(t)$ is obviously the analog to capital, K. The natural growth rate $F(\ )$ is the analog to the production function. Indeed we can refer to $F(\ )$ as the natural production function. The fixed aquatic environment $A = \overline{A}$ plays a role similar to the stationary labor force, while h is the analog to the rate of consumption c.

If the resource is unharvested, the fish population, or biomass, reaches an equilibrium $\overline{x}$ given by the equation:

$$F(\overline{x}, \overline{A}) = 0 \tag{9}$$

In practice, the growth function is often assumed to be logistic in form:

$$F(\overline{x}, \overline{A}) = rx(1 - \frac{x}{z}) \tag{10}$$

(so that $\overline{A}$ has two components, r and z) in which case we have simply

$$\overline{x} = z \tag{11}$$

The harvest function, in turn, is often assumed to be of the form:

$$h = qEx \tag{12}$$

where E denotes the rate of fishing effort, the flow of conventional capital and labor services devoted to harvesting, and q, a constant, the catchability coefficient.

As in the capital theory model, the stock of natural capital can either be depleted, by harvesting at a rate in excess of natural growth, or built up, by harvesting less than natural growth. The problem of optimal fishery management can therefore be conceptualized as a problem of optimal investment in a capital stock. This idea is the essence of the theory of natural capital.

Once again, the solution to the above dynamic optimization problem possesses an optimal equilibrium solution, $x=x^*$, given by the following modified Golden Rule equation:

$$F'(x^*) + \left. \frac{\partial R/\partial x^*}{\partial R/\partial h} \right|_{h=F(x^*)} = \delta \tag{13}$$

The optimal approach to x^* is either the most rapid or asymptotic depending upon circumstances.

The modified Golden Rule (equation 13) is very similar to the previous Golden Rule (equation 4), except for the inclusion of a second term on the L.H.S.

The second term is a reflection of the impact upon resource rents, R, of the stock level x. Note from equation 12 that the average product with respect to fishing effort, h/E, is qx. In essence, the denser the biomass, the lower the costs of harvesting. More will be said about this later.

Now, let us focus on A, the aquatic environment, which we see as the analog to L in our simple capital theory model. We have seen that, in modern capital theory models, L is assumed to be growing exponentially. We have seen further that, while there are arguments over whether diminishing returns can emerge with respect to capital on a per capita basis, the question of the possibility of excessive capital in absolute terms is a non-issue.

The aquatic environment, A, on the other hand, is decidedly finite, being bounded by the finite area of the oceans and the finite flux of solar energy. Human beings can attempt to enhance the productive power of the aquatic environment (e.g., through aquaculture), but the effects are minor at best. Indeed, what is almost certainly the most important impact that human beings have upon the productive capacity of the aquatic environment is negative. Pollution of the aquatic environment, resulting from human activity, serves to diminish significantly the environment's productive capacity.

In any event, the consequence of a finite aquatic environment, A, is that the law of diminishing returns does come into play with respect to the fishery biomass, in absolute terms, and that consequently, it is decidedly possible to have, from an economic perspective, an excessive amount of natural capital in absolute terms. To take one obvious example, consider an unexploited fish population with $x = \bar{x} = z$, which had hitherto been unprofitable to exploit, but which has now become commercially valuable because of changed market conditions. From equation 9 we have $F(\bar{x}, \bar{A}) = 0$, hence *any* exploitation will lead to a diminution of the resource. Thus the natural capital is unquestionably excessive from an economic perspective.

Negative investment, or disinvestment, in the resource thus becomes a central issue. In fact, under certain, admittedly extreme, circumstances it may appear optimal in economic terms to institute a program of resource disinvestment leading to the outright extinction of the resource.[2]

We are then led to discuss optimal disinvestment programs and to consider circumstances giving rise to over-disinvestment in the resource. It shall be seen that the risks of excessive disinvestment are particularly great in fisheries

---

2. For exhaustible resource stocks, the point is even more transparent. Utilization implies disinvestment, though not necessarily to the point of absolute exhaustion. Renewable resources are characterized by the possibility of providing a sustained yield, and in this way they more closely resemble the capital stock of classical theory than do exhaustible resources.

because of the common property nature of the resource. Moreover, as the result of environmental uncertainties, and the general lack of information about any given fish population, inadvertent overcapitalization becomes all the more likely.

Past mistakes can, of course, be recognized. Consequently, we must discuss optimal policies for positive resource investment when it is recognized that ineffective or misguided management in the past has led to undue depletion of the resource.

RESOURCE DISINVESTMENT PROBLEMS

We have now made the point that fishery resource *dis*investment can be entirely rational from an economic standpoint, and indeed is inevitable in the initial stages of fishery development. In this section we consider an optimal resource disinvestment strategy under ideal circumstances, and then go on to show how readily fishery resources can be subject to non-optimal, and at times devastating, disinvestment.

We shall continue to use the dynamic fishery model as set out in equations 6–12, but introduce some further simplifying assumptions. It will be assumed that both the demand for harvested fish and the supply of fishing effort, E, are perfectly elastic. Hence both the price of harvested fish and the unit cost of fishing effort can be treated as constants. Denote the price of harvested fish and unit cost of fishing effort as p and a respectively. Then $R(x,h)$ can be expressed simply as $ph - aE = (p - c(x))h$, where $c(x) = a/qx$.

Suppose that, with regards to the fishery resource in question, we initially have $p < c(\bar{x})$. It is unprofitable to exploit the resource at any stock level. As natural capital, the fishery resource is without value.

Consumer tastes now change, with the consequence that there is a once-and-for-all outward shift in demand for the harvested fish. We now find that $p \gg c(\bar{x})$. The previously valueless resource, becomes, from society's point of view, a valuable natural resource asset. Harvesting of the resource is now a profitable undertaking.

Exploitation now commences and does so, we shall suppose, under the iron control of a resource manager, who lives in a deterministic world. That is to say the manager has full and complete knowledge of the population dynamics of the resource and can predict the future with certainty.

Next observe that, given our newly introduced assumptions, the equilibrium equation, equation 13, can be re-written as:

$$\frac{1}{\delta} \frac{d}{dx^*} \{(p - c(x^*)F(x^*))\} = p - c(x^*) \tag{14}$$

The R.H.S. is marginal rent from current harvesting. The L.H.S. is the present value of sustainable resource rent gained from an incremental investment in the resource. It can also be interpreted as the *loss* to society of an incremental reduction in the resource and as such is referred to as the "user cost" of the resource. Thus if resource disinvestment is called for, the resource manager must be constantly balancing off the current gains from exploitation against the future losses to society resulting from resource depletion.

At $x = \bar{x}$, $F(\bar{x}) = 0$. Since we now have $p >> c(\bar{x})$, a program of resource disinvestment is called for. The stock of natural capital, in the form of the fishery resource, is indeed excessive. Resource disinvestment should continue until equation 14 holds, and we have $x = x^*$. Further resource depletion beyond this point would have the result that, at the margin, future losses to society would exceed current benefits from resource depletion.

We have assumed that the resource manager has perfect knowledge of the resource biology. In practice, biological modeling of a fishery resource is very difficult. If the resource manager has imperfect information about the size of the resource and the population dynamics, he may follow an exploitation program leading to inadvertent overexploitation. For example, when the New Zealand government introduced 200-mile Exclusive Economic Zones in 1979, it found that it had gained control over a valuable groundfish resource known as orange roughy. The authorities permitted a seemingly conservative resource depletion program, which took place under reasonably controlled circumstances. Subsequently, the authorities discovered that their initial biological parameter estimates were wrong and that serious overexploitation of the resource had been allowed to occur.[3]

While such examples of resource depletion due to overoptimism are not unusual, the real danger of excessive resource disinvestment lies in the fact that fishery resources are normally common property. Under common-property conditions, the individual fisherman has no control over the resource and is competing with other fishermen for harvests. Thus the individual fisherman will act as if $\delta = \infty$, i.e., his perception of the resource-user-cost is that it is equal to zero.[4]

Return to our example in which we suppose that a previously commercially valueless resource becomes valuable because of a shift in consumer tastes. Suppose now, however, that exploitation takes place under conditions of an uncontrolled, open-access, common-property fishery. Because of the high price level, the fishermen will enjoy a seeming bonanza. Fishing effort will be high, and the resource stock will be reduced. We can predict that the stock reduction will continue until we reach the stock level $x = x_\infty$, at which $p - c(x_\infty) = 0$, (i.e., resource rents are reduced to zero). Economists refer to x_∞ as bionomic equilibrium (Gordon 1954).

---

3. We later discuss problems arising from uncertainty in greater detail.
4. Similar common-property resource problems arise in many other circumstances, particularly in the field of environmental pollution. Failure to deal effectively with the "tragedy of the commons" can have serious implications in terms of sustainable development. See Hardin, (1968); Clark (1990).

Since the social rate of discount, δ, is certainly finite, we can say without hesitation that $x_\infty < x^*$. Excessive resource disinvestment has occurred because the individual fisherman has no incentive other than to regard resource user cost as equal to zero.

There are many examples of marine living resource stocks that have been driven to bionomic equilibrium. During 1930–1960, when the international whale resource was an ineffectively controlled common-property resource, the whalers extracted billions in profits. It is generally agreed, except by a few countries, that the whale stocks were seriously overexploited. Antarctic blue whales, for example, were reduced in number from some 150,000 to less than 10,000 (and perhaps less than 1,000) during this period.

The authorities can intervene in a common-property fishery to try to prevent the resource from being driven down to $x = x_\infty$, by restricting harvests. Consider how difficult the position of the authorities is, however. Because their perception of the benefits of resource conservation may differ radically from that of the authorities, the fishermen will be encouraged to fight the authorities every inch of the way. If the authorities enter the picture only after the initial highly profitable phase of resource reduction, the proposed harvest restrictions will lead to greatly reduced incomes for fishermen. Short-term needs of the fishermen may then outweigh any long-term considerations of resource conservation.

Moreover, it will not be sufficient for the authorities just to restrict harvests. If the authorities, in our example, attempt to bring the resource depletion to a halt by restricting harvests alone, the consequences of common property will manifest themselves in another way. Fishermen will compete for shares of the valuable restricted harvest. The inevitable result will be overcapitalization in the fishery. The amount of vessel capital will exceed by a wide margin the amount required to take the allowable harvest. The true fleet costs associated with the fishery will rise until the net economic benefits have again been fully dissipated.

Consider the implications of such a management strategy. Suppose that the authorities stabilize the resource at $x = x^*$, by controlling harvest rates, but make no attempt to control the fleet size. The resource user cost will be driven to zero as a result of overcapitalization. Resource conservation, from an economic standpoint, will be futile.

Indeed, resource conservation may be worse than futile. We have said nothing about the costs to society of attempting to manage the resource. Management costs can be extremely high (and are often considered to be beyond the means of underdeveloped countries). Once these costs are taken into account, we may be forced to conclude that the economic return to society from the fishery resource as natural capital is negative.

Even if the resource management authorities are able to address effectively the common property problem within the nation's waters, there remains yet another threat to resource conservation, one that has gained particular promi-

nence under the New Law of the Sea and the regime of 200-mile Exclusive
Economic Zones (EEZs).

As a consequence of the mobility of many species, coastal states, upon establishing 200-mile Exclusive Economic Zones, have found that some of the
encompassed fishery resources are transboundary in nature. The fishery resource is either shared with a neighboring coastal state or states and/or extends
beyond the boundary of the Zone into the remaining high seas, where it may
be subject to exploitation by so-called distant water fishing nations.

If a fishery resource is shared by say, two coastal states, the coastal states
may, or may not, be able to agree on a cooperative management regime.
Inability to cooperate can prove to be destructive.

It can be shown analytically through the application of differential game
theory, that, under non-cooperation, the two coastal states would be driven to
an outcome similar to that of an uncontrolled, open access, common property
fishery. Over-exploitation and excessive disinvestment in the fishery resource
would be all but guaranteed.

The United States and Canada are currently engaging in the joint management of an important shared fishery resource, Pacific salmon, and are doing so under a formal treaty arrangement. The negotiation of the treaty
proved to be enormously difficult and complex. Negotiations extended over
15 years and seemed likely to break down at several points during the 15-year
period. What drove the American and Canadian negotiators to persist was the
mounting evidence that a potentially highly destructive Pacific salmon "fish
war" was emerging between their two countries.

The establishment of effective cooperative resource management regimes
between, or among, joint coastal state owners of a fishery resource may be
difficult, but they are certainly feasible. The New Law of the Sea, as embodied
in the United Nations Law of the Sea Convention, makes the rights and obligations of coastal states as joint resource owners reasonably clear.

Much more difficult, however, are the cases of fishery resources found in
both the coastal state Exclusive Economic Zone and the adjacent high seas.
Perhaps, because at the close of the U.N. Third Conference on the Law of the
Sea, such stocks, commonly referred to as "straddling" stocks, were thought
to be unimportant, the Law of the Sea convention is exceedingly vague on the
rights and obligations of states exploiting the resources in the adjacent high
seas.

Straddling stocks are no longer seen as unimportant, but rather have become a major international resource management issue. If the coastal state
and the distant water fishing nations exploiting the straddling stock in the adjacent high seas are incapable of establishing effective cooperative management regimes, then the analysis which applies to non-cooperation by joint
coastal state owners of a fishery resource applies with full force. Excessive
disinvestment in the resource is virtually certain to occur.

At present, because of the vagueness of the Law of the Sea Convention on straddling stocks, the outlook is unpromising. Consider the following example.

Canada's major gains from the New Law of the Sea are to be found on the famous Grand Bank off Newfoundland. The 200-mile limit slices off two portions of the Grand Bank, one in the south and one in the east, known respectively as the Tail and Nose of the Bank. The straddling stocks extending into the Tail and Nose of the Bank are exploited by members of the European Community, Spain and Portugal in particular. When Canada implemented EEZs in 1977, it helped to establish an organization, the Northwest Atlantic Fisheries Organization (NAFO), to serve as a mechanism for the cooperative management of the aforementioned straddling stocks. For a time, the organization seemed to work reasonably well. Over the past several years, however, Canadian-EC fisheries' relations have deteriorated, with Canada accusing the EC of gross overfishing in the Tail and Nose of the Bank. The difficulty has created a major political crisis for the Canadian government.[5]

In June 1992, the U. N. Conference on the Environment and Development (UNCED) was held in Rio de Janeiro. It was agreed at the Conference that the United Nations should convene, as soon as possible, an intergovernmental conference on the difficult issue of management of high seas fisheries resources. At the time of writing, it is anticipated that the conference will be convened in late 1993.

RESOURCE INVESTMENT PROGRAMS

Resource management authorities, recognizing the effects of excessive resource disinvestment in the past, can attempt to correct the situation by rebuilding the resource (i.e., by engaging in positive resource investment). We take as our example a fishery resource confined to the waters of a single EEZ, which through past mismanagement had been reduced to bionomic equilibrium, $x = x_\infty$. It is supposed, for the time being at least, that our assumption of perfect elasticity of demand for harvested fish and supply of fishing effort is valid.

---

5. Since the first draft of this chapter was written, the situation in Canada's Atlantic fisheries has deteriorated substantially. Cod stocks in many areas had reached drastically low levels by mid-1992, forcing the Canadian government to close the fishery for at least 18 months, thereby throwing tens of thousands of fishermen and plant workers out of work. While the breakdown in Canada's fishery relations with the EC was a contributing factor to the collapse of the fishery, it was certainly not the only factor. It is also generally agreed that over-optimistic catch forecasts on the part of the Canadian authorities, leading to continual overfishing, were also instrumental in the collapse. This is particularly disturbing in view of the Canadian government's attempt to manage the fishery rationally. In December, 1992, the Canadian Minister of Fisheries and Oceans announced that revolutionary new management practices would be introduced into the Atlantic fisheries to prevent a recurrence of the problem.

Many years ago, before fisheries economists applied their theory of capital and investment seriously to fisheries management, bionomic equilibrium was commonly represented as in Figure 19.1. The level of fishing effort, E_∞, is that associated with bionomic equilibrium, and the complete dissipation of resource rent. Fishing effort is obviously excessive, and the correct prescription is that of reducing fishing effort.

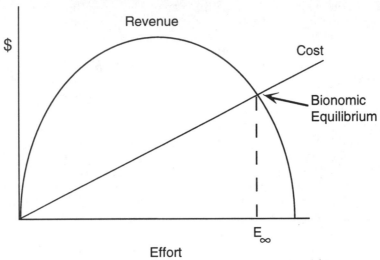

Figure 19.1. Bionomic equilibrium.

Figure 19.1 is not inherently incorrect, but it is misleading. It appears that, initially at least, a reduction of fishing effort will result in both an increase in total revenue from harvesting and a reduction of fishing effort cost. Everyone wins, apparently.

Yet if we consider the harvest production function as given in equation 12, $h = qEx$, any reduction in E must, initially, result in a proportionate fall in the harvest rate, h—something that is self-evident to every fisherman. A reduction in E will, *in time*, lead to an increase in h and in total revenues, but only after the resource stock, x, has had time to rebuild. Resource investment, like any other form of investment, requires a current sacrifice to be endured in the hope of future gain.

The resource investment target, the optimal biomass level, x^*, is given by equations 13 or 14. How long it takes for x^* to be achieved and for there to be a "payoff" to the resource investment depends entirely on the biological potential of the fish species. Some species, (e.g., skipjack tuna), have high intrinsic growth rates and are capable of rapid growth and recovery. Other species grow much more slowly. For example, when Canada implemented Exclusive Economic Zones in 1977, it maintained that several important groundfish stocks off its Atlantic coast had been subject to over-exploitation and would have to be rebuilt. One example was provided by redfish stocks. Canadian biologists estimated that the redfish stocks would require 25 years before achieving levels associated with Maximum Sustainable Yield even if catches were reduced to zero over that period. In actuality, recovery seems to have been achieved in about half the projected time, but then even 13 years is not a short period of time.

The fact that current sacrifice is unavoidable if the fishery resource is to increase, means that careful attention must be paid to the optimal rate of investment. The most rapid rate of resource investment will be achieved by letting $E = h = 0$ over the period of resource recovery.

Indeed, with our assumption of perfect elasticity of demand for fish and supply of fishing effort, our optimal control model is linear in the control variable h. This leads to the conclusion, given that certain other conditions are met, that a harvesting moratorium is the optimal policy during the resource investment phase:

$h(t) = 0$, as long as $x(t) < x^*$ (15)

One of the other conditions to be met is that conventional capital in the form of fishing vessels, processing plants, and fishermen's skills is perfectly *malleable*. The implication of perfect malleability of capital is that it can quickly and costlessly be shifted to other activities.

If the aforementioned capital is non-malleable, one can be assured that a fishing moratorium will bring forth a vehement reaction from the affected fishing industry. The reaction would be understandable since the harvest moratorium policy would wreak havoc in the industry. It can in fact be demonstrated that, if the aforementioned capital is non-malleable, a policy of slow, or gradual, resource investment may be called for.

We close this section with a warning. If the authorities succeed in curbing fishing effort sufficiently so that the stock size is, quickly or slowly, rebuilt, but do not limit the fleet size upon re-opening the fishery, over-capitalization is certain to emerge. The potential resource rent, resulting from the resource investment, will attract additional vessels competing for shares of the seemingly valuable harvests. The likely consequence will be that the *sustainable* return to society on the resource investment will again prove to be zero.

When Canada introduced an EEZ regime in 1977, it appeared to have some success in its groundfish resource investment program. The reduction of fishing effort was politically easy, as it involved ejecting distant water fishing nation fleets from the newly established EEZ. The Canadian Atlantic Coast fishing industry was to reap the benefits of the resource investment program. Yet in 1981–82 that industry was plunged into a deep crisis and required massive assistance from the Canadian federal government in order to survive. A Royal Commission established by the federal government (Canada 1982) to examine the situation revealed that the authorities had permitted their resource investment program to be accompanied by massive over-investment in both the harvesting and processing sectors of the fishing industry.

EFFECTS OF FLUCTUATIONS AND UNCERTAINTY

The basic fishery model used in the above discussion can be criticized on a number of counts. Its most important shortcoming is one to which we have already alluded at several points, namely the assumption of deterministic dynamics and perfect information regarding the stock level x and the biological production function $F(x)$.[6] In reality, every fish population undergoes random, unpredictable fluctuations. Estimating the current stock level of a given fish population (which may be spread over thousands or millions of sq. km. of the ocean) is extremely difficult and expensive. Attempting to estimate biological production functions with any degree of certainty is even more difficult.

A full discussion of these statistical problems, and their management implications, is beyond the scope of this article. Standard practice has been to base management decisions (e.g., annual quotas) on simple, "most likely" point estimates. This procedure is almost doomed to failure, it seems to us, given the vast degrees of uncertainty present in fishery systems. The approach is particularly fraught with danger when the conventional and human capital applied to the fishery are non-malleable. Books such as those of Walters (1986), Mangel (1985), and Savory (1988) and the article by Charles (1983), give some insight into basic management principles for uncertain resource assets.

The type of uncertainty which we have described with regards to the natural capital stock, x, and the key production function, $F(x)$, is assumed not to exist in the analysis of investment in conventional capital. It is rather assumed that the stock of conventional capital, K, can be measured, and the key production function $F(K)$, can be specified, with complete accuracy. The existence of the aforementioned uncertainty constitutes a second, and perhaps even more fundamental, difference between the application of capital theory to fisheries as capital and its application to conventional capital.

INDIVIDUAL TRANSFERABLE QUOTAS

A fishery resource, we have argued, is a stock of natural capital. As such, it possesses an intrinsic asset value, $V(x_0)$, which depends on the current size x_0 of the stock:

$$V(x_0) = \max \int_0^\infty e^{-\delta t} R(x,h)dt$$

$$\text{subject to } \frac{dK}{dt} = F(x) - h, \quad x(0) = x_0$$

---

6. Ecological factors, such as interactions between different fish species, and between fish and their environment, are also ignored in the simple model.

As we have seen, however, in the case of a common-property resource, this asset value is not apparent (or even of much interest) to the individual fisherman. Instead fishermen observe the marginal rent

$$MR(x_0) = p - c(x_0)$$

which represents their net revenue from catching one unit of fish. If this is positive, the fisherman is motivated to continue fishing.

Management strategies that merely attempt to achieve some "optimal" stock level $x^* > x_\infty$ have the effect of increasing the marginal rent $MR(x)$, and thereby increasing the fisherman's desire to catch fish in spite of the regulations. In order to maintain fishing at the desired level, the managers are forced to shorten the fishing season.

What management strategies could be envisioned that might overcome these undesirable incentives? Three such strategies have been discussed in the literature:

1. private ownership of the fishery;

2. landings taxes; and

3. individual transferable quotas (ITQs).

Each of these strategies has the potential for internalizing the user cost of the resource, thereby resulting in economically rational exploitation—at least in principle. The three strategies have quite different distributional implications, however. (The strategies can also be combined, with correspondingly different distributional implications.)

The private owner of a resource stock, it would appear, should be motivated to exploit the resource in a socially desirable manner. But several important provisos need to be mentioned. First, the resource should be free of externalities affecting other resources. Second, the resource owner should employ a discount rate approximately equal to the social rate of discount. Third, the resource owner should not occupy a monopoly position in the fish market. If these conditions are not met, the private owner's exploitation policy may differ sharply from the social optimum.

Although a few economists continue to argue vociferously for privatization of fishery resources, in most cases of commercial fisheries privatization is virtually out of the question, for various reasons (e.g., geographical extent of the fish population, past history of exploitation, etc.). Alternative approaches, therefore, need to be considered. Landings taxes are one theoretical possibility. If the government imposes a tax of τ per unit of landed fish, then the fisherman's net marginal revenue becomes

$$MR(x, \tau) = p - \tau - c(x)$$

By setting τ so that

$$p - \tau - c(x^*) = 0$$

the authorities can ensure that the (otherwise) common-property fishery achieves a taxed bionomic equilibrium equal to $x*$. At this equilibrium the fishermen receive zero rent, with the entire rents

$$(p - c(x^*))h = \tau\, h$$

accruing to the government. Needless to say, regulation by taxation has not proved to be politically feasible.

From a theoretical viewpoint, allocated quotas are known to be equivalent to taxes, at least under certain assumptions. How this works out in the fishery has been discussed in detail by Clark (1980). Provided that quotas are transferable, what happens is that quota units achieve a value m on the quota market. Any fishermen who use their quota, rather than selling it, therefore, face an opportunity cost equal to the value of the quota. Net marginal rent then equals

$$p - c(x) - m$$

which again equates to zero under competitive fishing. The authorities fix the total quota so as to achieve some target biomass objective $x = x^*$. Then the continual quota-market clearing and bionomic equilibrium conditions determine the quota price

$$m = p - c(x^*) = \tau^*$$

i.e., the quota price equals the optimal landings tax.

An advantage here is that the quota price is determined by market forces, rather than being set by the regulatory bureaucracy. Rents accrue to the original quota holders.

A problem may arise if some fishermen have lower discount rates than the rest. These fishermen will place a higher value on quota units, and will therefore tend to accumulate them by buying up the quotas of other fishermen. In the end, all of the quotas may end up in the hands of a single owner (as has actually happened in two fisheries in British Columbia). Of course such monopolization could easily be prohibited, if it was thought to be socially undesirable. Also, if it was thought that some of the rents ought to accrue to the public, a landings tax could be imposed in conjunction with the quota system.

In spite of these and other practical problems, ITQ's have the potential for greatly improving the efficiency of fishery management. First, assuring that catches are carefully monitored by the authorities (this is absolutely essential to the success of the ITQ system), the ITQ system ensures that the total annual catch is at an appropriate level. But equally important, the ITQ system eliminates the incentive for excessive capitalization of vessels and gear, which has been such a prevalent and troublesome feature of TAC ("total allowable catch"—i.e., unallocated catch quotas) management regimes. Holding an ITQ, the vessel owner has no incentive to exceed the quota; rather the incentive is to capture the quota at the least possible cost. If the cost of a larger vessel would be repaid by increased efficiency, the owner will contemplate the purchase, but not otherwise. He will not be forced by a competitive scramble to catch as many fish as he can at the beginning of the fishing season. In essence, the individual quota is a kind of property right to a portion of the total annual catch. The fisherman can leave the fishery and sell his quota, without serious financial loss. Likewise, new fishermen can enter the fishery by purchasing the necessary quotas.

ITQs have recently been introduced into a number of commercial fisheries in Canada, Iceland, New Zealand, and Australia. Following the implementation of ITQs in the Pacific halibut fishery in 1991, for example, fresh halibut became readily available in Canadian fish markets for the first time in fifty years. (Previously, the entire annual catch had been taken in a few days' fishing, requiring that most halibut be frozen.) Because of the high demand for fresh halibut, prices paid to fishermen doubled overnight.

Transferable quotas essentially turn the stock of natural capital into a jointly owned stock. Like common shares, the quotas attain a value that reflects the conditions of the capital stock and the prospects for future markets. But quotas also have the feature of permitting their owners to participate directly in the fishery. Of course, since fishing is now remunerative, there will exist an incentive for outsiders to catch fish, and for quotas holders to exceed their quotas. If the ITQ system is to be effective, catches must be carefully monitored and the quotas strictly enforced upon each fisherman. The quota owners will themselves have an incentive to ensure that these controls are properly administered, since the amortized value of their quotas depends on it.

Properly administered, individual transferable quotas have the potential for achieving a considerable degree of economic rationalization—and biological conservation. They explicitly recognize that a fishery resource needs to be treated as a valuable resource asset of natural capital, and managed accordingly.

REAL-WORLD COMPLEXITIES

Our simple model of optimal fishery investment/disinvestment strategies overlooks many practical complexities of actual fisheries management. We have already referred to problems arising from fluctuations and uncertainty. In this section we briefly discuss another source of complexities: the biological structure of fish populations.

Space precludes the discussion of other topics, such as user conflicts, marine pollution, and the like.

Treating a fish stock in terms of a single variable (x) is highly unrealistic from a biological standpoint. For example, in most fish populations age structure is important both biologically and economically (big, old fish are worth more, have greater fecundity, and may be easier to catch). Optimizing the age of capture may be as important as optimizing the total harvest rate. But aging is a well-known aspect of capital theory, the general principles of which still apply. The details, which are discussed in the bioeconomics literature, need not concern us here.

Equally important, but more difficult to deal with, is the fact that most species of fish eat, and are eaten by other species. Ecological interactions in the marine environment are often exceedingly complex and poorly understood. The depletion of a given species by commercial fishing may have unforeseen consequences for other species, possibly also of commercial importance. For example, pelagic schooling species, such as anchovies, sardines, and herring often support large fisheries, but are also primary food sources for cod, salmon, and whales. Where such ecological interactions are important, management based on single-species models may be inappropriate. An important example occurs in the Antarctic Ocean, where krill are harvested commercially, but are also the basis of the food chain for fish, birds, whales, and other species. The International Convention for Conservation of Antarctic Marine Living Resources (CCAMLR), signed in 1980, was revolutionary in explicitly recognizing ecological interconnections as part of its management principles (May et al. 1978).

A further bioeconomic interaction often results from multispecies catches by the same vessels. For example, a typical fishing vessel in tropical waters may return to port with upwards of 100 species of fish. In British Columbia more than 3,000 genetically distinct populations of salmon (six species) have been identified, many of which may be caught in the same nets. In such situations the more vulnerable populations may gradually be fished out completely. Determining and implementing an optimal investment-disinvestment strategy may be difficult indeed. The fact that different user groups (e.g., commercial vs. sports fishermen) may be interested in different species further complicates the problem.

Even though difficult, species interaction, either ecological or arising through fleet activity, does not require us to abandon our capital theoretic approach. Instead of thinking of the resource manager as managing a single resource asset, we now think of the resource manager managing a complex *portfolio* of resource assets. It becomes important to focus on the economic return on the *portfolio* as a whole, rather than on individual resource assets.

The heterogeneous structure of fish populations may cause difficulties in the implementation of ITQs. When individual catches consist of a mixture of fish types (species or size of fish), fishermen holding lumped quotas may discard less desirable types. Alternatively, if quotas for valuable species have been exhausted, the fisherman may discard those species. Reported catches may then misrepresent actual rates of fishing mortality. The severity of these problems will depend on particular circumstances; controlling the location and timing of fishing may help alleviate some of the difficulties.

In spite of a long history of errors and blunders, fishery management may yet provide a paradigm for the treatment of natural resources as forms of natural capital.

ACKNOWLEDGMENTS

Comments on the first draft of this chapter by Dr. Valentina Krysanova and an anonymous referee are gratefully acknowledged. Research expenses of CWC are partly covered by NSERC Operating Grant 83990.

REFERENCES

Burmeister, E. 1980. Capital Theory and Dynamics. Cambridge, U.K.: Cambridge Univ. Press.

Canada Ministry of Supply and Services. 1982. Report of the Task Force on Atlantic Fisheries. Ottawa: Ministry of Supply and Services.

Charles, A. 1983. Optimal fisheries investment under uncertainty. *Canadian Journal of Fisheries and Aquatic Science* 40: 2080–91.

Clark, C. W. 1980. Towards a predictive model for the economic regulation of commercial fisheries. *Canadian Journal of Fisheries and Aquatic Science* 37: 1111–129.

———. 1990. Mathematical Bioeconomics: the Optimal Management of Renewable Resources. New York: Wiley.

Clark, C. W., and G. R. Munro 1975. The economics of fishing and modern capital theory: a simplified approach. *Journal of Environmental Economists and Management* 2: 92–106.

Gordon, H. S. 1954. The economic theory of a common-property resource: the fishery. *Journal of Political Economy* 62: 124–42.

Hardin. G. 1968. The tragedy of the commons. *Science* 162: 1243–7.

Mangel, M. 1985. Decision and Control in Uncertain Resource Systems. New York: Academic Press.

May, R. M., J. R. Beddington, C. W. Clark, and S. J. Holt. 1979. Management of multispecies fisheries. *Science* 205: 267–77.

Munro. G. R. 1992. Mathematical bioeconomics and the evolution of modern fisheries economics. *Bulletin of Mathematical Biology* 54: 163–84.

Munro, G. R., and A. D. Scott. 1985. The economics of fishery management. In Handbook of Natural Resource and Energy Economics, eds. A. V. Kneese and J. L. Sweeny. Amsterdam: North-Holland.

Samuelson, P. A. 1965. A catenary turnpike theorem involving consumption and the golden rule. *American Economic Review* 55: 486–96.

Savory, A. 1988. Holistic Resource Management. New York: Island Press.

Walters, C. J. 1986. Adaptive Management of Renewable Resources. New York: Macmillan.

20 ECOLOGICAL FOOTPRINTS AND APPROPRIATED CARRYING CAPACITY: MEASURING THE NATURAL CAPITAL REQUIREMENTS OF THE HUMAN ECONOMY

William E. Rees and Mathis Wackernagel
The University of British Columbia
School of Community and Regional Planning
6333 Memorial Road
Vancouver, BC Canada V6T 1Z2

ABSTRACT

This chapter advances a novel approach to estimating the natural capital requirements of the economy based on considerations of human carrying capacity. We begin by contrasting conventional economic rationality with economic principles. Many economists reject the notion that carrying capacity imposes serious constraints on material growth on grounds that substitution and imports can overcome local resource constraints. However, monetary analyses are themselves incapable of producing useful estimates of the physical stocks and processes required to sustain defined human population. We therefore develop an empirical approach to this measurement problem based on a reinterpretation of carrying capacity that can account for technological advances and trade.

Carrying capacity is defined as the maximum population of an organism a given habitat can support indefinitely and could therefore be related to the ecological sustainability of a human population living in an isolated region. However, in an economy globally integrated through trade, regions can no longer be viewed as independent units. Therefore, rather than asking how many people a given region can support, the relevant question becomes: How much land/water, wherever it may be located, is required to produce the resources flows (consumption) currently enjoyed by that region's population? Our estimates of this quantity represent the total "ecological footprint" of the population on the Earth and provide a physical measure of its demand on natural capital.

Our data shows that typical urban industrial regions (>300 people per km²) appropriate the bioproduction of 10–20 times more land than is usually contained within the regions themselves. Extrapolation reveals that human material demand now exceeds the long-term carrying capacity of Earth.

These results suggest that while unrestricted trade may relieve local ecological constraints, it actually reduces long-term global carrying capacity. Ecological footprint analysis therefore provides a new framework in which to consider global development and North-South relationships. Indeed, in a world of expanding populations, rising material expectations, and deteriorating ecosystems, our model underscores the necessity of reevaluating the prevailing international development paradigm and diverting much of present consumption to investment in the maintenance of natural capital stocks.

INTRODUCING THE DIALECTIC

There are two competing visions of global economic reality. One of these, the *expansionist worldview,* is the dominant social paradigm (Taylor 1991; Milbrath 1989). Its confident logic shapes the macroeconomic policy of the world's major countries and provides the economic rationale driving mainstream international development efforts today. The other vision is an *ecological worldview*. Not fully formed and inherently less confident, this perspective has to date been little more than minor if increasingly persistent irritant snapping at the heels of its dominant rival.

Nothing is closer to the center of the tension than the question of whether the ecosphere imposes practical constraints on the material activities of humankind. Lawrence Summers, chief economist of the World Bank and among the most outspoken engineers of the expansionist vision, has said:

> There are no limits to carrying capacity of the Earth that are likely to bind at any time in the foreseeable future. There isn't a risk of an apocalypse due to global warming or anything else. The idea that the world is headed over an abyss is profoundly wrong. The idea that we should put limits on growth because of some natural limit is a profound error (George 1992).

By contrast, the ecological perspective holds that the "profound error" resides wholly in Summers' statement. As Garrett Hardin most succinctly put it: "carrying capacity is the fundamental basis for demographic accounting" (Hardin 1991).

The dominant perspective, as articulated by the World Bank (1992) and the 1987 U.N. World Commission on Environment and Development (WCED), acknowledges the ecological damage caused by development. However, it sees developing world problems such as soil erosion, and the lack of clean water and sewers (failing infrastructure generally) as the most pressing issues and poverty as the cause. It follows that to fix the environment, we have to fix poverty and "the cure for poverty is growth" (*The Economist* 1992). Indeed, the Brundtland Commission effectively equated sustainable development with "more rapid economic growth in both industrial and developing countries" and observed that "a five- to ten-fold increase in world industrial output can be anticipated by the time world population stabilizes some time in the next century" (WCED 1987).[1] However, by failing to assess the biophysical feasibility of this prescription, the Commission put carrying capacity at center stage in the evolving world development debate.

We suggest that an ecological perspective on carrying capacity is essential to any rational approach to the global development conundrum. There are three simple reasons for this. First, despite our technological wizardry and assumed mastery over the environment, humankind remains a creature of the ecosphere existing in a state of obligate dependency upon many products and processes of nature (Rees 1990). On the simplest level, our ecological relationships to the rest of the ecosphere are indistinguishable from those of the millions of other species with which we share the planet. Like all other organisms, we survive and grow by extracting energy and materials from those ecosystems of which we are a part. Like all other organisms, we "consume" these resources before returning them in altered form to the ecosphere. Second, the five-fold increase in the human economy in the post-war period has begun to induce ecological change on a global scale that simply can no longer be ignored. Finally, orthodox economic analysis is so abstracted from biophysical reality that its ability to detect, let alone advise on critical dimensions of carrying capacity, is severely compromised.

ECONOMICS AS ERRANT HUMAN ECOLOGY

Ecology is often defined simply as the study of the relationships between organisms and their environments. A more insightful definition is "the experimental analysis of distribution and abundance" [of plants and animals] (Krebs 1972). However, from an ecological economics perspective, ecology is best defined as the scientific analysis of the flows of energy, material, and information through ecosystems and of the competitive and cooperative mechanisms that have evolved for the allocation of resources among different species. This definition stresses the homology of ecology and economics, the latter commonly being defined as the scientific study of the efficient allocation of scarce resources (energy, material, and information) among competing uses in human society. From this perspective, ecology and economics are seen to share not only the same semantic roots, but also the same substantive focus. In fact, it could logically be argued that economics is really human ecology.

Or rather, it should be. The problem is that mainstream economics has deviated markedly from the theoretical foundations that still support its sister discipline. The material ecology of *other* species has roots in the chemical and thermodynamics laws that are the universal regulators of all transformations of energy and matter in the organic world. Economics, by contrast, had abandoned its classical organic roots by the end of the 19th Century. Neoclassical economics (which has recently enjoyed a remarkably uncritical renaissance

---

1. While this may seem like an extraordinary rate of expansion, it implies an average annual growth rate in the vicinity of only 3.5%–4.5% over the next 50 years. Growth in this range has already produced a near five-fold increase in world economic output since World War II.

the world over) is firmly based on methods and concepts borrowed from Newtonian analytic mechanics.

The result of this divergence is a dominant economic paradigm that "lacks any representation of the materials, energy sources, physical structures, and time-dependent processes basic to an ecological approach" (Christensen 1991). Prevailing theory therefore produces analytic models based on reductionist and deterministic assumptions about resources, people, firms, and technology that bear little relationship to their counterparts in the real world (Christensen 1991). In short, mainstream economists inevitably ignore critical elements of ecological theory, having sought refuge in the more theoretically tractable but environmentally less relevant realm of mechanical physics.

We are therefore confronted with a double irony in applied human ecology. Conventional economists, arguably the most influential of human ecologists, are also the most theoretically errant. Meanwhile, ecologists, who start from appropriate theory, have all but ignored humankind. *This severely limits the contribution of both disciplines to resolving the global ecological crisis.*[2]

Will the Myth Most Closely Approximating Reality Please Stand Up?

Four important consequences of this theoretical dichotomy will serve to illustrate the dilemma:

- Traditional economic models often represent the economy as essentially separate from and independent of "the environment." By contrast, an ecological economic perspective would see the human economy as an inextricably integrated, completely contained, and wholly dependent sub-system of the ecosphere.

- Economic theory treats capital and individual inputs to production as inherently productive, ignoring both their physical connectedness to the ecosphere and the functional properties of exploited ecosystems. By contrast, systems ecology emphasizes connectivity, particularly material and energy flows, in relation to the functional integrity of ecosystems.

- According to neoclassical theory, resource depletion is not a fundamental problem—rising prices for scarce resources automatically lead to conservation and the search for substitutes (Barnett and Morse 1963; Dasgupta and Heal 1979). As Nobel Laureate, Robert Solow, observes, "If it is very easy to substitute other factors for natural resources, then....The world can, in effect, get along without natural resources, so exhaustion is just an event, not a catastrophe" (Solow 1974).[3] Today, the conventional wisdom holds that substitution through

---

2. This critique is not aimed at entire disciplines but rather at the particular "brand" of economics that currently dominates the development policy arena and at mainstream academic ecology. Many economists do work with dynamic, ecologically realistic, multiple equilibrium models, and many systems ecologists do focus similar tools on the impacts of human beings.

3. Solow (1991) acknowledges, however, that "there is no reason to believe in a doctrinaire way" that "the goal...of sustainability can be left entirely to the market."

technological progress has, in fact, been more than sufficient to overcome emerging resource scarcities (Victor 1991).By contrast, ecological analysis reveals that humankind remains in a state of obligate dependency on numerous deteriorating biophysical goods and services with great positive economic value but for which there are no markets or feasible substitutes (e.g., the ozone layer). In the absence of markets, the already questionable scarcity indicators of conventional economics—prices, costs, and profits—fail absolutely.

• Finally, the mechanical metaphor describes an economy that is self-regulating and self-sustaining in which complete reversibility is the general rule (Georgescu-Roegen 1975). From this perspective, the starting point for economic analysis is the circular flow of exchange value (Daly 1990). By contrast, thermodynamic reality means the economy is sustained entirely by low entropy energy and matter produced "externally" by ecosystem and biophysical processes. Thus all economic production is actually consumption—the ecologically relevant material and energy flows through the economy are not circular but unidirectional and irreversible (Rees 1990).

This last factor is crucial to any attempt to account for the ecological effects of any economic process. Without reference to entropic throughput "it is virtually impossible to relate the economy to the environment," yet the concept is "[all but] absent from economics today" (Daly 1990).

NATURAL CAPITAL AND "LIVING ON THE INTEREST"

Some economists have accepted the argument that sustainability[4] requires the conservation of certain biophysical entities and processes. These "resources" may have immeasurable economic value, yet are often not even recognized as inputs to the economy. They maintain the life-support functions of the ecosphere, the risks associated with their depletion are unacceptable, and there are no technological substitutes. For these reasons, *"conserving what there is could be a sound risk-averse strategy"* (Pearce et al. 1990 [emphasis added]).

Ecological economists have begun to regard such assets as a special class of "natural capital," and are exploring various interpretations of a "constant capital stock" condition for sustainability (Costanza and Daly 1990; Daly 1990; Pearce and Turner 1990; Pearce et al. 1989, 1990; Pezzey 1989; Rees 1993).[5] The following interpretation is most relevant to the concept of carrying capacity:

---

4. Sustainable development is defined as positive socioeconomic change that does not undermine the ecological systems or basic social infrastructure upon which society is dependent. "Development" implies qualitative betterment or improvement, which may or may not involve material growth within ecosystems constraints. For political viability, specific measures for sustainability require the support of the people through their governments, their social institutions, and their private activities (Rees 1989).
5. The idea of inviolable resources stocks is anathema to conventional economists who argue that resources should be used to generate more wealth, including more productive substitutes for the original resource.

Each generation should inherit an adequate stock of natural assets *alone* no less than the stock of such assets inherited by the previous generation.[6]

This interpretation reflects basic ecological principles, particularly the multifunctionality of biological resources. It corresponds to Daly's (1990) definition of "strong sustainability" which recognizes that manufactured and natural capital "are really not substitutes but complements in most production functions" (Daly 1990). The constant natural stocks criterion also implies that, for the foreseeable future, humankind must learn to live on the annual flows—the "interest"—generated by remaining stocks of natural capital (Rees 1990). It is therefore related to Hicksian (or sustainable) income, the level of consumption that can be maintained from one period to the next without reducing wealth (productive capital). Of course, if populations or material standards increase, natural capital stocks would have to be enhanced to satisfy demand (see Figure 20.1).

Determining what mix and just how much ecosystems capital to preserve remains a major problem. Microeconomic theory suggests that *development* should proceed only to the point at which the marginal costs of natural capital depletion (diminished ecological services) begin to exceed the marginal benefits produced (additional jobs and income). Macroeconomics, however, has historically ignored the question of optimal aggregate scale. In any event, marginal analysis assumes that we can identify, quantify, and price all relevant life support functions and that any change in the properties of ecosystems under stress will be smoothly continuous (i.e., predictable) and reversible. Unfortunately, neither assumption holds (Rees 1993; Holling 1994 this volume).[7] This approach also leaves unanswered the questions of substitutability within natural capital and how global natural capital requirements should be allocated geographically. Here, too, conventional theory fails—prevailing economic rationality is indifferent to equity considerations or place, often reducing the whole economy to a single statistic. Thus, in the specification of optimal stocks, monetary analysis provides but a single critical insight: beyond some theoretical optimum, material growth is actually "anti-economic growth" that ultimately "makes us poorer rather than richer" (Daly 1990).

---

6. "Natural assets" encompasses not only material resources (e.g., petroleum, the ozone layer, forests, soils) but also process resources (e.g., waste assimilation, photosynthesis, soils formation). It also includes renewable as well as exhaustible resources. Our emphasis here is on the need to maintain adequate stocks of renewable biophysical resources. (Note that the depletion of non-renewables could be compensated for through investment in renewable assets.)

7. Economists regard cost-benefit analysis as the definitive tool in environmental decision making. However, "...difficulties with missing data, uncertainty, and too little time and resources for an exhaustive analysis combine with the theoretical difficulties to make ineffectual any serious claim that an applied study produces an optimal or theoretically justified outcome" (Lave and Gruenspecht 1991).

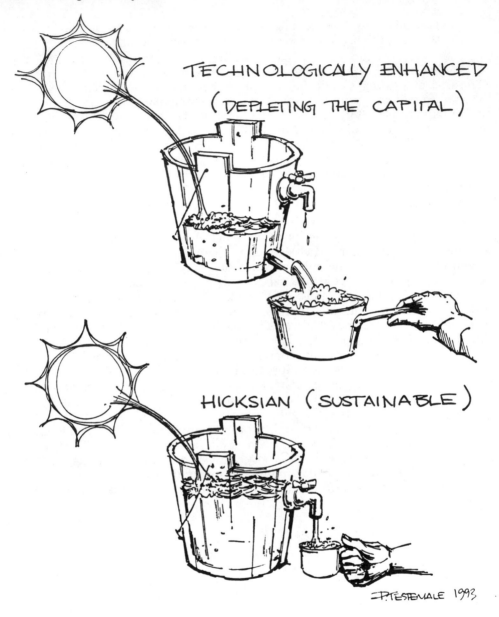

Figure 20.1. Rates of resource use: technology has progressively enabled humankind to exploit natural capital far beyond sustainable levels.

ECOLOGICAL FOOTPRINTS AND APPROPRIATED CARRYING CAPACITY

This section describes an alternative empirical approach to the optimal stocks question. We derive estimates of the actual physical stocks of natural capital necessary to sustain a given human population and compare this to carrying capacity of the population's home territory. This ecological approach avoids pricing problems altogether. While arguably central to ecology/economy integration, the ideas explored here are largely ignored in the mainstream policy arena.

In fact, many economists have totally rejected the concept of ecological carrying capacity (and remain unaware of the natural capital concept). Some do acknowledge that certain countries may face carrying capacity limitations "...even if the rest of the world is poised for sustainable growth indefinitely without significant environmental or resource constraints" (Muscat 1985). However, a 1986 economists' committee report on population growth and economic development for the U.S. National Research Council (1986) is more typical: "...neither the word *nor the concept* of 'carrying capacity' played a role" (Hardin 1991). The conventional wisdom seems to be that "the carrying capacity of the planet in terms of food (and other raw materials) appears to be well in excess of any likely human population magnitudes for the next century" (Muscat 1985).

Ecological analysis, however, reveals dimensions of the human population/resources problem that are invisible to conventional economic rationality. For example, economists see cities as loci for intense socioeconomic interaction among individuals and firms, and as engines of *production* and national economic growth. By contrast, ecology highlights the extended relationships among concentrated human populations, patterns of *consumption*, and the inward flows of usable energy and material. The latter approach shows that the common perception of the city as specific geographic location is illusory—urban areas can survive only if reliable supplies of low entropy material resources and surplus waste absorption capacity is being produced elsewhere in the ecosphere (Overby 1985). From the ecological perspective, this absolute dependency underscores the fact that cities and developed regions are mostly not where they appear to be!

Carrying Capacity Revisited

Ecologists define "carrying capacity" as the population of a given species that be supported indefinitely in a defined habitat without permanently damaging the ecosystem upon which it is dependent. However, because of our culturally variable technology, different consumption patterns, and trade, a simple territorially-bounded head-count cannot apply to human beings. Human carrying capacity must be interpreted as the maximum rate of resource consumption and waste discharge that can be sustained indefinitely without progressively impairing the functional integrity and productivity of relevant ecosystems *wherever the latter may be*. The corresponding human population is a function of *per capita* rates of material consumption and waste output or net productivity divided by *per capita* demand (Rees 1990). This formulation is a simple restatement of Hardin's (1991) "Third Law of Human Ecology":

(Total human impact on the ecosphere) = (Population) x (*Per capita* impact).

Early versions of this law date from Ehrlich and Holdren who also recognized that human impact is a product of population, affluence (consumption), and technology: I = PAT (Ehrlich and Holdren 1971; Holdren and Ehrlich 1974). The important point here is that a given rate of resource throughput can support fewer people well or greater numbers at subsistence levels.

Now the inverse of traditional carrying capacity provides an estimate of natural capital requirements in terms of productive landscape. Rather than asking what population a particular region can support sustainably, the question becomes: How much productive land and water area in various ecosystems is required to support the region's population indefinitely at current consumption levels?

Our preliminary data for developed regions suggest that *per capita* primary consumption of food, wood products, fuel, and waste-processing capacity co-opts on a continuous basis up to several hectares of productive ecosystem—the exact amount depends on the average levels of consumption (i.e., material throughput). This average *per capita* "personal planetoid" can be used to estimate the total area required to maintain any given population. We call this aggregate area the relevant community's total "ecological footprint" (see Figure 20.2) on the earth (Rees 1992).

This approach reveals that the land "consumed" by urban regions is typically at least an order of magnitude greater than that contained within the usual political boundaries or the associated built-up area. However brilliant its economic star, *every city is an entropic black hole* drawing on the concentrated material resources and low-entropy production of a vast and scattered hinterland many times the size of the city itself. Borrowing from Vitousek et al. (1986), we say that high density settlements "appropriate" carrying capacity from all over the globe, as well as from the past and the future (Wackernagel 1991).

The Vancouver-Lower Fraser Valley Region of British Columbia, Canada, serves as an example. For simplicity's sake consider the region's ecological use of forested and arable land for domestic food, forest products, and fossil energy consumption alone: assuming an average Canadian diet and current management practices, 1.1 ha of land *per capita* is required for food production, 0.5 ha for forest products, and 3.5 ha would be required to produce the biomass energy (ethanol) equivalent of current *per capita* fossil energy consumption. (Alternatively, a comparable area of temperate forest is required exclusively to assimilate current *per capita* CO_2 emissions (see "Calculating the Ecological Footprint"). Thus, to support just their food and fossil fuel consumption, the region's 1.7 million people require, conservatively, 8.7 million ha of land in continuous production. The valley, however, is only about 400,000 ha. Our regional population therefore "imports" the productive capacity of at least 22 times as much land to support its consumer lifestyles as it actually occupies (see Figure 20.3). At about 425 people/km² the population density of the valley is comparable to that of the Netherlands (442 people/km²).

Figure 20.2. The ecological footprints of individual regions are much larger than the land areas they physically occupy.

Calculating the Ecological Footprint

The ecological footprint concept is rooted in the idea that for most types of material or energy consumption, a measurable area of land and water in various ecosystems is required to provide the consumption-related resource flows and waste sinks. Thus, to estimate the ecological footprint of a particular community or region we must understand the land- use implications of each significant category of consumption. The sum of the land requirements for individual categories represents that community's ecological footprint, the total area "appropriated" from nature to support its particular pattern of consumption.

Since it is impossible to estimate the land required for the provision, maintenance, and disposal of every consumption good, we confine our calculations to major categories. To increase the policy relevance of our findings and to facilitate data collection, we adopt the consumption classifications and statistics used in official reports prepared for other purposes.

We have also adopted the six simple land-use categories proposed in the 1991 IUCN report, *Caring for the Earth*—consumed or degraded land (such as is occupied by the built environment), land currently used in gardens, crop land, pastureland (including grasslands), productive forest, and energy land. We have only begun to assess the role of the oceans.

"Energy land" can be defined in two ways. The first provides an estimate of the area of average productive land that would be required to produce a flow of high-quality biomass energy today (e.g., ethanol) equivalent to the present flow of commercial hydrocarbon energy. We use optimistic biomass-to-ethanol conversion efficiencies and therefore derive fairly conservative land-use equivalents. This approach assumes the eventual depletion of economic petroleum reserves or their abandonment as fuel for ecological reasons, such as the risk of climate change. It treats energy as a potentially renewable resource based on recycling carbon through the biosphere's active carbon pool, thereby reducing greenhouse gas accumulation.

An alternative estimate of fossil energy land requirements can be obtained by calculating the area of "carbon-sink" forests that would be required to sequester the CO_2 emissions released by contemporary hydrocarbon combustion. This approach assumes that non-renewable hydrocarbons will remain the dominant energy source for industrial societies in the foreseeable future and that fossil carbon extracted from an historically inactive pool should be kept out of circulation to avoid atmospheric change. Various estimates suggest that growing forests can assimilate up to 6 tonnes carbon/ha/year (net), with the rate generally declining from tropical to boreal forests. Thus, a highly productive forest of 2 million km^2 (200 million ha or about 5 times the area of Japan) could remove more than one gigatonne of carbon per year (or about one-fifth of the current level of fossil carbon emissions) for 10–40 years, depending on forest type (Karas and Kelly 1989; Worldwatch Institute 1988, 1989; Brown 1986). We should note that at current rates of soil depletion, deforestation, and fossil energy use there is a significant imbalance between CO_2 output and assimilation. About half of the world's carbon emissions (3 gigatonnes/year) are apparently accumulating in the atmosphere (Canada 1991).

Both definitions of energy land described above are conceptually related through the carbon cycle to the "phantom land" that Catton (1980) identified as land borrowed from the past and used today through fossil fuel consumption.

Paved over, built upon, or badly eroded land is considered to have been "consumed" as it is no longer biologically productive. An additional debit could be charged against such degraded lands by estimating the material, energy, and time that would be required to restore productivity. However, our current studies do not include this step.

We are also working with five major classes of consumption: food, housing, transportation, consumer goods, and the resources embodied in services received. We summarize the results for a particular study area or population in a simple two-way matrix. Each of the 30 cells (5 consumption classes • 6 land-use categories) converts a particular consumption rate into its corresponding appropriated land area. Presently we are developing a working handbook for communities to use in estimating their local ecological footprints. The manual contains ACC calculations for 22 categories of consumption in the five classes noted above (Wackernagel et al. 1993).

Our current approach almost certainly underestimates sustainable land-use requirements. For example, we have yet to attempt estimates of the land and water needed for waste assimilation (with exception of energy-related CO_2); our data assume that current forestry and agricultural practices are sustainable when in reality typical cultivated soils in North America are depleted 18 times faster than they self-renew (Pimentel et al. 1987); we use a biomass-to-ethanol conversion efficiency (NREL 1992) considerably higher than those accepted by other studies (e.g., Giampietro 1992; Giampietro and Pimentel 1990a, b; McKetta 1984). In short, however startling our results may appear, they are actually conservative estimates of the resource flows and productive land appropriated to sustain the human economy.

The following simplified examples illustrate how consumption can be translated into land-use:

Example 1: Energy Consumption

Question: How much land would be required to produce the biomass energy equivalent of current *per capita* fossil energy consumption in Canada?

Ethanol is the best current renewable substitute for fossil energy. We assume that conversion technology could produce 80 Gj/ha/yr net as ethanol (NREL 1992). Most other estimates fall below this. Indeed, studies that assume fossil energy use for ethanol processing consistently show net available energy losses.

$$\frac{277 \text{ [Gj/year/capita]}}{80 \text{ [Gj/ha/year]}} = 3.46 \text{ [ha]}$$

for energy production alone.

Example 2: Forest Area for Paper Consumption

Question: How much forest land is dedicated to paper production?

Canadians consume approximately 6,200,000 t of paper each year. For each tonne of paper, 1.8 m³ of wood are required (Policy and Economics Directorate 1991). Estimates suggest prime temperate coniferous forest land produces about 2.7 m³/ha of wood per year (average over 70 year rotation).

Therefore, the average Canadian requires

$$\frac{6,200,000 \ [t] \bullet 1.8 \ [t/m^3]}{25.6 \ \text{million [Canadians]} \bullet 2.7 \ [m^3/ha/yr]} = 0.16 \ [ha]$$

of prime forest in continuous production to provide the fiber for paper.

Example 3: Area Covered by Roads and Rail

Question: On how much land per person is covered by roads, highways, and railways (U.S. example)?

Roads and highways cover 21.5 million acres and railways occupy 7 million acres of land in the United States (Corson 1990). On average, U.S. residents therefore appropriate

$$\frac{28,500,000 \ [ac]}{2.47 \ [ac/ha] \bullet 245,000.000 \ [\text{Americans}]} = 0.05 \ [ha/capita]$$

for vehicular ground transportation.

Even with generally lower *per capita* consumption, European countries live far beyond their ecological means. For example, the Netherlands' population (see Figure 20.4) consumes the output of at least 14 times as much productive land as is contained within its own political boundaries (approximately 110,000 km² for food and forestry products and 360,000 km² for energy) (basic data from WRI 1992).[8]

IMPLICATIONS OF INTER-REGIONAL TRADE FOR GLOBAL CARRYING CAPACITY

The economists' case against carrying capacity stands in part on the twin pillars of technological development (which can increase the efficiency of resource use and substitute manufactured for natural capital), and inter-regional trade (which can relieve any locally significant constraints on growth). Next we examine the hidden ecological dimensions of trade.

---

8. The Reijksinstituut voor Volksgezondheid en Milieuhygiene in the Netherlands suggests that for food production alone that country appropriates 170,000 to 240,000 km² of agricultural land (Meadows et al. 1992).

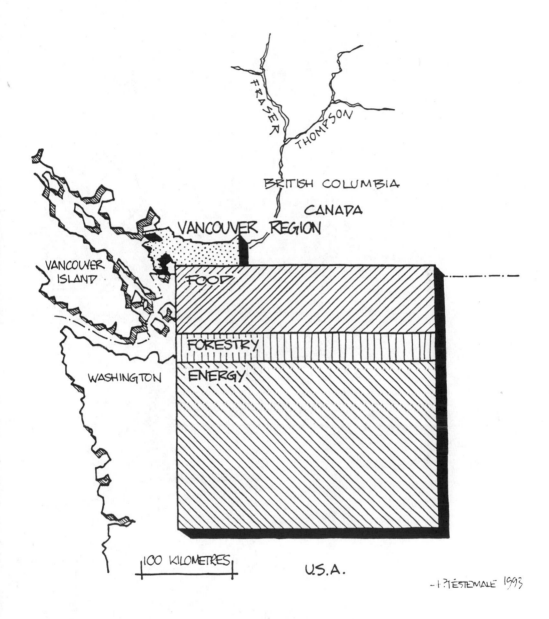

Figure 20.3. The Vancouver-Lower Fraser Valley Region appropriates from nature the ecological production of an area 22 times larger than the Lower Fraser Valley itself.

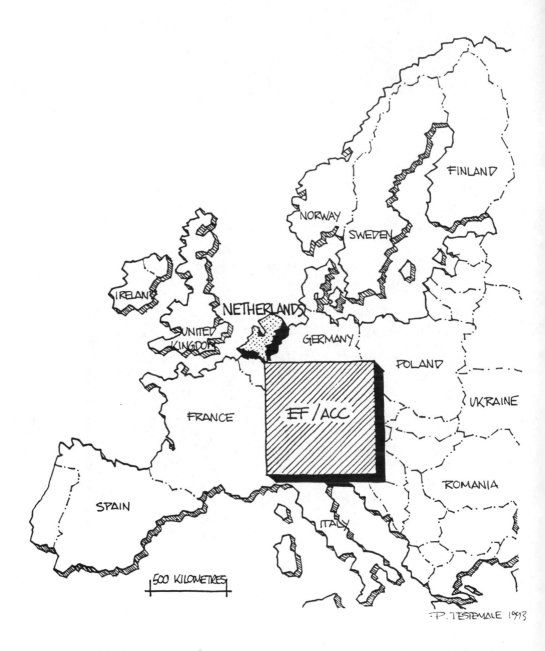

Figure 20.4. The Netherlands' population density is similar to that of the Lower Fraser
            Valley. To provide for their consumption, its people use an area 14 times
            larger than their country.

If all human populations were able to live within their own regional carrying capacities (i.e., on the continuous flows generated by local natural capital), the net effect would be global sustainability. However, no region exists as an independent unit—the populations of all urban regions and many whole nations already exceed their territorial carrying capacities and depend on trade and natural capital depletion for survival. Such regions are running an unaccounted ecological deficit—their populations are *appropriating carrying capacity from elsewhere or from future generations*.

Regional ecological deficits do not necessarily pose a problem if import-dependent regions are drawing on true ecological surpluses in the exporting regions. A group of trading regions remains within net carrying capacity as long as total consumption does not exceed aggregate sustainable production. The problem is that prevailing economic logic and trade agreements ignore carrying capacity and sustainability considerations. Commodity flows and the balance of trade are monitored in monetary, not ecological, units. In these circumstances, the terms of trade may actually accelerate the depletion of essential natural capital thereby undermining global carrying capacity.

Open Economies: The "Distancing" Effect of Urbanization and Trade

Urbanization and trade have the effect of physically and psychologically distancing urban populations from the ecosystems that sustain them. Access to bioresources produced outside home regions both undermine peoples' sense of dependency on the land and blinds them to the far-off social and ecological effects of imported consumption. These phenomena illustrate geographic or spatial discounting, distance effects that introduce systematic biases into peoples' valuation of place (Hannon 1992; Daily and Ehrlich 1992).

Importing Carrying Capacity

Because the products of nature can so readily be imported, the population of a given region can expand beyond local carrying capacity unknowingly and with apparent impunity.[9] In the worst cases, trade actually contributes to reduced local capacity. By eliminating negative feedback on population or lifestyles, trade reduces the direct incentive for participating regions to maintain adequate local stocks of productive natural capital. For example, the ability to import food makes people less averse to the risks associated with urbanization of locally limited agricultural land. Indeed, if low-priced imports undermine the viability of local agriculture, it becomes economically rational to convert the land to some other activity that yields a higher return.

Meanwhile, rising consumption in the expanding network of trading regions may be drawing down "surplus" natural capital stocks everywhere. As global markets gain access to locally limited resources, upward pressure on

9. Keep in mind that the economic role of trade is precisely to relieve local constraints, thereby increasing the mutual potential for and efficiency of economic growth.

prices may encourage over-exploitation. Ironically, in a competitive market, downward pressure on prices may have the same effect as sellers compensate for reduced revenues through increased volume. Paradoxically then, trade only appears to increase carrying capacity. In reality, by encouraging all regions to exceed local limits, by reducing the risks associated with local resource depletion, and by eroding the global safety net, it actually increases the long-term threat to everyone.

This situation applies not only to commercial trade but also to the unmonitored flows of goods and services provided by nature. Northern urbanites, wherever they are, are now dependent on the carbon sink, global heat transfer, and climate stabilization functions of tropical forests. There are many variations on this theme, from drift-net fishing to ozone depletion, each involving open access to, or shared dependency on, some form of valuable natural capital. Pollution externalities and the so-called "common property problem" increasingly erode the integrity of the global commons.

Exporting Ecological Degradation

The importing of carrying capacity by wealthy industrialized countries may equate to exporting ecological and social malaise to the developing world. Many impoverished countries have little other than agricultural, fisheries, and forest products to offer world markets and current development models encourage the "modernization" and intensification of production of specialized commodities for export. Unfortunately, the consolidation of land and other scarce resources to produce luxury export crops often jeopardizes domestic staples production, contributing to local food shortages and malnutrition. The process also displaces thousands of small farmers, farm workers, and their families from the better agricultural lands. This concentrates wealth in relatively few hands while forcing the dislocated peasants either to eke out a subsistence existence on inferior land that should never go under the plow, or to migrate to already overcrowded cities.[10] Erosion, desertification, and deforestation in the countryside and debilitating social and environmental problems in burgeoning squatter settlements around many Third World cities are the frequent result.

These trends are invisible to First World consumers who otherwise might not feel so sanguine about their imported coffee, tea, bananas, cocoa, groundnuts, sugar, cotton, etc. The distancing of consumers from the negative impacts of their consumption therefore only makes things easier for local elites, international capital, and transnational corporations, all of whom benefit from

---

10. In theory, the wealth generated by exporting local carrying capacity is supposed to "trickle down" to ordinary people, enabling them to purchase staple foods imported from more efficient producers at a lower cost than they could grow it themselves. However, trickle-down development theory is largely discredited and the benefits of the so-called green revolution are increasingly called into question. (See, for example, *The Ecologist*, March/April 1991, on the role of international development in "promoting world hunger.")

the expanding commodity trade. Meanwhile, the developing country itself may become increasingly dependent on export earnings, much of which must go to pay off the original "development" loans. This overall pattern has contributed to the net transfer of wealth and sustainability from the poor to the rich within developing countries and from the poorest countries to the richest among the trading nations. The latter transfers reached tens of billions of dollars annually in the late 1980s.

Ultimately, of course, the drain of debt servicing, downward pressure on prices in an increasingly competitive global market, import tariffs, and other market distortions, reduce the economic surpluses available for sound resource management in the developing world. This, combined with intensive cultivation methods, accelerates the erosion of the export regions' best lands, reducing both present productivity and the future potential for more sustainable forms of development.

DISCUSSION AND CONCLUSIONS: CARRYING CAPACITY, NATURAL CAPITAL, AND GLOBAL CHANGE

The dominant vision of the global economy is one in which "the factors of production are infinitely substitutable for one another" and in which "using any resource more intensively guarantees an increase in output" (Kirchner et al. 1985). In short, prevailing economic mythology assumes a world in which "carrying capacity is infinitely expandable" (Daly 1986).

By contrast, the ecological perspective holds that some biophysical resources and processes are irreplaceable and should not be "consumed" in the usual sense at all. Carrying capacity is ultimately constrained by the ability of self-renewing natural capital to continue providing ecological goods and essential life-support services. It should therefore be a fundamental consideration in demographic and resource analysis. Under prevailing assumptions, this conclusion simply eludes conventional analysis—certain of these critical ecological factors are not recognized by the economy at all and where markets do exist for ecological commodities, marginal prices may reveal nothing about the ecosystemic roles, the ultimate value, the necessary minimum quantities, or even the remaining volumes of relevant capital stocks.

Such structural flaws in conventional thinking now threaten the stability of our rapidly industrializing and increasingly urban world. Development policies generally ignore the fact that the sustainability of industrial regions and their wealth creation invariably depend on the continuous production of ecological goods and services somewhere else. In ecological terms, the city is a node of pure consumption existing parasitically on an extensive external resource base. While the latter may be spatially and temporally diffuse, "the relevant knowledge is that it must be somewhere, it must be adequate, it must be available, *and it must grow if the city grows*" (Overby 1985, emphasis added).

Consider this "relevant knowledge" in light of current trends pertaining to the state of global natural capital: encroaching deserts (6 million ha/year); deforestation (11 million ha/yr of tropical forests alone); acid precipitation and forest dieback (31 million ha damaged in Europe alone); soil oxidation and erosion (26 billion tonnes/yr in excess of formation); soil salination from failed irrigations projects (1.5 million ha/yr); draw-down and pollution of ground water; fisheries exhaustion; declining *per capita* grain production since 1984; ozone depletion (5% loss over North America [and probably globally] in the decade to 1990); atmospheric and potential climate change (25% increase in atmospheric CO_2 alone in the past 100 years). (Data from Brown et al. [Annual]; Brown and Flavin 1988; Canada 1988; WCRP 1990; Schneider 1990; U.S. Environmental Protection Agency [reported in Stevens 1991]). This partial list shows that far from growing with the expansion of the urban world, the resource base sustaining the human population is in steady decline. It should also remind us of an important corollary to Liebig's law of the minimum—carrying capacity is ultimately determined not by general conditions but by that single vital factor in least supply.

Catton has detailed the ecological history leading to such disquieting trends: the expansion of the human enterprise, particularly since the industrial revolution, has been sustained first by the "takeover" of other species' niches (energy and material flows), accompanied more recently by the "drawdown" of accumulated stocks of resources as illustrated here. He argues that humankind has long since "overshot" the permanent carrying capacity of the earth. His bleak prognosis is that we must now do whatever we can "to ensure that the inevitable crash consists as little as possible of outright die-off of *Homo Sapiens*" (Catton 1980). An important first step would be recognition that current levels of consumption in the industrialized countries are excessive—a significant portion of present "income" should be diverted to investment in rebuilding both national and global natural capital stocks.

This perspective also supports a disturbing interpretation of North-South geo-political tensions. The pollution, congestion, and land-use problems of Northern industrial regions stem largely from wealth and associated high levels of material consumption. By contrast, both rural ecological decline and the deplorable physical conditions and appalling public health standards of burgeoning southern cities stem from debilitating poverty and material deprivation. However, there is a connection. Much of the industrial countries' wealth came from the exploitation (sometimes liquidation) of natural capital, not only within their own territories, but also within former colonies. Colonial rule, with its *direct appropriation of extra-territorial carrying capacity* may have ended, but many of same resource flows continue today in the form of commercial trade. What used to require territorial occupation is now achieved through commerce!

Indeed, this persistent relationship is an inevitable consequence of thermodynamic law. The techno-economic growth and high material standards of developed countries require continuous net transfers of negentropy (exergy and available energy/matter) to the industrial center.[11] Conversely, less-developed regions and countries "must experience a net increase in entropy as natural resources and traditional social structures are dismembered" (Hornborg 1992).

To the extent that the restructuring of rural economies in the South to supply the North displaces people from productive landscapes to the cities, it is an direct cause of impoverishment, urban overpopulation, and local ecological decay. The people of developing countries cannot simultaneously live on carrying capacity exported to sustain the people of industrial countries. Moreover, to the extent that current development models and terms of trade favor net transfers of ecological wealth to the North and the continued depletion of natural capital in the South, poverty and ecological decline in the South are permanent conditions.

The Ecologist (1992) develops a similar argument in an issue dedicated to the proposition that the prevailing international development model represents the effective "enclosure" of the global commons: "The market economy has expanded primarily by enabling state and commercial interests to gain control of territory that has traditionally been used and cherished by others, and by transforming that territory—together with the people themselves—into expendable 'resources' for exploitation" (p.131). This argument is compatible with both the present analysis and Hornborg's view that "the ecological and socioeconomic impoverishment of the periphery are two sides of the same coin..." (Hornborg 1992).

We are suggesting that for all the ebullient optimism of development economists, many developing countries will not be able to follow northern industrial countries along their historical path to material well-being. There simply is not sufficient natural capital both to support the present world population at northern material standards and to maintain the functional integrity of the ecosphere (see "How Close to Practical Limits"). Figure 20.5 illustrates the dilemma by contrasting historic trends in land availability *per capita* with the productive area actually "appropriated" by wealthy consumers. It is not merely a question of adequate resources for direct consumption. The world must also now come to appreciate that many remaining stocks of natural capital as diverse as forests, fossil hydrocarbons, and the ozone layer are already fully committed *in place* providing unpriced life support services.

If development cannot safely be based on the further draw-down of natural capital, wealth transfers through, for example, the purchase of ecological services provided by nationally-held stocks of natural capital (e.g., tropical forests) may be the only peaceful alternative means to achieve greater global equity. Pearce (1991) shows that the purchase of such services by current "users" in the developed countries may well be a rational economic solution for both North and South.

---

11. Remember, manufactured capital can be sustained only by flows from natural capital.

Although never stated in quite these terms, the appropriation of most of the world's carrying capacity by the urban industrial North (and reluctance to give it up) and the insistence by the South of its right to a fair share (and the threat to seize what it can through sheer growth in numbers and inefficient technologies) were really the only issues at the Earth Summit in Rio in June 1992.

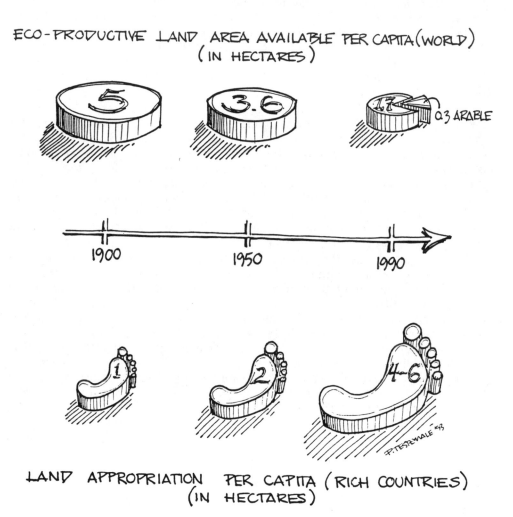

Figure 20.5. Trends in land availability per capita and actual ecological appropriation by wealthy consumers since 1900.

How Close to Practical Limits?

There is accumulating evidence that humanity may soon have to confront real carrying capacity constraints. For example, nearly 40% of terrestrial net primary productivity (photosynthesis) is already being used ("appropriated") by humans, one species among millions, and this fraction is steadily increasing (Vitousek et al. 1986). If we take this percentage as an index of the human carrying capacity of the earth and assume that a growing economy could come to appropriate 80% of photosynthetic production before destroying the functional integrity of the ecosphere, the earth will effectively go from half to completely full within the next doubling period—currently about 35 years (Daly 1991).

The significance of this unprecedented convergence of economic scale with that of the ecosphere is not generally appreciated in the current debate on sustainable development. Because the human impact on critical functions of the ecosphere is not uniform "effective fullness" may actually occur well before the next doubling of human activity. (Liebig's law reminds us that it takes only a single critical limiting factor to constrain the entire system.) Indeed, data presented in this chapter suggests that long-term human carrying capacity may already have been reached at less than the present 40% preemption of photosynthesis. If so, even current consumption (throughput) cannot be sustained indefinitely, and further material growth can be purchased only with the accelerated depletion of remaining natural capital stocks.

This conundrum can be illustrated another way by extrapolation from our ecological footprint data. If the entire world population of 5.6 billion were to use productive land at the rate of our Vancouver/Lower Fraser Valley example, the total requirement would be 28.5 billion ha. In fact, the total land area of Earth is only just over 13 billion ha, of which only 8.8 billion ha is productive cropland, pasture, or forest. The immediate implications are two-fold: first, as already stressed, the citizens of wealthy industrial countries unconsciously appropriate far more than their fair share of global carrying capacity; second, we would require an additional "two Earths," assuming present technology and efficiency levels, to provide for the present world population at Canadians' ecological standard of living. In short, there may simply not be enough natural capital around to satisfy current development assumptions. The difference between the anticipated ecological footprint of the human enterprise and the available land/natural capital base is a measure of the "sustainability gap" confronting humankind.

Homer-Dixon et al. (1993) provide evidence that generally deteriorating ecological conditions and growing scarcities of renewable resources will contribute to social instability and precipitate civil and international strife. We suggest that such geopolitical tensions may be exacerbated by the dependencies created through foreign trade. Populations that have come to rely on imports have no direct control over the natural capital stocks that sustain them from afar. What is the inherent stability of trading relationships in an era of global change? Are external sources secure if climate change, ozone depletion, or soil erosion reduce biological production or if local populations lay claim to the land or export flows? Do prevailing management practices ensure adequate maintenance or enhancement of natural capital stocks? To the extent that interregional dependency threatens geopolitical security, we have an argument for policies to enhance regional economic diversity, independence, and self-reliance, and restrict trade to the exchange of true ecological surpluses. (In this context, bioregionalism stands out as an appropriate ecopolitical philosophy.) Such prescriptions counter prevailing development rhetoric that calls for economic specialization, concentration of capital, unrestricted access to resources, freer markets, and expanded trade as the route to future prosperity.

The pace of global ecological change suggests that the productive capacity of some forms of natural capital has already been breached on the scale of the ecosphere. While economic analysis properly treats individual urban regions as open to exchange, it does not recognize that the ecosphere in aggregate is materially closed and therefore ultimately limiting. Clearly, not all countries or regions can run ecological deficits indefinitely—some must remain surplus producers if the net effect is to be global balance and stability. (We have yet to acquire the means to import carrying capacity from off-planet!) Inter-regional relationships, equity, and the role of trade in appropriating carrying capacity should be re-examined in this light. (See "EF/ACC" below for some of the other policy implications of the EF/ACC concept).

AN ECOLOGICAL, ECONOMIC, AND ETHICAL RATIONALE FOR THE CONCEPTS OF ECOLOGICAL FOOTPRINTS AND APPROPRIATED CARRYING CAPACITY (EF/ACC)

Ecological Relevance

The EF/ACC concept:

- recognizes land as a valid proxy for several key functions of terrestrial ecosystems essential for human existence (thermodynamic transformations in resource production and waste absorption, biogeochemical cycling, self-production and renewal [natural capital formation]).

- emphasizes the connectedness of all participants in the web of life through global energy and material flows. As a sub-system of the ecosphere, the human economy is inseparable from those of all other species and is still fundamentally dependent on nature.

- underscores from the human perspective the need for adequate stocks of renewable and replenishable natural capital as a *necessary condition* for sustainability. (Humane existence also requires social equity, economic security, and personal freedom.)

- helps to determine the ecological constraints within which society operates and to establish criteria (benchmarks or targets) for sustainability.

- gives economic stability a new ecological dimension: uninterrupted access to the required "carrying capacity" (the continuity of resource flows) is a precondition for a stable economy.

- provides a measure of current (or expected future) consumption against which to contrast current (or likely future) production, thereby revealing a "sustainability gap." For example, Canadians' current land requirement of 5 ha *per capita* is 3 times the 1.6 ha of productive land *per capita* actually "available" on Earth. (A global population of 10 billion—expected by 2030—would leave us with only .88 ha *per capita* assuming no land degradation.)
- reveals the overshoot of local (and global) carrying capacity by industrialized societies.

Contributions to Economic Analysis

- highlights the ecological and thermodynamic basis of economic process.
- recognizes productive natural capital as the basis for human-made wealth.
- helps distinguish, in contrast to current economic analysis, between sustainably and non-sustainably produced renewable flows.
- helps define "Hicksian natural capital" and provides an area-based indicator of the physical limits to material growth.
- promotes the necessary shift from consumption to investment (in natural capital) which is at the heart of developing sustainably.
- provides a biophysical (not price dependent) interpretation of resource scarcity and compensates for other blind spots of monetary assessments.
- provides a new measurement unit or "currency" for sustainability analysis.
- encourages the extension of traditional economic cost/benefit and marginal analyses to the macro level. Recognizes the economy's biophysical requirements and constraints forces consideration of the cumulative effects of growth, the notion of optimal scale, the implications of ecological inequity, etc. at the regional, national, and global levels.

Implications for Political Discourse and Decision-Making

- increases ecological awareness; in particular, it raises to consciousness the dependency of the human enterprise on biophysical resources and life-support services (natural capital).
- serves as a powerful heuristic tool for communicating the sustainability concept. It aggregates complex information into a single easily understood ecological indicator.
- explores the contribution of both population and material consumption to global ecological decline emphasizing the need for policies to control both.
- underscores the global imperative for local action. It demonstrates an inter-regional ecological multiplier effect of industrial levels of consumption on the welfare of human populations and other species everywhere.
- gives decision makers a physical criterion for ranking policy, project, or technology options according to their impact on ecological sustainability.
- shows who depends on which resource stocks (all over the world).
- reveals the extent to which wealthy people and countries have already "appropriated" the productive capacity of the ecosphere (through both commercial trade and unaccounted demand on open access source and sink functions). This challenges the most

basic assumptions of growth-oriented international development models and helps shift the focus to equity considerations (see below).

Ethical Dimensions

•   provides a new context for the sustainability debate. By raising to consciousness the moral, ethical (and ecological) consequences of conventional development patterns, it forces over-consumers to face the otherwise implicit trade-off made between their own consumption levels and the poverty and human suffering that results "elsewhere."

•   helps to quantify both intra-and inter-generational inequities. In a just world this should impose greater moral accountability on the wealthy and give the poor greater leverage in bargaining for development rights, technology transfers, and other equity measures.

•   strengthens the case for international agreement on how to share the Earth's productive capacity more equitably, by showing that not everyone can become as materially rich as average North Americans or Europeans without undermining global life support systems.

•   underscores intertemporal and interspatial interdependence of all living things, adding a practical plank to the extensionist platform for granting moral standing to non-human species.

A Cautionary Note

We admittedly make no allowance for potentially large efficiency gains or technological advances. Even at carrying capacity, further economic growth is possible (but not necessarily desirable) if resource consumption and waste production continue to decline per unit GDP (Jacobs 1991). We should not, however, rely exclusively on this conventional rationale. New technologies require decades to achieve the market penetration needed to significantly influence negative ecological trends. Moreover, there is no assurance that savings will not simply be directed into alternative forms of consumption. Efficiency improvements may actually increase rather than decrease resource consumption (Saunders 1992). We are already at the limit in a world of rising material expectations in which the human population is increasing by 94 million people per year. The minimal food-land requirements alone *each year* for this number of new people is $18.8 \cdot 10^6$ ha (at 5 people/ha, the current average productivity of world agriculture)—the equivalent of all the cropland in France.

EPILOGUE: COMPETING MYTHS, INACCESSIBLE REALITY

The complex issues arising from population-resources-development are characterized by a confusing mix of science, myth, and ideology. Even optimists admit that we may never again have sufficient confidence in our knowledge to embark on any global development strategy with the same arrogant certainty we enjoyed in the past. Two things do seem assured, however. First, globalization of the economy means globalization of the ecological crisis. Neither

North nor South can act in isolation of the other nor be unscathed by the consequences of ill-considered development programs wherever they may be implemented. Material wealth may provide a temporary buffer from the initial consequences, but ultimately we are all part of the same ecosphere and destined henceforth to share the same macro-ecological fate.

Second, whatever one's ideological persuasion, there is a biophysical reality "out there" that is totally indifferent to human habit or preferences, important dimensions of which will always remain inaccessible to understanding. Our models and paradigms can only ever be more or less reliable descriptions of mere particles of nature.

Of course, not all paradigms are created equal. In the struggle for sustainability, that development model whose internal logic is most compatible with the behavior of the biophysical world has the highest probability of success. Consider just one key cybernetic theorem, Ashby's Law of Requisite Variety: the internal variety (diversity, complexity) of a management system must correspond to the variety of the managed system if the manager is to maintain control. (Or, in Stafford Beer's more colorful terms: "We cannot regulate our interaction with any aspect of reality that our model of reality does not include because we cannot by definition be conscious of it" [Beer 1981]).

This theorem acquires particular poignancy when we consider that the primary decision rules determining humankind's increasingly uneasy relationship with nature come from ecologically empty economic theory. The human economy is becoming increasingly coincident with the ecosphere. But how can we hope control the "internal variety" of the ecosphere when there is no analog in the internal variety of prevailing management models? Humankind has already much reduced the latitude for trial and error on the earth. We can ill afford any approach to global sustainability based on the assumption that the economic future will resemble our exuberant past, only more so.

ACKNOWLEDGMENTS

We would like to thank Phil Testemale, who prepared the figures. Parts of an earlier version of this chapter appear in Rees (1992).

REFERENCES

Barnett, H., and C. Morse. 1963. Scarcity and Growth: The Economics of Natural Resource Scarcity. Baltimore: Johns Hopkins Univ. Press.

Beer, S. 1981. I said, You are gods. *Teilhard Review* 15 (3): 1–33.

Brown, L. et al. Annual. State of the World. Washington, DC: Worldwatch Institute.

Brown, L., and C. Flavin. 1988. The Earth's vital signs. In State of the World 1988, eds. L. Brown et al. Washington, DC: Worldwatch Institute.

Brown, S. 1986. Biomass of tropical tree plantations and its implications for the global carbon budget. *Canadian Journal of Forest Research* 16: 2.

Catton, W. 1980. Overshoot: the Ecological Basis of Revolutionary Change. Urbana: Univ. of Illinois Press.

Christensen, P. 1991. Driving forces, increasing returns and ecological sustainability. In *Ecological Economics: The Science and Management of Sustainability*, ed. R. Costanza. New York: Columbia Univ. Press.

Corson, W. 1990. *The Global Ecological Handbook*. Boston: Beacon Press.

Costanza, R., and H. Daly. 1990. *Natural Capital and Sustainable Development*. Paper prepared for the CEARC Workshop on Natural Capital, Vancouver, BC, March 15–16, 1990. Ottawa: Canadian Environmental Assessment Research Council.

Daily, G., and P. Ehrlich. 1992. Population, sustainability, and Earth's carrying capacity. *BioScience* 42 (10): 761–71.

Daly, H. 1986. Comments on "Population Growth and Economic Development." *Population and Development Review* 12: 583–585.

———. 1990. Sustainable development: from concept and theory towards operational principles. *Population and Development Review* (special issue 1990).(Also published in Daly, H. 1991. Steady-State Economics. 2d. ed. Washington, DC: Island Press.

———. 1991. From empty world economics to full world economics: recognizing an historic turning point in economic development. In *Environmentally Sustainable Development: Building on Brundtland*, eds. R. Goodland, H. Daly, and S. El Serafy. Washington, DC: The World Bank.

———. 1991. *Steady-State Economics*. 2d ed. Washington, DC: Island Press.

Dasgupta, P., and D. Heal. 1979. *Economic Theory and Exhaustible Resources*. Cambridge; New York: Cambridge Univ. Press.

Economics brief: a greener bank. 1992. (Review of the World Bank's World Development Report 1992). *The Economist* 323 (May 23): 79–80.

The Ecologist. 1991. 21 (March/April): 2.

Ehrlich, P., and J. Holdren. 1971. Impact of population growth. *Science* 171: 1212–17.

Environment Canada. 1988. *The Changing Atmosphere: Implications for Global Security*. Conference Statement, Toronto, June 27–30, 1988. Ottawa: Environment Canada.

———. 1991. *The State of Canada's Environment* (Chapter 22). Ottawa: Environment Canada.

George, S. 1992. Comment on the World Bank's "World Development Report 1992" and the June 1992 Earth Summit in Rio. *The Guardian* (Reprinted as Perils of a fat bank in a poor world. In *The Globe and Mail*, Toronto, May 29, 1992).

Georgescu-Roegen, N. 1975. Energy and economic myths. *Southern Economic Journal* 41 (3): 347–81.

Giampietro, M. 1992. Personal communication.

Giampietro, M., and D. Pimentel. 1990a. Energy analysis models to study biophysical limits for human exploitation of natural processes. In *Ecological Physical Chemistry*, eds. C. Rossi and T. Tiezzi. Amsterdam: Elsevier.

———. 1990b. Alcohol and biogas production from biomass. *Critical Reviews in Plant Science* 9 (3): 213–33.

Hannon, B. 1992. Sense of Place: Geographic Discounting by People, Animals and Plants. Paper presented to the Second Meeting of the International Society for Ecological Economics, "Investing in Natural Capital." Stockholm, August 3–6, 1992.

Hardin, G. 1991. Paramount positions in ecological economics. In *Ecological Economics: The Science and Management of Sustainability*, ed. R. Costanza. New York: Columbia Univ. Press.

Holdren, J., and P. Ehrlich. 1974. Human population and the global environment. *American Science* 62: 282–92.

Homer-Dixon, T., J. Boutwell, and G. Rathjens. 1993. Environmental change and violent conflict. *Scientific American* (February): 38–45.

Hornborg, A. 1992. Codifying Complexity: Towards an Economy of Incommensurable Values. Paper presented to the Second Meeting of the International Society for Ecological Economics, "Investing in Natural Capital." Stockholm, August 3–6, 1992.

Jacobs, M. 1991. The Green Economy. London: Pluto Press.

Karas, J., and P. Kelly. 1989. The Heat Trap: the Threat Posed by Rising Levels of Greenhouse Gases. Briefing document prepared for Friends of the Earth. Norwich: Univ. of East Anglia.

Kirchner, J., G. Leduc, R. Goodland, and J. Drake. 1985. Carrying capacity, population growth, and sustainable development. In Rapid Population Growth and Human Carrying Capacity: Two Perspectives, ed. D. Mahar. Staff Working Papers #690, Population and Development Series #15. Washington, DC: The World Bank.

Krebs, C. 1972. Ecology: The Experimental Analysis of Distribution and Abundance. New York: Harper & Row.

Lave, L., and H. Gruenspecht. 1991. Increasing the efficiency and effectiveness of environmental decisions: benefit-cost analysis and effluent fees. *Journal Air Waste Management Association* 41 (5): 680–93.

McKetta, J. 1984. Encyclopedia of Chemical Processing and Design, Vol. 20. New York and Basel: Marcel Dekker.

Meadows, D. H., D. L. Meadows, and J. Randers. 1992. Beyond the Limits. Toronto: McClelland & Stewart.

Milbrath, L. 1989. Envisioning a Sustainable Society: Learning our Way Out. Albany: State Univ. of New York Press.

Muscat, R. 1985. Carrying capacity and rapid population growth: definition, cases, and consequences. In Rapid Population Growth and Human Carrying Capacity: Two Perspectives, ed. D. Mahar. Staff Working Papers #690, Population and Development Series, #15. Washington, DC: The World Bank.

National Research Council Committee on Population. 1986. Report of the Working Group on Population Growth and Economic Development. Washington, DC: National Academy Press.

National Renewable Energy Laboratory. 1992. Internal data sheets. Cited by Barbara Goodman, personal communication. Golden, CO: NREL.

Overby, R. 1985. The Urban Economic Environmental Challenge: Improvement of Human Welfare by Building and Managing Urban Ecosystems. Paper presented in Hong Kong to the POLMET 85 Urban Environment Conference. Washington, DC; The World Bank.

Pearce, D. 1991. Deforesting the amazon: toward an economic solution. *Ecodecision* 1: 40–9.

Pearce, D., A. Markandya, and E. Barbier. 1989. Blueprint for a Green Economy. London: Earthscan.

Pearce, D., and R. Turner. 1990. Economics of Natural Resources and the Environment. New York: Harvester Wheatsheaf.

Pearce, D., E. Barbier, and A. Markandya. 1990. Sustainable Development: Economics and Environment in the Third World. Aldershot: Edward Algar.

Pezzey, J. 1989. Economic Analysis of Sustainable Growth and Sustainable Development. Environment Department Working Paper No. 15. Washington, DC: World Bank.

Pillet, G. 1991. Towards an Inquiry into the Carrying Capacity of Nations: What Does Over-Population Mean? Ecosys, S.A. Report to the Coordinator for International Refugee Policy. Berne, Switzerland: Federal Department of Foreign Affairs.

Pimentel, D. 1991. Ethanol fuels: energy security, economics, and the environment. *Journal of Agriculture and Environmental Ethics* 4 (1): 1–13.

Pimentel, D., et al. 1987. World agriculture and soil erosion. *BioScience* 37: 277–83.

Policy and Economics Directorate. 1991. Selected Forestry Statistics for Canada 1991. Information Report E-X-46. Ottawa: Forestry Canada.

Rees, W. 1989. Defining "Sustainable Development." CHS Research Bulletin. Center for Human Settlements. Vancouver: Univ. of British Columbia, .

———. 1990. The ecology of sustainable development. *The Ecologist* 20 (1): 18–23.

———. 1991. Economics, ecology, and the limits of conventional analysis. *Journal Air Waste Management Association* 41 (10): 1323–27.

———. 1992. Ecological footprints and appropriated carrying capacity: what urban economics leaves out. *Environment and Urbanization* 4 (2): 121–30.

———. 1993. In press. Understanding sustainable development: natural capital and the new world order. *Journal of the American Planning Association* .

RIVM. 1991. National Environmental Outlook, 1990–2010. Bilthoven, The Netherlands: Rijksinstituut voor Volksgezondheid en Milieuhygiene.

Saunders, H. 1992. The Khazzoom-Brooks Postulate and neoclassical growth. *The Energy Journal* 13 (4): 131–48.

Schneider, S. 1990. The science of climate-modeling and a perspective on the global warming debate. In Global Warming: The Greenpeace Report, ed. J. Leggett. New York: Oxford Univ. Press.

Solow, R. 1974. The economics of resources or the resources of economics. *American Economics Review* 64: 1–14.

———. 1991. Sustainability: An Economist's Perspective. The Eighteenth J. Seward Johnson Lecture, June 14, 1991. Marine Policy Center, Woods Hole Oceanographic Institution. Woods Hole, MA: Woods Hole Oceanographic Institution.

Stevens, W. 1991. Ozone layer thinner, but forces are in place for slow improvement. *New York Times* (April 9): B2.

Taylor, D. In press. Sustaining development or developing sustainability?: Two competing world views. *Alternatives*.

Victor, P. A. 1991. Indicators of sustainable development: some lessons from capital theory. *Ecological Economics* 4: 191–213.

Vitousek, P., P. Ehrlich, A. Ehrlich, and P. Matson. 1986. Human appropriation of the products of photosynthesis. *BioScience* 36: 368–74.

Wackernagel, M. 1991. Using "Appropriated Carrying Capacity" as an Indicator: Measuring the Sustainability of a Community. Report for the UBC Task Force on Healthy and Sustainable Communities. Vancouver: UBC School of Community and Regional Planning.

Wackernagel, M., J. McIntosh, W. Rees, and R. Woollard. 1993. How Big is our Ecological Footprint? A Handbook for Estimating a Community's Appropriated Carrying Capacity (draft). Vancouver: Univ. of British Columbia Healthy and Sustainable Communities Task Force.

World Bank. 1991. Urban Policy and Economic Development: An Agenda for the 1990s. Washington, DC: The World Bank.

———. 1992. World Development Report 1992. Washington, DC: The World Bank.

World Climate Research Program. 1990. Global Climate Change. World Meteorological Organization and International Council of Scientific Unions.

Whose Common Future? 1992. *The Ecologistt* 22 (July/August): Special issue on the enclosure of the global commons.

WCED. 1987. Our Common Future. Oxford: Oxford Univ. Press.

Worldwatch Institute. 1988. State of the World 1988. Washington, DC: Worldwatch.

———. 1989. State of the World 1989. Washington, DC: Worldwatch.

World Resources Institute. 1992. World Resources. Washington, DC: WRI.

PART THREE

Environmental Management
and Policy Implications:
Adjusting Economic,
Technical, Socio-Political,
and Cultural Systems

21 THREE GENERAL POLICIES TO ACHIEVE SUSTAINABILITY

Robert Costanza
Director, Maryland International Institute for Ecological
Economics and
Professor, Center for Environmental and Estuarine Studies
University of Maryland
Box 38, Solomons, MD 20688

ABSTRACT

Sustainability is a long-term goal over which there is broad and growing consensus. Establishment of this goal is fundamentally a social decision about the desirability of a survivable ecological economic system. It entails maintenance of (1) a sustainable *scale* of the economy relative to its ecological life-support system; (2) a fair *distribution* of resources and opportunities between present and future generations, as well as between agents in the current generation, and (3) an efficient *allocation* of resources that adequately accounts for natural capital. We can only be certain we have achieved sustainability in retrospect. Sustainable policies and instruments are therefore those that we *predict* will lead to the achievement of the goal. Like all predictions, they are uncertain. In designing sustainable policies and instruments, one would like to maximize the likelihood of success, while acknowledging and minimizing the remaining uncertainty.

This chapter describes three broad, mutually reinforcing policy instruments that have a high likelihood of assuring that economic *development* (as distinct from economic *growth*) will be ecologically sustainable. They utilize incentives to produce the desired results (sustainable scale and efficient allocation). They address only that aspect of the distribution issue having to do with distribution between current and future operations. Other aspects must be handled politically. They are:

1. A *natural capital depletion tax* aimed at reducing or eliminating the destruction of natural capital. Use of non-renewable natural capital would have to be balanced by investment in renewable natural capital in order to avoid the tax. The tax would be passed on to consumers in the price of products and would send the proper signals about the relative sustainability cost of each product, moving consumption toward a more sustainable product mix.

2. The *precautionary polluter pays principle (4P)* would be applied to potentially damaging products to incorporate the cost of the uncertainty about ecological damages as

well as the cost of known damages. This would give producers a strong and immediate incentive to improve their environmental performance in order to reduce the size of the environmental bond and tax they would have to pay.

3. A system of *ecological tariffs* aimed at allowing individual countries or trading blocks to apply 1 and 2 above without forcing producers to move overseas in order to remain competitive. Countervailing duties would be assessed to impose the ecological costs associated with production fairly on both internally produced and imported products. Revenues from the tariffs would be reinvested in the global environment, rather than added to general revenues of the host country.

INTRODUCTION

The integration of ecology and economics has begun to provide new insights about the linkages between ecological and economic systems, and to suggest some broad policies concerning how to achieve sustainability (Daly 1990; Costanza 1991; Young 1992). In this chapter three fairly broad, interdependent proposals are described and discussed. Taken together, they are comprehensive, and may be sufficient to achieve ecological sustainability, a necessary prerequisite to total system sustainability. Ecological sustainability implies maintaining the economy at a scale that does not damage the ecological life-support system (i.e., safe minimum standards) and a fair distribution of resources between present and future generations. The market incentive-based instruments suggested to implement the policies are intended to do the job with relatively high efficiency and effectiveness. They are not the only possible mechanisms to achieve these goals, but there is considerable evidence that they would work rather well. By focusing on specific policies and instruments, we can also address the essential changes that need to be made in the system and begin to build a broad enough consensus to implement these changes.

Various aspects of the proposals have appeared in other forms elsewhere (Pearce and Turner 1989; Daly 1990; Cropper and Oates 1990; Perrings 1991; Costanza 1991; Costanza and Daly 1992; Costanza and Cornwell 1992; Young 1992; and Bishop 1993). This chapter represents an attempt to synthesize and generalize them as the basis for developing an "overlapping consensus." (Rawls 1987). According to Rawls, a consensus that is affirmed by opposing theoretical, religious, philosophical, and moral doctrines is most likely to be fair and just, and is also most likely to be resilient and to survive over time. A key overlapping consensus that has emerged recently is the goal of sustainable development, a form of economic development that maintains the ecological processes and functions that underpin it and reaps the benefits of improving the quality of life now without denying future generations a similar opportunity (WCED 1987; Young 1992; *Agenda 21* 1992).

The proposals are interrelated and interdependent. A natural capital depletion tax assures that resource inputs from the environment to the economy are sustainable in a general and comprehensive way (Costanza and Daly 1992), while giving strong incentives to develop new technologies and processes to minimize impacts. The precautionary polluter pays principle (4P) assures that the full costs of outputs from the economy to the environment are charged to the polluter in a way that adequately deals with the huge uncertainty about the impacts of pollution, and encourages technological innovation (Costanza and Cornwell 1992). A system of ecological tariffs is one way (short of global agreements that are difficult to negotiate and enforce) to allow countries to implement the first two proposals without putting themselves at an undue disadvantage (at least on the import side) relative to countries that have not yet implemented them.

NATURAL CAPITAL AND SUSTAINABILITY

A minimum necessary condition for sustainability is taken to be maintenance of the total natural capital (TNC) stock at or above the current level (Pearce and Turner 1989; Costanza and Daly 1992). This condition is sometimes referred to as *strong sustainability* as opposed to *weak sustainability* which requires only that the total capital stock (including both human-made and natural capital) be maintained (Costanza and Daly 1992). Since natural and human-made capital are, in general, complements rather than substitutes, the strong sustainability condition is more appropriate. While a lower stock of natural capital *may* be sustainable, given our uncertainty and the dire consequences of guessing wrong, it is best to at least provisionally assume that the we are at or below the range of sustainable stock levels and allow no further decline in natural capital. This "constancy of total natural capital" rule can thus be seen as a prudent minimum condition for assuring sustainability, to be abandoned only when solid evidence to the contrary can be offered.

In the past only human-made stocks were considered as capital because natural capital was superabundant. Human activities were at too small a scale relative to natural processes to interfere with the free provision of natural goods and services. Expansion of human-made capital entailed little or no opportunity cost in terms of the sacrifice of services of natural capital. Human-made capital was the limiting factor in economic development, and natural capital was a free good. But we are now entering an era, thanks to the enormous increase of the human scale, in which natural capital is becoming the limiting factor. Human economic activities can significantly reduce the capacity of natural capital to yield the flow of ecosystem goods and services upon which the very productivity of human-made capital depends.

Of course, the classical economists (Smith, Malthus, Ricardo) emphasized the constraints of natural resources on economic growth, and several more recent economists, especially environmental and ecological economists, have explicitly recognized natural resources as an important form of capital with major contributions to human well-being (Scott 1954; Daly 1968, 1973, 1977; Page 1977; Randall 1987; Pearce and Turner 1989). But environmental economics has, until now, been a tiny subfield far from the mainstream of neoclassical economics, and the role of natural resources within the mainstream has been de-emphasized. If we are to achieve sustainability, the economy must be viewed in its proper perspective as a subsystem of the larger ecological system of which it is a part, and ecological economics needs to become a much more pervasive approach to the problem (Costanza 1991).

WHY IS ACCOUNTING FOR NATURAL CAPITAL SO IMPORTANT?

Natural capital produces a significant portion of the real goods and services of the ecological economic system, so failure to adequately account for it leads to major misconceptions about how well the economy is doing. This misconception is important at all levels of analysis, from the appraisal of individual projects to the health of the ecological economic system as a whole. This chapter will concentrate on the level of national income accounting, however, because of the importance of these measures to national planning and sustainability.

There has been much recent interest in improving national income and welfare measures to account for depletion of natural capital and other mismeasures of welfare (Ahmad et al. 1989). Daly and Cobb (1989) have produced an index of sustainable economic welfare (ISEW) that attempts to adjust GNP to account mainly for depletions of natural capital, pollution effects, and income distribution effects. Figure 21.1 shows two versions of their index compared to GNP from 1950 to 1986. What is strikingly clear from Figure 21.1 is that while GNP has been rising over this interval, ISEW has remained relatively unchanged since about 1970. When depletions of natural capital, pollution costs, and income distribution effects are accounted for, the economy is seen to be not improving at all. If we continue to ignore natural capital, we may well push the economy down while we think we are building it up.

NATURAL CAPITAL DEPLETION (NCD) TAX

One way to implement the sustainablility constraint of no net depletion of natural capital is to hold throughput (consumption of TNC) constant at present levels (or lower truly sustainable levels) by taxing TNC consumption, especially energy, very heavily. Society could raise most public revenue from such a natural capital depletion tax, and compensate by reducing the income tax, especially on the lower end of the income distribution, perhaps even financing a negative income tax at the very low end. Technological optimists who believe that efficiency can increase by a factor of ten should welcome this policy, which raises natural resource prices considerably and would powerfully encourage just those technological advances in which they have so much faith. Skeptics who lack that technological faith should nevertheless be happy to see the throughput limited since that is their main imperative to conserve resources for the future. The skeptics would be protected against their worst fears; the optimists would be encouraged to pursue their fondest dreams. If the skeptics are proven wrong and the enormous increase in efficiency actually happens, then they will be even happier. They will get what they wanted, but it will cost less than they expected and were willing to pay.

The optimists, for their part, can hardly object to a policy that not only allows, but offers strong incentives for the very technical progress on which their optimism is based. If they are proven wrong, at least they should be glad that the rate of environmental destruction has been slowed.

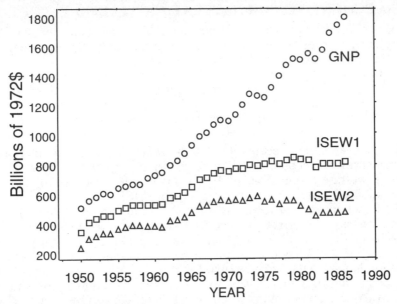

Figure 21.1. U.S. GNP compared to the Index of Sustainable Economic Welfare (ISEW, from Daly and Cobb 1989) for the interval 1950 to 1986. ISEW 2 includes corrections for depletion of non-renewable resources and long-term environmental damage; ISEW 1 does not.

Implementation of this policy does not hinge upon the *precise* measurement of natural capital, but the valuation issue remains relevant in the sense that the policy recommendation is based on the perception that we are at or beyond the optimal scale. The evidence for this perception consists of the greenhouse effect, ozone layer depletion, acid rain, and the general decline in many dimensions of the quality of life. It would be helpful to have better quantitative measures of these perceived costs, just as it would be helpful to carry along an altimeter when we jump out of an airplane. But we would all prefer a parachute to an altimeter if we could take only one thing. The consequences of an unarrested free fall are clear enough without a precise measure of our speed and acceleration. We would need at least a ballpark estimate of the value of natural capital depletion in order to determine the magnitude of the suggested NCD tax. This, we think, is possible, especially if uncertainty about the value of natural capital is incorporated into the tax itself, using, for example, the refundable assurance bonding system discussed below.

The political feasibility of this policy is an important and difficult question. It certainly represents a major shift in the way we view our relationship to natural capital, and would have major social, economic, and political implications. But these implications are just the ones we need to expose and face squarely if we hope to achieve sustainability. Because of its logic, its conceptual simplicity, and its built-in market incentive structure leading to sustainability, the proposed NCD tax may be the most politically feasible of the possible alternatives to achieving sustainability.

We have not tried to work out all the details of how the NCD tax would be administered. In general, it could be administered like any other tax, but it would most likely require international agreements or at least national ecological tariffs (as discussed later) to prevent some countries from flooding markets with untaxed natural capital or products made with untaxed natural capital (as discussed later). By shifting most of the tax burden to the NCD tax and away from income taxes, the NCD tax could actually simplify taxation administration while providing the appropriate economic incentives to achieve sustainability.

DEALING WITH TRUE UNCERTAINTY

One of the primary reasons for the problems with current methods of environmental management is the issue of scientific uncertainty—not just its existence, but the radically different expectations and modes of operation that science and policy have developed to deal with it. If we are to solve this problem, we must understand and expose these differences about the nature of uncertainty and design better methods to incorporate it into the policy-making and management process.

To understand the scope of the problem, it is necessary to differentiate between *risk* (which is an event with a *known* probability, sometimes referred to as "statistical uncertainty") and *true uncertainty* (which is an event with an *unknown* probability, sometimes referred to as "indeterminacy"). Every time you drive your car, you run the *risk* of having an accident because the probability of car accidents is known with very high certainty. We know the risk involved in driving because, unfortunately, there have been many car accidents on which to base the probabilities. These probabilities are known with enough certainty that they are used by insurance companies to set rates that will assure those companies of a certain profit. There is little uncertainty about the risk of car accidents. If you live near the disposal site of a newly synthesized toxic chemical, you may be in danger as well, but no one knows even the *probability* of getting cancer or some other disease from this exposure, so there is true uncertainty. Most important environmental problems suffer from true uncertainty, not merely risk.

One can think of a continuum of uncertainty where zero represents certain information, moving to intermediate levels representing information with statistical uncertainty and known probabilities (risk), and ending with high levels for information with true uncertainty or indeterminacy. Risk assessment has become the central guiding principle at the U.S. EPA (Science Advisory Board 1990) and other environmental management agencies, but true uncertainty has yet to be adequately incorporated into environmental protection strategy.

Science treats uncertainty as a given, a characteristic of all information that must be honestly acknowledged and communicated. Over the years scientists have developed increasingly sophisticated methods to measure and communicate uncertainty arising from various causes. It is important to note that the progress of science has, in general, uncovered *more* uncertainty rather than leading to the absolute precision that the lay public often mistakenly associates with "scientific" results. The scientific method can only set boundaries on the limits of our knowledge. It can define the edges of the envelope of what is known, but often this envelope is very large and the shape of its interior can be a complete mystery. Science can tell us the range of uncertainty about global warming and toxic chemicals, and maybe *something* about the relative probabilities of different outcomes, but in most important cases, it cannot tell us which of the possible outcomes will occur with any degree of accuracy.

Our current approaches to environmental management and policy making, on the other hand, abhor uncertainty and gravitate to the edges of the scientific envelope. The reasons for this are clear. The goal of policy is to make unambiguous, defensible decisions, often codified in the form of laws and regulations. While legislative language is often open to interpretation, regulations are much easier to write and enforce if they are stated in clear, black and white, absolutely certain terms. For most of criminal law this works reasonably well. Either Mr. Cain killed his brother or he didn't; the only question is whether there is enough evidence to demonstrate guilt beyond a reasonable doubt (i.e., with essentially zero uncertainty). Since the burden of proof is on the prosecution, it does little good to conclude that there was an 80% chance that Mr. Cain killed his brother. But many scientific studies come to just these kinds of conclusions because that is the nature of the phenomenon. Science defines the envelope while the policy process gravitates to its edges—generally the edge that best advances the policy maker's political agenda. We need to deal with the whole envelope and all its implications if we are to rationally use science to make policy.

The problem is most severe in environmental policy making. Building on the legal traditions of criminal law, policy makers and environmental regulators desire absolute, certain information when designing environmental regulations. But much of environmental policy is based upon scientific studies of the likely health, safety, and ecological consequences of human actions. Information gained from these studies is therefore only certain within their epistemological and methodological limits (Thompson 1986). Particularly with the recent shift in environmental concerns from visible, known pollution to more subtle threats like radon, regulators are confronted with decision making outside the limits of scientific certainty with increasing frequency (Weinberg 1985).

Problems arise when regulators ask scientists for answers to unanswerable questions. For example, the law may mandate that the regulatory agency come up with safety standards for all known toxins when little or no information is available on the impacts of these chemicals. When trying to enforce the regulations after they are drafted, the problem of true uncertainty about the impacts remains. It is not possible to determine with any certainty whether the local chemical company contributed to the death of some people in the vicinity of the toxic waste dump. One cannot *prove* the smoking/lung cancer connection in any direct, causal way (i.e., in the courtroom sense), only as a statistical relationship. Similarly, global warming may or may not happen after all.

As they are currently set up, most environmental regulations, particularly in the United States, *demand certainty* and when scientists are pressured to supply this nonexistent commodity, there is not only frustration and poor communication, but mixed messages in the media as well. Because of uncertainty, environmental issues can often be manipulated by political and economic interest groups. Uncertainty about global warming is perhaps the most visible current example of this effect.

The "precautionary principle" is one way the environmental regulatory community has begun to deal with the problem of true uncertainty. The principle states that rather than await certainty, regulators should act in anticipation of any potential environmental harm in order to prevent it. The precautionary principle is so frequently invoked in international environmental resolutions that it has come to be seen by some as a basic normative principle of international environmental law (Cameron and Abouchar 1991). But the principle offers no guidance as to what precautionary measures should be taken. It "implies the commitment of resources now to safeguard against the potentially adverse future outcomes of some decision," (Perrings 1991) but does not tell us how many resources or which adverse future outcomes are most important.

This aspect of the "size of the stakes" is a primary determinant of how uncertainty is dealt with in the political arena. The situation can be summarized as shown in Figure 21.2, with uncertainty plotted against decision stakes. It is only the area near the origin with low uncertainty and low stakes that is the domain of "normal applied science." Higher uncertainty or higher stakes result in a much more politicized environment. Moderate values of either correspond to "applied engineering" or "professional consultancy," which allow a good measure of judgment and opinion to deal with risk. On the other hand, current methods are not in place to deal with high values of either stakes or uncertainty, which require a new approach—what might be called "postnormal" or "second order science." (Funtowicz and Ravetz 1991). This "new" science is really just the application of the essence of the scientific method to new territory. The scientific method does not, in its basic form, imply anything about the precision of the results achieved. It *does* imply a forum of open and free inquiry without preconceived answers or agendas aimed at determining the envelope of our knowledge and the magnitude of our ignorance.

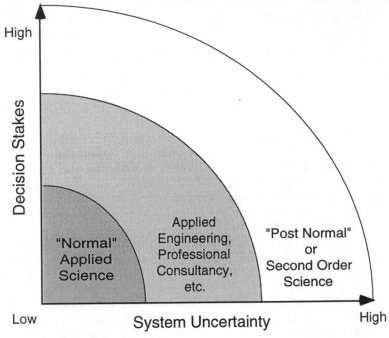

Figure 21.2. Three kinds of science (S. O. Funtowicz and J. R. Ravetz 1991).

THE PRECAUTIONARY POLLUTER PAYS PRINCIPLE (4P)

Implementing this view of science requires a new approach to environmental protection that acknowledges the existence of true uncertainty rather than denying it, and includes mechanisms to safeguard against its potentially harmful effects while at the same time encouraging development of lower impact technologies and the reduction of uncertainty about impacts. The precautionary principle sets the stage for this approach—the real challenge is to develop scientific methods to determine the potential costs of uncertainty, and to adjust incentives so that the appropriate parties *pay* this cost of uncertainty and have appropriate incentives to reduce its detrimental effects. Without this adjustment, the full costs of environmental damage will continue to be left out of the accounting (Peskin 1991), and the hidden subsidies from society to those who profit from environmental degradation will continue to provide strong incentives to degrade the environment beyond sustainable levels.

Over the past two decades there has been extensive discussion about the efficiency that can theoretically be achieved in environmental management through the use of market mechanisms (Brady and Cunningham 1981; Cropper and Oates 1990). These mechanisms are designed to alter the pricing structure of the present market system to incorporate the total, long-term social and ecological costs of an economic agent's activities. Suggested "incentive-based" mechanisms include pollution taxes, tradable pollution discharge permits, financial responsibility requirements and deposit-refund systems. Dealing with the pervasive uncertainty inherent in environmental problems in a precautionary way is possible using some new versions of these incentive-based alternatives.

An innovative incentive-based instrument currently being researched to manage the environment for precaution under uncertainty is a *flexible environmental assurance bonding system* (Costanza and Perrings 1990). This variation of the deposit-refund system is designed to incorporate *both* known and uncertain environmental costs into the incentive system, and to induce positive environmental technological innovation. It works in this way: in addition to charging an economic agent directly for known environmental damages, an assurance bond equal to the current best estimate of the largest potential future environmental damages would be levied and kept in an interest-bearing escrow account for a predetermined length of time. In keeping with the precautionary principle, this system requires the commitment of resources *now* to offset the potentially catastrophic future effects of current activity. Portions of the bond (plus interest) would be returned *if and when* the agent could demonstrate that the suspected worst case damages had not occurred or would be less than originally assessed. If damages did occur, portions of the bond would be used to rehabilitate or repair the environment, and possibly to compensate injured parties. Funds tied up in bonds could continue to be used for other economic activities. The only cost would be the difference (plus or minus) between the interest on the bond and the return that could be earned by the firm had they invested in other activities. On average one would expect this difference to be minimal. In addition, the "forced savings" which the bond would require could actually improve overall economic performance in economies like that of the United States, which chronically undersave.

By requiring the users of environmental resources to post a bond adequate to cover uncertain future environmental damages (with the possibility for refunds), the burden of proof (and the cost of the uncertainty) is shifted from the public to the resource user. At the same time, agents are not charged in any final way for uncertain future damages and can recover portions of their bond (with interest) in proportion to how much better their performance is than the worst case.

Deposit/refund systems, in general, are not a new concept. They have been successfully applied to a range of consumer, conservation, and environmental policy objectives (Bohm 1981). The most well-known examples are the systems for beverage containers and used lubricating oils that have both proven to be quite effective and efficient.

Another precedent for environmental assurance bonds are the producer-paid performance bonds often required for federal, state, or local government construction work. For example, the Miller Act (40 U.S.C. 270), a 1935 federal statute, requires contractors performing construction contracts for the federal government to secure performance bonds. Performance bonds provide a contractual guarantee that the principal (the entity which is doing the work or providing the service) will perform in a designated way. Bonds are frequently required for construction work done in the private sector as well.

Performance bonds are frequently posted in the form of corporate surety bonds that are licensed under various insurance laws and, under their charter, have legal authority to act as financial guarantee for others. The unrecoverable cost of this service is usually 1%–5% of the bond amount. However, under the Miller Act (FAR 28.203-1 and 28.203-2), any contract above a designated amount ($25,000 in the case of construction) can be backed by other types of securities, such as U.S. bonds or notes, in lieu of a bond guaranteed by a surety company. In this case, the contractor provides a duly executed power of attorney and an agreement authorizing collection on the bond or notes if they default on the contract (PRC Environmental Management 1986). If the contractor performs all the obligations specified in the contract, the securities are returned to the contractor, and the usual cost of the surety is avoided.

Environmental assurance bonds would work in a similar manner (by providing a contractual guarantee that the principal would perform in an environmentally benign manner), but would be levied for the current best estimate of the *largest* potential future environmental damages. Funds in the bond would be invested and would produce interest that could be returned to the principal. An "environmentally benign" investment strategy would probably be most appropriate for a bond such as this.

These bonds could be administered by the regulatory authority that currently manages the operation or procedure (for example, in the United States, the Environmental Protection Agency could be the primary authority). But a case can be made that it is better to set up a completely independent agency to administer the bonds. The detailed design of the institutions to administer the bond is worthy of considerable additional thought and analysis, and will depend on the requirements of an individual situation).

The bond would be held until the uncertainty or some part of it was removed. This would provide a strong incentive for the principal to reduce the uncertainty about their environmental impacts as quickly as possible, either by funding independent research or by changing their processes to ones that are less damaging. A quasi-judicial body would be necessary to resolve disputes about when and how much refund on the bonds should be awarded. This body would utilize the latest independent scientific information on the worst-case ecological damages that could result from a firm's activities, but with the burden of proof falling on the economic agent who stands to gain from the activity, not the public. Protocol for worst-case analysis already exists within the U.S. EPA. In 1977 the U.S. Council on Environmental Quality required worst-case analysis for implementing NEPA (National Environmental Policy Act of 1969). This act required the regulatory agency to consider the worst environmental consequences of an action when scientific uncertainty was involved (Fogleman 1987).

One potential argument against the bond is that it would favor relatively large firms that could afford to handle the financial responsibility of activities potentially hazardous to the environment. This is true, but it is exactly the desired effect, since firms that *cannot* handle the financial responsibility should *not* be passing the cost of potential environmental damage on to the public. In the construction industry, small "fly-by-night" firms are prevented (through the use of performance bonds) from cutting corners and endangering the public in order to underbid responsible firms.

This is not to say that small businesses would be eliminated. Far from it. They could either band together to form associations to handle the financial responsibility for environmentally risky activities, or, preferably, they could change to more environmentally benign activities that did not require large assurance bonds. This encouragement of the development of new environmentally benign technologies is one of the main attractions of the bonding system, and small, start-up firms would certainly lead the way.

The individual elements of the 4P system have broad theoretical support, and have been implemented before in various forms. The precautionary principle is gaining wide acceptance in many areas where true uncertainty is important. Incentive-based environmental regulation schemes are also gaining acceptance as more efficient ways to achieve environmental goals. For example, the U.S. Clean Air Act reauthorization contains a tradable permit system for controlling air pollution. Both the precautionary and the polluter pays principles are incorporated in *Agenda 21*, the final resolutions of the 1992 United Nations Conference on Environment and Development (*Agenda 21* 1992). By linking these two important principles, we can begin to effectively deal with uncertainty in an economically efficient and ecologically sustainable way.

In a sense, we are already moving in the direction of the 4P system. As strict liability for environmental damages becomes more the norm, far-sighted firms have already begun to protect themselves against possible future lawsuits and damage claims by putting aside funds for this purpose. The 4P system is, in effect, a *requirement* that all firms be far-sighted. It is an improvement on strict liability because it:

(1) explicitly moves the costs to the present where they will have the maximum impact on decision making,[1]

(2) provides "edge-focused, second order scientific" assessments of the potential impacts from a comprehensive ecological economic perspective in order to ensure that the size of the bond is large enough to cover the worst case damages, and

(3) insures that appropriate use of the funds are made in case of a partial or complete default.

Because of its logic, fairness, efficiency, ability to implement the precautionary and polluter pays principles in a practical way, and use of legal and financial mechanisms with long and successful precedents, the 4P system promises to be both practical and politically feasible. We think it can do much to help head off the current environmental crisis before it is too late.

ECOLOGICAL TARIFFS: MAKING TRADE SUSTAINABLE

If all countries in the world were to adopt and enforce the 4P system and NCD taxes, there would be no problem (at least from an ecological point of view) in allowing "free" trade. Given recent commitments of the global community to the idea of sustainable development (*Agenda 21* 1992), it does not seem totally out of the question that a global agreement along these lines could someday be worked out. But in the meantime, there are alternative instruments that could allow individual countries or trading blocks to apply the 4P system and NCD taxes in their local economies without forcing producers overseas. It is within at least the spirit of the GATT guidelines to allow countervailing duties to be assessed to impose the same ecological costs on internally produced and imported products. The key is fairness. A country cannot impose duties on imports that it does not also impose on domestically produced products. But if a country chose to adopt the 4P and NCD tax systems domestically, it could also adopt a system of ecologically based tariffs that would impose equivalent costs on imports. This is a different use for tariffs than the usual one. In the past, tariffs have been used to protect domestic industries from foreign competition. The proposed (and more defensible) use of tariffs (in conjunction with the 4P and NCD taxes) is to protect the domestic (and global) environment from private polluters and non-sustainable resource users, regardless of their country of origin or operation. The mechanisms for imposing tariffs are well-established. All that changes is the motive and the result. The proposed ecological tariffs (ETs) would result in patterns of trade that do not endanger sustainability.

Revenues from the tariffs could (and should) be reinvested in natural capital. It would be particularly attractive to reinvest in natural capital in the country on which the tariff was imposed. This would "close the loop" and

---

1. Several studies of "social traps" have shown that the timing of information about costs is more important than the actual expected magnitude (Costanza and Shrum 1988; Costanza 1987; Platt 1973; Cross and Guyer 1980; Teger 1980; Bockner and Rubin 1985).

prevent trade from inducing net destruction of natural capital in less-developed countries, as it does now.

SYSTEMS OF RESOURCE RIGHTS

The success of either command and control regulation or market-based incentives for sustainable management are predicated on having an adequate, legally viable resource use and property rights system. Young (1992) argues that by designing overlapping, conditional systems of resource use and property rights to cover the many different aspects of natural capital within a region, the stage can be set for sustainable use. *Changing* these systems of explicit and implicit rights are likewise often the most difficult and neglected step in implementing (e.g., the former Soviet Union) or enlarging the scope (e.g., western Europe and the United States) of market-based systems of allocation. The resource use and property rights system, laws, and regulations set the stage and largely determine the goals for an economy, while competitive markets are efficient tools to help society achieve its goals. As Young points out, "competitive markets are excellent servants but bad masters." Our challenge is to employ these powerful servants in the service of sustainability.

CONCLUSIONS

Taken together, the three policy instruments suggested here (Natural Capital Depletion (NCD) taxes, the Precautionary Polluter Pays Principle (4P), and Ecological Tariffs (ETs) would go a long way toward assuring ecological sustainability while at the same time taking advantage of market incentives to achieve this result at high efficiency. The time for action is running short, but the political will to implement significant changes seems to be finally at hand. The three instruments suggested embody the mix of environmental protection and economic development potential necessary to make them politically feasible. The next steps are to further elaborate and test the instruments, and to build a broad, overlapping consensus to allow their ultimate implementation. It is not too late to protect our natural capital and achieve sustainability.

ACKNOWLEDGMENTS

Partial funding was provided by the Ford Foundation, the Moriah Fund, and the U.S. EPA contract #CR-815393-01-0, S. Farber and R. Costanza, Principal Investigators, titled: "A Flexible Environmental Cost Charging and Assurance Bonding System for Improved Environmental Management." This paper was originally presented at the Second Conference of the International Society for Ecological Economics (ISEE), "Investing in Natural Capital," Stockholm, Sweden, August 3–6, 1992. I thank D. King, T. O'Riordan, S. Funtowicz, J. R. Ravetz, and J. Bartholomew for their helpful comments on earlier drafts. Particular thanks are due to H. Daly and L. Cornwell, co-authors on previous papers that were the basis for many of the ideas discussed here.

REFERENCES

Ahmad, Y. J., S. El Serafy, and E. Lutz. 1989. Environmental Accounting for Sustainable Development. A UNEP-World Bank Symposium. Washington, DC: The World Bank.

Bohm, P. 1981. Deposit-Refund Systems. Resources for the Future. Baltimore, MD: Johns Hopkins Univ. Press.

Brady, G. L., and R. D. Cunningham. 1981. The economics of pollution control in the U.S. *Ambio* 10: 171–5.

Brockner, J., and J. Z. Rubin. 1985. Entrapment in Escalating Conflicts: A Social Psychological Analysis. New York: Springer-Verlag.

Cameron, J., and J. Abouchar. 1991. The precautionary principle: a fundamental principle of law and policy for the protection of the global environment. *Boston College International and Comparative Law Review* 14: 1–27.

Costanza, R., ed. 1991. Ecological Economics: The Science and Management of Sustainability. New York: Columbia Univ. Press.

Costanza, R., and L. Cornwell. 1992. The 4P approach to dealing with scientific uncertainty. *Environment* 34: 12–20.

Costanza, R., and H. E. Daly. 1992. Natural capital and sustainable development. *Conservation Biology* 6: 37–46.

Costanza, R., S. C. Farber, and J. Maxwell. 1989. The valuation and management of wetland ecosystems. *Ecological Economics* 1: 335–62.

Costanza, R., and C. H. Perrings. 1990. A flexible assurance bonding system for improved environmental management. *Ecological Economics* 2: 57–76.

Costanza, R., and W. Shrum. 1988. The effects of taxation on moderating the conflict escalation process: an experiment using the dollar auction game. *Social Science Quarterly* 69: 416–32.

Costanza, R., F. H. Sklar, and M. L. White. 1990. Modeling coastal landscape dynamics. *BioScience* 40: 91–107.

Cropper, M. L., and W. E. Oates. 1990. Environmental Economics: A Survey. Resources for the Future, Discussion Paper: Q E90-12. Washington, DC: Resources for the Future.

Cross, J. G., and M. J. Guyer. 1980. Social Traps. Ann Arbor: Univ. of Michigan Press.

Daly, H. E. 1968. On economics as a life science. *Journal of Political Economy* 76: 392–406.

———. 1973. Toward a Steady-State Economy. San Francisco: W. H. Freeman.

———. 1977. Steady-State Economics: the Political Economy of Biophysical Equilibrium and Moral Growth. San Francisco: W. H. Freeman.

———. 1990. Toward some operational principles of sustainable development. *Ecological Economics* 2: 1–6.

Daly, H. E., and J. B. Cobb, Jr. 1989. For the Common Good: Redirecting the Economy Toward Community, the Environment, and a Sustainable Future. Boston: Beacon.

El Serafy, S. 1989. The proper calculation of income from depletable natural resources. In Environmental Accounting for Sustainable Development, eds. Y. J. Ahmad, S. El Serafy and E. Lutz. A UNEP-World Bank Symposium. Washington, DC: The World Bank.

Fogleman, V. M. 1987. Worst case analysis: a continued requirement under the National Environmental Policy Act? *Columbia Journal of Environmental Law* 13: 53.

Funtowicz, S. O., and J. R. Ravetz. 1991. A new scientific methodology for global environmental problems. In Ecological Economics: the Science and Management of Sustainability, ed. R. Costanza. New York: Columbia Univ. Press.

Gever, J., R. Kaufmann, D. Skole, and C. Vörösmarty. 1986. Beyond Oil: the Threat to Food and Fuel in the Coming Decades. Cambridge, MA: Ballinger.

Hall, C. A. S., C. J. Cleveland, and R. Kaufmann. 1986. Energy and Resource Quality: the Ecology of the Economic Process. New York: Wiley.

Lovins, A. B. 1977. Soft Energy Paths: Toward a Durable Peace. Cambridge, MA: Ballinger.

Lovins, A. B., and L. H. Lovins. 1987. Energy: the avoidable oil crisis. *Atlantic* (December): 22–30.

MacNeil, J. 1990. Sustainable Development, Economics, and the Growth Imperative. Workshop on the Economics of Sustainable Development, Background Paper No. 3. Washington, DC.

Nordhaus, W., and J. Tobin. 1972. Is Growth Obsolete? National Bureau of Economic Research. New York: Columbia Univ. Press.

Page, Talbot. 1977. Conservation and Economic Efficiency: an Approach to Materials Policy. Washington, DC: Resources for the Future.

Pearce, D. W., and R. K. Turner. 1989. Economics of Natural Resources and the Environment. Brighton: Wheatsheaf.

Pearce, D. W., A. Markandya, and E. B. Barbier. 1989. Blueprint for a Green Economy. London: Earthscan.

Perrings, C. 1991. Reserved rationality and the precautionary principle: technological change, time and uncertainty in environmental decision making. In Ecological Economics: the Science and Management of Sustainability, ed. R. Costanza. New York: Columbia Univ. Press.

Peskin, H. M. 1991. Alternative environmental and resource accounting approaches. In Ecological Economics: the Science and Management of Sustainability, ed. R. Costanza. New York: Columbia Univ. Press.

Platt, J. 1973. Social traps. *American Psychologist* 28: 642–51.

PRC Environmental Management. 1986. Performance Bonding. A final report prepared for the U.S. Environmental Protection Agency, Office of Waste Programs and Enforcement. Washington, DC: EPA.

Randall, A. 1987. Resource Economics. 2d ed. New York: Wiley.

Rawls, J. 1987. The idea of an overlapping consensus. *Oxford Journal of Legal Studies* 7: 1–25.

Science Advisory Board. 1990. Reducing risk: setting priorities and strategies for environmental protection. SAB-EC-90-021. Washington, DC: EPA.

Teger, A. I. 1980. Too Much Invested to Quit. New York: Pergamon.

Thompson, P. B. 1986. Uncertainty arguments in environmental issues. *Environmental Ethics* 8: 59–76.

UNEP. 1992. *Agenda 21*. New York: United Nations Environment Program.

Young, M. D. 1992. Sustainable Investment and Resource Use: Equity, Environmental Integrity and Economic Efficiency. Paris: UNESCO and Parthenon.

WCED. 1987. Our Common Future: Report of the World Commission on Environment and Development. Oxford: Oxford Univ. Press.

Weinberg, A. M. 1985. Science and its limits: the regulator's dilemma. *Issues in Science and Technology* 2: 59–73.

22 IMPLEMENTING ENVIRONMENTAL POLICIES IN CENTRAL AND EASTERN EUROPE

Tomasz Zylicz[1]
Economics Department
Warsaw University
Ul. Dluga 44/50
Warsaw, Poland 00–241

ABSTRACT

The chapter identifies the causes of environmental policy failure in former centrally planned economies in Europe, and discusses ways to make these policies more successful in the period of transition to market economies. The countries of central and eastern Europe have embarked on a process of unprecedented political changes and structural adjustment. While economic reforms are expected to remove some of the factors responsible for excessive environmental pressure in the past, a careful design of far-sighted environmental policies is inevitable in order to let the new market systems prove their potential.

This chapter reviews alternative economic instruments developed for environmental management purposes, two of which have attracted particular interest (i.e., pollution charges and marketable pollution permits). In the past, policy makers in central and eastern Europe relied on the former too much. It is argued, however, that the latter possess several characteristics that make them a preferred tool in many circumstances. In particular, they are an excellent tool to add flexibility to regulations and other non-market solutions which are inevitable in the transition period when markets are weak and structural problems inherited from the past are immense. The preference for marketable permits does not preclude using pollution charges as a politically feasible method of raising funds to co-finance environmental projects when strict adherence to the Polluter Pays Principle is difficult. Finally, the chapter discusses ways to mobilize the private sector to contribute to environmental improvement and recovery in central and eastern Europe.

1. This chapter was written while the author was Valfrid Paulsson Guest Professor at the Beijer International Institute of Ecological Economics, The Royal Swedish Academy of Sciences, Stockholm. Critical comments from Ralph C. d'Arge, Faye Duchin, Christian Koelle, Valentina Krysanova, and Bedrich Moldan on an earlier draft of the chapter are gratefully acknowledged.

The author draws from the most recent experience in reforming environmental policies in the central/eastern European region to mark progress, identifies typical traps faced in this process, and illustrates all essential arguments. However, it was not the author's aim to give a detailed description of how reforms proceed in central/eastern Europe or how they affect the state of the environment. The three-year time perspective is simply too short to validate any such conclusions.

THE PREDICAMENT

The restructuring of the European post-communist economies over the next several decades will require massive investment dependent upon both domestic and international resources. The general objectives are to raise the average material standard of living while continuing to provide critical social services and to reduce the pressures on the environment. The national governments need to play a critical role in determining specific objectives and in transferring increasing responsibility for decision making to lower levels of government and to the private sector. They need to encourage domestic entrepreneurs as well as the inflow of international capital without compromising environmental requirements.

Heritage of the Old System

It is well documented that centrally planned economies used to exert a significantly greater pressure on their environmental resources than comparable market ones. For instance, there is a clear distinction between the European Community (EC) and the centrally planned economies of central and eastern Europe with respect to energy and resource intensities of their domestic products (see Table 22.1).[2] A more detailed analysis (Krawczyk 1991; Hughes 1991) reveals that these intensities tend to decline when material welfare increases in either group. Even the wealthiest centrally planned economies (German Democratic Republic, Hungary, and Czechoslovakia) had in general higher intensities than the least wealthy EC states (Portugal, Greece, and Ireland). This suggests that the two groups of economies did not follow the same pattern of development. Even though the central planning mechanism has been abandoned in Europe, problems remain and should be taken into account when designing clean up and recovery policies.

A recent study done at Warsaw University (Krawczyk 1991)—using some of the most up-to-date econometric techniques—confirms that not the level of industrial development, but rather the economic system adhered to is crucial in explaining differences in environmental performance among European countries. However, what can be concluded from sophisticated statistical re-

---

2. The table offers very rough estimates of indicators prevalent in the late 1980s. The quality of the input prevents more than two digit accuracy. For data availability reasons, the table excludes the former Soviet republics. Readers may, however, wish to refer to Salay (1993) to check that the general pattern portrayed here was replicated in Estonia, Latvia, and Lithuania.

search is also apparent by comparing the air in Rome or Paris to the air in any of the central/eastern European capitals. Atmospheric emissions provide visible proof of environmental protection failure in centrally planned economies. But this could only partially be attributed to an ill-designed environmental policy as it primarily resulted from running inefficient economies. General inefficiency of a non-market allocation results in an excessive (relative to output) use of any resources. This is accompanied by several additional factors, leading to excessive use of natural resources in particular.

Table 22.1. Economic Pressure on the Environment in the late 1980s

	Central/Eastern Europe–6	European Community–12
GDP per capita, $1000/person	4.8	10.1
Energy intensity of GDP, TOE/$1000	0.77	0.23
Energy intensity of GDP, TOQE/$1000	0.56	0.21
Water intensity of GDP, m³/$1000	153	82
Pollution intensity of GDP:		
—solid waste, tonnes/$	1.0	0.4
—wastewater, m³/$1000	83	24
—gases, kg/$1000	51	24
—dust, kg/$1000	13	1

Central/eastern Europe–6 = Bulgaria, Czechoslovakia, GDR, Hungary, Poland, and Rumania
European Community–12 = European Community of the Twelve.
TOE = tonne of oil equivalent.
TOQE = tonne of "oil quality" equivalent: identical with TOE in the case of oil and gas; 2 TOE of coal are assumed to be 1 TOQE; 1 TOE of hydro- and nuclear-electricity is assumed to be 3 TOQE.
Source: Author's estimates based on Krawczyk (1991) and Olszewski et al. (1987).

The Anatomy of Environmental Policy Failure

A bias towards heavy industry in national product mix and the resulting under-representation of the service sector are perhaps the most important reasons for the continuing pressure on natural resources in centrally planned economies and their heirs. The position of heavy industries has historical roots and proved to be stable despite modernization efforts. Industry interest groups still exercise some power and contribute to the inertia of the economy they are part of, because of the weakness of democratic institutions that could effectively represent social preferences. The dominance of heavy industries is coupled with weakness of environmental enforcement inherited from the past.

Furthermore, there are some additional misconceptions regarding environmental policy in central/eastern Europe. For instance, in the 1980s, policy in Poland was (*de facto*, but sometimes even *de iure*) left almost entirely to market forces hardly existing at that time that were supposed to achieve the protection through a system of financial incentives. Unfortunately, many of the central/eastern European economists advocated emission charges rather than alternative tools, overlooking the fact that even in the highly developed market economies, environmental management has not been left to uncontrolled market forces (Opschoor and Vos 1989). The excessive pollution is then linked to an inappropriate level of taxes or fines, and thus the attention of policy makers is distracted from the real issues. It should be clearly acknowledged that in the system where all essential inputs were allocated administratively and where the firms were operating under so-called soft budget constraints (Kornai 1986), there was no need to pay much attention to price stimuli.

The pure effect of policy failure—as opposed to the effects of general system inefficiency—becomes evident when comparing energy, water, and pollution intensity in Table 22.1. In general, the European non-market economies required roughly 2–3 times more inputs to produce a given effect, and this explains much of the difference in their environmental performance. In particular, the hard-to-abate gaseous emissions resulting from fuel combustion can be explained in terms of excessive energy use which is typically 2–3 times higher than in the West. Energy intensities of GDP are either directly or indirectly attributable to system inefficiency. Central/eastern European countries—especially Czechoslovakia and Poland—use a fuel mix of a much inferior quality than the European average. Taking into account quality differentials, the energy efficiency gap is smaller, but still exists and represents a *direct* effect of system inefficiency (accounted for in energy intensity of GDP measured in oil quality equivalent, TOQE). Given their inability to compete in international markets, centrally planned economies had to rely on domestic resources as much as possible. Their inability to upgrade the fuel mix can be viewed as an *indirect* effect of the system inefficiency (accounted for in energy intensity of GDP, TOE). In sum, the excessive gaseous emissions in centrally planned economies result (either directly or indirectly) from general system inefficiency, and to a limited extent, could have been mitigated through better environmental protection practices. The particulate matter emissions (dust) offer a sharp contrast to this pattern. Here the gap between centrally planned economies and market economies is enormous and cannot be explained in terms of inefficiency. Simultaneously, the abatement techniques are less costly and have been known and practiced for decades. Thus it was only inability or reluctance to enforce the abatement that allowed central/eastern European industries to become "dirty" in the most literal sense.

The comparison of gases and dust in Table 22.1 thus illustrates a major difference between environmental performance of centrally planned economies and market economies. The latter did not do a perfect job with respect to SO_2 abatement and even less so with respect to NO_x throughout the 1980s. Consequently, the contrast between both groups of countries is not so apparent as in the case of particulate matter emissions. These could by no means be explained in terms of (excessively) consumed energy. Because of strict environmental policies, the developed market economies can now enjoy low levels of dust emissions which are fairly easy to abate. Despite moderate abatement costs and moderate level of technological sophistication, central/eastern European economies have failed to solve their dust emission problem due to the lack of adequate policy measures.[3]

COMBINING ECONOMIC AND ENVIRONMENTAL REFORMS

In their struggle for a friendlier, safer environment, the central/eastern European nations have to overcome two barriers: general system inefficiency and inadequate policy principles. Economic reforms quite vigorously promoted by most of the new governments will remove the first obstacle. Given all the recent developments in the region, it is unlikely that a sudden comeback of the *ancien régime* may threaten the future of economic and political reforms. Nevertheless, preventing social disturbances or dealing with their effects may postpone achieving efficiency goals of the reform and make environmental improvements slower. Thus environmental policy makers cannot take the beneficial results of market reforms for granted.

The other barrier is not so easy to overcome either. Quite commonly in central/eastern Europe, the failure of environmental policies has so far been linked to an inadequate level of charges and to the lack of markets. Thus there is a natural temptation to conclude that with the advent of a market economy some "optimally calculated charges" will finally start to work. This feeling is occasionally reinforced by Western authors (Sand 1987; Anderson et al. 1990) who uncritically refer to East European financial instruments in environmental protection.

Pollution Charges *Versus* Marketable Pollution Permits

The rationale for emission charges—or, more specifically, so-called *Pigouvian taxes*[4]—stems from the fact that in a perfect information competitive world, the same socially desirable effect can be achieved both through setting emission quotas and a per-unit charge. If the marginal costs of abatement and the marginal benefits from it are known, then the point where they meet defines the "socially right" level of emissions. It minimizes the sum of total costs of environmental damages (or foregone benefits) plus total costs of protective activities. Environmental policy makers can reach that level by simply not allowing emissions to be higher than optimal. Under standard microeconomic assumptions[5] the same effect can be achieved through establish-

---

3   In some cases, the non-availability of adequate environmental information can be identified as a source of policy failure. According to an OECD survey (World Bank 1992b), a wealth of data has already been available, at least in Czechoslovakia, Hungary, and Poland. However, even in these countries, the quality of information needs to be improved to form the sufficient basis for more effective policies.

4. A *Pigouvian tax* is a charge levied at those who adversely affect other parties without reimbursing them for the damage. The charge is intended to give intruders an incentive to treat the damage as if this was their own problem, without any additional regulation.

5. Most importantly including the convexity of cost functions.

ing a charge (corresponding to the level of the marginal abatement costs and marginal benefits) per every unit of emission released into the environment. The equivalence of both solutions is only a formal one. In the real world, however, the superiority of one system over the other is contingent upon specific assumptions (Weitzman 1974). For practical purposes there is often a strong, a priori preference for the first, i.e., quantity-based approach (Adar and Griffin 1976). Unfortunately, neither the seminal paper of Weitzman, nor the subsequent discussions have ever been referred to in the East European and Soviet literature on emission charges, which represent the price-based approach.

There is a way to combine firm, quantity-based regulations with flexibility of market response. The way—invented with the institution of marketable permits—perfectly conforms with Herman Daly's (1984) well-known principle of separating scale and allocation decisions in environmental policy. To regulate the environment means to decide to what extent a given resource—say, the carrying capacity of the atmosphere or an aquifer—should be used, and what portion of the resource should be allocated to any of its potential users. Only the second problem is suitable for market regulation. When it comes to the question of scale, the market may easily fail as a guide, and public debate should determine the solution and finally, an administrative standard. This theoretical background and an enormous cost-reducing potential make marketable permits a preferred instrument of environmental policy.

It should be stressed that the quantity-based approach is not just a more understandable and acceptable option for those exposed to non-market mechanisms. It is not that the preference for marketable permits may result from central/eastern European managers' familiarity with physical rather than financial targets (Levin 1989). Also, it is not just a preferred tool in an economy with soft budget constraints. This is an instrument which combines the certainty and effectiveness of a standard with the flexibility of a market incentive. Its viability was first demonstrated in a developed market economy.[6] Nevertheless, unlike charges and fines, it may also work well in a poorly developed institutional context even though here its cost-saving potential can not be fully utilized.

Environmental and Economic Policy Challenges

Striking the balance between sectoral and/or regional *versus* financial (macroeconomic) approaches is, perhaps, the major challenge environmental and industrial policies meet. On one hand, any discretionalism has to be avoided, in order to provoke the necessary psychological breakthrough in the

---

6. To date, most of the many thousand market transactions of emission permits took place in the United States. In some countries (e.g., in Germany) there exist legal provisions for arranging permit transfers between various emission sources within the same plant. In some other Western European countries (e.g., in Sweden), plans for applying marketable permits are currently being debated.

attitudes of industrial managers in the former centrally planned economies, who are used to the excessive paternalism of state administration. Since the very beginning of the present reform, they have been seeking individual or sectoral subsidies, exemptions, tariff protection, etc. To yield to such pressures would lead to the instant decay of the emerging market system in central/eastern Europe. Meanwhile, the over-reliance on financial measures—in the absence of well-established market structures and entrepreneurial behavior—has lead (at least in Poland) to an economic collapse much deeper than initially expected (and, perhaps, deeper than absolutely unavoidable). Finally, "economies of scale" can be achieved, if government intervention provides services that can be used by groups of firms. Strategic studies for industrial restructuring make a prime example of activities that are often too expensive to be carried out by firms—especially those barely surviving the transition period.

Environmental policy makers in central/eastern Europe must seek solutions to reconcile the need for sectorally and regionally customized policies while requiring firm financial rigors. The answer cannot simply point at the positive OECD experience, since those countries could combine discretionary measures with the functioning of well-established markets. Eastern European economies do not operate under the same circumstances.

There is, however, a possibility to gradually integrate market choices into sectoral and regional physical plans. The option is provided precisely by marketable permits which can be applied both at regional and sectoral scales. Successful examples of *regional* implementation of these instruments are provided by numerous "bubbles"[7] found to be the only practical and cost-effective way to meet ambient standards in many non-attainment areas in the United States. The textbook example of a successful *sectoral* program is phasing-out lead additives in American refineries. Finally, combined *sectoral/regional* effects were observed in cases where industry-specific pollutants (e.g., volatile organic compounds released from laundries) were tackled on a municipal basis (Tietenberg 1993).

In their transition to a greener economy, the central and eastern European societies and governments are offered assistance of the best western European, North American, and Japanese experts. Environmental policies for former centrally planned economies have become a part of general environmental policy studies (Howe 1991). They have also become a separate target of some analyses both in the East and West (Zylicz 1990, 1991; Dudek, Stewart, and Wiener 1992). Post-communist Europe is thus given a chance to learn from the Western experience and to avoid some of its traps and mistakes. One study of economic instruments concludes with a statement whose optimism is, however, somewhat premature:

---

7. The "bubble" refers to the concept of treating emission sources as if they were under an imaginary dome (bubble), and as if only the sum of the emissions originating under the dome mattered.

Market-based environmental policies are an ideal tool for nations nurturing new market-based economies. Employing market-based approaches to environmental policy from *the outset* (emphasis added) can avoid some of the costly mistakes made in the U.S., and propel Eastern Europe into the forefront of environmental protection. (Dudek, Stewart, and Wiener 1992).

The assumption about "the outset" is by far not an obvious one. Although it is true that environmental reforms in central/eastern Europe ought to be comprehensive, this does not mean that their architects may start from scratch. The countries of the region have had extensive legislative systems which cannot be ignored without putting at risk environmental objectives (at least in the short run).

Nevertheless, the need to reform environmental institutions is quite obvious in central/eastern Europe. It is thus both desirable and probable that the post-communist economies will replicate the American experience with market-oriented instruments sooner than the rest of the continent. Indeed, in Poland there are already two marketable permit projects under way (Dudek, Kulczynski, and Zylicz 1992). These have been prepared in the absence of a fully satisfactory legal framework, and therefore, even though they receive support both from the government and local environmental groups, they can be carried out as "experiments" only. Replicating such projects on a larger scale requires major amendments to environmental legislation. The Hungarian and Polish governments introduced provisions for marketable permits into their new draft Environmental Protection Acts, but it may take time before these are approved by the Parliaments.[8]

Setting Environmental Priorities

Even though the emerging market structures in central/eastern Europe are likely to stimulate more environmentally sound developments in the long run, firm yet flexible government policies are needed in order to make it happen. A consistent environmental policy is an absolute must and should determine a set of short- and long-term priorities, and a rational mix of market-oriented tools and direct implementation controls, as either of the instruments has its merits and should be relied on in certain circumstances. Setting priorities is a delicate issue, as it implies not only inclusions in, but also exclusions from, the collection of objectives to be addressed. This is why several time horizons should be adopted: whatever cannot be accomplished in the short run ought to be taken care of later, but within a specified time frame. Otherwise, the policy may become either unrealistic if it tries to integrate too many

---

8. Judged by Western standards, the new Parliaments in central/eastern Europe are quite productive. However, the number of fundamental problems which have to be settled increases. Because the environment is not high on the agenda, the legislative process can be disappointingly long. Even studies like the one prepared by the World Bank (1993), which confirm the cost-saving potential of market-oriented instruments to address Poland's air pollution problems, are not likely to strengthen politicians' interest in the subject.

"justified" goals, or it may lack social support if the public is disappointed with too many painful exclusions. A high ratio of immediate benefits derived from a project to its cost is a useful indication of whether it should be considered a priority. However, there are also projects lacking obvious short-term benefits that have serious adverse long-term effects if not undertaken early. The substantial benefits of such ventures (in terms of keeping future options open, or avoiding losses in the future) may justify their inclusion on priority lists even if the time factor and uncertainty are taken into account.

A list of environmental priorities makes an irreplaceable component of any national policy. It highlights key issues, helps to focus the attention of media and environmental groups, and coordinates public agency activities. However, such lists do not necessarily imply specific instruments to implement policy objectives. The choice of appropriate instruments always poses a challenge, but nowhere has this challenge been more profound than in economies in transition. The difficulty is caused by the need to base post-communist economic policies on firm, system solutions, while the system itself is *in statu nascendi*, and certain discretionary measures are unavoidable. A careful balance between the necessary state intervention and a general system development is thus of paramount importance. Otherwise, environmental "interventionism" could be an obstacle to a successful therapy of the economic system. Is it possible to design a set of instruments which serve the purpose of improving environmental policy and supporting the market logic simultaneously? As suggested in this section, economic instruments have a particularly important role to play here, but one should not expect miracles.

Can Environmentalists Rely on Pro-Market Reforms?

It is commonly assumed that the return to market logic will cause economic actors to be much more sensitive to price signals. At the same time, they will be forced to seek every opportunity for cost reduction by way of energy and material efficiency comparable to levels in western Europe, thus curbing a fundamental source of pressure on the environment and its resources. Achieving market flexibility and efficiency will not solve the problem automatically. Moreover, as a World Bank (1992a) study for Poland shows, the short-run structural adjustments of a post-communist economy may, in fact, *increase* rather than decrease pressure on the environment. This paradox can be explained by a lack of abatement equipment in the existing light industries, whose output is most likely to expand in response to market stimuli, before environmental authorities have even had an opportunity to react. This clearly points to the necessity of an extremely careful design of environmental policies matching the macroeconomic ones.[9]

The establishment of competitive markets will unleash tendencies to save on costs. This in turn will reduce energy and material intensities of produc-

---

9. Bochniarz (1991) discusses how practically to design coordinated financial, industrial, and environmental policies for economies in transition.

tion, thus attacking the main single cause of environmental disruption. This same tendency, however, will also create incentives to save on environmental protection costs, thereby calling on the need for environmental authorities to be particularly effective in enforcement programs. Following the example of the developed market economies, the post-communist ones should make use of market-oriented instruments. Simultaneously, the governments should implement direct regulation wherever necessary and where commonly applied worldwide.

THE ROLE FOR ECONOMIC INSTRUMENTS

The Rationale of Economic Instruments

It is the standard wish (and expectation) of environmentalists that economic actors allow ecological considerations the same standing in their calculations as considerations of profit. This means that each kilogram of emissions and each tree cut can be priced in such a way that firms will only emit pollutants or disturb nature when it is socially optimal. Unfortunately, this is unrealistic as a policy option, and it is important to make the public and decision makers in central/eastern Europe aware as quickly as possible that in none of the market economies are firms regulated through the use of this type of signal. Perhaps in the future economists will be able to accurately enough calculate value of natural capital to allow environmental management to rely solely on fees. So far the usefulness of fees for the use of nature has been as a means of generating revenues for environmental management and less so as a means of stimulating reduced pressure on the environment. Hence, it is necessary to employ other tools for setting limits on use of the environment.

Economic instruments include, first of all, charges and marketable permits. These are supplemented by subsidies (distributed in the form of grants, tax allowances, soft-loans); budget-neutral instruments (e.g., tax differentiation and deposits on harmful products); and finally, non-compliance fees.

Adhering to the following four policy principles while applying economic instruments is likely to enhance the chance for a success (Ministry 1990):

- Seeking economic efficiency to be achieved through "harnessing" market forces in order to minimize the total cost of policy implementation;

- Setting limits so that the overall scale of use of the environment not be exceeded despite economic agents' willingness to pay a non-compliance fee or other form of financial reimbursement;

- Regionalism understood as orienting environmental requirements to regional circumstances and providing local authorities the freedom to choose instruments (to meet these requirements) including the right to use effluent charges as a way of co-financing local environmental programs; and

- Polluter Pays Principle (PPP) understood as the polluter's responsibility for carrying out all the expenditures necessary to comply with environmental regulations in order to free the authorities from excessive responsibility for the implementation of abatement programs.

Economic instruments perform a number of useful functions. The most important task of these instruments is to minimize overall costs of environmental protection through an efficient differentiation of control requirements. Namely, those agents with the lowest abatement costs should be given the most stringent requirements. In practice, only marketable permits can perform this role in an environmentally safe way. Theoretically, the same goal can be achieved by setting effluent charges at sufficiently high (Pigouvian) levels; however, almost nowhere is this solution actually relied upon due to political and methodological obstacles. Another function performed by economic instruments is to raise funds to be spent on environmental management needs. Charge systems throughout the world serve this purpose, and their implementation in the post-communist economies is expected to do the same. Charges, as well as other economic instruments, may also create incentives for environmental conservation and improvement, but this is their secondary role for the time being.

Marketable permit systems and charges raised to their "biting" (Pigouvian) levels can often be viewed as substitutes. While the previous argument points at marketable permits as a tool more appropriate for economies in transition, it does not exclude pollution charges used as a revenue-raising instrument. They can perform this role quite satisfactorily even with low rates. Thus, in this context they can be seen as complements rather than substitutes for marketable permits (or any other instrument, for that matter).

Specific Recommendations

Effluent charges should be made available for regional or local authorities who are responsible for regulating the effluent. On the other hand, product charges (e.g., a surcharge on fuel prices) do not have to be decentralized, and, therefore, can be used to raise environmental funds at the national level.[10] Such centralized funds are an indispensable tool to deal with problems that transcend regional boundaries and/or have accumulated for a long time, so that leaving them eventually to the local jurisdictions would be neither effective nor equitable. The central/eastern European countries have an abundance of such cases.

Pollution charges applied in the central/eastern European countries cover a wide range of pollutants, usually much more than in the developed market

---

10. The pattern prevailing in central/eastern Europe is quite different now. The revenues raised from effluent charges are shared by central and regional authorities in certain proportions: e.g., in Lithuania the "central" share is 30% (Semeniene 1993); in Poland it is 40% in most cases (Zylicz 1993). This causes endless frictions, as regions demand that 100% be retained at the local level. In Estonia local authorities are free to change the national rates by up to 20% only. Other central/eastern European countries do not even have that flexibility.

economies. Such schemes contrast sharply with, for example, the Dutch fuel tax introduced in the 1980s to substitute for a number of other charges, and thus simplify administration (Opschoor and Vos 1989). The Estonian system also provides a rare example of "ambient taxes" (as opposed to more typical emission taxes), since charge rates vary with the location of a polluter to account for its ambient environmental impact (Kallaste 1993). Because inflation makes it difficult to maintain a high level of rates, the Estonian system—as well as a similar Lithuanian one—has never become a really significant instrument. One can expect, however, that once the charges start to bite, the polluters will raise difficult questions. Consequently, the system will need to be simplified; otherwise, the rates will remain much below their incentive levels.

Charges are used in particular in water resources management (for water intake and sewage discharge), in waste management (for disposal and storage), in air quality management (for emissions), in the protection of living resources, and in the protection against noise and radioactivity. Their rates should meet the budgets devoted to respective sectors of environmental protection or resource management.[11] These budgets should be self-financing with the possibility of receiving moderate support from the general budget of the national or local governments—especially during the next few years when the backlog is enormous.

Extractive activities should also be subject to licensing procedures and charges whose rates may vary with the quality of the deposit. The charge should be paid into the national budget (on behalf of the owner, i.e., the state) or to a fund dedicated to the development of the extractive sector. After the competitive market for geological services and mining activities has developed, the charge should be substituted with a concession fee arrived at through a bidding procedure. This will ensure that the budget receives the full rent due. In order that the geological/mining sector operates smoothly, exploration and documentation activities should be carried out continuously. This sector had been previously sponsored by the state which should continue until the market fully develops for geological services.

Forest resources—which for the most part in central/eastern Europe are under state administration—also are an element of the national wealth which can raise revenues (in the form of a rent) for the budget. At the same time, given that the forestry sector is adversely affected by environmentally disruptive economic activities, the governments should provide this sector with subsidies. In addition, these subsidies can support the development of new forest lands. The entitlements of these subsidies should be established by law.

---

11. Pollution charges are typically earmarked for environmental protection despite objections both by theorists and state budget ministers. Even when they are formally not, an informal earmarking often takes place—as in Lithuania with respect to the centralized 30% of charge revenues (Semeniene 1993)—and consequently, environmental protection budgetary expenditures correspond to the amount collected on pollution.

The scope of subsidies for environmental protection paid from public funds should be retained and even broadened during the period of eliminating the backlog. The subsidies may be continued in the form of direct grants from the national or local budgets for environmental investment. Tasks pertaining to the creation of a (minimum) municipal infrastructure (e.g., facilities providing water supplies, or basic sewage and solid waste treatment), which is the responsibility of the municipality, should be financially secured in the local budgets. In municipalities with low revenues, this share should be warranted by means of allocations from the national budgets. (The approach to dealing with environmental protection should be comparable to that of meeting the basic needs of education, health care, and public safety). Subsidies for water resources management budgets—now widespread in central/eastern Europe—should be gradually eliminated so that in the near future each river basin would be self-financed through increased user fees. Subsidies for industrial environmental protection programs should also be employed, to a limited extent, as a part of the governments' programs to restructure the economy.

A portion of the subsidies for environmental projects may take the form of tax allowances.[12] Their performance—wherever applied—should be analyzed in order to work out detailed guidelines for an efficient use of these budgetary means. In Poland, to date, only the National Fund for Environmental Protection and Water Resources Management, and the (commercial) Bank of Environmental Protection offer soft-loans for environmental projects. The preferential terms are made possible by revenues incurred by the Fund from effluent charges and non-compliance fees. As a result of International Monetary Fund consulted structural adjustment and stabilization policies, soft-loans are not widely used throughout the central/eastern Europe region now. However, the model developed in Poland was considered consistent with these policies, and proved to be effective. The governments may also wish to consider a possibility for allotting budgetary funds to be channeled through (selected) commercial banks in the form of reduced interest rates and extended grace periods for environmental protection loans.

Budget-neutral instruments can be introduced to provide incentives for speeding up pro-ecological changes in the economy and society. Deposits on commodities, chemical compounds, or containers that are not environmentally friendly serve as examples here. The environmental lobbies can also plea for tax and tariff differentiation to penalize manufacturers and users of environmentally harmful products, regardless of whether the difference stays in the general budget or gets earmarked for environmental projects.

Non-compliance fees based on fixed (perhaps even high) rates contradict the principle stating that a predetermined, overall scale-of-use of the environment must not be exceeded in exchange for a fine. For this reason, non-compliance fees, which have existed in Poland, can be accepted only as a

---

12. Applied or at least debated in many central/eastern European countries. In Poland discontinued since 1991, perhaps somewhat prematurely, as economically inefficient.

means of discouraging accidental, one-time spillovers. With respect to the polluters being consistently in non-compliance, other enforcement measures should be applied.

As this brief outline suggests, the scope for economic instruments in environmental policy is quite extensive, and most of them have been in use in central/eastern Europe in one form or another. Their application, however, is inconsistent. As a result, their cost-saving potential has not yet been realized. The post-communist economies must still learn how to "harness" market forces safely; how to take advantage of these in allocation processes while protecting the overall scale of environmental resource use from reckless expansion; and how to effectively address economic restructuring problems without excessive centralization of authority and finance.

FINANCING ENVIRONMENTAL PROTECTION INVESTMENT

There is a widespread irrational belief in central/eastern Europe that environmental protection does not require expenditures, provided that the polluters are given the "right" price incentives. As a result, people often resist the idea of bearing the burden of abatement. Meanwhile, the communists are expected to pay for the clean-up of "the mess for which they are responsible." Alternatively, it is expected that the West will bail out the newly emerging market economies by providing the necessary finances.[13] As a result, the fundamental principle that the financial resources should be mobilized domestically first has been ignored. Unfortunately, international financial institutions did not help to rectify the fallacy when they linked the low share of foreign aid to insufficient "capacity to absorb" such aid (World Bank 1992b). Both in Hungary and Poland—the two central/eastern European countries most successful at attracting international assistance—approximately 95% of environmental investment expenditures in 1991 was financed from purely domestic sources. Even though the institutional capacity to absorb assistance is indeed far from satisfactory (Zylicz 1992), its improvement would not change this picture dramatically.

Environmental recovery and abatement costs in central/eastern Europe are probably not much higher than they used to be in the OECD countries which made their policies more effective in the 1970s (Hughes 1991). Still they are in the order of billions rather than millions of dollars per year. This is in sharp contrast with available international environmental assistance funds whose order of magnitude is tens or, at best, hundreds of millions of dollars

---

13. As a rule, environmental groups in the central/eastern European region—quite visible and vigorous at the time of revolutions in 1989—lost their popular support significantly once it became apparent that the Western democracies would not pay for the clean-up of the new allies. The only reason why *Eco-Glanzost* in Bulgaria survived as a political force longer than its counterparts elsewhere in the post-communist Europe was that the Western redemption myth was more widespread there (Dickman 1990).

per year.[14] Thus the major part of the burden will be born by the countries of the central/eastern European region themselves. This leaves the open question of how adequate resources could be mobilized internally. Of course, the *Polluter Pays Principle* could be a perfect policy guide, but its applicability is constrained by several factors.

According to the PPP—as interpreted by the OECD—the polluter is financially responsible for meeting the standards set by the respective environmental authority. The OECD countries, however, acknowledge the need to bypass this principle under special circumstances, especially if polluters are to pay for the backlog of neglect for which they may not have been fully responsible. PPP is not considered to be violated if the polluters are financially supported through funds collected from charges. Here the role of the charges is to raise funds which can be used by environmental authorities to share the costs of the most urgent projects.

Two major reasons justify the departure from a strict version of the PPP in the post-communist Europe:

- The polluter may not exist any more. This is the case in thousands of inefficient plants—mostly state-owned—which are about to terminate their operations (or have already done so), leaving devastated sites and unpaid bills.

- The polluter may be too poor to meet its environmental requirements, and yet its continued operation cannot be questioned. A number of municipalities functioning without adequate sewage and waste-disposal infrastructure provide prime examples.

In some OECD countries, whenever a strict responsibility of a given polluter is hard to enforce, an alternative *Polluters Pay Principle* (note the plural) is applied. The direct relationship between environmental stress and payment is relaxed, but considering the alternative of charging the general taxpayer, the modified principle offers a compromise. In Poland, the Polluters Pay Principle is implemented through the National Fund and 49 Regional Funds for Environmental Protection. Both the user and emission charges provide an important source of environmental project financing in Poland. They proved instrumental in increasing the share of environmental investment expenditures in GDP above 1%, to the relative level corresponding to what is currently spent in the OECD countries.[15] Emission charges and their "recycling"

---

14. Sweden—the single most generous international donor with a strong interest in the region—commits $140 million per annum (in 1993–95) to projects in central/eastern Europe. However, only a minor fraction of these funds will be spent on the environment, which competes with a number of other (political, educational, cultural, medical, and commercial) objectives (Ministry 1992).

15. In absolute terms, it corresponds to the *per capita* level the OECD countries started from in the 1960s when they began their environmental recovery programs. It is widely accepted that maintaining environmental investment expenditures at the level of 1%–1.5% of their GDPs for more than two decades is responsible for a significant improvement of ambient conditions.

method through the National and Regional Funds make a visible contribution to the environmental recovery process and are a crucial part of the financial mechanism. Operation of such funds should be encouraged throughout central/eastern Europe.

Another important role for environmental funds in central/eastern Europe is their potential contribution to the emerging system of environmental institutions. In most of the OECD countries, environmental policies are shaped in a creative process of interactions between public agencies and non-government organizations (NGOs). For historical reasons, both the public sector and environmental NGOs are weak in post-communist Europe. To the extent that such environmental funds could enjoy a relative independence from the political instability of new government structures, their existence positively contributes to the capacity build-up. It helps to train cadres and to retain environmental professionals who would have otherwise left the public sector, discouraged by the political process (Zylicz 1992).

It is ironic that while the communist system was notorious for its "adverse selection" mechanism, also the new democracies do little to upgrade the quality of its administration. Government departments are plagued by political appointments reaching down to the medium level staff. At the same time, professionals employed there are paid salaries corresponding to 10%–20% of what they could make in the thriving private sector. Certainly not many can resist such a temptation.

In addition to their technical role in implementing the *Polluters Pay Principle*, environmental funds may thus become a means to strengthen the domestic capacity build-up process in central/eastern Europe. Their overall impact, however, crucially depends on the ability of the charge systems to keep pace with inflation. To this end, it is necessary that they are adjusted according to the anticipated inflation rather than to an official (*ex post*) price index. Such a procedure (successfully applied in Poland 1990–1991) involves a risk that anticipations are incorrect. Nevertheless, as long as the charging rates are kept much below their Pigouvian levels, overestimating the inflation rate is less harmful than underestimating it.

WILL PRIVATE INVESTORS CONTRIBUTE?

Environmental recovery, coupled with economic restructuring of central/eastern Europe implies a massive investment process expected to take place within the next thirty years. The necessary investment expenditures can be reduced if environmental improvements are properly combined with economic reforms. Further cost savings are possible if countries take the advantage of market-oriented policy instruments. Part of the financial burden can be spread over a large number of economic agents by applying the *Polluters Pay Principle*. In the end, however, this will take a major effort and requires the mobilization of private resources, not only domestically, but internationally. Will private investors significantly contribute to this process?

Making the Rules of the Game Clear

For foreign investors, the most striking feature of the environmental manage-ment in central/eastern Europe is the disparity between strict regulation and lax enforcement. With its 32 $\mu g/m^3$ ambient standard for SO_2 (mean annual concentration), Poland exceeds the EC (where the 80–120 $\mu g/m^3$ standards apply), Switzerland, and the United States. Obviously, no one takes such a standard seriously. Poland is not an exception among the central/eastern European countries, and its SO_2 standard is not an isolated case. The Estonian maximum *daily* concentration of 50 $\mu g/m^3$ is even stricter, as it does not allow for compensating "dirty" days of, say, 70 $\mu g/m^3$ with "clean" ones of, say, 30 $\mu g/m^3$; whereas, in fact, even larger variations in ambient conditions are quite common.

Prospects for adopting the EC standards perhaps indicate the likely evolu-tion of environmental policy in central/eastern Europe. The expected harmo-nization of standards offers a valuable reference point for both national and foreign investors. However, a number of problems are to be sorted out in or-der to determine when and how these (reasonable) standards will be enforced. To some degree environmental policies can be financed through government budgets and dedicated funds, but the success ultimately depends on the ability to mobilize polluters' (or users') own resources, innovativeness, and creativ-ity. This mobilization will not take place unless economic agents receive clear and reliable signals regarding environmental requirements to be enforced.

Pollution control policy provides an apt illustration of the role economic instruments can play in policy implementation process. First, national ambient standards (e.g., concentrations of toxic substances in the air, water quality classification, soil contamination limits) should be binding on government bodies at the regional or other levels legally responsible for environmental regulation. Achievement of standards—within the framework of timetables set in conjunction with government agencies (e.g., Ministries of the Environment, regional administrators) and formalized in Parliamentary resolutions or some other high-level acts—is the responsibility of the government body specifi-cally in charge of environmental permitting.[16] Individual firms, in turn, are re-sponsible for preparing timetables laying out their abatement programs and submitting them to the appropriate authorities. These sub-timetables are an important means for meeting ambient standards regionally.

Neither regional nor municipal government authorities are necessarily re-sponsible for current ambient conditions, which are beyond their control as a result of interregional or international flows of pollutants. It is the responsi-bility of the national government (e.g., through Parliamentary action on pro-

---

16. The *State Implementation Plans* (SIP) prepared under the United States Clean Air Act are, perhaps, the best known example of an arrangement clearly dividing the responsibility for meeting environmental objectives between various levels of administration and polluters. It is also worth noting that SIPs triggered the application of marketable permits as an instrument of meeting environmental policy objectives for the lowest cost.

posals prepared by the Ministry of the Environment) to incorporate into the regional timetables constraints on interregional "exports" of pollutants. These constraints can also follow from a given country's international obligations.

Within the framework of the above-mentioned constraints, the agency empowered to issue emissions permits should have full flexibility to plan and implement pollution control policies. In particular, the agency, based on its informational, technical, and staff capabilities, may choose one or more of the following four methods:

- Demonstrate attainment of national ambient standards by way of compliance by individual sources with their respective emissions standards as established by national authorities. (Currently there exist emission standards for a limited number of source types in the central/eastern European countries.)

- Implement stricter emissions or discharge limits for individual sources which are not exchangeable for charges or fines.

- Regulate the use of environment by means of a system of charges and fines similar to the one currently in use in some central/eastern European countries, which combines modest financial incentives to abate with raising funds to co-finance priority projects.

- Institute permit markets in which individual users trade all or part of their rights to emit or discharge to other users of the same water- or air-basin, subject to the approval of the pertinent permitting authority.

The approach proposed here relies on two postulates: (1) that environmental protection cannot rely on a system of charges and fines as the sole stimulus for required activities, and (2) that it is necessary to create possibilities for entities responsible for implementing these activities to generate funds for this purpose. In the case of firms, who are the main target group of environmental protection requirements, this is made possible by the abandonment of administrative price control. In the case of national and local governments, which are co-responsible for implementation of environmental policy, this is made possible by allocating the ability to raise funds at precisely the level where governmental responsibility for a given portion of this policy rests. Alternately, establishing the level of subsidies for environmental protection derived from general tax revenues represents a decision to be politically determined, taking into account both actual progress in achieving policy objectives and the state's international obligations in this area.

Despite recent reforms, the legislative systems governing environmental protection in central/eastern Europe are far from introducing clear division of responsibility between central governments, local authorities, and polluters (environmental resource users). Still there exist ambient standards—often too strict to be taken seriously—without an authority responsible for their implementation. As a result, not only investors, but also administrators are confused as to what requirements should be imposed.

Combining Privatization with Environmental Improvement

In addition to their environmental recovery programs, the countries of central/eastern Europe are engaged in an ambitious privatization effort. Its success will largely depend on whether domestic and foreign investors and their financing banks can accurately calculate the costs of investment, and resolve elements of uncertainty and risk. The costs of meeting environmental requirements are an important element of this calculation. Several environmental issues must be addressed and resolved, if substantial privatization is to be accomplished in a reasonable time period.

If, therefore—as commonly expected—international investors are to contribute to environmental recovery in central/eastern Europe, the governments must create clear and stable legal conditions for business activities. For those countries who are in the process of associating with the European Communities, the EC environmental standards may offer a valuable reference point. However, even in this case, the question arises as to the requirements for the interim period and with respect to these areas where EC directives leave their member countries certain freedom. In any event, the expected contribution will depend on the ability of central/eastern European governments to articulate their priorities, design consistent policy instruments, and create adequate financial mechanisms by enforcing Polluter Pays and User Pays Principles, as well as by operating public environmental funds.

A largely unresolved question facing investors in central/eastern Europe is their liability for the past contamination of the site, or—in general—for the past environmental performance of the plant they purchase. The German model of virtually exempting new owners of the former GDR firms from this responsibility is made possible by the enormous potential of the West German economy to absorb the resulting liability of the federal government.[17] At the other extreme, there is the Polish Privatization Act of 1990, which transfers the entire responsibility onto purchasers (new owners). However, even the most rigorous legislation in this respect, does not necessarily have to act as a deterrent for investors, since they can always negotiate the price of a purchase so as to subtract their expected liability costs. In fact, substantial rebates were negotiated in several privatization deals between multinational companies and Poland's Ministry of Privatization. Thus, the question boils down to how accurately these costs can be predicted, and what financial instruments to rely on to take them into account (Górka and Peszko 1992).

Apparently only in the Czech Republic has this question been satisfactorily dealt with on the government side. Comprehensive environmental audits must precede any privatization agreements there (Wajda 1992). This routine procedure offers as much comfort for investors, as for privatization authorities who are then in a better position to negotiate any possible rebates and/or compliance schedules. In all the other central/eastern Europe countries, excluding

---

17. *Treuhandanstalt*, the federal entity responsible for the privatization of the former GDR economy, absorbs 90% of environmental liability (Wajda 1992).

Germany, environmental audits are performed on an ad hoc basis (if at all), which slows down the mobilization of private capital and is most likely to provoke conflicts in the future.

This myopic attitude results from the lack of appropriate legal tradition, as well as from the current division of authority between government departments. On the one hand, Ministries of Privatization see their mission as the one to sell the state property as quickly as possible. On the other hand, Ministries of Environment see the privatization process as a unique opportunity to settle (at once) many of the problems resulting from the past neglect. Consequently, their efforts are perceived by privatization officers as an impediment to transactions rather than something which—in the long run—may improve the environment and the financial predicament of the state as well. Both privatization and environmental recovery processes throughout the central/eastern European region would benefit from learning from the Czech experience.

SUMMARY AND CONCLUSIONS

This chapter identifies the causes of poor environmental performance in the former centrally planned economies. Their excessive resource use has its roots both in the general inefficiency of the non-market allocation mechanism, and in inappropriate environmental policies. While it can be expected that the introduction of market logic will gradually eliminate one source of problems, a proper choice of policy priorities and instruments poses a challenge. Even though central/eastern Europe is witnessing a major political breakthrough now, it will be a long time until it fully reaches market coordination and efficiency. Meanwhile many of the old problems will continue, and the emerging markets are likely to surprise with an increase rather than decrease in the pressure on the environment.

In the past, despite their clear ineffectiveness, the policy makers and their advisors in central/eastern Europe favored emission charges rather than alternative nonfinancial instruments. There is an obvious tendency to adhere to this approach, especially now, and to wait until it allegedly produces the right results in the new market context. However, administrative solutions and other direct regulations in environmental policy are widely used by the OECD countries, and there is every reason to use them in the reformed economies of central/eastern Europe. This does not preclude developing permit markets wherever appropriate.

This chapter emphasized the importance of setting policy priorities. These should coordinate various conflicting goals within the environmental sector and between the environment and other fields. They should also offer a guidance for authorities and the private sector regarding areas of particular concern that are likely to focus attention. A number of instruments can be applied to implement priority programs. Permit markets have several interesting characteristic features which make them a promising environmental, sectoral, and regional policy tool. They combine effectiveness with flexibility, and are likely to assist policy makers in central/eastern Europe in addressing their concerns in a least-cost manner.

The Polluter Pays and User Pays Principles have been discussed in the context of circumstances prevailing in central/eastern Europe. In some countries (e.g., in Poland), emission charges have already become very substantial—they rank among the highest in the world—and it is unlikely that governments will be able to raise them further. However, even these very high charges are still lower than marginal abatement costs. That is why environmental policies have to assume that available instruments—at least within the next couple of years—will not include the Pigouvian taxation. Thus, if one seeks least-cost solutions, one has to rely on other economic instruments, and, most importantly, permit markets.

This chapter also addressed environmental aspects of the privatization process, and their implications for future foreign direct investment in the central/eastern European region. The disparity between strict regulation and lax enforcement is identified as one of the serious obstacles to any deeper involvement of private investors. In general, prospective buyers or their financing institutions are not concerned as much with their liability for what previous owners were responsible for; neither are they concerned with the growing costs of environmental control (which is inevitable). They are concerned because of the uncertainty regarding (1) whether these strict regulations will be sustained despite their inconsistency with EC measures, and (2) when a revision may happen.

The future of environmental policy reforms in central/eastern Europe depends on the governments' willingness and ability to resolve the following key issues:

- Progressing with pro-market reforms to increase economic efficiency and thus remove excessive environmental resource use;

- Creating a "critical mass" of market institutions to let economic agents assume responsibility for their actions, remove an important psychological barrier to overcoming "state paternalism," and pave the way for more effective enforcement;

- Learning to combine direct regulations (whenever unavoidable for historical reasons) with giving economic agents a choice of adjustment strategies;

- Applying innovative policy instruments—like marketable pollution permits—to bring environmental compliance costs to a minimum;

- Setting environmental priorities—to protect both human health and natural capital from further decay—so that public agencies and regional authorities receive clear signals indicating where to concentrate their efforts;

- Establishing mechanisms to secure a reasonably high share of environmental protection expenditures in Gross Domestic Products;

- Maintaining or even increasing the role of public funds in co-financing environmental investment projects during the period of economic transition; and

- Channeling private entrepreneurship into commercial investment projects which embody environmentally friendly technologies and/or help clean up contaminated environments.

REFERENCES

Adar, Z., and J. M. Griffin. 1976. Uncertainty and choice in pollution control instruments. *Journal of Environmental Economics and Management* 3: 178–88.

Anderson, R. C., L. A. Hofman, and M. Rusin. 1990. The Use of Economic Incentive Mechanisms in Environmental Management. Research Paper #051, American Petroleum Institute.

Bochniarz, Z., ed. 1991. Capacities and Deficiencies for Implementing Sustainable Development in Central and Eastern Europe. UNDP Regional Assessment Report. H. H. Humphrey Institute. Minneapolis: Univ. of Minnesota.

Daly, H. E. 1984. Alternative strategies for integrating economics and ecology. In Integration of Economy and Ecology: An Outlook for the Eighties, ed. AM. Jansson. Proceedings from the Wallenberg Symposium. Stockholm: Stockholm Univ.

Dickman, S. 1990. Western help comes slowly. *Nature* 348 (December 6): 472.

Dudek, D. J., Z. Kulczynski, and T. Zylicz. 1992. Implementing Tradable Rights in Poland: a Case Study of Chorzów. Paper presented at the Third Annual Conference of the European Association of Environmental and Resource Economists, Cracow.

Dudek, D. J., R. B. Stewart, and J. B. Wiener. 1992. Environmental policy for Eastern Europe: technology-based versus market-based approaches. *Columbia Journal of Environmental Law*.

Górka, K., and G. Peszko. 1992. Privatization, Environmental Liabilities and Credit Insurance Fund for Foreign Investments in Central and Eastern Europe against Environmental Risk. Paper presented at the Third Annual Conference of the European Association of Environmental and Resource Economists, Cracow.

Howe, C. W. 1991. Lessons from the U.S. Experience for the Single European Market and Eastern Europe. Wageningen: Agricultural Univ. Mimeo.

Hughes, G. 1991. Are the costs of cleaning up Eastern Europe exaggerated? Economic reform and the environment. *Oxford Review of Economic Policy* 7 (4): 106–36.

Kallaste, T. 1993. Economic instruments in Estonian environmental policy. In Economic Policies for Sustainable Development, ed. T. Sterner. Dordrecht: Kluwer.

Kornai, J. 1986. The soft budget constraint. *Kyklos* 39: 3–30.

Krawczyk, R. 1991. Wzrost gospodarczy a srodowisko - wzajemne zaleznosci. (Economic growth versus environment—mutual relationships). Unpublished Ph.D. thesis, Economics Department. Warsaw: Warsaw Univ.

Levin, M. H. 1989. An environmental manifesto for Poland. *Wall Street Journal*, October 24.

Ministry of Environment. 1990. National Environmental Policy: Outline of Economic Instruments. Warsaw.

Ministry of Finance. 1992. The Swedish Budget 1992/93. Stockholm: Allmänna Förlaget.

Olszewski, D., W. Bielanski, and M. Kaminska. 1987. Economic Pressure on the Environment in the European CMEA Countries: Statistical tables. EKO^2 Series No. 8716. Warsaw: Warsaw Univ.

Opschoor, J. B., and H. B. Vos. 1989. Economic Instruments for Environmental Protection. Paris: OECD.

Salay, J. 1993. Environment and energy in the Baltic countries. In Economic Policies for Sustainable Development, ed. T. Sterner. Dordrecht: Kluwer.

Sand, P. H. 1987. Air pollution in Europe: international policy responses. *Environment* 29 (10): 16–20.

Semeniene, D. 1993. Economic Instruments of Environmental Protection Management in Lithuania. Vilnius: Environmental Protection Department.

Tietenberg, T. 1993. Market-based mechanisms for controlling pollution: lessons from the U.S. In Economic Policies for Sustainable Development, ed. T. Sterner. Dordrecht: Kluwer.

Wajda, S. 1992. Problemy ekologiczne w procesie prywatyzacji na terenie b. NRD, Czechoslowacji i Wegier (Environmental problems in privatization process in former GDR, Czecho-Slovakia and Hungary). Warsaw: Ministry of Environment. Mimeo.

Weitzman, M. 1974. Prices vs. quantities. *Review of Economic Studies* 40: 447–91.

World Bank. 1992a. Poland: Environmental Strategy. Report No: 9808-POL. Washington, DC: World Bank.

———. 1992b. Setting Environmental Priorities in Central and Eastern Europe. Discussion document on Analytical Approaches. Draft, December 3. Washington, DC: World Bank.

———. 1993. Alternative Policy Instruments for the Control of Air Pollution in Poland. Washington, DC: World Bank.

Zylicz, T. 1990. Environmental Policies for Former Centrally Planned Economies. Keynote address at the First Annual Conference of the European Association of Environmental and Resource Economists, Venice.

———. 1991. Environmental reform in Poland: theories meet reality. *European Environment* 1 (Part 1): 16–18.

———. 1992. Debt for Environment Swaps: The Institutional Dimension. Beijer Discussion Papers #18. The Beijer Institute. Stockholm: The Royal Swedish Academy of Sciences.

23 ENERGY PRODUCTION, CONSUMPTION, ENVIRONMENT, AND SUSTAINABILITY

James J. Zucchetto[1]
National Academy of Sciences
Energy Engineering Board
2101 Constitution Ave., N.W.
Washington, DC 20418

ABSTRACT

This chapter addresses issues of sustainability and the production and consumption of energy resources in human economies. Energy resources in and of themselves do not pose a constraint on economic activity over the coming decades. However, environmental, institutional, and social constraints will probably be the most important factors limiting the production and use of energy. To this end, the chapter addresses energy efficiency, environmental costs of energy production and use, the increasing importance of interdependence among nations and regions on the acceptability of energy technologies, and energy-related strategies for sustainability.

INTRODUCTION

Sustainability, sometimes referred to as sustainable growth or development, concerns the ability of world economies to sustain a reasonable level of economic output and standard of living in the face of widespread stress on the natural environment (Brown 1981; World Commission on the Environment 1987; Odum 1988).

This chapter explores the relationship between energy supply, consumption, and sustainability, and suggests some strategies to maximize the likelihood of achieving sustainable economic activity. The relationship between energy production and use, and the sustainability of natural capital (such as natural resources, forests, lakes, oceans, and the variety of ecosystems on the earth) includes:

---

1. The views of the author expressed here do not represent the views of the National Academy of Sciences, the National Research Council, or any of its constituent units.

1. husbanding non-renewable energy resources (e.g., coal, oil, natural gas) in ways that allow them to last to the maximum extent possible;

2. managing renewable resource systems so that they provide both a sustaining level of energy and materials as well as other natural services contributing to the quality of the environment; and

3. producing and using energy in ways that reduce associated environmental effects both locally and globally, thus minimizing impacts on the ecosphere and its constituent living systems.

Other broad scale trends affecting long-term sustainability, such as soil erosion, land and water pollution, overfishing, deforestation, or rapid population growth are not addressed here. Current and anticipated trends in world population growth and the economy are taken as given over the coming decades without pronouncements regarding their suitability or desirability. These, of course, may impact sustainability and change with time.

ENERGY SUPPLY AND SUSTAINABILITY

A source of energy is necessary for both natural ecosystems and human economies. Humans have realized great benefits from the exploitation of energy resources and the development of technology to harness them (Cottrell 1955; Odum 1971, 1983).

A first consideration regarding sustainability involves several questions: To what extent can global energy resources sustain the continuing increase in economic growth and world population experienced over the last few centuries? Can the world economy maintain existing levels of economic activity, much higher levels, or much lower levels? For how long? Such questions remind one of the "energy crisis" of the 1970s, which was primarily an oil crisis. Many analysts predicted shortages of energy and a period of high energy prices that would constrain growth and become a central problem for the economies of the world. Prices of oil did indeed rise rapidly in the late 1970s and early 1980s, but declined sharply in the mid-1980s. Similar results occurred for natural gas. The high prices of the 1970s and early 1980s (1) encouraged conservation and more efficient use of fuels, (2) stimulated a shift towards greater energy efficiency, (3) changed behavior patterns, (4) encouraged exploration and development of energy resources, (5) led to additions to energy reserves, and (6) stimulated fuel switching. As a result, oil supplies increased while demand decreased, leading to a sharp decline in the price of oil in the 1980s.

Considering a rough estimate of world energy resources (Table 23.1) and the annual levels of consumption (Table 23.2), the free energy stored in mineral resources can last for many decades and even centuries; however, other constraints may prevent their complete use. Even if the net energy yield of the fossil resources were as low as one-half of these values (which is unlikely), the quantities are still large. In the case of fossil fuels, the useful energy produced minus that to mine, transport, and process the resource into useful forms, is positive and been shown to be substantial (Odum et al. 1976; Hall et al. 1988). The energy yield ratio (ratio of useful energy produced to the energy cost) tends to decline from natural gas to petroleum to coal but, of course, depends on the quality of the resource, its accessibility, and its distance from points of end use. From a resource point of view, there is also no constraint on uranium supplies for fueling nuclear power plants over this period of time.

Table 23.1.   Estimated Levels of World Energy Resources (Gross Energy Values are Indicated in Parentheses, not Net Energy Available to Society)

NON-RENEWABLE:	
1. OIL	
a. proved and indicated	800 Bbbl (4,600 Quads)
b. undiscovered	380 Bbbl (2,200 Quads)
2. HEAVY OIL	
a. proved and indicated	450 Bbbl (2,600 Quads)
b. undiscovered	90 Bbbl (520 Quads)
3. IN-PLACE TAR SANDS	4000 Bbbl (23,000 Quads)
4. RECOVERABLE OIL SHALE	1000 Bbbl (5,800 Quads)
5. NATURAL GAS PROVED RESERVES	3,800 TCF (3,800 Quads)
6. RECOVERABLE COAL	1,000–1,700 billion tons (22 to 38,000 Quads
7. URANIUM (U_3O_8)	3.6 million tons
Estimated Additional	5.5 million tons
RENEWABLE:	
1. BIOMASS RESIDUES (not including dedicated energy crops)	134 Quads/year
2. SOLAR ENERGY FLUX OVER LAND	ca. 500,000 Quads/year

Source: OTA (1992b); Brower (1990); NRC (1990); EIA (1988, 1989); Riva (1988); Hall et al. (1986); Alberta Oil Sands and Tar Research Authority of Canada; Considine (1977).

Note: 1 Quad   = 1 quadrillion British Thermal Unites (BTU)
          Bbbl   = 1 billion barrels
          TCF   = thousand cubic feet
          These estimates do not include other additional resources and speculative resources

Note: 1 Quad   = 1 Quadrillion Btus of Energy

Table 23.2. World Consumption of Selected Resources, 1989

Resource	Quantity
Oil	25 Bbbl/yr
Natural Gas	70 billion cu ft/yr
Coal	5.2 billion tons/year
Uranium	60,000 metric tons/yr of U_3O_8
Biomass	35 Quads/yr
Hydroelectric	2,080 billion kwh/yr

Source: EIA (1991). Uranium consumption estimated as 190 metric tons uranium ore per year per Gigawatt Light-Water Reactor. Installed global capacity of fission reactors about 310 GW. Biomass estimates from OTA (1992b).

Many analysts believe (1) world oil prices will remain moderate for a number of decades, (2) large finds of natural gas have materialized, and (3) coal is noteworthy for its abundance. If scarcity occurred in the coming decades, rising prices could again stimulate additions to these fossil fuel reserves.

Direct solar conversion technologies may have the most problematic net energy concerns because of the dilute nature of the solar flux, its intermittency, and its lack of storage. Biomass conversion may also have marginal net energy yields, although this will depend on the use of the biomass and the conversion technologies employed; for example, fermentation of corn to produce ethanol is marginal from a net energy point of view (Ho 1989). However, the solar energy flux is vast, and the development of economic technologies that could only convert 10% of this flux to useful energy would still represent enormous quantities.

Recent analysis of world energy demand projects consumption of ca. 1,400 Quads/year in 2060; in the event of significant increases of efficiency on the order of 50%, then 810 Quads/year consumption is projected for 2060 (Starr et al. 1992).[2] One would expect oil prices to rise with time and as they reached approximately $35 to $40/bbl (1988 dollars), conversion of coal and shale into liquid fuels would become attractive (NRC 1990). Furthermore, over the coming decades, it is expected that petroleum-based liquid fuels will be supplemented with alcohol fuels from biomass and natural gas.

However, as discussed later, a number of environmental, social, and institutional constraints may prevent the use of these fuels at the rates projected by Starr et al. (1992). If the world achieves such levels of energy consumption, economic activity, and population, can they be sustained? Thus, beyond 2050, levels of energy consumption will primarily depend on whether technology can produce energy supplies from such sources as nuclear power and solar energy in safe, environmentally acceptable, economic ways. Otherwise, fossil fuels are finite, and although human ingenuity may extend such reserves for quite some time, they will inevitably be exhausted. As exhaustion occurs and prices rise without other energy sources to substitute, economic activity would decline and society would adjust to lower levels of consumption. Consequently, because of the anticipated difficulties in exploiting, converting, and using energy, attention should be paid to energy efficiency, full social cost pricing, institutional issues, and strategies for sustainability.

REDUCING ENERGY CONSUMPTION

Opportunities exist for reducing energy consumption either through improving the technical efficiency with which energy is produced, converted, or used,

---

2. Note that 1 Quad is equal to a quadrillion British Thermal Units (Btus), which is equivalent to about 170 million barrels of crude oil. World energy consumption in 1989 was about 340 Quads.

or by changing the manner in which members of societies behave and organize their activities. It must be emphasized, however, that improvements in energy efficiency alone are insufficient for long-term sustainability. In the long term, external energy resources are required to fuel economic activity, even if the efficiency with which energy is used greatly increases in the future.

Energy Efficiency Opportunities Through Technology

There are numerous opportunities to improve the efficiency with which energy is either produced or used in an economy. Examples include more efficient drilling for oil, better insulation in housing, improved fuel economy in vehicles, and more efficient lighting. Although, in many instances, the improvements will require some initial cost or investment, both energy and economic life-cycle savings will result. In addition, they can be substantially less expensive and entail fewer environmental costs than trying to build additional energy supply capacity.

Estimates of efficiency improvements in different sectors of the economy have ranged widely; advocates of energy efficiency claim large gains are possible, while other analysts generate more sanguine estimates (Goldenberg et al. 1987; Chandler et al. 1988; OTA 1992 a, b). Opportunities differ from nation to nation because of historical energy prices and investments in energy efficient equipment, as well as in the degree of industrialization. For example, historically high prices of transportation fuel in Europe and Japan in comparison with the United States, have resulted in higher on-the-road fuel economies of the passenger car, although in recent years the fuel economy of new vehicles has declined in Japan and some countries in Europe to levels similar to those in the United States. On the other hand, in countries that have lower standards of living, bicycles and mass transit may be used more commonly than in industrial countries with higher per capita levels of automobiles. Such societies can not improve their transportation energy efficiency very much; in fact, they might want to use more energy by incorporating automobiles or modern mass transit into their economies because of the advantages these types of transportation might confer on their economies and lifestyles.

A recent study on the potential for improved U.S. automotive fuel economy suggests that, by the year 2006, with automobiles and light trucks having the same performance and attributes as 1990 models, significant gains in fuel efficiency are possible depending on vehicle size class and with technologies that are already in mass production somewhere in the world; however, such gains will entail significant increases in vehicle prices (NRC 1992).[3] The de-

3. For passenger cars, low and high estimates are as follows: subcompact cars, 39–44 mpg; compact cars, 34–38 mpg; midsize cars, 32–35 mpg; large cars, 30–33 mpg. The Environmental Protection Agency's (EPAs) composite average fuel economy for Model Year 1990 passenger cars are as follows: subcompact, 31.4 mpg; compact, 29.4 mpg; midsize, 26.1 mpg; and large, 23.5 mpg (Heavenrich et al. 1991).

gree of price increase is uncertain but varies from about \$500 to \$2,500 on the initial price of the car (consider that current car prices average \$16,000). Whether such investments would turn out to be cost-effective would depend very much on the price of fuel in the future. If one assumes reduced levels of performance, greater gains could be achieved. In the longer term, other analysts see greater gains with new engines (e.g., greater penetration of the direct injection diesel and/or two-cycle engine with greater engine efficiencies), materials (e.g., lightweight plastics and composites), and design (e.g., aerodynamics) (Chandler et al. 1988; Bleviss 1988; OTA 1992c). Of course, significant gains in fuel economy of cars could be realized by increasing the proportion of lightweight cars in any given market, but such strategies must be undertaken in light of the economic, social, employment, and safety implications that would accrue, considering the demand for such vehicles by consumers. For example, in the United States, consumers are increasingly showing preference for "light trucks" such as vans and utility vehicles, which, on average, have substantially lower fuel economy than automobiles. Higher fuel costs would probably change such consumer preferences.

There are also opportunities for efficiency improvements in other sectors of the economy. In U.S. buildings, improvements in space conditioning (such as heating and cooling), lighting, water heating, and refrigeration and freezing, could result in about 33% less energy consumption by the year 2015 than would occur under a business-as-usual scenario (OTA 1992a). There are broad opportunities in developing countries for cost-effective, energy-efficient technologies as well (OTA 1992b). For example, a number of technologies could cost-effectively reduce electrical consumption by 50%; such technology also produces savings in required capital investments, perhaps one of the most important constraining factors for developing countries.

The improvement of efficiency in producing and using energy can have direct benefits on the environment and reduce stress on the natural capital both locally and globally. If a given amount of electricity can be produced with less fuel, whether it is oil or coal, then fewer emissions of oxides of nitrogen (NO_x), sulfur dioxides, particulates, greenhouse gases, and other pollutants can result. However, improvements in energy efficiency will not always result in lower emissions of pollutants. If, as in the U.S., automotive emissions are regulated on a grams/mile basis, emissions of NO_x and hydrocarbons would depend on the amount of driving regardless of the efficiency of fuel use. However, improving energy efficiency is an important strategy for reducing greenhouse gas emissions because of the reduced consumption of energy required to provide a given level of service and, hence, reduced emissions of carbon dioxide. Again, significant technological opportunities have been identified for all sectors of the U.S. economy and are probably applicable to other economies as well (NRC 1990; NAS 1991).

Systems Issues

Aside from technologies for conserving energy, policies may be needed to manage societies in a broader fashion. In the transportation sector, approaches to reduce petroleum use include not only vehicles with higher fuel economy, but also: (1) improved traffic flow management, perhaps with the use of intelligent highway vehicle systems; (2) encouragement of ride sharing and mass transit through various incentives; and (3) in the longer term, the design of cities in such a fashion that places less reliance on the automobile.

Urban forms that reduced fuel use might result from deliberate urban planning policies (these would only work in those societies that have fairly strong central authorities, land-use planning and zoning approaches) or arise through adaptation to higher transportation fuel prices. They could also have significant economic impacts if they were to severely constrain the access and freedom of people to participate in the economic and social affairs of any given society. The sprawl patterns that are so prevalent in U.S. society, and to a greater or lesser degree in other countries, have arisen during the era of cheap petroleum and the development of the automobile—a means of transportation that, for all its problems, also confers great advantages and flexibility of mobility. If prices of fuel were to rise, we might see a slow evolution of urban form that was less sprawling possibly consisting of suburban centers surrounding an urban core (Zucchetto 1983). But, of course, the flexibility of some form of personal vehicle might be retained by improving fuel economies as petroleum-based fuel prices went up, or developing vehicles based on alternative fuels or electricity.

In the industrial sector, it may be possible to link up industries with increased material recycling of "wastes" and consequent energy savings (NRC 1990). Recycling, to some extent, from residential and commercial sectors of metals such as aluminum, as well as producing products with the goal of waste minimization, can also improve energy efficiency as well as improve environmental quality—but the potential savings probably differ from case to case. Extensive implementation of such practices requires systemic changes in the organization of industrial societies, realized through both regulations and appropriate pricing policies.

Society may also completely reorganize with the advent of the communications and computer revolution. Although talked about for some years, computers and telecommunications links throughout the world might at least replace the need for part of the transportation that occurs now for the purposes of work. Whether such systems replace "face-to-face" interaction remains to be seen, but the potential for important sectors of the economy to allow their employees to work from home part of the time remains. Just how this "information society" will be organized and the extent to which it can improve energy efficiency has yet to be determined.

ENVIRONMENTAL COSTS OF ENERGY SUPPLY AND USE

The environmental effects of energy production and use have been under intensive study for only a few decades. As alluded to above, issues include (1) local disruptions and pollutant emissions, (2) acid rain precursors, and (3) increasing atmospheric concentrations of greenhouse gases with possible implications for climate change (see Table 23.3). In the case of nuclear energy, possible large releases of radioactivity or proliferation of nuclear weapons are of concern. A complete understanding of such effects requires examination of the entire fuel cycle, from the ore in the ground to its transport, processing, conversion into useful fuels, and end use. Such fuel cycle analysis was quite active in the late 1970s during the period of rapid rises in energy prices and a perception that there was an "energy crisis." Some recent studies are being undertaken (e.g., by the U.S. Department of Energy) but are not yet published. One recent study has explored the environmental costs of electricity generation from different energy sources, as we discuss later (Ottinger et al. 1990; Hohmeyer 1988).[4]

Table 23.3. Estimates of Contributions of Different Sectors to Global Warming

Sector	Percent Contribution
Industrial energy use	25
Transportation energy use	18
Agriculture	15
Residential/commercial energy use	14
CFC-12	10
Forestry	8
CFC-11	4
Other CFCs	3
Other industry	3

Source: Lashof and Tirpak (1990).

Establishing the costs of these environmental emissions from these fuel cycles, the so-called externalities, is a difficult proposition which raises many questions. For example, even if we knew the human lives lost for each kilowatt-hour of electricity produced from a given electric energy power producer, how do we value the cost of that human life? Several methods have been proposed, (e.g., the lost income of the remaining years of a human life),

---

4. The exploitation of energy sources engender a series of environmental, health, and social impacts, as energy is mined, transported, processed and delivered to its point of end use. This is particularly true of energy resources that must be mined, but also holds for petroleum, natural gas, and solar-based technologies, but to different degrees. For example, the use of natural gas is a relatively clean approach to the generation of electricity, but it also produces its share of NO_x and CO_2 (see Table 23.4 for a summary of emissions from different types of power plants, considering the entire fuel cycle). Different transportation fuels will have fuel cycle impacts depending on the raw feedstock and the fuel produced. For example, Deluchi (1991) has estimated the greenhouse gas emissions for different transportation fuels considering the entire fuel cycle (see Table 23.5).

Table 23.4. Estimates of the Direct and Indirect Emissions from a Variety of Electrical Power Plants (tons/Gwh)

	Technology*				
	Conventional	AFBC	IGCC	BWR	PV Central Station
Carbon Dioxide	1,058	1,057	824	8.6	5.9
Nitrogen Oxides	2.99	1.55	0.25	0.034	0.008
Sulfur Oxides	2.97	2.97	0.34	0.029	0.023
Particulates	1.63	1.63	1.18	0.003	0.017
Carbon Monoxide	0.267	0.267	—	0.018	0.003
Hydrocarbons	0.102	0.102	—	0.001	0.002
Aldehydes	0.008	0.008	0.006	—	—

Source: Meridian Corporation (1989).
*AFBC: atmospheric fluidized bed combustion; IGCC: integrated gasification combined-cycle; BWR: boiling water reactor; PV: photovoltaic.

Table 23.5. Emissions Associated with Different Transportation Fuels in the United States.

	Gasoline from Petroleum	Nat. Gas	Methanol from Nat. Gas	Methanol from Coal	Methanol from Wood	Ethanol form Corn	Ethanol from Wood
Carbon Dioxide equivalent emissions (g/MMBtu)	25,000	11,100	42,000	131,000	28,000	135,000	7,000
Carbon Dioxide (g/mile)	400	300	400	600–800	100	350	30
Nitrogen Oxides (g/mile)	32	34	47	—	34	76	44

Source: Based on DeLuchi (1991). Carbon dioxide for methanol from coal is estimated assuming about 1.5–2 times that of gasoline from petroleum (NRC 1990).

but there is not general agreement regarding this approach or the manner in which evaluations should be conducted. How should we value the discomfort to those members of the population affected by high concentrations of urban ozone? How should the lost productivity from an ecosystem or the loss of species diversity or rare species be evaluated? What is the cost (or the benefit) of a one degree rise in the earth's average temperature? These and other questions are not easy to answer in the attempt to generate quantitative answers to the evaluation of costs for policy purposes. Nevertheless, such evaluations, although they may be imprecise, can give us at least some rough idea of relative costs among energy fuel cycles.

In fact, there have been several attempts at economic estimates of the "social costs" of electricity generation. Reviewing a variety of studies, Ottinger et al. (1990) have summarized the costs associated with emissions of sulfur and nitrogen oxides, particulates, and carbon dioxide from various electricity fuel cycles and technologies. Tables 23.6 and 23.7 demonstrate these costs are substantial and, if directly internalized, would raise the cost of

Table 23.6. Postulated Externality Costs for Pollutants

Pollutant	Externality Cost	"Typical" Value
Carbon Dioxide	$17–$120/ton of C	$50/ton C
	($0.0085–$0.06/lb C)	$.025/lb C
Sulfur Dioxide	$240–$9,000/ton	$2.03/lb
	($0.12-$4.5/lb	
NO$_x$	$2–$3,460/ton	$0.82/lb
	($0.011–$1.73/lb)	
Particulates	$0–$11,000/ton	$1.19/lb
	($0–$5.50/lb)	

Source: Ottinger at al. (1990).

Table 23.7. Postulated Externality Costs for Electricity Generation

Technology	Cost (cents/kwh)
Coal-Fired	2.5 to 5.8
Oil-Fired	2.5 to 6.7
Natural Gas	0.7 to 1.0
Nuclear	2.91
Direct Solar	0 to 0.4
Wind	0 to 0.14
Biomass (Wood)	0 to 0.7
Waste to Energy	4

Source: Ottinger at al. (1990).

electricity significantly. For example, the average U.S. cost of electricity is about 5.6 cents/kwh for nuclear and 3.1 cents/kwh for coal-fired power plants (NRC 1992). Thus, the externality costs estimated by Ottinger (1990) could increase costs for electricity by about 50% and 100% for nuclear and coal respectively. For transportation fuels, the typical externality values (see Table 23.6) for CO_2 and NO_x would cost about 2.5 cents and 5.8 cents per mile respectively, compared to a U.S. gasoline cost of about 5 cents/mile.[5] These costs do not include impacts on ecosystems, which could also be high, although some of these emissions could increase the productivity of different

5. Assuming gasoline costs about $1.40 per gallon and the average U.S. fuel economy is 27.5 mpg, the cost is about 5 cents/mile. The carbon dioxide and oxides of nitrogen costs are derived from Tables 23.5 and 23.6.

ecosystems depending on the concentrations and levels of pollution imping-
ing on a given ecosystem.

It is not clear how these costs should be incorporated into decision making.
Several advocates of least-cost energy planning suggest adding these costs
onto the price of electricity (Wiel 1991). In fact, several public utility com-
missions in the United States have ordered utilities to incorporate environmental cost accounting when acquiring of new sources of electricity. It is uncertain, however, whether this approach results in an equating of the marginal
costs with the marginal benefits (Jaskow 1992). If not, resources are being inefficiently expended on control. Also, current environmental regulations and
laws are already burdening energy systems with some of these costs so that
double counting must be carefully considered.

Another approach to keeping emissions below a certain level but letting the
market act to sort it out and allow flexibility is to establish *an emissions permit
trading scheme*. Thus, within a given regional area, a certain amount of a
pollutant would be allowed. Determining the appropriate level of a given
pollutant would require a scientific understanding of the natural environment
impacts of given levels of emissions; in the absence of such information, political decisions might determine the levels. Individual polluters within a given
region would be allowed to trade their rights to pollute within the given constraint of a total allowable emissions load for the region. Such a scheme might
also be considered globally for greenhouse gas emissions.

Incorporating taxes on fuels in some way tied to their external costs would
also engender other effects in economies. For example, it would reduce demand for a given fuel type that is priced relatively higher than others, stimulate conservation, and encourage switching to energy sources that have less
impact. But a decrease in use would also tend to lower prices, possibly encouraging use of the fuel that was initially taxed; an equilibrium would eventually be reached.

In assigning some price to externalities, an attempt is made to value the
damage that might be done from a given level of emissions. Using dollar values through willingness-to-pay measures biases the results based on what humans perceive to be of value. Using damage functions focuses, to a greater
extent, on what actually is required to replace a given loss. Some argue that a
more objective measure of nature's contributions, or of natural losses incurred, can be obtained by using thermodynamic estimates of work contributions, in the process accounting for the ability of different forms of energy
and information to perform work. H. T. Odum (1992) has continued to develop this line of thought in what he refers to as "Emergy analysis." Whether
it can more accurately capture the so-called externalities is still being researched. Nevertheless, to make it intelligible to individuals and political decision makers, it would need to be translated into some unit that is readily understood in social transactions.

The incorporation of the cost of externalities, whatever the manner employed, allows the social and environmental implications of given technologies to be reflected in the price. Such a scheme signals directly to the purchasers of energy both direct and indirect costs. To the extent that such costs are accurate, they allow the market to act on choosing the most economic energy sources. In such a way, those technologies with significant impact on the environment will be burdened to a greater degree. However, it is unclear whether such technologies would be ignored. For example, coal-fired electricity with a large emissions assessment for greenhouse gases might turn out to be more expensive than wind-generated electricity for a given size of operating power plant. However, if demand for electricity is at such a level that the wind resource cannot meet it, then the wind-based technology will only be used to the extent that it can provide cost-effective supplies. The balance of supply and demand in a free market would drive the price of the wind technology up. The supply of coal-fired electricity would still dominate because of the large supply of coal. Nevertheless, such social cost-pricing can help in the transition to increasingly environmentally compatible energy sources.

INTERDEPENDENCE AND THE MANAGEMENT OF THE COMMONS

One characteristic of modern societies is the increasing congestion arising from increased population, standard of living, and economic growth, as well as the increasing democratization of society. Societies are becoming more complex with ever more regulations and laws, and citizens groups can exert a greater and greater influence on decisions to locate facilities. Another trend is the increasing economic and environmental interdependence among nations. These both impact the energy systems that can prosper and compete in the future and must be factored into our thinking of energy and sustainability.

Resistance to Siting

Within modern industrial societies with democratic social institutions, as congestion increases, it becomes more difficult to locate facilities. The "Not in My Backyard" (NIMBY) syndrome surfaces as individuals become wealthier and take more interest in the protection of their localities and environment. In the United States, it is a well known aspect of modern development. Examples include the: (1)opposition to building waste-to-energy plants because of the perceived risks of incineration; (2) difficulty of building new rights-of-way for electrical transmission lines because of land use constraints in congested areas and some concerns about the effects of electromagnetic radiation (EMF), and (3) siting of toxic and radioactive waste dumps.

Nuclear power plants may particularly suffer from the problems of siting if there is general concern about accidents involving the release of radioactivity. For example, in the United States, a recent poll indicated that about 75% of the respondents said that nuclear energy was the most dangerous way to generate electricity (NRC 1992). Yet 81% said it was very important or somewhat important to the nation's future national energy strategy. However, when asked whether they would favor, oppose, or reserve judgment for a new nuclear plant in their area, 59% said they would reserve judgment, and 23% said they would oppose a nuclear plant.

International events, such as the Chernobyl nuclear power plant accident, only exacerbate general concerns about nuclear power even though the Chernobyl reactor's design was different from U.S. reactors, and the United States has never experienced a significant release of radioactive material over a cumulative commercial reactor operation time of 1,400 years (NRC 1992). In 1989, nuclear power accounted for about 19% of the electricity supply in the United States, 77% in France, 26% in Japan, and 33% in the former West Germany. A new nuclear power plant has not been ordered in the United States since 1978. When new plants will be built in the United States, and whether a slowing of the nuclear option occurs in other countries, remains to be seen.

Other countries may not experience the NIMBY aspect to the same degree, but environmental awareness and opposition seems to be occurring throughout the industrialized world. That is not to say that these problems cannot be overcome, but they do engender delays, costs, discussion, and consensus-building, and they reflect the struggle over the commons. We should recognize that they are a part of modern societies and will strongly influence the development of future energy systems.

International Interdependence

With regard to trade among nations, one can think of the energy embodied in the production of a good as it goes through its various stages of upgrading to become a finished product.[6] A certain amount and type of energy will be consumed per unit of output, whether expressed in monetary or physical units. This consumption of energy, as discussed above, also entails a certain cost to the ecosphere because of disturbance of habitat and pollutant emissions. Thus, we can think of an environmental cost associated with energy consumption also embodied in a finished product. (There are also other environmental costs that accrue strictly from processing and manufacturing, and not necessarily directly tied to energy per se.) If externality calculations are done correctly and reflected in the end price of the product (e.g., through a full social cost-pricing of the energy used), then these environmental costs are reflected in the price. If some countries burden their products with externality costs while others do not, they will be at a disadvantage in the international

---

6. Zucchetto (1984) analyzed energy to economic ratios for different countries considering the embodied energy of trade among these countries. The differences in the ratio of total energy consumption to gross economic output among countries decreases when such indirect effects of interdependence are taken into account. We might expect similar results when considering environmental impact, but it would depend on the nature of the trading relationships and the products involved.

marketplace. Thus, it is important that consistent rules be established across nations for valuing such externalities. Otherwise, incentives will be created for countries to protect their own markets through, for example, tariffs if they are implementing environmental protective strategies or relaxing environmental protection.

Not only has an increased level of international trade made the world "smaller," but so have the technologies of transportation and communication. They have made nations more interdependent, but so also has the scale of technology employed by modern economies. The Chernobyl nuclear accident is a stark reminder of how one nation's energy system can affect the entire world. The emissions of greenhouse gases conjures up visions of dramatic global change. The building of large dams to divert rivers can change the hydrologic cycle of large regions, the harvesting of large parts of tropical forests for energy and goods can affect the global carbon cycle, the pollution emitted by one country's energy systems in the world's oceans or seas can affect the health and well-being of other countries, and so on. Thus, everyone begins to see that they have a stake in the energy affairs of others.

Consequently, major investments in one type of energy system or another may increasingly depend on possible global effects. Widespread use of coal-fired electrical generation by one nation may conflict with the concerns of other nations. Recent protests regarding the shipment of plutonium by the Japanese for their breeder reactor program illustrate such interconnectedness. These concerns may be expected to increase in the future.

DISCUSSION AND STRATEGIES FOR SUSTAINABILITY

Conclusions from the preceding discussion may be summarized as follows:

- Global energy resources *in and of themselves* are quite sufficient to support economic growth well into the 21st century and probably into the 22nd century. However, the availability of energy will increasingly be more constrained by environmental, congestion, social, and institutional matters than with the size of energy resources per se.

- There are great uncertainties regarding the viability of long-term sustainable energy technologies. Over the next few decades, a diverse energy system will probably exist much as today, perhaps with a greater proportion of solar-based technologies. In the longer term, if concerns about climate change dictate, a global economy based on an extensive use of non-fossil sources may be required, but it is unclear which technologies will be viable.

- There are extensive opportunities to cost-effectively improve the efficiency with which energy is converted in the energy supply sector and used to provide end-use services.

- There are significant externalities associated with the energy supply, distribution, and end-use sectors that need to be factored into the full social cost-pricing of energy.

- Global interdependence and the management of the global commons will increasingly dictate international cooperation and consideration of externalities and safety in the development of the world's energy resources.

- In the long term, social reorganization may represent an effective approach to reducing the use of energy while retaining a desirable quality of life.

One certainty is that the future is uncertain. The discussion so far has, to some extent, assumed that the world develops with no startlingly great change or conflagrations. If climate and environmental changes occur, they are assumed to occur gradually; although if high enough concentrations of atmospheric greenhouse gases are realized, dramatic change might result, as the climate system could "flip" into new patterns. Thus, strategies are needed that are flexible and resilient to changing future circumstances and uncertainty, including:

- **A Diverse Portfolio of Energy Research and Development.** Since it is uncertain what technologies will be needed in the future, R&D support is needed across several different technology options. Even with the threat of global warming, fossil fuels will probably continue to play an important role in economic growth well into the 21st century because of the economic needs of countries such as China and India. If, for economic reasons, these energy sources are the ones of choice, they should function in efficient and clean ways. Carbon dioxide capture might even be necessary, although it is unlikely such an approach would be economic and would be very energy costly. Production of liquid transportation fuels from fossil and biomass resources, advanced nuclear reactor concepts, and a host of energy-efficient and renewable technologies should be included in such an R&D portfolio (NRC 1990a, b; 1992). Considering the expense of energy technology R&D and the importance of advanced energy systems for the entire world, international cooperation should be encouraged.

- **A "No Regrets" Path of Implementing Energy-Efficient Technologies.** There are numerous opportunities to produce and use energy with significant improvements in efficiency in both the industrialized countries and the developing world. If those technologies that are cost-effective are implemented, economic gains will be realized that reduce negative environmental impacts. This makes good sense from the perspective of economics, the environment, conservation of energy resources, and sustainability. Such a strategy is beneficial anyway whether or not, for example, global climate change is occurring. If it turns out that drastic reductions in greenhouse gases are required over the next two or three decades, processes to implement these technologies would already be in play. One of the problems of implementing such technology has been the high first cost, even though the life-cycle costs imply net life-cycle savings. Thus, incentives and policies are required to overcome such first-cost hurdles. As fuels become scarce and begin to increase in price, and externalities are factored into energy prices, energy-efficient technologies will become increasingly attractive.

- **Energy Prices That Reflect All Environmental and Social Costs.** Much work is needed on fuel-cycle analysis to understand the full impacts of energy production and consumption on the environment and humans. Although there are many estimates and projections, there is still great uncertainty. Not only should monetary measures be used, but many attempts have been made to understand the impact on nature through an analysis of the work functions loss of the natural systems. These provide another view of environmental impact and can also be incorporated in any "net energy" analysis of new options that can give insight on whether proposed energy sources will truly make a contribution in the future. Further understanding of the best approach to incorporate externalities is also needed.

- **International Trade That Reflects Embodied Environmental Costs.** Full incorporation of externalities can again reflect to purchasers of goods the environmental effects of what they are purchasing. Development and enforcement of rules on externalities could help prevent some countries from gaining in international competitiveness at the expense of their own environment, and people, as well as the global commons. Transfer of clean technologies to the developing world, perhaps subsidized through an international environmental fund, might be an approach to help the poorer countries implement such technologies more rapidly.

- **Reorganizing Economic Systems to Improve Energy Efficiency.** In the long term, as environmental constraints become more severe, and energy resources become more scarce, a better understanding is needed of how societies can organize using emerging technologies and new settlement patterns for enhanced energy efficiency. The appropriate mix of market mechanisms, planning, zoning, incentives, and regulations should be analyzed for their implications.

- **Ameliorating the Impacts of Greenhouse Gas Emissions.** In the event that it becomes too problematic and costly to eliminate fossil fuels to any great extent from the energy systems of economies, an understanding of more effective approaches to either capture carbon dioxide in economic and environmentally acceptable ways, or adapt economic systems and human activity to global climate change will be required.

REFERENCES

Bleviss, D. L. 1988. The New Oil Crisis and Fuel Economy Technologies. Westport, CT: Quorum Books.

Chandler, W. U., H. S. Geller, and M. R. Ledbetter. 1988. Energy Efficiency: A New Agenda. American Council for an Energy Efficient Economy. Washington, DC (July).

Considine, D. M., ed. 1977. Energy Technology Handbook. New York: McGraw-Hill.

Cottrell, F. 1955. Energy and Society. New York: McGraw-Hill.

Energy Information Administration (EIA). 1991. Annual Energy Review 1990. DOE/EIA-0384 (90). Washington, DC: U.S. Department of Energy.

———. 1988. International Energy Outlook 1987. DOE/EIA-0219. Washington, DC: U.S. Department of Energy.

————. 1989. Annual Energy Outlook with Projections to 2000. Washington, DC: U.S. Department of Energy.

DeLuchi, M. 1991. Emissions of Greenhouse Gases from the Use of Transportation: Fuels and Electricity. Vol. 1. ANL/ESD/TM-22. Center for Transportation Research. Argonne, IL: Argonne National Laboratory.

Goldenberg, J., T. B. Johansson, A. K. N. Reddy, and R. H. Williams. 1987. Energy for a Sustainable World. Washington, DC: World Resources Institute.

Hall, C. A. S., C. J. Cleveland, and R. Kaufmann. 1986. Energy and Resource Quality: The Ecology of the Economic Process. New York: Wiley .

Ho, S. P. 1989. Global Warming Impact of Ethanol versus Gasoline. Amoco Oil Company, Naperville, IL. Presented at the 1989 National Conference "Clean Air Issues and America's Motor Fuel Business," October 3–5, Washington, DC.

Joskow, P. L. 1992. Weighing Environmental Externalities: Let's Do It Right. *The Electricity Journal*, (May): 53–67.

Lashof, D. A., and D. A. Tirpak, eds. 1990. Policy Options for Stabilizing Global Climate. Report to Congress. Washington, DC: U.S. Environmental Protection Agency (December).

Meridian Corporation. 1989. Energy System Emissions and Material Requirements. Report prepared for the Deputy Assistant Secretary for Renewable Energy, U.S. Department of Energy. Alexandria, VA: Meridian Corp.

National Academy of Sciences (NAS). 1991. Policy Implications of Greenhouse Warming. Washington, DC: National Academy Press.

National Research Council (NRC). 1990a. Fuels to Drive our Future. Washington, DC: National Academy Press.

————. 1990b. Confronting Climate Change: Strategies for Energy Research and Development. Washington, DC: National Academy Press.

————. 1992a. Automotive Fuel Economy: How Far Should We Go? Washington, DC: National Academy Press.

————. 1992b. Nuclear Power: Technical and Institutional Options for the Future. Washington, DC: National Academy Press.

Odum, E. P. 1988. Ecology and Our Endangered Life-Support Systems. Sunderland, MA: Sinauer.

Odum, H. T. 1994. The Emergy of Natural Capital. In Investing in Natural Capital: the Ecological Economics Approach to Sustainability, eds. AM. Jansson, M. Hammer, C. Folke and R. Costanza. Washington, DC: Island Press.

————. 1983. Systems Ecology. New York: Wiley .

————. 1971. Environment, Power and Society. New York: Wiley.

Odum, H. T. et al. 1976. Net Energy of Alternatives for the United States. United States Energy Policy Trends and Goals, Part V. U.S. Congress, 2nd Session, Subcommittee on Energy and Power of Committee on Interstate and Foreign Commerce: 253–302.

Office of Technology Assessment (OTA). 1992a. Building Energy Efficiency. U.S. Congress. OTA-E-518. Washington, DC: U.S. Government Printing Office (May).

————. 1992b. Fueling Development: Energy Technologies for Developing Countries. U.S. Congress. OTA-E-516. Washington, DC: U.S. Government Printing Office (April).

————. 1992c. Improving Automotive Fuel Economy: New Standards, New Approaches. U.S. Congress. OTA-E-504. Washington, DC: U.S. Government Printing Office (October).

Ogden, J. M., and R. H. Williams. Solar Hydrogen: Moving Beyond Fossil Fuels. Washington, DC: World Resources Institute.

Ottinger, R. L. et al. 1990. Environmental Costs of Electricity. New York: Oceana Publications, Inc.

Riva, J. P., Jr. 1987. Fossil Fuels. In Encyclopedia Britannica, Vol. 19: 588–612.

———. 1988. Oil Distribution and Production Potential. *Oil and Gas Journal* 86 (3): 58.

Starr, C., M. F. Searl, and S. Alpert. 1992. Energy Sources: A Realistic Outlook. *Science* 256 (May 15): 981–7.

United States Department of Energy (DOE). 1988. Energy Technologies: Environmental Information Handbook. DOE/EH-0077. Springfield, VA: National Technical Information Service.

Wiel, S. 1991. The New Environmental Accounting: A Status Report. *The Electricity Journal* (November): 46–55.

The World Commission on Environment and Development. 1987. Our Common Future. New York: WCED.

Zucchetto, J. 1983. Energy and the Future of Human Settlement Patterns: Theory, Models and Empirical Considerations. *Ecological Modelling* 20: 85–111.

———. 1984. Global Energy/Economic Measures, 1960–80. *Journal of Resource Management and Conservation* 13 (3): 147–53.

24 ECONOMIC CONVERSION, ENGINEERS, AND SUSTAINABILITY

Martha W. Gilliland
Graduate College
Departments of Hydrology and Water Resources,
Civil Engineering and Engineering Mechanics
University of Arizona
Tucson, AZ 85721

Jeffrey P. Kash
Department of Political Science
University of Arizona
Tucson, AZ 85721

ABSTRACT

Economic conversion is now underway; it is being driven by both political and economic factors. It offers the opportunity to redefine National Security in terms of sustainability. Economic conversion is most commonly understood to be the changeover in defense-related facilities and activities. That conversion can take many directions. If it is to move us toward sustainability, it requires proactive participation by the technical workforce, more specifically, by engineers. Unplanned economic conversion is likely to cause negative economic impacts, such as job and income loss, at least in the short term and especially among engineers. Both because of negative impacts and the inherent lack of involvement of the technical workforce in social change, that workforce is likely to impede economic conversion. Yet their involvement in the development of sustainable technologies is critical to conversion that moves us toward sustainability. Engineers have a history of following, if not resisting social change; they have generally not provided leadership. Thus, encouraging their proactive involvement will be difficult. It may best be accomplished by appealing to what motivates engineers and by attempting quite consciously to produce a shift in values. Engineers are motivated primarily by the nature of the technical challenge; they are problem solvers. They have not been exposed to, involved in, or motivated by discourse on the social consequences of technology. Therefore, any statement on the purpose of sustainable technology (e.g., to both facilitate economic growth and reduce pressures on the natural environment) must be translated into specific technical challenges, if it is to provide motivation. A shift in values is required for most people living in industrialized countries, if development and implementation of sustainable technologies is to be successful. Thus, any programs aimed at shifting the values of engineers will also be applicable to other groups. Values are very difficult to change; the literature suggests, however, that they can be shifted via education used in conjunction with personal experimental involvement in the issue. Engineers will have to be provided knowledge about conversion, the impacts of technology, and sustainability through their professional societies as well as through the formal education process. But, more importantly, any shift in values will require opportunities to "experience" the issue of sustainability, to work in teams, and to work with the end users of technology.

INTRODUCTION

Economic conversion offers the opportunity to redefine National Security for the United States in terms of sustainability. Economic conversion is the planned shift of significant resources from military to civilian applications. Both the "from" and the "to" in that definition must exist for conversion to be a reality. The world is beginning to address the "from" component; both the United States and the former Soviet Union are investing fewer resources in military applications. In truth, however, at least in the United States, the rhetoric greatly exceeds the reality. Some real cuts in the $300 billion military budget have occurred, but most of the action has been at the discussion level. Even less progress has been made on the "to" side of the conversion equation. Hard questions (i.e., what are we going to convert to?; what ought to be the focus of government spending?; and where should those military dollars be directed?) need to be answered if progress is to be made. In short, debate and discussion about both the "from" and the "to" must be increased. A window of opportunity to focus that debate on sustainability exists. This chapter suggests how to do that relative to the technical workforce—more specifically, to the engineering workforce. A key component of successful conversion is the expertise of engineers. That expertise is especially important for conversion towards sustainable systems. Yet, engineers in general are not participating in the discussion, are certainly not leading the effort, and in fact, could hinder and slow the effort. More specifically, then, this chapter suggests how to bring engineers into the process. Many of the ideas also apply to bringing other sectors of society into the process. Consequently, this chapter addresses a social component of the problem of moving toward sustainability. It identifies defense engineers as a major source of technical knowledge that can be used to develop a sustainable society.

Concepts of economic conversion range from the mundane to the philosophical and from the technical to the ethical. This chapter first summarizes these concepts. The next two sections delineate, respectively, the evidence of economic conversion and the forces driving it. The fourth section summarizes the impact that conversion is likely to have on engineers and their likely response. Based on those impacts and responses, the final section identifies strategies that will encourage the participation and even the leadership of engineers in conversion from a military focus to a focus on sustainability in the world's economies.

Throughout this chapter, the discussion of economic conversion and the role of engineers applies to the current economic and political situation in the United States. However, many of the social and economic factors that affect the situation in the United States may also occur on an international level. All countries need to refocus engineering talent from designing better defense technology to developing sustainable systems.

CONCEPTS OF ECONOMIC CONVERSION

What is "conversion"? Conversion is the substantial shift of resources—labor, equipment, facilities, skills, and money—*from* military *to* civilian production and use. Three perspectives on conversion, each with different implementations for public policy, are provided here.

- First, the most limited definition of conversion simply includes a reduction of military expenditures—a reduction in annual United States defense department purchases and services as measured in dollars and corrected for inflation. The decline in the number of defense products being manufactured results in job losses in affected industries. Within these parameters, a decline in military expenditures will lead to the loss of skills and capital in defense industries. This conception of conversion is too narrow. Conversion is not just a reduction in expenditures—it requires a *shift* of resources to other productive uses. The important issues are those addressing how to redirect resources saved from defense cuts to the civilian sector—and how to direct them toward the design and development of sustainable systems.

- Second, some people envision conversion as having its most positive impact with only minimal direct public sector involvement and minimal planning. Conversion, according to this point of view, is a business matter, not an ethical or public policy matter. It should be left to market forces, which will determine the products that should replace weapons systems in the economy. Resourceful engineers will retrain and seek new positions; flexible companies will retool and create new products for the civilian economy. The problem with leaving conversion to market forces is that many engineers do not have the resources necessary to find new jobs. Furthermore, U. S. defense companies cannot easily shift to civilian-oriented production because the equipment and skills used in defense manufacturing are too specialized and not applicable to the civilian market (Mosley 1985). These conditions result in fewer job opportunities for defense workers who lose their jobs. The large numbers of people employed in the U. S. defense industries magnifies the effects of cuts in defense spending that may result in widespread unemployment. Leaving conversion to market forces, therefore, could have a destabilizing effect on the U. S. economy.

  An alternative viewpoint maintains that reductions in defense spending will negatively impact the national economy and will dramatically impact local communities; therefore, planning the conversion is critical to its success. The nature of this planning—business, economic, and public policy—then becomes central to the discussion of conversion. Planning can occur at the national level and focus on new emphases in the economy (i.e., rebuilding the nation's infrastructure, cleaning up hazardous wastes, creating new manufacturing technologies, and generating education reform, social programs, and more broadly, sustainability). In addition, planning needs to take place at the local level to mitigate the impact on local communities.

- Third, the broadest conception of conversion is that associated with changes in the United States' philosophy of national security. Traditionally, the U.S. defense philosophy rests on five propositions: (1) a strong military is the key component of U.S. security in the world order, (2) large scale military production is essential for economic welfare, (3) U.S. weapons sales generate important foreign exchange, (4) the United States must be able to protect its energy resources abroad using its military, and (5) a major standing army and weapons inventory are needed for rapid deployment because a converted defense structure would be difficult to mobilize rapidly enough if needed (Jameton 1993).

A philosophy of conversion would challenge all five of these claims. With respect to the first, U.S. economic production and exchange, and more specifically, a sustainable economy can be seen as more fundamental to U.S. security than military might. On the second point, it has often been argued that non-military production is more productive than military production in creating jobs and maintaining general material welfare. Indeed, those who point to the examples of Japan and Germany argue that the United States' emphasis on military development and manufacture can be seen as one source of economic stagnation. Third, the use of weapons as an important element of foreign exchange fosters armed conflict among trading partners and hinders the peace necessary for economic development. Fourth, long-term health and environmental problems create a critical need to reduce the world's dependency on consuming non-renewable natural resources such as oil. It is thus more important to reduce dependency on oil and other raw materials than to secure availability with force. Fifth, economic strength can be seen as the key to the ability to mobilize military forces. If keeping a large standing army and weapons inventory undermine economic strength, then a large military force weakens the ability of the United States to respond flexibly and militarily to sustain resistance over an extended period.

These changes in defense philosophy must be linked to changes in economic philosophy. At its most basic level, a commitment to planned economic conversion is a commitment that long-range planning and deliberate public decisions regarding the use of resources can aid economic development. Moreover, a philosophy of conversion must recognize that the world is facing a crisis of overproduction, exhaustion of non-renewable resources, waste disposal, and overpopulation. To plan for conversion is to plan toward a redesigned national economy that can produce goods more efficiently, that can reuse and recycle goods, and that can reduce dependency on commodities to fuel material welfare (e.g., a plan for sustainability).

In this scenario, the philosophical shift is toward support for economic development that focuses on people rather than on economic growth, and that harbors remaining global resources, reduces damage to the environment, and focuses more on equality and human welfare. Under this view, "conversion" is a philosophical shift—a shift in values. People must make a fundamental shift in their perspectives on the world—a major shift that involves personal identity. Such a fundamental shift provides the motivation needed to achieve successful conversion.

Technology is central to the changes associated with any type of conversion. It provides the physical means of changing from military to civilian production. Since engineers are directly involved in the design and implementation of technology, engineers will be involved in any conversion process. They provide the link between technology's capabilities and society's needs. Technology cements values because it defines what products and services society wants; technologies shape work patterns and the nature of products. Thus, values are inevitably involved in choosing technological directions. The search for a more efficient automobile, a better bicycle, a transit system among major cities, or a fleet of faster Concords embodies different social and economic priorities. On a larger scale, a focus on sustainable technologies or on technologies that are perceived to increase economic competitiveness embodies different values.

Society's best chance of improving the welfare of the planet and its populations depends on an open public and professional debate on what values might best be employed in the future. Without the debate, no "great" project can be undertaken. "Appropriate technology" in the largest sense has the potential to create a different integration of civilian and military production. Engineers must participate in the design of the technologies; thus, their active participation in the debate about conversion is essential.

IS ECONOMIC CONVERSION UNDERWAY?

A review of trends in four categories of activity suggest the "from" half of the conversion equation is imminent in the United States and may have begun. Specifically, the momentum is building toward substantial cuts in the military budget. These early trends in cuts, however, could be reversed relatively easily and quickly. The "to" half of the conversion equation has not begun and is rarely discussed. No public or private effort is directed at how best to use the resources that are freed by reductions in the military budget. Those involved in issues of sustainability on planet Earth have not presented their case within the framework of economic conversion.

Cuts in U.S. Defense Spending

Discussions in Congress and the rhetoric of President Bush differed substantially from defense cuts that have actually occurred. The estimated defense budget authority for FY 1992 was the same as the authority for FY 1993 at $290.8 billion (Cain 1991). That is, when adjusted for inflation, there is a decline in real dollars, but the decline is minimal. In contrast, both the President and the Congress talk about major cuts in defense spending. Current discussions are centered on the goal of reducing the defense budget from 5% of the GNP to 3% of the GNP over a five-year period.

Scheduled Base Closings

In 1988 Congress selected 86 military bases for closure (Wake 1990). Bases from this first round are now involved in conversion planning. The best known example is the Presidio Base in California. In addition, the 1990 Base Closure Act led to the formation of the Defense Base Closure and Realignment Commission. The Commission identified another 35 domestic bases for closure (Wake 1991). Few closures have actually occurred.

Conversion of U.S. Defense Contractor Companies

Currently, no situation exists in which a major U.S. defense contractor is converting a large-scale military production facility to civilian production; however, limited examples of smaller defense contractors' marketing products for civilian consumption do exist. For example, the Kavlico Corporation in California makes sensors for military aircraft and is now selling its products to the Ford Motor Company. Major companies, rather than converting, are scaling back and laying off employees. Major defense companies likely to be impacted by cutbacks include General Dynamics, Northrop Corporation, Lockheed, Boeing, and United Technologies. Any impact on these major companies also has a ripple effect through their subcontractors. Hughes Aircraft in Tucson, Arizona announced a layoff of 1600 employees. At the same time, Hughes is aggressively pursuing diversification, relying on markets where, because of patent rights, it may have a competitive advantage (Hughes 1991). Examples of technologies under development at Hughes are character-recognition systems, smart highways, smart cars, and automated manufacturing systems. No defense contractor, to our knowledge, is focusing on "clean" or "green" technologies.

Federal Legislation

In addition, to the "Defense Economic Adjustment, Diversification, Conversion, and Stabilization Act of 1990" and the "Defense Base Authorization and Realignment Act of 1990," which have been enacted, other legislation reallocating defense spending to the civilian sector has been proposed (e.g., Senator Tom Harkin's proposed amendment to the "Health and

Human Services Appropriations Bill for 1992"). The amendment would have reallocated $3.1 billion from unobligated Department of Defense funds to biomedical research, breast cancer screenings, research on mental illness, educational grants, AIDS research, child immunization programs, the pre-school programs for low income families, the low income home heating program, and a program to educate immigrants in the Southwest (Kash 1991). Other proposed legislation has focused on industrial policy and technological competitiveness. One bill suggested that the National Security Council be expanded to include the Secretary of Commerce, the Secretary of the Treasury, and the U.S. Trade Representative. Senator Joseph Liebermann's bill would have set aside 10% of the projected defense cuts to pay for industrial diversification. None of these or other bills that deal with the "to" side of conversion equation has yet passed. And none had dealt with the issues of sustainability, sustainable technologies, or global environmental issues.

WHAT IS DRIVING ECONOMIC CONVERSION?

Economic conversion in the United States is being driven by a combination of interlinked political and economic factors. It is not being driven by an organized move toward environmental protection or sustainability.

Political Factors

The threat of global war is widely perceived to have diminished. The fall of the Berlin Wall, START II (an arms reduction treaty that deactivates a large number of U.S. medium-range nuclear missles and cuts the number of nuclear warheads by about 30%), Soviet democratization, the failed Soviet coup, and the dissolution of the Warsaw Pact and the Soviet Union itself are all perceived to contribute to a reduced worldwide military threat.

In August 1991, President Bush stunned the world by unilaterally taking strategic bombers and intercontinental ballistic missile (ICBM) silos off the 24-hour alert they had endured since the 1950s. Arms negotiator Gerald Smith said, "We are rapidly running out of an enemy in the form of the Soviet Union, and I don't know what we're going to need 20,000 warheads for."

The future role of the U.S military is now the subject of great debate. Military actions such as the Persian Gulf war are now seen as anomalies. General Colin Powell was quoted in the *Army Times* as saying "I would be very surprised if another Iraq occurred," (Budiansky and Auster 1991). In the same interview, General Powell stated: "I am running out of demons, I am down to Castro and Kim IL Sung." Pentagon planners are searching for a new set of threats to support the United States' billion-dollar weapons development programs; both the B-2 stealth bomber and the Block III tank were designed as responses to Soviet threats in Europe (Budiansky and Auster 1991). The implications for defense spending of a redefinition of the role of

U.S. military is not clear, but an overall smaller defense budget and troop reductions are likely.

Economic Factors

Three interelated economic factors are important drivers of economic conversion in the United States :

- *The Federal Debt*: Over the last decade the United States' national debt has tripled from $1 trillion to approximately $3.2 trillion. The debt places pressure on the Federal government to find ways to balance the budget; military spending is a prime target. The underlying assumption, of course, is that a more balanced budget would stimulate economic growth and productivity.

- *U.S. International Trade Competitiveness:* U.S. competitiveness in world markets, as measured by the trade balance, has been declining substantially. The trade surplus that characterized the period from 1900 to 1970 was replaced by a deficit that amounted to approximately $100 billion by 1991 (Council 1991). One view is that economic conversion could make talented engineering resources available for the design and production of civilian products. In turn, such a redirection of talent might re-establish the role of U.S. products in foreign markets.

- *Military Research and Development (R&D) and Economic Productivity:* Widely accepted is the fact that research and development play an important role in product development and in the nation's economic growth. But there is also evidence that the engineering resources for military-oriented R&D has an overall negative impact on economic productivity. Dumas shows that the focus of engineering resources on military technology represents a scientific "brain drain" from the civilian sector (Dumas 1986). Although positive spin-offs into the civilian sector do occur, they are inconsequential compared to the negative effects of brain drain. The best engineers gravitate to the military sector because the wages are higher, producing an overall negative economic impact on the production of consumer goods and services.

HOW WILL ECONOMIC CONVERSION IMPACT ENGINEERS?

Four categories of impacts are summarized here; the severity of these impacts depends on (1) the extent of defense cuts, (2) the extent to which conversion is planned, (3) the "to" side of the equation, and (3) the extent to which conversion is decentralized. Planning (not leaving conversion to market forces) will lessen the severity of the negative impacts. If the planning is also decentralized, negative impacts can be further mitigated. Decentralized planning involves the development of conversion plans for specific military installations and specific weapons productions facilities at the local level. The retraining of employees and retooling the facility is most effective when it occurs at the local level and is directed at the specific community or specific facility.

Jobs

Economic conversion will create a situation in which more defense engineers are seeking fewer defense positions. This oversupply situation will be particularly acute for certain categories of engineers (e.g., aerospace engineers), older engineers, engineers doing highly technical work, and engineers doing work that requires a security clearance (Yudken and Markusen 1993).

The U.S. Congress Office of Technology Assessment estimates that 1.5 million Americans will lose their defense-related jobs by 1995 (OTA 1992). Since one-third of all engineers work on military projects, engineers can be expected to experience major job dislocation. Unemployment rates in the first half of 1991 averaged 6.8% for all civilian workers and 2.8% for managerial and professional specialties (Braham 1991). Although engineers have been insulated from unemployment in the past, their unemployment rate doubled from 1.4% in 1989 to 2.2% in 1990 (Labor 1992). The Institute of Electrical and Electronics Engineers predicts that 55,000 engineers in the defense sector could lose their jobs by 1995.

Income

Clearly, when fewer jobs are available for the same number of qualified people, salaries and benefits decline. This will be true, at least in the short-term, during economic conversion. The magnitude of the decline depends on how rapidly defense cuts occur, on what the economy converts to, on the number of graduates from engineering schools, and the extent to which planning occurs. Relocation of former defense engineers may be required. This will exacerbate the problem of limited resources since relocation is expensive, and the location from which an engineer is moving may have a depressed housing market because other unemployed defense engineers are also trying to leave the same area.

Transferability

Issues associated with the transferability of engineers from military to civilian sectors will become highly visible during economic conversion. More specifically, the principle focus of military work is the development of products that will perform in combat situations. Cost is the primary consideration. In contrast, civilian work is aimed at developing products within severe cost restraints. Retraining military engineers to consider cost, in addition to performance, in the design of a product is difficult. Consequently, the skills of engineers from the defense sector are not always transferable to the civilian sector. This could be a major hurdle to a smooth transition. Some researchers expect it to be particularly difficult for engineers working in the national laboratories whose main focus has been the development of defense technology (Hughes 1993).

Job Satisfaction

Engineers enjoy technical work and work that is well defined and challenging. They also find satisfaction in good salaries and benefits. Of secondary importance to job satisfaction for most engineers is the value framework in which the work is carried out. Maintaining national security through defense-related work has long been an important value for engineers, but it is probably secondary to job satisfaction, the focus on making something happen, and the solving of interesting technical problems.

Engineering jobs outside the defense sector are not as technical, nor do they pay as well. Thus, job satisfaction could decline. While it may be possible to convince engineers that their work is socially valuable because, for example, it increases economic competitiveness, protects the environment, or enhances energy security, if it is also not technically challenging, dissatisfaction could occur. Landfill design, for example, is simply not as technical as missile design.

HOW WILL ENGINEERS IMPACT ECONOMIC CONVERSION?

A variety of common behavioral patterns, values, and cultural norms are held by engineers. Information about these characteristics provides some insight into the role engineers may play in economic conversion, as well as their willingness to contribute to a sustainable society.

Political Orientation

Engineers tend to be members of conservative political organizations and to avoid political issues. They like to be challenged by solvable, discrete problems. Consequently, the compromises associated with the solutions to social problems and the open-ended nature of public policy issues is unappealing to engineers. Engineers who lack the interest in social change may oppose economic conversion simply because of the change it represents. These same attitudes may discourage engineers from participating in the political debate surrounding a societal shift towards sustainability. Therefore, increasing the impact engineers will have on developing sustainable systems may mean appealing to them from an apolitical standpoint.

The Corporate Culture

Historically, company managers have chosen the problems on which engineers work, resulting in a lack of autonomy in the work environment for the technical workforce. Moreover, this structure provides no opportunity for engineers to participate in policy making, even inside the corporation. In short, if decisions about conversion are made by managers, engineers will respond but are unlikely to be part of the decision.

Values

Little research exists about engineers' personal values or about their beliefs regarding justice; values and beliefs together influence their opinions about economic conversion. If engineers value conversion and believe it is just, they will support the process, even if some negative consequences accrue. A common belief among engineers is that absolute truths can be found if given enough time and effort for the search. Engineers have a deeply rooted optimism about the effects of technology on society, regardless of whether technology is defense- or civilian-oriented (Gagne 1993). Their time is spent on the pursuit of technology, often at the expense of the pursuit of personal or community social interaction. Their focus is on things, not on society's role in progress or on the definition of progress in our society. Thus, involving engineers and society in conversion requires building bridges of values and meanings that connect the seemingly utopian goals of economic conversion to present workplace realities.

Engineering Societies

In general, engineers do not actively seek membership in political organizations that influence their values systems; however, many engineers are members of professional societies. Engineering professional societies have not participated in debating values or societal issues, rather they are a reflection of engineers' values and of the focus on technology. They have been important in defining the professional role of engineers, and they serve as a clearinghouse for information. They do not provide a forum for the discussion of broader issues.

For these reasons, engineers and engineering societies should not be expected to be dynamic advocates for economic conversion or to provide leadership. The conservatism and focus on discrete problems suggest that engineers will tend to support the status quo and resist change. Engineers will tend to oppose or not become involved in the debate about conversion. If they follow historical trends, they may even be expected to obstruct conversion. Neither extreme is likely. Most likely, engineers will tackle company or societal problems without value-laden judgments about the problems.

WHAT FACTORS WILL ENCOURAGE PARTICIPATION?

Engineering talent is needed for economic conversion at any level. Although our knowledge of what motivates engineers is limited, it does suggest some factors that hold widespread appeal to most engineers. The following could encourage participation of engineers if not lead to their active support of economic conversion.

Personal Resources

Engineers who cannot afford to relocate, or who lack the skills required for other employment will feel particularly vulnerable when faced with the job displacement associated with conversion. Much of defense work involves highly technical skills, and some of it is classified. Skills acquired in defense work are not easily applicable in designing consumer-oriented technology, and the classified nature of defense technology creates legal barriers for engineers who want to use their skills in the civilian market. Both of these factors inhibit transferability to the civilian sector. Clearly, any assistance by government or firms in retraining and relocation will encourage participation.

Knowledge about Economic Conversion

Currently, little information about economic conversion is available. In fact, no commonly held definition of the term even exists. Without adequate information on the conversion process, engineers will not participate in the process because they are unable to visualize the end result. Information concerning the economic, political, technical, and moral issues associated with conversion need to be made available to engineers if discussions about the opportunities and benefits associated with the design of sustainable technologies are going to occur.

The Extent of Planning

If conversion is planned, it will help identify and create civilian engineering jobs for unemployed defense engineers. Such planning can involve engineers in solving the technical problems associated with converting defense manufacturing facilities to civilian production and with designing new products for a sustainable society. Involvement of engineers in the planning will greatly increase support and participation.

Action by Respected Leadership

Engineers will be influenced by the leadership within their profession. Engineers in leadership positions can motivate fellow engineers in three ways: (1) leaders can serve as a reliable sources of information about the conversion process, (2) leaders who participate in the conversion process can serve as role models for defense engineers, and (3) leaders can be liaisons between defense engineers and other groups participating in the conversion process.

Changing the Nature of the Workplace

New management techniques and a philosophy of worker involvement are changing the nature of the workplace. The team approach and the cooperative approach to productivity embedded in Total Quality Management, Quality Circles, and Employee Stock Ownership Plans, as well as new literature associated with productivity in the private sector (Senge 1990), lends itself to the participation of engineers. To the extent these approaches become the norm in the private sector, the process of conversion will be easier.

Providing the Personal Connection

The literature on the psychology of empowerment suggests that participation depends on a personal connection to the issues (Everett 1993). If engineers feel personally connected to economic conversion, they may be more willing to promote it, participate in it, and make the changes in their work lives dictated by the conversion process. Identifying what defines personal connection is difficult. At least three elements are involved in the creation of personal connections for engineers: (1) interest in the technology, (2) short-term satisfaction in the work environment, and (3) long-term goals that satisfy values. More specifically, one of the primary attractions of defense work is the engineers' involvement in cutting edge technologies that defense work provides. Creating a personal connection for engineers will require the application of cutting edge technologies to the commercial sector. The high technology nature of sustainable systems, such as some alternative energy sources, may provide engineers with the technical challenge needed to generate greater participation. Providing job satisfaction in the commercial sector in the short-term requires attention to comparable salaries, increased freedom to be creative, and increased responsibility. And finally, the personal connection will be enhanced to the extent that conversion is linked to long-term goals important to engineers. Long-term goals valued by engineers include the challenges of solving complex technological problems, the design of products that solve societal problems efficiently (e.g., sustainable technologies), and the challenge of making U.S. commercial products competitive worldwide.

In short, engineers' participation in economic conversion will depend on: their knowledge about the opportunities and the moral issues, their personal resources, opportunities for retraining, the extent of planning at corporate and societal levels, their involvement in the decision-making apparatus within corporations, and the extent to which engineers feel personally connected to issues.

PUBLIC POLICY: PLAYERS IN PROVIDING THE MOTIVATION

Existing institutions can provide motivation for economic conversion. Potential roles for the federal government, engineering societies, and industry are summarized.

Federal Government

The federal government is the most important institution for fostering conversion policy. The federal government influences conversion policy in at least four ways: (1) it controls vast financial resources that can be used to support conversion policy, (2) it can disseminate information supporting conversion policy to all levels of society through government agencies, (3) it can pass legislation that provides incentives for businesses and individuals to support conversion policy, and (4) it controls defense expenditures and can help initiate conversion by diverting those expenditures. Five types of federal activities that support conversion are identified.

- First, the federal government can take the lead role in planning conversion. With planning, potential negative impacts will be less severe and prolonged. Some consensus on the direction of the shift (e.g., infrastructure, manufacturing, the environment) is required, and that consensus must develop at the federal level.

- Secondly, the federal government must provide leadership in refocusing the priorities of the national laboratories. The multi-program national laboratories that are funded through the U.S. Department of Energy conduct research on technologies with applications in defense. These include the design, development, and testing of weapons systems. These multi-program laboratories include Argonne, Brookhaven, Lawerence-Berkeley, Oakridge, Pacific Northwest, Idaho Engineering, Lawerence-Livermore, Los Alamos, and Sandia. Total budgets for all of these laboratories have been in the range of $5–$10 billion per year. The laboratories are immersed in the defense engineering culture; if they can shift to research and development of products for the civilian sector, they can model the elements for such a shift for the private sector. The labs, in fact, may be pivotal in demonstrating how defense engineers can become civilian engineers. During the last five years, there has been a move toward involving the national labs more directly in pre-commercialization of civilian technology and in technology transfer activities. An interagency committee for federal laboratory technology transfer exists to promote such efforts. Perhaps more encouraging—in July 1992 the Senate passed the Department of Energy Laboratory Partnership Act (S.2566). The bill directs the Secretary of the Department of Energy to ensure that the labs enter partnerships with private companies and educational institutions to develop technology in critical areas such as energy efficiency, energy supply, high performance computing, environmental protection, advanced manufacturing, advanced materials, and transportation.

- Thirdly, the federal government can become involved directly in establishing joint ventures among federal laboratories, universities, and private companies. Such joint ventures can be directed at developing technology for solving problems in the public domain.

- Fourthly, the federal government can provide direct support for retraining. Retraining is required to refocus the national laboratories and to establish joint ventures. Fellowships for engineers to update their skills at universities are one approach. Funding and support of retraining programs developed by professional societies is another.

- Lastly, the federal government can offer incentives to defense companies such as investment tax credits. Such incentives would be directed at companies that implement innovative programs for the development of civilian technologies. Obviously, such incentives raise the issues of government interference in the marketplace, but such issues are resolvable.

Engineering Societies

The professional engineering societies are in an excellent position to provide leadership for their members. The societies have established communications networks, administrative support, and legitimacy from the perspective of the public and political leaders. The societies can involve the membership in the conversion debate by including forums at the local, chapter, and national meetings; they can act as a clearinghouse providing information about conversion from congressional staff to their membership; they can design and implement retraining efforts; they can lobby for congressional bills on conversion that provide assistance to engineers; they can involve the membership in discussions about the engineering code of ethics as it applies to conversion; and they can identify specific technical challenges associated with solving problems in the civilian sector.

Industry

Industry itself plays a critical role in educating working engineers. Individual corporations must identify the civilian products on which they are going to focus. In their diversification program, Hughes Missile Systems defines the differences in market development between defense and commercial markets as follows: for commercial markets, the requirements for products must be delineated internally; however, for defense production, military specifications define products. Success in the commercial sector also requires being first to the market and requires system designs that produce payoffs for customers. Hughes selects markets in which it has unique, sustainable, competitive advantage (e.g., where it owns patents in important technologies).

SUMMARY

Economic conversion involves a shift *from* the use of resources for military applications *to* some other focus of activity or combinations of activities. ISEE may want to ensure that the primary focus is the creation of sustainable systems. Discussions about economic conversion in the United States are focused almost entirely on a reduction in military spending; little debate is occurring on how to redirect that spending. An opportunity to define sustainability as a focus of the redirection now exists. Part of successful economic conversion is the ability to think of science and technology in entirely different terms—not as mechanisms for weapons production or as mechanisms to increase wealth and comfort, but as sources of innovation that can lead to less consumption, less pollution, less depletion of resources, and lower rates of population growth.

Substantial engineering talent is now directed at weapons research, development, design, and manufacture. That talent needs to be redirected toward research, development, design, and manufacture of sustainable technologies. Sustainability for the earth cannot happen without the assistance of the technical workforce. How do we access and redirect that engineering talent?

1. Identify and define the technological challenges associated with sustainability. The technological challenges will stimulate engineering interest and creativity.

2. Involve engineers in the debate with their professional societies and their employers. Within corporations, this means providing for more participation in decision making and a recognition of the role of teams.

3. Increase engineers' involvement with the end users of their products. If engineers can see the results of their work, as well as explain the uses of technology to the public, they will feel that their work is impacting society.

4. Provide monetary incentives for engineers to design sustainable systems. These incentives can be provided through government grants or at the company level. Positive incentives create a reward system for engineers which may increase engineers' conception of self-worth through the recognition of individual effort.

5. Develop retraining programs for engineers in the area of sustainability. Retraining can occur at universities or at companies that are planning to manufacture sustainable technologies. The government could make available loans for engineers who are seeking continuing education in the field of sustainability.

6. Create jobs in the area of sustainability. The most critical factor for getting engineers to design sustainable systems is to create jobs in this field. This may involve government tax and financial incentives for businesses interested in manufacturing sustainable systems. The U.S. government could also help create global markets for sustainable technologies by providing companies with information about what sustainable technologies are in demand internationally.

7. Redirect government research and development funds from defense projects to sustainable systems. The effect of increased spending on the development of sustainable systems will be a value shift in support of sustainability. In addition, engineers, once employed in defense research, will see that designing sustainable systems is a field for future job security.

8. Publicize the role of engineers in society. Greater media attention towards engineers will force them into the debate over the worth of defense industries. This may change engineers' attitudes towards sustainable systems because it forces them to deal with the ethical questions of designing defense technology. Under these conditions both working engineers and future engineers may choose jobs in developing sustainable technologies, because the public perceives these technologies as more beneficial than weapons systems.

9. Redirect the work of the National Laboratories toward research development of sustainable technologies.

If the talents of engineers are going to be applied toward designing sustainable systems, engineers must accept the goals of sustainability. Economic conversion offers a window of opportunity for engineers to replace their current belief system, which places great value on developing defense technology, with a new system that values the financial and societal benefits of sustainability.

ACKNOWLEDGMENTS

This project was sponsored by the National Science Foundation, Ethics and Values Studies in Science, Technology, and Society Program (Grant No. DIR–9013961). The original chapter and a book to be published by Springer-Verlag are a synthesis of ideas generated at a workshop; we thank all of the workshop participants.

REFERENCES

Braham, J., 1991. A pink slip among the blueprints. *Machine Design* (April 11): 35–39.

Budiansky, S., and B. B. Auster. 1991. Missions implausible: the evaporating communist threat has challenged many of the assumptions that still guide United States military strategy. *U.S. News and World Report* (October 14): 28.

Cain, S. A. 1991. Analysis of the FY 1992–93 Defense Budget Request: With Historical Defense Budget Tables. Defense Budget Project, Washington, DC, (February): 1–3.

Council on Competitiveness. 1991. Competitiveness Index 1991, Council on Competitiveness, Washington, DC, (July): 6.

Dumas, L. J. 1986. The Over-Burdened Economy: Uncovering the Causes of Chronic Unemployment, Inflation and National Decline. Los Angeles: Univ. of California Press.

Everett, M. 1993. Engineers and economic conversion: a psychological perspective. In Engineers and Economic Conversion: From the Military to the Marketplace, eds. P. L. MacCorquodale, M. W. Gilliland, J. Kash and A. Jameton. New York: Springer-Verlag.

Gagne, E. 1993. Economic conversion in perspectives: the values and ethics of engineers. In Engineers and Economic Conversion: From the Military to the Marketplace, eds. P. L. MacCorquodale, M. W. Gilliland, J. Kash and A. Jameton. New York: Springer-Verlag.

Hughes K. H. 1993. National technology priorities and economic conversion. In Engineers and Economic Conversion: From the Military to the Marketplace, eds. P. L. MacCorquodale, M. W. Gilliland, J. Kash and A. Jameton. New York: Springer-Verlag.

Hughes Inc. 1991. Focus on Hughes, Volume III, Fall 1991. A videotape from Hughes Aircraft Company.

Jameton, A. 1993. Economic conversion and global justice: the moral issues. In Engineers and Economic Conversion: From the Military to the Marketplace, eds. P. L. MacCorquodale, M. W. Gilliland, J. Kash and A. Jameton. New York: Springer-Verlag.

Kash, J. 1991. Conversation with Senator Tom Harkin's staff member, Sandy Thomas, on September 10, 1991. Graduate College, Univ. of Arizona, Tucson, AZ.

Mosely, H. H. 1985. The Arms Race: Economic and Social Consequences. Lexington, MA: Lexington Books.

OTA. 1992. After the Cold War: Living with Lower Defense Spending. Congress of the United States, Office of Technology Assessment, OTA-ITE-524. Washington, DC: U.S. Government Printing Office, (February).

Senge, P. M. 1990. The Fifth Discipline. New York: Doubleday.

U.S. Dept. of Labor. 1992. Table A-39: Selected Unemployment Indicators, Employment and Earnings Vol. 39, No. 4, U.S. Department of Labor, Bureau of Labor Statistics, (April): 44

Wake, J., ed. 1990. Basically confused: government gridlock on next round of base closings. *Base Conversion News*, 1 (1): 4–5. Mountain View, CA: Center for Economic Conversion.

———. 1991. Around the country. *Base Conversion News* (Winter): 6–7. Mountain View, CA: Center for Economic Conversion.

Yudken, J., and A. Markusen. 1993. The labor economics of conversion: prospects for military-dependent engineers and scientists. In Engineers and Economic Conversion: From the Military to the Marketplace, eds. P. L. MacCorquodale, M. W. Gilliland, J. Kash and A. Jameton. New York: Springer-Verlag.

25 PUBLIC POLICY: CHALLENGE TO ECOLOGICAL ECONOMICS

Stephen Viederman
Jessie Smith Noyes Foundation
16 East 34th Street
New York, NY 10016

ABSTRACT

Ecological economists are challenged to become more involved in the policy process. This chapter reviews the defining characteristics of ecological economics as a post-normal science, focusing attention on a view of sustainability that includes equity as well as economics and ecology. The need to create a vision of a sustainable future and to understand the processes of change is underlined, and examples of policy issues for ecological economists to address are outlined.

INTRODUCTION

How can ecological economics inform and change political and policy debates and outcomes? What does ecological economics have to contribute now, and what must we do in the future to participate more actively and fruitfully in the debates over sustainability in our country, and in the world as a whole?

We have a historic opportunity, and an obligation to participate actively in these debates. Whether we choose to do so is a question that we must address now. Is ecological economics really a problem-oriented, policy-relevant, "post-normal" science, as its leaders suggest, or is it going to become an elegant and largely irrelevant discipline, not too unlike the conventional economics and ecology that ecological economists have found wanting? Though I welcome proof to the contrary, I fear that it is in the latter direction that we are moving.

POLITICS, ECONOMICS, AND VALUES

What is the problem that confronts the field? I offer the following from George Orwell's 1945 essay, *Catastrophic Gradualism:*

> The practical men have led us to the edge of an abyss, and the intellectuals in whom acceptance of political power has first killed the moral sense, and then the

sense of reality, are urging us to march rapidly forward without changing direction.

Ecological economists are similarly guilty of having allowed conventional economics to become the language of politics, a language without a moral sense (because of its origins), and with a limited relevance to real economic circumstances. In doing so we have effectively ignored values and ethics. Yes, conventional economics has an ethics of sorts—the devil take the hindmost. We talk of costs and benefits without reference to values. But it is politics, not economics, that must reflect what we value in society. And it is a new politics, as well as a new economics, that must connect the economy and the environment, people and nature.

Recall that "politics" is not, and need not be, "A strife of interests masquerading as a contest of principles. The conduct of public affairs for private advantage," as suggested by Ambrose Bierce, the nineteenth century American journalist. Rather, "politics," as defined by the Oxford English Dictionary, "is the science and art of government (and of) public or social ethics, that branch of moral philosophy dealing with the state or social organism as a whole."

Ecological economics can become the knowledge base, or at minimum make vital contributions to the knowledge base, for the new politics we must build. We must clearly demonstrate, in our words and in our actions, our commitment to policy relevance. To do so, however, will require self-conscious attention to where we are and where we are going.

DEFINING CHARACTERISTICS

The principles of ecological economics suggest that addressing the real world issues at the intersection of economics and ecology—the human household and nature's household—are our reason for existence. Therefore, we ought not to—and cannot—shy away from or apologize for engaging in political dialogues.

What do we say about ourselves?

The original officers of International Society for Ecological Economics (ISEE), in mapping long-range goals for ecological economics, emphasized the importance of policy, both to be addressed directly and as part a framework for research and training and curriculum development. They identified the need to develop mechanisms "to ensure that an ecological economics perspective is introduced into the relevant policy debates (and to) facilitate communication and cooperation between ecological economics professionals and the legislative and executive branches of the government at federal and, eventually, at state levels ... encouraging the use of ecological economics perspectives in the policy development process." (Viederman 1991).

The potential policy importance of ecological economics was highlighted in the first issue of *Ecological Economics* by Costanza and Proops, among others, and in a number of the papers presented at the first international meeting of the Society in 1990 (Costanza 1991).

Among the defining characteristics of ecological economics with particular relevance to policy are "a holistic view of the environment-economic system (and) wider values than those encompassed by environmental economists (i.e., utility) and greater concern for moral obligations towards future generations" (Pearce 1992). Ecological economics is also characterized by a problem focus, pragmatism, and concern for sustainability. It takes a comprehensive global view, over the long term and in context. ecological economics is also concerned with the process of institutional change, expressing a tolerance for uncertainty, and a willingness to explore questions for which we do not at present know the answers. While efficiency is the policy objective of conventional economics, ecological economics is concerned with resiliency, safety, and common sense.

HOW ARE WE DOING?

Policy involvement is clearly an oft-stated goal of ecological economics, and our defining characteristics support that goal. Unless, however, we make our words a reality, we face the distinct possibility of becoming irrelevant. To what extent have we been able to influence policy?

In the United States, the Congress has shown little interest in the alternative view that ecological economics presents of economics and the environment. With the important exception of Vice-President Al Gore, probably no other elected representative has thought seriously about the issues that ecological economics raises (Gore 1992). Rarely does the Congress, the executive branch, or the media turn to one of our ranks for an opinion, either because we are still invisible, or perhaps worse and equally likely, because we have not persuaded them we are relevant. In addition, ecological economics is largely scorned by the so-called leading economists. For example, Daly and Cobb's *For the Common Good*, which to many of us is a seminal work, has been largely ignored by the economics profession, in much the same way that they ignored Daly's earlier *Steady State Economics*. In brief, whether in government, in academia, or in the public eye, ecological economics is hardly a household word.

WHY IS THIS?

To begin with, our numbers are small. There are only 1400 ISEE members in the world, compared with the tens of thousands of members of the American Economics Association. To argue for a greater involvement in policy and politics does not imply that we must mobilize the entire membership. Not all ecological economists are by temperament or intellect drawn to policy issues. But there appear to be too few of us who are truly committed to political dialogue in our roles as ecological economists.

In addition, until very recently there were few places where one could be trained as an ecological economist. And getting trained did not, and still does not, guarantee a job, since demand for ecological economists is still limited. There is, as yet, no effective institutional setting through which ecological economics can exert an influence on the policy process.

Change does not occur easily, nor quickly. The institutions whose support would be important for a new economics—the universities and research institutes, the various branches of governments, business and industry, the media, among others—have vested interests in the status quo.

HOW HARD HAVE WE TRIED?

Allowing for these institutional realities, the question arises, how hard have we tried? Evaluating our influence on the policy process presents problems.

My review of the six volumes of the journal *Ecological Economics* since its inception in 1989, suggests that we are still more comfortable as a "normal science" rather than as a "post-normal science," as described by Funtowicz and Ravetz (1991):

> We adopt the term "post-normal" to mark the passing of an age when the norm for effective scientific practice could be a process of problem-solving in ignorance of the wider methodological, societal, and ethical issues raised by the activity and its results. The scientific problems which are addressed can no longer be chosen on the basis of abstract scientific curiosity or industrial imperatives. Instead, scientists now tackle problems introduced through policy issues where, typically, facts are uncertain, values in dispute, stakes high, and decisions urgent...Here we find decisions that are "hard" in every sense, for which the scientific inputs are irremediably "soft." (p.138)

We might each review the contents of these volumes and ask ourselves; how many of these articles could have been published elsewhere? How many of them are truly reflect ecological economics, according to our stated goals and defining characteristics? And within that smaller number, how many were motivated by a concern for policy, *and* are potentially usable in the political/policy process? I venture that the number that are "truly" ecological economics, as I am using the term, is significantly under 25%.

Another possible measure is the audience ecological economists address. Is it other ecological economists, or scientists in other disciplines? Or is it politicians and policy makers, advocates and organizers? I hasten to emphasize, again, that not all ecological economists need to be doing the same thing, but I want to suggest that in our present form we are closer to becoming just another discipline or field, rather than living up to our self-definition. Our chosen audience still seems to be our colleagues. I wonder if we are willing to get our hands dirty?

THE TASKS AT HAND

How can we move closer to our goals for ourselves? What are the tasks at hand, both over the long term and more immediately? The former leads to further redefinition of who we are, what we are, and how we work: an extension of ecological economics as a post-normal science. The latter is directed to what we can and must do now.

THE LONGER TERM

What is a "Problem"?

Over the longer term, we must first be clear about what constitutes a problem: the *Oxford English Dictionary* defines a problem as a "difficult question proposed for solution." In common usage, "our problem" is the distress we feel at the moment—a statement of where we are. The problem then becomes global warming, ozone depletion, the loss of biodiversity, or toxic waste dumps. This description of where we are is necessary information, but it is not sufficient to develop a solution. What is needed is a vision of the future that states where we want to be. The real problem is not where we are—the immediate situation that has generated a feeling of societal or individual suffering—but rather the gap between where we are, where we are going, and where we want to be.

Creation of a Vision

Here then is the beginning of the contribution to the policy process of ecological economics: leadership in creating a vision of a sustainable society, in and among our individual countries. As the late American social critic Michael Harrington has argued, "there must be a vision in the sense of purpose, of aspiration.... " (Harrington 1963).

We do have the beginnings of a vision, when we speak of sustainability and sustainable development. These provide the organizing principles for this new vision. Much, of course, needs to be done to make these concepts clear and more operational. But we should not linger too much on definitions, for as De Vries has insisted, "Sustainability is not something to be defined, but to be declared. It is an ethical guiding principle." This principle is our standard and our goal, yet it too can change as we learn more.

A Broader View of Sustainability

Our declarations, however, must include a broader view of "sustainability" than is evident in much of the work that we have done to date. Too much of our concern has focused too narrowly on "ecological" or "physical" sustainability. For example, Daly's three principles—(1) not to use renewable resources at rates that exceed their capacity to renew, (2) not to use non-renewable resources at rates that exceed the capacity to substitute for them and (3) not to use any resources beyond the earth's capacity to assimilate the wastes associated with their use—speak to "ecological" sustainability alone, which is necessary but not sufficient.

While accepting the inherent limitations in defining complex concepts and ethical principles, the following demonstrates the breadth of our concern:

> A sustainable society is one that ensures the health and vitality of human life and culture *and* of nature's capital, for present and future generations. Such a society acts to *stop* the activities that serve to destroy human life and culture and nature's capital, and to encourage those activities that serve to *conserve* what exists, *restore* what has been damaged, and *prevent* future harm (Viederman 1992).

Equity

Implicit, and to be made much more explicit, is the assertion that there can be no real sustainability without equity, within and among the nations of the world. Without equity there is no political stability. Therefore, to fulfill its potential role, ecological economics must incorporate equity as a third "E": equity between and among peoples and nations, between present and future generations, and between the human and the natural worlds. The three "Es"—Ecology, Economics, and Equity—together form the ethical dimension of our task as ecological economists, which thereby gives us a place in the political and policy debate.

Trickle-Up

Our concerns for equity, as well as our desire to be effective, oblige us to forgo topdown approaches. "Trickle-up" approaches can be achieved by working with others to reasonably ensure that our "expertise" is balanced by the "experience" of people most in need, who reflect their own forms of "expertise." Too many "visions" have been statements of conventional expertise alone, and have ended up gathering dust on library shelves. Participatory does not mean that each of us will need to work, for example, with grassroots organizations directly, although there could be significant benefits in that. Rather it suggests that we must make every effort, and ensure success, in communicating what we believe is important, while at the same time listening carefully to hear what activists and others are saying is important. All are leaders, and all are followers. There are issues that must be resolved at the macro level, but dialogue and inputs from the micro level—the community—are necessary if these issues are to be resolved successfully.

Our goal is not to create unworkable and unattainable "utopias," but rather to describe desirable and attainable futures, and the alternative paths to those futures. Listening can help recast or refine the agenda of ecological economics in policy-relevant ways.

Processes of Change

Participation in creating a vision requires that we clearly understand both the processes of change and the substance of change. We are all more comfortable with the latter: what needs to be done? But to be effective we must also address the processes of *behavior* change at both the individual and institutional levels: how does change occur? And I emphasize *behavior* because that must be our goal; changes in knowledge, values, and attitudes may be useful, but these changes must be measured by how they affect behavior. For example, it is too simple to argue, as is often the case, that education and training change behavior.

For us as ecological economists, this means that we must look at the processes of changing the way economists and ecologists do their work and define their goals, because they serve as important gatekeepers in the policy process. This is in addition to our efforts to effect changes in the economy and the society itself.

We must also look at the way that society's institutions change. The most important of these for ecological economics include, but are not limited to, the executive and legislative branches of governments at all levels, business and industry, non-governmental organizations and informal institutions, universities and research institutions, and the media.

If we look at the policy process, the importance of making explicit our conception of the processes of change, before beginning work on the substance of the changes needed, becomes clear. For example, an assumption that policy making is a rational process places a high value on research of a normal kind. Normal science lends itself to rational appraisal. If, however, we assume that policy making is a process of muddling through with overtly political choices, the role of a post-normal science, as opposed to a normal science, becomes more apparent, and values, advocacy, uncertainty, quality, and timelines take on a higher order of priority.

Roles of Knowledge

As a post-normal science, ecological economics also requires that its practitioners consider, as part of the research process, how the research will be used and to what end. The focus changes from dissemination of science (where will the work be published?), to the utilization of knowledge (how can the use of the work be ensured, and by whom?). This places great emphasis on our need to understand the system into which our research enters. It also suggests the need for greater awareness of who our partners are, or should be, in effecting change, including those who will be affected by the changes.

We must also question the very nature of our emphasis on research as part of a process of envisioning, and of policy participation. The philosopher Peter Brown has recently observed that a call for more research is an almost conditioned reflex among American progressives when we are confronted with problems. However, without a vision of where we are going, we are likely to be spinning our wheels. He notes that the so-called Reagan Revolution was put together by some conservative think-tanks—especially the Heritage Foundation—with little concern for research and much concern for a vision of a society that would be different from that which proceeded it. He further notes the conflict which emerged between Robert Brookings, who in 1927 established the institution that bears his name in Washington, and the economists who ran it. Their needs were for "objectivity," not vision, and the die was cast (Brown 1993). As Vaclav Havel observed recently, "traditional science, with its usual coolness, can describe the different ways that we might destroy ourselves, but it cannot offer us truly effective and practical instructions on how to avert them" (Havel 1992). There is a lesson here for ecological economics.

Yes, we should seek for a form of "objectivity," but as we have already seen, facts are uncertain and values are disputed. For millennia people believed that for all questions there was an answer—one founded in truth at that. But, as Isaiah Berlin has observed,

> After the Romantics had done their work it began to be believed that some answers were not to be discovered, but created; that moral and political values are not found but made.... We inherit both these traditions, objective discovery and subjective creation, oscillate between them, and try vainly to combine them, or ignore their incompatibility (Berlin 1992).

Individual Change

Attention to the processes of change at the individual level is equally important. Neoclassical economics tends to assume rationality on the part of "economic agents" with complete information, which means they efficiently pursue their individual agendas. Obviously that assumption or assertion leads to policy prescriptions that differ widely from prescriptions that assume, say, that "people" also act on the basis of altruism, out of concern for community, and for a public rather than a solely private agenda—and that they do so with less-than-perfect information. Similarly, conventional economics assumes that competition is the natural state of affairs in relations between people and entities, whether they are nations or businesses. What has happened to cooperation, which ecologists and psychologists know is also "natural"?

Agenda for Change

Therefore, four questions relating to the processes of social change should become the key elements of the ecological economics research agenda:

- How does the environmental behavior of individuals change?

- How does an issue rise and get acted upon on as part of the public agenda?

- How do public institutions, including governments, businesses and industry, non-governmental organizations, and the media, change their environmental behavior over the long term? How do these institutions influence the behavior of individuals?

- How do training and research institutions change so that they can respond to the societal need of citizens and professionals who can deal with the problems that must be addressed, now and in the long term?

For the present, these are, to varying degrees, questions for which we do not have the answers. They will necessitate relationships with persons in other fields, with whom we have not yet seen a common agenda.

A problem focus requires vision, and attainment of a vision requires an understanding of and participation in the processes of change. In order to have the greatest impact on the substance of the changes needed, we must be clear about our destination, have a road map, and a sense of the vehicles necessary to overcome any of the barriers to our achieving our goal.

THE SHORTER TERM

In the shorter term, the first policy task is the description and ultimately the implementation of a system that balances the needs of nature's household and the human household, while recognizing that the former is the basis for the latter. The lack of understanding of the systemic, structural nature of our problems, is most likely the single important failure of policy. If we think, and act, systemically, we are led beyond "tinkering" to seek real change.

Many people are recognizing that the problems we are confronting today are significantly different from those that they have dealt with in the past. For example, responding to Robert Solow's observation that this generation may be the first in America that will leave its children poorer than itself, Daniel Bell concluded: "the economic foundation for culture is beginning to show cracks" (Bell 1992). Just as Hayek observed in his Nobel address in 1974, we are now being called upon to recommend solutions for problems that we have created.

But the political process continues to respond in a piecemeal fashion. Witness, for example, some of the issues being debated during 1993 in the U.S. Congress:

- Enterprise zones and other packages of incentives designed to entice businesses back to Los Angeles after the 1992 riots include reductions of pollution standards, despite the fact that a poor environment is often cited as one of the causes of the riots, and that Los Angeles is the most polluted city in the United States.

- Loans and grants are being made available to the former republics of the Soviet Union and the countries of Eastern Europe, without reference to the kinds of "free market" activities they will support, almost assuring serious and continuing environmental destruction. As we know from the experiences of Chile under Pinochet, and of Eastern Europe, the "free market" can be as environmentally destructive as central planning.

- Proposals for carbon taxes are being discussed. But what is needed is a revision of the whole taxation system.

The list could be expanded with ease.

The Immediate

How does ecological economics contribute to the policy debate?

Let us imagine, for the moment, that an ecological economist has been called to a seat of political power. The President or Prime Minister looks at our colleague and says: "What are we going to do now? How do we get the prices right? What do I do about jobs, trade, and equity? What is the distinction between growth and development? How do we achieve sustainability?" Once outsiders, playing the comfortable role of critic, we are now insiders, with responsibility.

Is this farfetched? Perhaps. But recall that in 1916 Lenin said that he would never live to see the Revolution. The next year he was in power, and he was not prepared.

The issues raised here relate mainly to the United States. They are comparable to those existing in other countries.

Growth

"Growth," even the oxymoron "sustainable growth," is seen by all countries as the way out of our economic dilemmas. Political strategists in the U.S. with whom I have spoken say that to even suggest you are against growth, (so deeply held is the belief in its saving graces), is to commit political suicide.

What does ecological economics have to offer as an alternative that is ecologically and economically appropriate and politically possible? The Meadows' have spoken of the need for "targeted growth" to meet the needs of poor and attack problems of inequity (Meadows 1992). Do ecological economists know what that means and how to implement it?

Jobs and the Environment

This is another politically sensitive issue with high stakes in the U.S. and goes to the heart of questions of equity. Former President Bush stated that he had given too much weight to the environment over jobs. On the other hand, the environmental movement has tended to be insensitive to employment issues, or to approach them as problems in the aggregate without reference to individual and community needs. It is acknowledged that a certain number of jobs will be lost as a result of an environmental intervention, and asserted that a larger number of jobs will be created. Or that the jobs will be lost at some time in the near future, anyway, because of the depletion of the resource, and therefore, why not bite the bullet now. These have been the arguments used by American environmental groups to argue for the protection of the old-growth forests of the Pacific Northwest, and the fish stocks of the New England coast. But these responses are not sufficient, devoid as they are of any reference to the quality and nature of the jobs to be created, and of the immediate and present impacts upon communities and individuals. Who wins and who loses? Who pays, and when?

With the end of the Cold War and the likely reduction of military budgets, opportunities for new investments in the economy will arise. What kinds of jobs will be available to meet nature's needs, as well as the needs of workers and communities?

Capital flight and job loss are being blamed increasingly on environmental regulations in the United States. Businesses often have "problems" meeting air quality regulations, while circumventing "responsibilities" to comply with regulations to protect human health and the health of the environment.

How do we create well-paying jobs, which people want, in settings that honor their integrity, as well as the integrity of the environment and community? What do we say in the United States to the loggers of the Northwest and the fisherman of the Northeast? What are the ecologically and economically appropriate uses for monies re-allocated from military spending? How do we contribute to equity for present generations, while reflecting our concern for equity with future generations? And as difficult as these issues are within the so-called developed world, how do we approach the same issues in less developed countries?

Trade and the Global Economy

The Wall Street Journal editorialized on July 29, 1992 that "... It is crucial that the world's leaders get the trade issue right at this point in the world's economic history." With this, we can all agree. But their "right" is that "once again we're offered a false choice between hygienic poverty and toxic growth, unless free trade is made less free with all manner of environmental regulations, (but) it is precisely economic freedom, including the freedom to trade and invest across borders, that makes people rich enough to pay for clean air and water."

My "right," informed by the questions of ecological economics, is quite different.

The potentially negative impacts of so-called free trade on the environment are myriad. The trading system that encourages specialization can result in the significant loss of biodiversity and cultural diversity. More specific assaults may arise from the migration of dirty industries to countries with less environmental regulation, thereby affecting jobs in the sending country and pollution in the receiving country. Efforts at "harmonization" of environmental standards are likely to favor weaker rather than stronger regulation. Yet we are told by conventional economists that increases in trade are necessary to create the wealth that will allow countries to invest in environmental protection, the so-called *rising tide theory*.

What do we have to say about the ways that trade could contribute to, rather than destroy, sustainability? Can a competitive trading system maintain and enhance natural capital, human life, and culture? Can a redefined trading system contribute to greater equity within and among nations, while preserving the environment?

CONCLUSION

This list of issues could be expanded, but is sufficient to indicate the range of specific challenges that ecological economics must address in order to contribute to the political/policy process. In short, ecological economics, as a post-normal science grounded in values as well as knowledge, has to develop critiques, and underline positive alternatives that incorporate concerns for nature and the human household, while embracing equity and ethics.

Is this asking too much of ecological economics? That is for each of us to decide. We have choices to make. But as a whole, I see no alternative other than to accept a policy agenda as central to ecological economics. It mirrors what we have said about ourselves, if we listen carefully to what we have said. We have a historic opportunity. We have an obligation because of our concerns for the ecosystem and future generations.

ACKNOWLEDGMENTS

(Thanks to Patricia Bauman, David Orr, Jael Silliman, Dan Viederman, and the two reviewers for very helpful comments on an earlier draft.)

BIBLIOGRAPHY

Bell, D. 1992. Into the 21st Century, Bleakly. *New York Times Week in Review*, July 26: E17.
Berlin, I. 1992. Conversation with.... New York: Scribners. (As quoted in the *New York Times* Book Section, July 19, 1992: 35).
Brown, P. 1993. Restoring the Public Trust: A Fresh Vision for Progressive Government in America. Boston: Beacon Press.
Costanza, R., ed. 1991. Ecological Economics: The Science and Management of Sustainability. New York: Columbia Univ. Press.
Funtowicz, S., and J. R. Ravetz. 1991. A new scientific methodology for global environmental issues. In Ecological Economics: The Science and Management of Sustainability, ed. R. Costanza. New York: Columbia Univ. Press.
Gore, A. 1992. Earth in Balance: Ecology and the Human Spirit. New York: Houghton Mifflin.
Harrington, M. 1963. The Other America: Poverty in the United States. Baltimore, MD: Penguin Books.
Havel, V. 1992. Cited in SCOPE Newsletter, 39 (March).
Meadows, D. H., D. L. Meadows, and J. Randers. 1992. Beyond the Limits. Post Mills, VT: Chelsea Green.
Pearce, D. 1992. Private communication.
Viederman, S. 1991. Long-range goals for ISEE and ecological economics, *ISEE Newsletter* 3 (May): 2.
———. 1992. A Sustainable Society: What Is It? How Do We Get There? Paper presented to the Poynter Center, Indiana Univ., April 24.

26 INVESTING IN NATURAL AND HUMAN CAPITAL IN DEVELOPING COUNTRIES

Olman Segura and James K. Boyce
Universidad Nacional de Costa Rica
Apdo. 555-3000 Heredia
San José, Costa Rica

ABSTRACT

Natural capital, like man-made capital, can depreciate through use or be augmented through investment. While the industrialized countries have begun to invest in natural capital within their own borders, in the developing countries natural capital is often still treated as a free good.

In this chapter we argue that a necessary condition for investment in natural capital in developing countries is investment in human capital through improvements in the nutritional well-being, health, and education of their poor majorities. In part, this is because people struggling for sheer survival cannot afford to worry about long-term environmental costs. In addition, such investment can strengthen the poor in their struggles against forms of environmental degradation of which they are victims, but not perpetrators. This latter effect is especially important if, as we argue, resource abuse by the rich (of the industrialized and developing countries alike), and not by the poor, is responsible for most of the environmental degradation in the world today.

Investment in human capital also can increase and spread knowledge of the relationships between economic activity and the environment, further enhancing investment in natural capital. Finally, insofar as population growth exacerbates environmental stresses, investment in human capital brings an added benefit by fostering voluntary reductions in birth rates.

INTRODUCTION

Until recently, economics and ecology were treated as distinct and unrelated subjects. But today the recognition is growing that the ways human beings manage nature are closely related to the ways we manage relationships among ourselves.

Threats to the ecosystems that underpin economic activity are forcing us to reconsider our treatment of natural resources in our daily lives and in our fields of study. The assumption that technical progress is a perfect substitute for natural resources is losing credibility. The inadequacy of myopic time horizons in the face of long-term environmental degradation is becoming ever more apparent. These insights have given impetus to the idea of sustainable development.

The treatment of natural resources as natural capital is an important element of the new sustainable development economics. Natural resources can no longer be regarded as free goods. In microeconomic decisions and in national accounts, the depletion of the natural resource base must be treated as a cost. At the same time, it is possible to invest in natural capital, for example, through reforestation, soil restoration, and pollution control. In other words, some natural capital can be, if not "man-made," at least helped along by purposeful human endeavor.

This chapter considers the prospects for sustainable development in the so-called Third World, (i.e., in the agrarian and semi-industrialized economies of Asia, Africa, and Latin America). Our central argument is that a necessary condition for sustainable development in these countries is a radical improvement in the nutritional well-being, health, and education of their poor majorities. In other words, investment in natural capital requires investment in human capital.

ECONOMIC MODELS

Classical economics distinguished three factors of production: land, labor, and capital. Today the terms "natural capital," "human capital," and "man-made capital" have replaced the original triad. Natural capital, like other forms of capital, can depreciate. Yet economic models and economic policies long ignored this basic fact. In recent decades this has begun to change in the industrialized countries. It has been estimated, for example, that air and water pollution controls implemented in the United States in the 1970s caused that country's measured GNP to be 5% lower in 1990 than it would have been without controls (Koop 1992). This cost represents a form of investment in natural capital. In the developing countries of the Third World, however, investment in natural capital remains meager indeed.

The economic models applied in Third World countries in the name of development have typically neglected environmental concerns. The import-substitution model developed after World War Two had important negative environmental impacts (Segura 1992). Selective trade barriers built industrial sectors highly dependent on imported capital goods and petroleum. Industrial concentration in urban centers contributed to multiple environmental ills. In agriculture, the "green-revolution" technology increased yields of food grains, but at the price of intensive use of agrochemicals and losses of genetic diversity.

More recently, an outward-oriented model of development has become fashionable. In Central America, this model emphasizes exports of non-traditional agricultural products. The new export-led model typically calls for less state intervention in the economy, taking it to be virtually synonymous with inefficiency, and instead embraces open markets, international trade, and comparative advantage. Like the import-substitution model before it, however, this model does not consider the social costs of natural resource depletion. The correct allocation of natural resources, as of other commodities, is simply left to the market (Stanfield 1991). Since market prices generally fail to capture the full cost of depreciation of natural capital, the spontaneous action of markets cannot be expected to lead to sustainability (Daly 1991; Kaimowitz 1992).

A central tenet of the new ecological economics is the proposition that, far from being a perfect substitute for natural capital, investment in man-made capital requires natural capital (Pearce, Barbier, and Markandya 1990; Daly 1991). In this chapter, we advance an analogous thesis: investment in natural capital requires human capital. Specifically, we believe that dramatic improvements in the nutritional well-being, health, and education of the poor majority in the Third World are crucial to the goal of sustainable development. The linkage between human capital and natural capital has far-reaching implications, for it means that issues of the distribution of wealth and power are fundamental to the quest for sustainable development.

A TYPOLOGY OF ENVIRONMENTAL DEGRADATION

A common feature in diverse strands of economic theory in the latter half of the twentieth century has been the tendency to downplay the importance of distributional issues. Neoclassical economics focuses on the goal of efficiency—defined in Paretian terms as a situation in which no one can be made better off without making someone else worse off. The division of the economic pie is left to politicians and others uninhibited by the neoclassical economists' professed aversion to "value judgments." Development economics has focused on the goal of economic growth—defined as rising gross national product. In the 1970s, in response to the perceived failure of the benefits of growth to "trickle down" to the poor, mainstream development economics embraced the goal of "redistribution with growth," but this meant merely an effort to secure a more equitable distribution of new increments to national income, rather than a redistribution of the existing pie (Chenery et al. 1974).

While economists tend to downplay distribution, ecologists tend to ignore it altogether. This is not surprising, since social differentiation is generally absent in the non-human populations studied by ecologists. Consider, for example, aquatic weeds growing in a pond. If volume doubles every day, and the weeds fill the entire pond in 30 days, the pond will, of course, be half-full on the 29th day. Ecologists have used this example to illustrate the perils of exponential growth of human population (Ehrlich and Ehrlich 1990, pp. 15–16). A notable feature of this metaphor is the total absence of distribution as an issue. Pond weeds are not differentiated by wealth or power. Hence neither the pond weeds' rate of nutrient depletion nor their rate of growth is affected by differences between rich and poor, or powerful and powerless. Moreover, the symptoms of ecological stress will affect equally all the weeds in the pond.

By contrast, inequalities of wealth and power are a noteworthy feature of human societies. In our view, these inequalities are of crucial importance in understanding the causes and consequences of environmental degradation. Why do people engage in economic activities which degrade the environment? And why, if they choose to do so, should anyone else worry about it? The answer to the first question, of course, is that some people reap net benefits from the activities, or at least think that they do. The answer to the second is that other people bear net costs as a result of these same activities. Leaving aside for the moment the possibility of ignorance (when people think that they will reap net benefits, but really will bear net costs), the winners and losers are different people.

In analyzing the causes and consequences of environmental degradation, therefore, we can pose three further questions:

- Who reaps the benefits?
- Who bears the costs?
- Why are the winners able to impose costs on the losers?

In neoclassical environmental economics, these questions are resolutely pushed aside. Negative externalities result from impersonal market failures; distribution has no bearing on their causes. As to consequences, neoclassical environmental economics attempts to sidestep distributional value judgments by means of the "compensation test": could the winners, in theory, compensate the losers, and remain better off? As Sen (1987, p. 33) observes, however, even by the Pareto-efficiency criterion, this device is either unconvincing (if compensation is not in fact paid) or redundant (if it is).

As a step towards redressing this neglect of distributional issues, Figure 26.1 shows a simple four-fold classification of environmental degradation, based on the relative wealth positions of the winners and losers. In Type I environmental degradation, both winners and losers are rich. In Type II, the rich bear the costs of environmental degradation caused by the poor. In Type III the reverse occurs—the poor bear the costs of environmental degradation caused by the rich. Finally, in Type IV the poor are both winners and losers.

A few examples will illustrate these possibilities. If a rich homeowner instructs his gardener to burn yard waste in the back yard, and the smoke pollutes the air breathed by his affluent neighbors, this is Type I environmental degradation. If a poor person tosses a bag of refuse on the rich person's lawn, this is Type II. If the rich person sends his trash to a landfill or incinerator which pollutes an adjacent community inhabited by poor people, this is Type III. And if a poor person burns his trash in a metal drum behind his dwelling, polluting the air breathed by his comparably poor neighbors, this is Type IV.

		Winners	
		Rich	Poor
Losers	Rich	I	II
	Poor	III	IV

Figure 26.1. A typology of environmental degredation

The relative frequency and importance of these four types of environmental degradation—with society appropriately partitioned into rich and poor—is an interesting topic for research. As a preliminary hypothesis, we offer the following: Type III environmental degradation is more prevalent than Types I and IV, and these in turn, are more prevalent than Type II. The rich are thus better able to impose environmental costs on the poor than vice-versa.

There are several reasons we expect this to be true. First, since environmental degradation is, *ceteris paribus*, an increasing function of consumption and production, the fact that the rich consume more implies that they generate more environmental degradation. Second, since money can be spent to reduce or avoid bearing the costs of pollution, for example, by residing and vacationing in relatively uncontaminated ecosystems, the rich can more readily escape the costs of environmental degradation. Finally, wealth is positively correlated with power, and power increases one's ability to impose negative externalities on others and to resist having them imposed on oneself (Boyce 1992).

The direct income effects of wealth disparities are captured in the recent assertion by World Bank Chief Economist, Lawrence Summers, that "the economic logic behind dumping a load of toxic waste in the lowest-wage country is impeccable" (*The Economist* 1992). The indirect power effects are hinted at by Becker (1983, p. 384) when he remarks that "an analysis of noncooperative competition among pressure groups can unify the view that governments correct market failures and what has seemed to be a contrary view that governments favor the politically powerful." The latter helps to explain why the remedies prescribed by environmental economists—regulations, taxes, marketable pollution permits, and so on—are implemented in some times and places, but not in others. In our view, then, both the extent and direction of the "negative externalities" arising from environmental degradation can be explained in terms of the relative wealth and power of the winners and losers.

HUMAN CAPITAL AND NATURAL CAPITAL

We now turn to the linkages between human capital and natural capital. We discuss these under four headings: time horizons, power, knowledge, and population growth.

Time Horizons

People cannot be expected to cease activities that degrade the environment but are essential to the sustenance of their families. Immediate survival is the paramount objective. In the countryside, many of the poor are driven to cultivate fragile environments—steep hillsides, semi-arid lands, thin tropical forest soils—where erosion and nutrient depletion follow swiftly. In the cities, they often live and work in highly precarious and contaminated environments. These choices are the results of desperation, which compels the poor to reap small immediate gains even at the price of large future costs.

The poor themselves are often all too well aware of the damaging long-run consequences of their activities. But to see tomorrow, one must survive today. The end result can be a vicious circle, in which poverty accelerates the depreciation of natural capital, which leads in turn to further impoverishment (Durning 1989; World Bank 1992). According to our categories in Figure 26.1, this is Type IV environmental degradation.

By lessening their desperation, improvements in nutrition, health, and education of the poor would permit them to invest more in natural capital: to protect and improve the physical environments, both rural and urban, which are essential to their own long-run well-being. In effect, human capital investments can lengthen the time horizons of the poor.

Power

Investments in the human capital of the poor can also strengthen their ability to combat environmental degradation of which they are victims, not perpetrators—Type III environmental degradation in Figure 26.1. This is potentially the most important linkage between human capital and natural capital. For much of the pollution and natural resource depletion in the developing countries, as elsewhere, is driven not by the desperation of the poor, but by the greed and negligence of the rich.

Tropical deforestation illustrates this point. In Central America, where very rapid deforestation took place in the 1960s and 1970s, the main cause was the clearing of lands for cattle ranching, stimulated by favorable access to subsidized credit and to the protected U.S. beef market. The main beneficiaries of this privately profitable (but socially costly) process were the large hacenderos and owners of meat-packing plants, both exemplified by former Nicaraguan dictator Anastasio Somoza. The main losers were poor peasants, who were denied access to previously cultivated as well as forested lands, and for whom extensive cattle grazing provided meager employment opportunities (Edelman 1985; Williams 1986; Ascher and Healy 1990).

In Southeast Asia, the main cause of rapid deforestation has been logging for tropical hardwood exports. The main beneficiaries are logging concessionaires—who are often military officers and the political cronies of top government officials—who capture rights to cut the public forests as a form of political patronage. Once again, the main losers are the poor, including displaced forest dwellers (often ethnic minorities) and downstream peasants whose crops depend on the forest's hydrological "sponge-effect" (Repetto and Gillis 1988; Hurst 1990; Kummer 1992; Boyce 1993).

In the Brazilian Amazon, the driving forces and main beneficiaries in deforestation have been generals, land speculators, and large-scale cattle ranchers. The principal victims are the 200,000 indigenous people of the Amazon, and the 2 million other Brazilians who earn their living by gathering rubber, nuts, resins, palm products, and medicines from the forest (Guppy 1984; Hecht and Cockburn 1990).

Third World elites have not pursued these assaults on the world's tropical forests on their own. On the one hand, they have many times enjoyed the avid support—economic, political, and military—of the governments and international financial institutions of the industrialized countries. On the other hand, they often employ their poorer countrymen to do the hard work. In Costa Rica, for example, landless farmers are often contracted to clear the land in return for permission to grow crops on it for two or three years, after which the cattle rancher converts it to pasture (Edelman 1985). In Brazil, temporary laborers, sometimes including Indians, are hired at minimal wages to cut and burn the forest. But as Barraclough and Ghimire (1990, p. 13) remark, "To blame poor migrants for destroying the forest is like blaming poor conscripts for the ravages of war."

Investment in the human capital of the poor can increase their leverage to oppose such instances of Type III environmental degradation. With better nutrition, health, and education, the poor become better able to resist the economic and political pressures of the rich; better able to analyze the causes and consequences of environmental degradation; and better able to score victories in the political arenas where conflicts are ultimately resolved. The complementarity between human capital and natural capital is here mediated by what can be termed the "political capital" of the poor.

Knowledge

Knowledge of the environment is itself a form of human capital. While it would be quite erroneous to assume that the poor are entirely ignorant of their environment—on the contrary, they are frequently more knowledgeable about environmental matters than the rich—it would be equally erroneous to assume romantically that their environmental knowledge is so profound as to be incapable of further advance. Environmental education can be particularly important in situations where people confront new environments, or where they face radically new stresses in old ones.

By improving natural resource management by the poor, education can help to reduce Type IV and Type II environmental degradation. No less important, education can enhance their ability to contest Type III environmental degradation. The latter effect comes about for two reasons. First, education can improve knowledge of environmental costs whose impacts and sources otherwise remain obscure. It is one thing to know that your child is sick, but something else to trace that sickness to a specific source of environmental contamination. Second, education can improve knowledge of how to wage successful political struggles—how to lobby government officials, initiate legal actions, and mobilize public opinion.

In sum, by providing environmental knowledge to people with an incentive to use it, education can alter the balance of power and thereby tilt environmental outcomes towards greater protection of natural capital.

Population Growth

In our view the impact of population growth upon natural capital is more complex than is often assumed. Obviously, the number of human beings on our planet cannot grow forever. It is equally obvious that holding any variable per capita constant, more people mean more of that variable. These truisms, however, do not logically imply the proposition that at this moment in human history the world is overpopulated, nor that population growth is today the principal cause of worldwide environmental degradation.

Consider the simple formula advanced by Ehrlich and Holdren (1971):

$$I = P \times F,$$

where I = the total negative impact on the environment; P = population; and F = per capita negative impact on the environment. This mathematical identity is often taken to prove that population growth necessarily accelerates environmental degradation. Let us define I_n as that subset of environmental degradation which has no causal relation whatsoever with population growth. An example might be the environmental degradation caused by the manufacture, deployment, and disposal of weapons for the U.S. military, which reportedly generates more toxic waste every year than the world's five largest multinational chemical companies combined (Seager 1993). Let I_p represent that subset of environmental degradation which is causally related to population growth. Let F_n and F_p represent these two subsets per capita. Now we can write the identity $I_n = P \times F_n$: total nonpopulation-related negative environmental impact equals population multiplied by such impact per capita. Yet it would be utterly fallacious to conclude, on the basis of this formula, that population growth is a driving force behind contamination of the environment by the U.S. military.

This is not to argue that population growth has no environmental effects. The point is that the formula, $I = P \times F$, tells us nothing about the importance of population growth as a cause of environmental degradation. That is, it tells us nothing about the magnitudes of the ratios I_n/I and I_p/I. The formula would be equally correct no matter whether $I_n/I = 1$ (and population growth had no effect whatsoever on environmental degradation), or $I_n/I = 0$ (and all environmental degradation was inexorably multiplied by population growth), or anything in between. In other words, it is an empty tautology.

Part of the appeal of this defective analysis, especially among natural scientists, may lie in the analogy it makes between humans and other animals. More caterpillars, for instance, mean more leaves must be eaten if they are to survive and reproduce. In the absence of predators or pesticides, their numbers will grow until famine or disease precipitates a population crash. Unlike animals which merely consume, however, people both produce and consume. To characterize babies born in Bangladesh, the Philippines, or elsewhere as "mouths" (Ehrlich and Ehrlich 1990, pp. 72, 75) is to ignore their brains and hands. Animals cannot invest in natural capital. Humans can.

Improvements in the human capital of the poor majority, notably in health and education, historically have been associated with declining fertility for a number of reasons: women gain greater access to employment opportunities outside the home; lower infant and child mortality mean fewer births are needed to attain a given probability of survival to adulthood; the importance of children's labor in family income diminishes; access to birth control and supporting health services improves; and so on (Cassen 1976; Repetto 1979; Caldwell 1982). Insofar as population growth exacerbates environmental degradation—and there are undoubtedly settings where, all else equal, it does so—this demographic transition constitutes a further link between human capital and natural capital.

We have argued above that investment in the human capital of the poor can increase their relative power. The possible effects on birth rates of this shift in the balance of power merits comment. Greater power for the poor—especially for women—increases the political effectiveness of their demand for access to birth control and reproductive health services. By the same token, however, it strengthens their capacity to resist unwanted birth control measures pushed upon them by governments in the name of population control. For example, improvements in living standards in Bangladesh might reduce the number of women who accept sterilization in return for "incentive" payments in the form of cash and clothing (Hartmann 1987). Empowerment of the poor thus could cause a short-run increase in fertility in settings where population-control incentives and coercion are now in use. However, this would also reduce the risk of a popular backlash against family planning programs, as occurred in India after the coercive sterilization drive of the mid-1970s. At the same time, it would give governments and international agencies greater incentive to provide more "user-friendly" birth control options, enhancing the long-run prospects for fertility decline.

CONCLUSIONS

In this chapter we have presented a case for investment in human capital as a necessary condition for investment in natural capital in the developing countries. Improvements in the nutrition, health, and education of the poor majority would lengthen their time horizons, strengthen their power to oppose environmental degradation caused by others, enhance their knowledge of environmental costs and how to reduce them, and contribute to voluntary diminution of population growth.

The foregoing analysis implies a dual role for the state. On the one hand, state interventions are needed to correct market failures, for example, through regulations on the use of toxic chemicals, incentives for reforestation, taxes on pollution, and depletion quotas for non-renewable resources. On the other hand, the state can directly and indirectly promote investment in the human capital of the poor.

Today we find a contradictory policy in place in many developing countries. In the name of macroeconomic adjustment, governments are slashing social expenditure, cutting public investment, and tightening credit. Although reductions in public expenditure are often unavoidable, it is important to establish priorities. Policies should be avoided which increase poverty, adversely affect public health, undermine the state's capability for environmental protection, or reduce financing for the development and diffusion of more environmentally sound technologies.

Many serious threats to sustainability in the Third World originate elsewhere, in the industrialized countries. In the case of carbon dioxide emissions, for example, per capita output from industrial processes in the United States is 19.7 metric tons per year, compared to 0.9 ton for the average Costa Rican, or 0.1 for the average Bangladeshi (World Resources Institute 1992). In addition to the industrialized countries' role in global environmental threats, their trade and financial policies often operate to the detriment of both the economic and environmental well-being of the poor majority in the Third World.

In the industrialized countries as well as the developing countries, therefore, complementary investments in natural capital and the human capital of the poor will require profound institutional changes. If sustainable development is to be more than a passing slogan, it will demand a reshaping not only of human relationships with nature, but also of our relationships with each other. This is the great challenge facing not only economists and ecologists, but humankind as a whole.

REFERENCES

Ascher, W., and R. Healy. 1990. Natural Resource Policymaking in Developing Countries. Durham; London: Duke Univ. Press.
Barraclough, S., and K. Ghimire. 1990. The Social Dynamics of Deforestation in Developing Countries. United Nations Research Institute for Social Development, Discussion Paper No. 16, Geneva.
Becker, G. S. 1983. A theory of competition among pressure groups for political influence. *Quarterly Journal of Economics* 48 (3): 371–400.
Boyce, J. K. 1992. La Degradación Ambiental y la Economía: Hacia una Economía Política del Desarrollo Sostenible. Paper presented to the Seminar on the Impact of Economic Policy on Sustainable Development in the Agricultural Sector, Inter-American Institute for Cooperation on Agriculture, Coronado, Costa Rica, November 1992.

————. 1993. The Philippines: The Political Economy of Growth and Impoverishment in the Marcos Era. London: Macmillan; Honolulu: Univ. of Hawaii Press; and Quezon City: Ateneo de Manila Univ. Press.

Caldwell, J. 1982. The Theory of Fertility Decline. New York: Academic Press.

Cassen, R. H. 1976. Population and development: a survey. *World Development* 4 (10/11): 785–830.

Chenery, H. et al. 1974. Redistribution with Growth. Oxford: Oxford Univ. Press.

Daly, H. 1991. Steady-State Economics. 2d ed. Washington, DC: Island Press.

Durning, A. B. 1989. Poverty and the Environment: Reversing the Downward Spiral. Worldwatch Paper No. 92. Washington, DC: Worldwatch Institute.

Edelman, M. 1985. Land and labor in an expanding economy: agrarian capitalism and the hacienda system in Guanacaste Province, Costa Rica. Ph.D. dissertation. Columbia Univ.

Ehrlich, P., and J. Holdren. 1971. The impact of population growth. *Science* 171 (March 26): 1212–7.

Ehrlich, P., and A. Ehrlich. 1990. The Population Explosion. New York: Simon & Schuster.

Guppy, N. 1984. Tropical deforestation: a global view. *Foreign Affairs* 62 (4): 928–65.

Hartmann, B. 1987. Reproductive Rights and Wrongs: The Global Politics of Population Control and Contraceptive Choice. New York: Harper & Row.

Hecht, S., and A. Cockburn. 1990. The Fate of the Forest: Developers, Destroyers and Defenders of the Amazon. New York: Harper Collins.

Hurst, P. 1990. Rainforest Politics: Ecological Destruction in South-East Asia. London: Zed Books.

Ives, J. 1985. Mountain environments. *Progress in Physical Geography* 5: 427–33.

Kaimowitz, D. 1992. La Valorización del Futuro: un reto para desarrollo sostenible en América Latina. In Desarrollo Sostenible y Políticas Económicas en América Latina, ed. O. Segura. San José, Costa Rica: Editorial DEI.

Koop, R. J. 1992. The role of natural assets in economic development. *Resources* (Winter). 7–10.

Kummer, D. M. 1992. Deforestation in the Post-War Philippines. Chicago: Univ. of Chicago Press.

Let them eat pollution. 1992. *The Economist* (February 8): 66.

Martinez-Alier, J. 1991 Ecology and the poor: a neglected dimension of Latin American history. *Journal of Latin American Studies* 23: 621–39.

Pearce, D., E. Barbier, and A. Markandya. 1990. Sustainable Development: Economics and Environment in the Third World. Aldershot, U.K.: Edward Elgar.

Repetto, R. 1979. Economic Equality and Fertility in Developing Countries. Baltimore: Johns Hopkins Univ. Press.

Repetto, R., and M. Gillis, eds. Public Policy and the Misuse of Forest Resources. Cambridge: Cambridge Univ. Press.

Seager, J. 1993. Earth Follies. London: Routledge.

Segura, O. 1992. El Desarrollo Sostenible y la Liberalización del Comercio Internacional. In Desarrollo Sostenible y Políticas Económicas en América Latina, ed. O. Segura. San José, Costa Rica: Editorial DEI.

Sen, A. K. 1987. On Ethics and Economics. Oxford: Blackwell.

Stanfield, D. 1991. La Liberalización del Comercio International y la Agricultura Sostenible: El Impacto del GATT. Panama, September. Unpublished paper.

Williams, R. G. 1986. Export Agriculture and the Crisis in Central America. Chapel Hill: Univ. of North Carolina Press.

World Bank. 1991. World Development Report 1992. New York: Oxford Univ. Press.

World Resources Institute. 1992. World Resources 1992–93. New York: Oxford Univ. Press.

INDEX